The Holy Fox: A Life of Lord Halifax

Eminent Churchillians

The Aachen Memorandum

Salisbury: Victorian Titan

Napoleon and Wellington

Hitler and Churchill: Secrets of Leadership

What Might Have Been (editor)

Waterloo: Napoleon's Last Gamble

*The Correspondence Between Mr Disraeli and
Mrs Brydges Willyams* (editor)

A History of the English-Speaking Peoples Since 1900

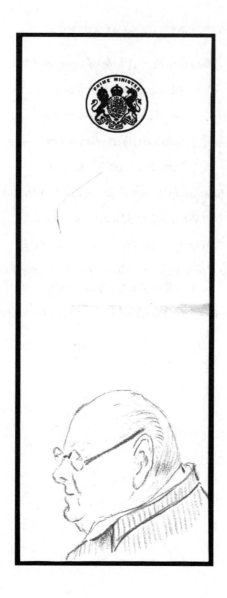

Sketch of Churchill by General Brooke on No. 10 writing paper, made during
a War Cabinet meeting in March 1942

ANDREW ROBERTS

Masters and Commanders

*How Roosevelt, Churchill, Marshall and
Alanbrooke Won the War in the West*

ALLEN LANE
an imprint of
PENGUIN BOOKS

ALLEN LANE

Published by the Penguin Group

Penguin Books Ltd, 80 Strand, London WC2R ORL, England

Penguin Group (USA) Inc., 375 Hudson Street, New York, New York 10014, USA

Penguin Group (Canada), 90 Eglinton Avenue East, Suite 700, Toronto, Ontario, Canada M4P 2Y3
(a division of Pearson Penguin Canada Inc.)

Penguin Ireland, 25 St Stephen's Green, Dublin 2, Ireland
(a division of Penguin Books Ltd)

Penguin Group (Australia), 250 Camberwell Road, Camberwell, Victoria 3124, Australia
(a division of Pearson Australia Group Pty Ltd)

Penguin Books India Pvt Ltd, 11 Community Centre, Panchsheel Park, New Delhi – 110 017, India

Penguin Group (NZ), 67 Apollo Drive, Rosedale, North Shore 0632, New Zealand
(a division of Pearson New Zealand Ltd)

Penguin Books (South Africa) (Pty) Ltd, 24 Sturdee Avenue, Rosebank, Johannesburg 2196, South Africa

Penguin Books Ltd, Registered Offices: 80 Strand, London WC2R ORL, England

www.penguin.com

First published 2008

2

Set in PostScript Adobe Sabon
Typeset by Rowland Phototypesetting Ltd, Bury St Edmunds, Suffolk
Printed in England by Clays Ltd, St Ives plc

A CIP catalogue record for this book is available from the British Library

ISBN: 978-0-713-99969-3

www.greenpenguin.co.uk

For my wife, Susan

Contents

PART III

Estrangement

List of Illustrations

Frontispiece: A sketch of Churchill by Alan Brooke (Trustees of the Liddell Hart Centre for Military Archives, King's College, London – Ref: Alanbrooke 6/4/1–5. Reproduced by kind permission of The Viscount Alanbrooke)

Preface: A page from Lawrence Burgis' account of the War Cabinet meeting of 10 December 1941 (Churchill Archives Centre, Papers of Laurence Burgis, BRGS 2/10, 10 December 1941)

1. The Masters and Commanders at the Casablanca Conference, January 1943 (reproduced by kind permission of Mrs Joan Bright Astley)

2. Pershing and Marshall, 1919 (courtesy of the George C. Marshall Research Library, Lexington, Virginia)

3. Alan Brooke in the uniform of the Royal Horse Artillery, 1910 (Trustees of the Liddell Hart Centre for Military Archives, King's College, London – Ref: Alanbrooke 13/1)

4. Churchill arriving at Downing Street, 15 May 1940 (Getty Images)

5. Roosevelt addressing Congress, 8 December 1941 (Getty Images)

6. Churchill and Roosevelt on board USS *Augusta*, 9 August 1941 (Topfoto)

7. Churchill and Roosevelt on board HMS *Prince of Wales*, 14 August 1941 (AP/PA Photos)

8. Marshall, Churchill and Henry L. Stimson, 24 June 1942 (Getty Images)

9. Alan Brooke's lunch for Marshall at the Savoy Hotel, July 1942 (David E. Scherman/Time & Life Pictures/Getty Images)

10. Harry Hopkins, Mark Clark, Roosevelt and Eisenhower in North Africa, 31 January 1943 (Bettmann/Corbis)

11. Eisenhower and Marshall in Algiers, 3 June 1943 (Corbis)

12. Churchill recuperating in Carthage, Christmas Day 1943 (Bettmann/Corbis)

13. Patton, Bradley and Montgomery in France (Corbis)

14. The Combined Chiefs of Staff at Casablanca, January 1943 (US Army Military History Institute)

15. Churchill, Eden and others at Allied HQ in North Africa, 8 June 1943 (Getty Images)

16. Combined Chiefs of Staff meeting at the First Quebec Conference, August 1943 (US Army Military History Institute)

17. Second Quebec Conference, September 1944 (reproduced by kind permission of Mrs Joan Bright Astley)

18. Churchill and Roosevelt at the Second Quebec Conference, September 1944 (reproduced by kind permission of Mrs Joan Bright Astley)

19. John Dill, Andrew Cunningham, Alan Brooke, Charles Portal and Hastings Ismay at Quebec, 1944 (reproduced by kind permission of Mrs Joan Bright Astley)

20. British Joint Planning Staff, Second Quebec Conference, September 1944 (reproduced by kind permission of Mrs Joan Bright Astley)

21. Allen Tupper Brown (courtesy of George Marshall Research Library, Lexington, Virginia)

22. Alan Brooke and Barney Charlesworth, October 1941 (Trustees of the Liddell Hart Centre for Military Archives, King's College, London – Ref: Alanbrooke 13/3)

23. Churchill and Jan Smuts in Cairo, August 1942 (Bettmann/Corbis)

24. Hastings Ismay, 1942 (George Karger/Time & Life Pictures/Getty Images)

List of Maps

The North African Littoral

SOCIALIST REPUBLICS

Barents Sea

Murmansk

Archangel

• Novgorod

Leningrad

FINLAND

Helsinki

Tallin

ESTONIA

• Kalinin

Gulf of Bothnia

SWEDEN

Riga
Sigulda Line
LATVIA
Riga Line

Memel

LITHUANIA

*Baltic
Sea*

N

300 miles
400 kilometres

xvi

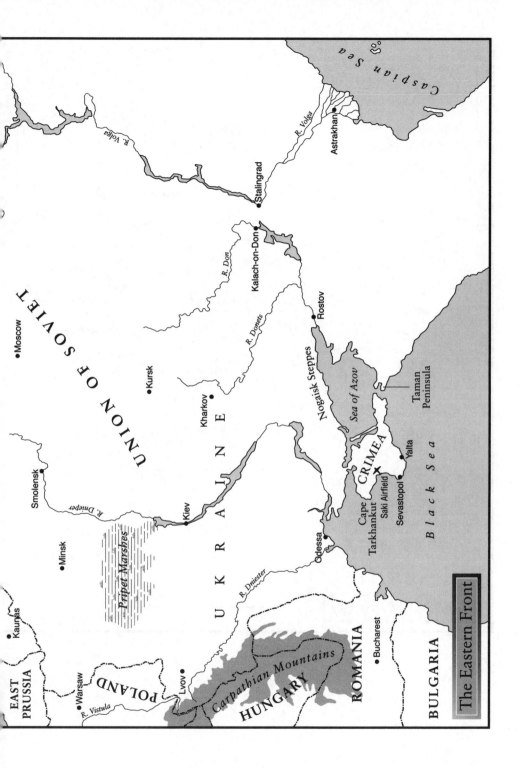

The Mediterranean Theatre

The Far East: Approaches to Japan

U S S R

Sea of Okhotsk

MONGOLIA

MANCHURIA

Vladivostok

HOKKAIDO

Kuril

Peking

Tientsin

KOREA

Sea of Japan

JAPAN

Seoul

Tokyo

Yellow Sea

Hiroshima

Kyoto

CHINA

Nagasaki

Shanghai

Ryuku Islands

Okinawa

Iwo Jima

FORMOSA

Hong Kong

500 miles

500 kilometres

Batan Islands

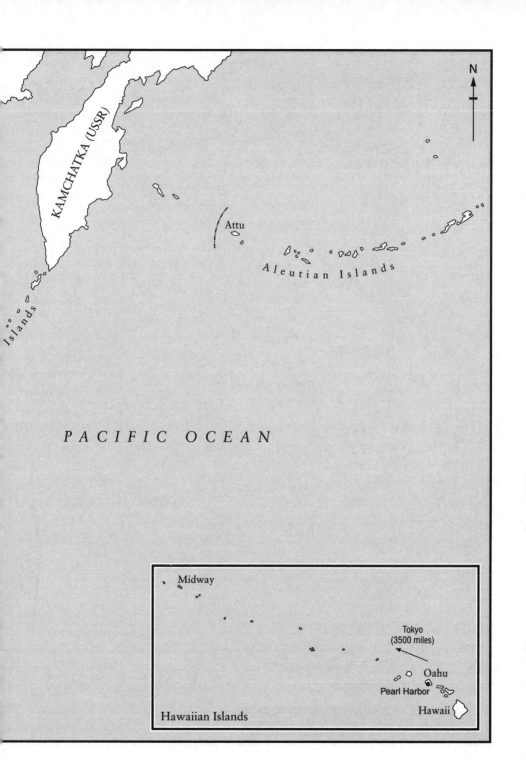

N

KAMCHATKA (USSR)

Islands

Attu

Aleutian Islands

PACIFIC OCEAN

Midway

Tokyo
(3500 miles)

Oahu

Pearl Harbor

Hawaii

Hawaiian Islands

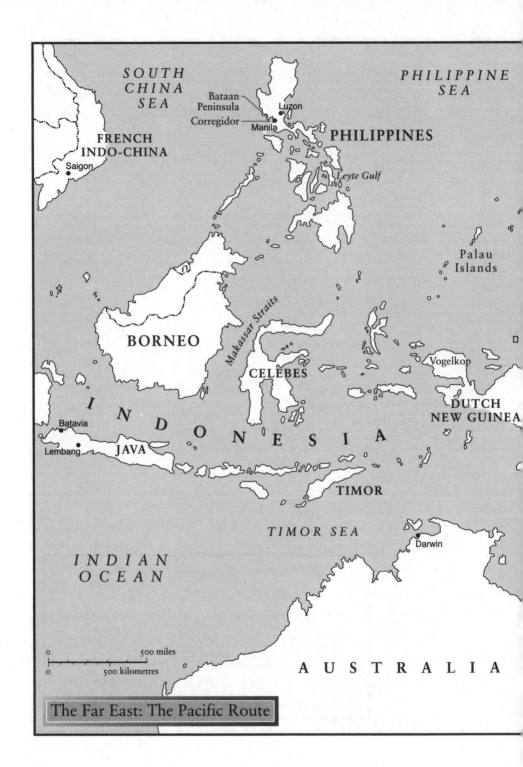

The Far East: The Pacific Route

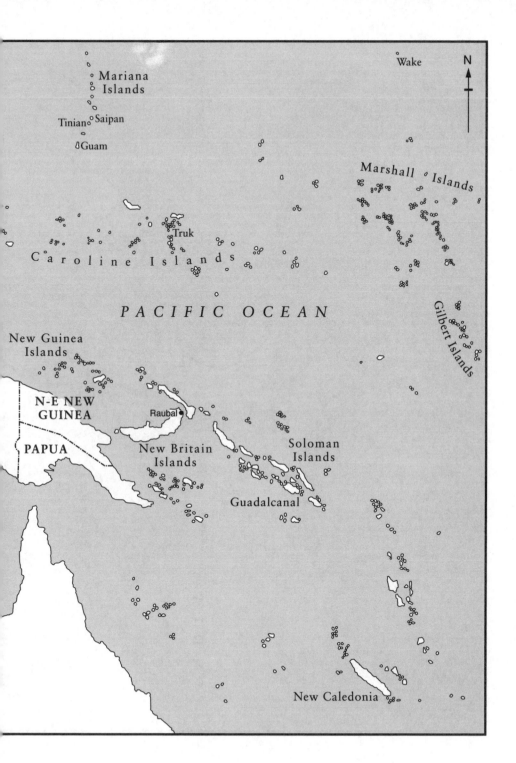

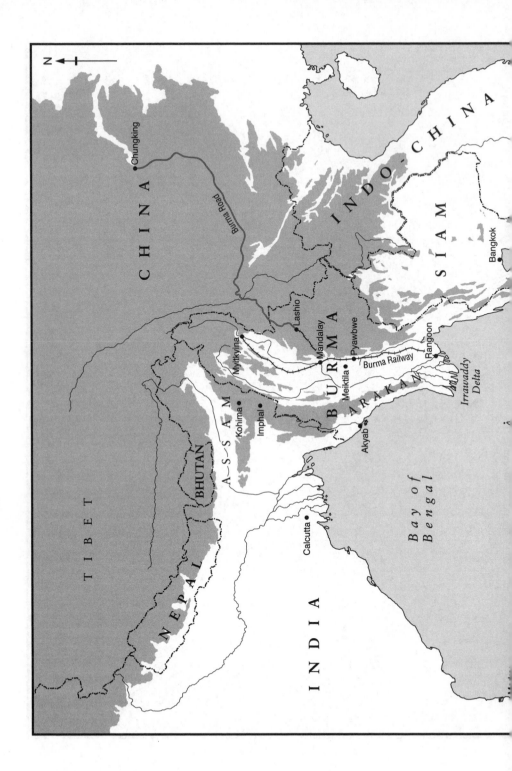

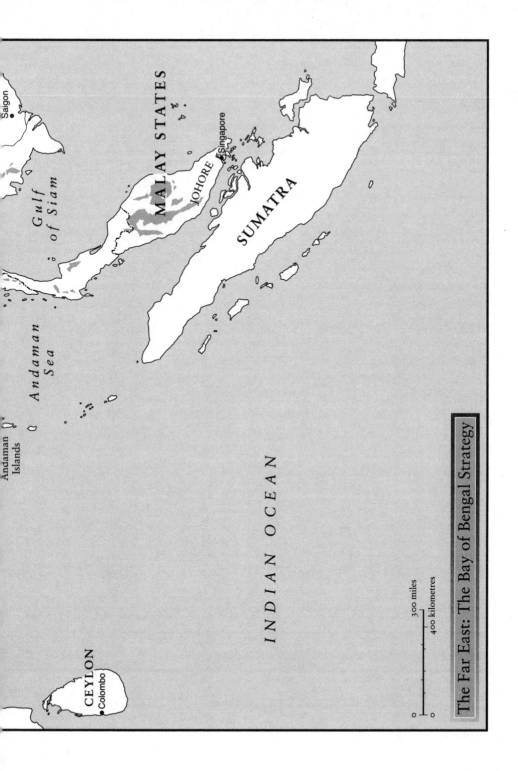

The Far East: The Bay of Bengal Strategy

Saigon

MALAY STATES

Gulf
of Siam

JOHORE

Singapore

SUMATRA

Andaman Sea

Andaman
Islands

INDIAN OCEAN

CEYLON
Colombo

300 miles
400 kilometres

0
0

Acknowledgements

In the three years that it has taken me to research and write this book, there have been a large number of people who have been tremendously generous to me, especially with their time, and I would like to take this opportunity to thank them.

Michael Crawford allowed me to quote from his father Sir Stewart Crawford's unpublished diary of the Yalta Conference, Joan Bright Astley gave me free rein in her fascinating archive, and Conrad Black permitted me to view his collection of President Roosevelt's private correspondence. Other people who have been immensely helpful for this book include Hugh Lunghi for his memories of translating for Churchill and Lord Alanbrooke at the Teheran, Casablanca, Yalta and Potsdam Conferences; the indefatigable Colonel Patrick Mercer MP who very kindly showed me around the battlefields of Monte Cassino, Salerno and Anzio, and accompanied me to the grave of General Marshall's stepson at the Sicily–Rome American Cemetery at Nettuno; Professor Sir Michael Howard for his unrivalled knowledge of the grand strategy of the war; Lord Alanbrooke's biographer General Sir David Fraser and Lady Fraser for memories of the field marshal; Geneviève Parent for opening up the Salon Rose at the Château Frontenac in Quebec for me; Professor Alex Danchev, the acknowledged expert on Anglo-American Staff relations and the co-editor of the Alanbrooke diaries, for many insights; Philip Reed for private tours of the Cabinet War Rooms; Lawrence Rees for videotapes of Alanbrooke's BBC television programmes; my aunt and uncle Susan and David Rowlands for letting me stay at their farmhouse in the Dordogne while I was writing this book; Victoria Hubner for showing me around FDR's home at Hyde Park, New York; James, Lisa and Helen-Anne Gable for making me feel so welcome in Virginia; the always exuberant Governors of the Other Other Club of

Madison, Wisconsin; and Sam Newton for showing me around General Marshall's house, Dodona Manor. Paul B. Barron of the George C. Marshall Library very generously invited me to Thanksgiving Dinner with his charming family, for which very many thanks. I should also like to thank profusely Campbell Gordon, who found my word processor – with the only copy of this book on it – after I moronically left it in the back of a taxi coming home from the London Library. Three years of research would have been wasted had it been lost. If Mr Gordon will please get in touch, I would like to give him lunch.

A large number of people have kindly discussed one or more of the four Masters and Commanders with me, often from their personal knowledge of them, and I should like to thank Joan Bright Astley, the Countess of Avon, Antony Beevor, Lord Black, Field Marshal Lord Bramall, Professor Donald Cameron Watt, Lord Carrington, Winston S. Churchill Jr, Lady de Zulueta, Colonel Carlo D'Este, Professor Sir Lawrence Freedman, Professor Sir Martin Gilbert, Field Marshal Lord Inge, Professor Warren Kimball, Paul Johnson, Sir John Keegan, Richard Langworth, Dr Anthony Malcolmson, Jon Meacham, Sir Anthony Montague Browne, Professor Richard Overy, Kenneth Rose, Celia Sandys, Lady Soames, Anne Sharp Wells and Lady Williams of Elvel.

I have encountered much friendliness and help in the archives and libraries I have visited in the course of researching this book, and would particularly like to thank Paul B. Barron, Peggy L. Dillard and the late, greatly lamented Dr Larry I. Bland at the George C. Marshall Foundation in Lexington, Virginia; Mark Renovitch at the Franklin D. Roosevelt Presidential Library in Hyde Park, New York; Katherine Higgon at the Liddell Hart Centre for Military Archives, King's College London; Allan Packwood, Andrew Riley and the staff of the Churchill Archives in Churchill College, Cambridge; Dr Richard Sommers, Bob Mages, David Keough and Paul Lynch at the USA Military History Institute in Carlisle, Pennsylvania; Natalie Milne at the Heslop Room of Birmingham University; Janet McMullin at Christ Church Library, Oxford; Simon Gough at the Parliamentary Archives in the Palace of Westminster, Frederick Augustyn at the Library of Congress, Washington, as well as the staffs of the Bodleian Library, London Library, National Archives at Kew and the Manuscripts Room of the British Library.

In Stuart Proffitt, Georgina Capel and Peter James, I know that I'm very fortunate to have a fantastically talented team for my publisher, literary agent and copy-editor; my profound thanks go to them. Various other friends, family and experts have read my manuscript, and although all errors in it are of course mine alone, I would like to thank for their advice and invaluable suggestions John Barnes; Paul Courtenay, the new chairman of the International Churchill Society (UK); Jeremy Elston; my wife Susan Gilchrist; Roger Jenkin; Hugh Lunghi; John McCormack; Sir Anthony Montague Browne, Sir Winston Churchill's former private secretary; Stephen Parker; Eric Petersen; my father Simon Roberts; Antony Selwyn and Allan Taylor-Smith.

I dedicate this book to my darling wife Susan, who in the course of my researches has accompanied me to many of the places that appear in the book, including Marrakesh, the Mena House in Giza, the Château Frontenac in Quebec, Bletchley Park, Stalingrad (now Volgograd), the Oval Office of the White House and the Cabinet Room at 10 Downing Street, Auschwitz-Birkenau, the battlefields of Kursk, Moscow, Anzio, Rome and Monte Cassino, Mussolini's execution spot above Lake Como, the Hadtörténeti Müzeum and the Holocaust museum in the Dohány Utca synagogue in Budapest, the Heeresgeschichtliches Museum in Vienna and, on our honeymoon last year, the Kanchanaburi death camp on the River Kwai.

She is the woman I have been seeking all my life.

Andrew Roberts
www.andrew-roberts.net
May 2008

Preface

I put all this aside. I put it on the shelf, from which the historians, when they have time, will select their documents to tell their stories.

Winston Churchill, House of Commons, 18 June 1940

Type 'strategy second world war' into the Google search engine and you will get no fewer than 1.64 million hits, so why am I trying to add to that figure? One aspect that I hope will differentiate this book from the hundreds already published on the subject is the inclusion of some hitherto unpublished material, including an extensive set of verbatim reports of Winston Churchill's War Cabinet meetings, previously quoted from only on the internet. In trying to reconstruct the intimacy of the often daily exchanges between my four principals, I was fortunate, through pure serendipity, also to chance upon the verbatim notes taken of the War Cabinet meetings by someone who was hitherto virtually unknown to history, Lawrence Burgis.

Burgis (pronounced Burgess) was, according to the diarist James Lees-Milne, 'the last serious attachment of Lord Esher's private life'.[1] When Esher and Burgis first met – it is not known how – Burgis was a seventeen-year-old schoolboy at Ing's School, Worcester, and the fifty-seven-year-old Reginald, second Viscount Esher, was a former courtier to Queen Victoria, a member of the Committee of Imperial Defence and the man who had introduced the idea of a General Staff for the Army in 1904, as well as being perhaps the best socially connected man of Edwardian England.

After leaving school, 'Thrushy' Burgis worked as Esher's private secretary, even though Esher's eldest son Oliver thought him 'plain and lower middle-class with a cockney accent'. Esher's relationship with Burgis was described by Lees-Milne as 'the most satisfactory of his love affairs, because it is unlikely that it was ever more than Socratic'. (He presumably meant 'Platonic'; Socrates was altogether more hands-on.)

Thrushy was 'alert, intelligent and eager to learn', and took down dictation very fast in his own private shorthand. 'It was wonderful [for Esher] to have once again a very young man to instruct,' explained Lees-Milne, 'to enrich with anecdotes of all the famous people he had known, to mould in his ways.' Burgis was heterosexual and married at the age of twenty-two, although this proved no 'impediment to their intimacy. There is no reason to suppose that Lorna Burgis resented Regy's love for her husband.'[2] Lawrence Burgis and Esher were due to lunch together at Brooks's Club on the day that Esher died in January 1930.

Esher was actuated by a strong desire to keep those he loved out of the fighting in the 1914–18 War, and by getting Burgis a post as aide-de-camp to Brigadier-General John Charteris, Lord Haig's intelligence chief in the Great War, saved him from service in the trenches. It was also down to Esher that Burgis secured a place on the staff of the Cabinet Office before the war ended. It is therefore due to this physically unconsummated love of Lord Esher for the lad he called 'My Thrush' that we today have verbatim reports of the War Cabinet meetings held during the Second World War, for by 1940 Burgis had risen to the post of assistant secretary to the Cabinet Office, and was thus one of the few people whose job it was to take down word for word whatever ministers said there.

There were strict rules against officials keeping diaries, but Burgis' practice of retaining the verbatim notes he made of War Cabinet meetings was far more serious. It was not simply a sackable offence; if he had been caught, he would have faced prosecution under the 1911 Official Secrets Act. That he knew he was breaking the law is evident from his unpublished autobiography, in which he explicitly stated that he kept his actions secret from the Cabinet secretary, Sir Edward Bridges, and his deputy Norman Brook. The Cabinet Office rules were unambiguous: all notes, after being used to draw up the official minutes, were to be burnt in the office grate in Whitehall. Instead, Burgis stashed them away. He had an eye for great events, and fully appreciated how fortunate he was to be present when history was being made. 'To sit at the Cabinet table at No 10 with Churchill in the chair was something worth living for,' he wrote. 'Perhaps some would have paid a high price to occupy *my* seat, and *I* got paid for sitting in it!'[3] He was proud to have been the only person besides Churchill and Field Marshal Jan Christian Smuts to have been present at the War Cabinet meetings of both world wars.

By the time of the Second World War, Lawrence Burgis was, according to his friend Leslie 'Joe' Hollis, who also worked in the Cabinet secretariat, 'a short, rotund and rubicund person, who loved a good story and a glass of wine'.[4] In later life he became an authority on judging gymkhanas in Oxfordshire, where he retired. He hugely admired Churchill, and was certain that had the Germans invaded Britain in 1940 the Prime Minister 'would have mustered his Cabinet and died with them in the pill-box disguised as a WH Smith bookstall in Parliament Square'. He recalled Churchill in the Cabinet Room:

sitting in his chair at the long table in front of the fire, either in his siren suit, or, if some engagement or attendance at the House of Commons followed the meeting, immaculately dressed in a short black coat, striped trousers, silk shirt and bow-tie with spots. Wonderful hands too – so well kept. He gave the impression that he had just dressed after a bath and had used talcum powder with liberality. As one entered that historic room one could generally tell from the expression on Churchill's face if the meeting was set for fine, fair, or wet and stormy . . . though, as with the uncertainty of our weather prophets, one could not be absolutely sure that an unexpected storm would not blow up from somewhere.[5]

After the Australian Prime Minister Robert Menzies sent Churchill a stuffed flat-billed platypus as a present, it was put on view to the left of the lobby at No. 10. A group of people, including Burgis, were waiting there one day when Churchill arrived and, 'beaming all over', pretended to be the showman at a fairground, crying: 'This way to the flat-billed platypus, gentlemen!'[6]

Sir Edward Bridges' instructions for the writing of Cabinet minutes insisted on their being '(a) brief (b) self-contained (c) in the main, impersonal, and (d) to the full extent the discussion allows – decisive'.[7] Often this was the very opposite of what had actually happened in meetings that were prolix, open ended, highly personal and indecisive. Official Cabinet minutes are therefore opaque documents, usually deliberately so. As one War Cabinet secretariat clerihew put it:

A page from Lawrence Burgis' account of the War Cabinet meeting of
10 December 1941

And so while the great ones depart to their dinner,

The secretary stays, growing thinner and thinner,

Racking his brains to recall and report

What he thinks that they think they ought to have thought.[8]

Sometimes the Cabinet minutes adopted a form of code for the initiated, similar to the Foreign Office euphemism whereby 'a full and frank discussion' meant a blazing row. When at the Cabinet Defence Committee of 2 March 1942, for example, Churchill and General Sir Alan Brooke clashed over the problems caused by the fan-belt drive and the lubrication system of the Cruiser tank, and the minutes record, 'Some discussion then took place on the subject of these defects, in the course of which surprise was expressed that they should not have been detected earlier,' one can be fairly sure that there was a hard-fought and possibly ill-tempered argument.[9]

By reading the original, contemporaneous, handwritten notes that Burgis took, one can see who said precisely what at the meetings. From his jottings it is now possible, six decades later, to recreate the exact discussions that took place. Burgis' very extensive papers have lain almost completely unexamined in the Churchill Archives in Cambridge since they were deposited in 1971. As he was a comparatively minor official, he has not so far excited any interest among historians, although admittedly his calligraphy and private shorthand is more hieroglyphic than easily interpreted English. Nonetheless the hundreds of yellow secretarial sheets do contain the record of what was actually said at those crucial meetings. Readers can if they wish check on my website – www.andrew-roberts.net – how I have reconstructed the sentences of speech from Burgis' shorthand notes.

Also appearing here for the first time in book form are the verbatim reports of Cabinet meetings made by Norman Brook (later Lord Normanbrook). These were released by the British National Archives in 2007 and provide a similar treasure trove of what precisely was said by ministers. Some of the more sensational revelations – such as Churchill's scheme to execute Hitler by the use of the electric chair – were reported in the press, but huge amounts of fascinating information were not, and appear here with the source notes CAB 195/1, 195/2 and 195/3.

Of course verbatim records, however well reported, can tell us next to nothing about the all-important aspects of exchanges besides the mere choice of words used. Swiftness of reply, absence of normal courtesies, tempo of speech, tone of voice, body-language, sheer decibel level, veins standing out on foreheads, clenching of fists, snapping of pencils and everything else that went to make up the expression of the arguments over wartime grand strategy simply cannot be conveyed in an account recording in cold print what was agreed, or even what was actually said. Attempting to reconstruct the scenes of wartime meetings from committee minutes and verbatim reports is like trying to rebuild a Roman villa from a handful of tiny floor mosaics. Nevertheless, a couple of sentences from a diarist who was present can sometimes be far more useful than pages of official documentation. It is therefore very fortunate for historians that there were so very many diarists among the primary actors of the Western Allies and among their best-placed spectators. Sir Alan Brooke, Chief of the Imperial General Staff (CIGS), was the only one among the four principal actors of this book, but a remarkable number of other senior figures kept diaries, 'a vast cloud of witnesses' as one of them put it, even though it was expressly forbidden in Britain on security grounds.

Britons who ignored the strict official regulations against keeping a journal included Churchill's private secretary Jock Colville, Lord Louis Mountbatten and his chief of staff Lieutenant-General Sir Henry Pownall, the Foreign Secretary Anthony Eden and his private secretary Oliver Harvey, Field Marshal Lord Wavell, Colonel Ian Jacob of the War Cabinet secretariat, the British Ambassador to Washington Lord Halifax, the permanent under-secretary at the Foreign Office Sir Alec Cadogan, Brigadier Vivian Dykes of the Joint Staff Mission in Washington, Harold Nicolson MP, the Minister Resident in North-west Africa Harold Macmillan MP, Churchill's doctor Sir Charles Wilson (later Lord Moran), Air Chief Marshal Sir Charles Portal's private secretary Stewart Crawford, the Secretary of State for India Leo Amery, General Sir Edmund Ironside, and even King George VI himself and his private secretary Sir Alan Lascelles. American diarists, who were admittedly under no such official strictures, included Dwight D. Eisenhower and his aide Harry Butcher, Vice-President Henry Wallace, the War Secretary Henry L. Stimson, the Chief of the Joint Chiefs of Staff Admiral William Leahy, the head of the US Army Air Force General Henry 'Hap' Arnold, the Treasury Secretary Henry Morgenthau and General Joseph Stilwell. In Canada, the Prime Minister William Mackenzie King also kept one.

These men knew they were making history, and as the official records can be extremely opaque, we must be grateful that they did. I have drawn extensively on these diaries, and on the unpublished papers of more than sixty confidants and contemporaries of the four principals, in order to try to recreate the drama and passion that went into the formation of Allied grand strategy.

Anyone who was shocked by the attacks on Churchill contained in Brooke's unexpurgated diaries that were published in 1994 – and serialized in the *Sunday Telegraph* under the headline 'Britain's Wartime Military Chief Thought Churchill "A Public Menace"' – ought to read the journals of the equally peppery Admiral Lord Cunningham in the British Library, which I have drawn on particularly in the second half of the book. Yet in Cunningham's 710-page autobiography, *A Sailor's Odyssey*, it is hard to spot a sentence of criticism of Churchill, who was prime minister at the time of publication.

Similar self-censorship took place in 1957 when Brooke's former director of military operations, Major-General John Kennedy, published *The Business of War*, an autobiography based on his daily diaries, at a time when many of the senior Allied wartime figures were still alive and in senior positions (Eisenhower was president for example, and Macmillan prime minister). Born in 1893, and thus ten years younger than Brooke, though sharing many experiences during their careers, Kennedy was educated at Stranraer Academy and Woolwich and entered the Royal Navy in 1911. He was commissioned into the Royal Artillery in January 1915 and served on the Western Front from 1916 to 1918, including at the Somme. Wounded in August 1916, he nonetheless fought at the battle of Ancre in 1917, becoming an acting major. He then served on the British military mission during the Russian Civil War, working with the White commanders-in-chief Denikin and Wrangel, which he 'looked upon as an adventure . . . when I was getting bored'. At the end of a decade spent at the Staff College and the War Office, he became director of plans in 1939.

John Kennedy receives relatively little attention today – possibly because attempting to locate him on internet search engines results in more than sixty million hits relating to someone else of the same name – but his testimony from the very heart of the military decision-making process is compelling. In June 1940 he commanded the Royal Artillery section of the 52nd Division in France under Brooke, and between 1940 and 1943 was director of military operations (DMO), the senior War Office Planner, before becoming assistant CIGS for the rest of the war. He was thus a central eyewitness, but *The Business of War* excised many of the most caustic comments that he had originally written in his diaries, which have never been published *in extenso*. The handwritten daily journals now in the Liddell Hart Centre for Military Archives at King's College London show what this exceptionally well-placed officer genuinely thought at the time, and are an invaluable, though by no means entirely objective, source for both the strategic thinking of the British War Office and the machinations between the principals in this story.[10]

In the decades after the war ended, with self-serving autobiographies and diaries, admiring biographies and slanted histories being published en masse, and with the fear of resurgent Communism revising the story of Yalta for political purposes in the West, it was difficult to arrive at an objective judgement about Allied grand strategy. History was often written in a partisan way, perhaps inevitably because of the immediacy, importance and sheer immensity of the subject. One of the quartet of power – President Roosevelt – never had the chance to tell his own tale, as Brooke did in the sulphurous diary extracts edited by Sir Arthur Bryant, published as *The Turn of the Tide* in 1957 and *Triumph in the West* in 1959, and as his American opposite number General George Marshall did to his biographer Forrest C. Pogue between 1956 and 1959. Churchill himself published no fewer than six beautifully written but highly subjective, not to say in many respects misleading, volumes of war memoirs. Today we can see that the real story was far subtler than the one that emerged shortly after the conflict, and than any of the surviving three represented it. As I hope this book will help show, historical truth tends to defy easy explanations, and is all the more fascinating for it.

Introduction

Franklin Roosevelt, Winston Churchill, George Marshall and Alan Brooke met for the first time in the Oval Office of the White House at noon on Sunday 21 June 1942. Scheduled as a routine strategy session, it was to turn into one of the most significant moments of the Second World War.

Roosevelt and Churchill had arrived in Washington on the presidential train from Hyde Park, FDR's family estate in upstate New York, soon after 9 a.m. Having breakfasted and read the newspapers and official telegrams in the White House, at 11 a.m. the Prime Minister summoned Britain's senior soldier, General Sir Alan Brooke, to come over from the Combined Chiefs of Staff offices on nearby Constitution Avenue. Lieutenant-General Sir Hastings 'Pug' Ismay, Military Secretary to the War Cabinet, who was as usual with the Prime Minister, warned Brooke that Churchill was 'very upset' by some recent decisions taken in his absence by the Combined Chiefs – that is, by the British Chiefs of Staff and their American counterparts the Joint Chiefs of Staff sitting in a powerful new Allied committee. But when he got to the White House Brooke found the Prime Minister 'a bit peevish, but not too bad and after an hour's talk had him quiet again'.[1]

Since Brooke had not expected to visit the White House that day, he was wearing an old suit, and asked to be allowed to change into uniform before he met the President for the first time, but Churchill would not hear of it. They went to the Oval Office together and found Roosevelt, who had been afflicted with poliomyelitis since 1921, seated behind the large desk that had been given to his predecessor Herbert Hoover by the Grand Rapids Furniture Manufacturers Association.

The desk itself was cluttered with knick-knacks and mementoes, many of which can be seen at Hyde Park today. There was a half-dollar commemorative coin in its box, a Lions Club International lapel pin, a stuffed elephant toy and carved wooden donkey, a capstan-shaped paperweight, a tape measure, a novelty figurine of an ostrich, a nail file, an enamelled copper ashtray made in Buffalo, NY, and a bullet about which nothing is known. It seemed more like a bric-a-brac store than the desk of the chief executive of the United States of America, and visiting a year later Brooke 'tried to memorize the queer collection', which also included a blue vase lamp, a bronze bust of Mrs Roosevelt, another small donkey made of hazelnuts, a pile of books, a large circular match stand, an inkpot and a jug of iced water. Colonel Ian Jacob, Ismay's assistant, while admitting that the President's study was 'a delightful oval room, looking south', uncharitably equated Roosevelt's 'junk of all sorts piled just anyhow' with a 'general lack of organization in the American Government'.[2]

After being introduced to the President, Brooke began by apologizing for his informal dress. 'What's wrong with you?' Roosevelt replied jovially. 'Why not take off your coat like I have, you will feel far more comfortable.' It was an oppressively hot day, and the flinty Ulsterman was understandably charmed, later writing in his diary: 'I was much impressed by him – a *most* attractive personality.'[3] The Chief of Staff of the US Army, the courtly but steely Pennsylvanian General George C. Marshall, then arrived, and talks began over the various alternative strategies for a major Allied attack against the Germans in 1942.

Discussions stopped for lunch with Mrs Roosevelt at one o'clock, at which the President reminisced that Brooke's father and brother had stayed at Hyde Park half a century earlier, which the general had not known. Sir Victor Brooke had visited America looking for investment opportunities, and had written to his wife of the 'glorious, wooded cliffs and rolling forests' of the Hudson Valley, as well as of the Roosevelts' kindness in putting them up for three days in their 'dear little house, with a verandah all around it'. Brooke confided to his diary that night that he 'could not help wondering what father would have thought if he had known then the circumstances in which Roosevelt and his youngest son would meet in the future!'

Back in the Oval Office after lunch, as they returned to their deliberations, a pink slip of telegraph paper was brought in and handed to the President, who read it and, without saying a word, gave it to the Prime Minister. It announced that the Mediterranean port of Tobruk, the

British Eighth Army's stronghold in Libya that had for months been a potent symbol of resistance to Field Marshal Erwin Rommel's Afrika Korps, had surrendered without warning to the 21st Panzer Division. Tobruk's garrison – including two South African brigades and one from a British Guards regiment, as well as sixty tanks – had been captured en masse, and German radio broadcasts were claiming twenty-five thousand prisoners-of-war. (Rarely for him, Dr Goebbels had underestimated; the true figure turned out to be almost thirty-three thousand.)

'This was a hideous and totally unexpected shock,' recalled Ismay, 'and for the first time in my life I saw the Prime Minister wince.' Neither Churchill nor Brooke had foreseen what Brooke called this 'staggering blow'. Marshall later spoke of how 'terribly shaken' Churchill looked.

Ismay, whose fifty-fifth birthday it was, left immediately to try to get confirmation of the news from London. As he walked down the corridor, he remembered that it was also the birthday of his friend General Sir Claude Auchinleck, Commander-in-Chief in the Middle East. 'Poor Claude,' he later recalled thinking to himself. 'What a horrible anniversary!' He soon returned with a copy of the message that the Commander-in-Chief of the Mediterranean Fleet, Admiral Sir Henry Harwood, had sent to the Admiralty, stating: 'Immediate. Tobruk has fallen and situation deteriorated so much that there is a possibility of heavy air attack on Alexandria in near future and in view of approaching full moon period I am sending all eastern Fleet units south of [the Suez] Canal to await events.'

Worse was to come: a telegram from Richard Casey, the British Government's Minister Resident in the Middle East, marked 'Most Secret. Most Immediate', reported that although it had been proposed 'to fight as strong a delaying action as possible' on the Egyptian border, it was concluded that 'The forces at our command in this theatre are inadequate to enable us to cope with the enemy.' There was every prospect, therefore, that Egypt might fall to the Axis powers of Germany and Italy. It later also transpired that the great bulk of stores for Tobruk's defence – vast quantities of oil, petrol, aviation fuel, ammunition and food – had inexplicably not been destroyed, but had fallen virtually intact into the hands of the Germans, who would now be using them for their march on Cairo.

A year earlier, when Tobruk had previously been under siege, Churchill had sketched out to Roosevelt's special representative Averell Harriman 'a world in which Hitler dominated all Europe, Asia and Africa and left the United States and ourselves no option but an unwilling peace'. He argued that this was only preventable because Tobruk 'still resists valiantly', for if Egypt and therefore the Suez Canal were to fall to the Nazis, then the whole of the Middle East would collapse, after which Spain, Vichy France and Turkey would embrace the Axis powers and Hitler's 'robot new order' would inevitably triumph. Tobruk was thus far more than a strategically important Mediterranean arsenal for Churchill: it was a shibboleth of survival, and its fall correspondingly dire.

At this point in the war, Britain had been defeated by the Germans wherever the two had fought on land: in Norway in April 1940, in France and Belgium the following month, in Greece in April 1941 and in Crete the following June. In May and early June 1942, Lieutenant-General Sir Neil Ritchie had been defeated by Rommel in the Gazala area, forcing a withdrawal towards Egypt and leaving Tobruk to defend itself. Alongside this debilitating series of defeats on land, Allied shipping losses in the Atlantic had doubled since January 1942; the Arctic convoys were coming under heavy pressure from the Kriegsmarine and Luftwaffe in northern Norway; the convoy route around southern Africa was increasingly threatened by U-boats, and the expansion of Bomber Command seemed to have stalled. Seven years later, Brooke summed up the global situation they had faced by saying: 'German Forces were through the Caucasus, Japanese forces were threatening Australia and India, the Mediterranean was closed, and Persia had been entirely depleted of forces to save threatened points. The whole of the oil reserves in the Middle East in Iraq and Persia were at Hitler's mercy.'[4]

Furthermore, Churchill knew he would now come under renewed political pressure back in London, and a motion of no confidence in his government was indeed tabled in the House of Commons soon afterwards. 'I am ashamed,' he confided to his doctor at the time. 'I cannot understand why Tobruk gave in. More than thirty thousand of our men put their hands up. If they won't fight . . .' The Prime Minister then 'stopped abruptly', since what followed was 'too ghastly to articulate'. As Churchill himself recalled in his memoirs: 'This was one of the heaviest blows I can recall during the war . . . Defeat is one thing; disgrace is another.'[5]

*

It was at this desperate juncture that there began the three-year relationship between the four chief strategists of the Western Allies, the quartet of power that ultimately crafted the victories that were to come. Although it is taken for granted that emotion, persuasiveness and charisma have a large part to play in politics, the same is not generally thought to be true of grand strategy. Intelligence reports, weather forecasts, hard facts about opposing forces and objective military assessments are believed to decide when, where, why and how great offensives are launched. Yet, as I hope this book will show, the two political Masters and two military Commanders of the Western powers who ultimately took these decisions together were flesh and blood, working under tremendous stress, and prey to the same subjective influences as everyone else.

Why, if the USA was attacked by the Axis in the Pacific Ocean, did she devote such effort to counter-attacking in North Africa? Why, if the most direct route to Germany from Britain was via north-west France, did the Western Allies march to Palermo and Rome? Why, if Operation Overlord was intended to drive into Germany via north-west France, did four hundred thousand men land 500 miles to the south more than two months later? Why did the Allies not take Berlin, Vienna or Prague, but allow the Iron Curtain to descend where it did? One of the aims of this book is to show the degree to which the answers to these questions, and many more, turned on the personalities and relationships of the four key figures who are its central focus: Franklin D. Roosevelt, Winston Churchill, George C. Marshall and Sir Alan Brooke.

The lives of hundreds of thousands of soldiers and civilians ultimately rested on the deliberations of these four: two Americans and two Britons, two politicians and two soldiers. Each of the four men was strong willed, tough minded and certain that he knew the best way to win the war. Yet, in order to get his strategy adopted, each needed to ensure that he could persuade at least two of the other three. Occasionally the politicians would side together against the soldiers, and vice versa. (Up in Hyde Park the day before the Tobruk news arrived, for example, Roosevelt and Churchill had agreed to oppose Marshall's plan for an attack on France in 1942.) More often the Britons and Americans would take up positions according to nationality, but sometimes alliances were formed across both professional *and* national lines; just as politicians had to master strategy, so the soldiers were forced to become political.

Once made, such groupings were always likely swiftly to reconfigure, as the four Masters and Commanders danced their complicated minuet, each fearing the potentially disastrous consequences of getting out of step with the others. When that happened to any one of the four – as it did to Churchill, Marshall and Brooke at different stages of the war – his views were overruled by the opposing trio. Each Master and Commander was thus constantly manoeuvring for position vis-à-vis the other three, and only one of them never found himself isolated.

Both real and feigned anger was seen at their many wartime meetings, as well as immense moral and political pressure, threats and cajolery, deliberate misleading of each other on occasion, high rhetoric masking low politics, shouting matches followed by last-minute compromises, mutual suspicion and exasperation, and even one near nervous breakdown. Yet charm, humour and good-fellowship could sometimes lift the mood at key moments too. There were titanic rows and emotional reconciliations, and at the end of it all there was, of course, Victory. This then is the story of how the four Masters and Commanders of the Western Allies fought each other over how best to fight Adolf Hitler.

PART I

Enchantment

I

First Encounters:
'I had heard a good deal about him!'
1880–June 1940

War is a business of terrible pressures, and persons who take part in it must fail
if they are not strong enough to withstand them.
                    Winston Churchill, *The World Crisis: 1915*[1]

Winston Churchill, a man who was said to have 'won the decathlon
of human existence', did not impress any of his fellow Masters and
Commanders on first acquaintance.[2] On Monday 29 July 1918, Franklin
D. Roosevelt, then Assistant Secretary of the US Navy, was asked to
speak impromptu at a dinner of Allied war ministers at Gray's Inn, one
of London's ancient legal Inns of Court, and years later he recalled that
Churchill had 'acted like a stinker' and was 'one of the few men in public
life who was rude to me'.[3] They then did not see each other again until
August 1941, when – to Roosevelt's evident chagrin – Churchill had to
admit to having completely forgotten the occasion. He later remembered
it for the benefit of his war memoirs, however, writing of how he had
been 'struck' by Roosevelt's 'magnificent presence in all his youth and
strength'.[4]

George Marshall was similarly underwhelmed by Churchill on their
first contact in 1919, at a great Allied victory parade in London,
and twenty-two years later regaled a Sunday luncheon party at the
British Embassy in Washington with the story. There had been three
thousand American troops present, 'all picked men of about 6'2", with
every kind of decoration', yet every time that Marshall tried to make
any observations to Churchill about them, all he elicited was gruff
silence. Prohibition had been ratified by the US Congress that year and
finally, after all the dignitaries, including King George V, had pro-
cessed around the rear rank and back up the flank of the parade,
Churchill turned to Marshall to make his only remark of the day: 'What

a magnificent body of men, and never to look forward to another drink!'[5]

Alan Brooke's first personal encounter with Churchill came down a crackling telephone line between his headquarters at Le Mans in France and 10 Downing Street in June 1940, and was to be the worst by far.

By contrast with Churchill's behaviour at the parade, the one adjective constantly employed to describe George Catlett Marshall was 'gentlemanly'. Good-natured, charming, with fine manners, Marshall was nonetheless a tough man, and knew it. 'I cannot afford the luxury of sentiment,' he once told his wife Katherine about his job as US Army chief of staff, 'mine must be cold logic. Sentiment is for others.'[6] She agreed, writing in her autobiography, *Together: Annals of an Army Wife*, of how she had read many articles and interviews that mentioned her husband's retiring nature and modesty, but she added: 'Those writers have never seen him when he is aroused. His withering vocabulary and the cold steel of his eyes would sear the soul of any man whose failure deserved censure. No, I do not think I would call my husband retiring or overly modest. I think he is well aware of his powers.'

There was self-effacement nonetheless. Marshall's friend and diligent biographer Forrest C. Pogue noticed that Marshall deprecated the use of the word 'I' and tended to adopt the first person plural in describing the actions of the War Department or the Joint Chiefs of Staff, even when he had been the driving force behind them. In a passage accusing Anthony Eden, Bernard Montgomery and others of vanity, Churchill's doctor Sir Charles Wilson (later Lord Moran) wrote: 'To remain gentle and self-effacing after climbing to the top of a profession', as Field Marshal Lord Wavell and George Marshall had done, 'is to me an endearing trait.'[7] It is one thing to be thought of as self-effacing, but altogether another to be regarded as an exemplar of it.

Alone among the four subjects of this book, Marshall – born in Uniontown, Pennsylvania on the last day of 1880 – did not come from the upper classes. His father was a prosperous co-owner of coke ovens and coalfields, at least until December 1890 when an unwise investment in a Shenandoah Valley land promotion brought him to the brink of bankruptcy. Marshall nonetheless had a happy childhood, and his family could still just about find the $375 per annum (plus $70 for uniforms) to send him to the prestigious Virginia Military Institute (VMI) in Lexington, Virginia.

Years afterwards, Marshall recalled that he had overheard his elder brother Stuart, who had himself graduated from VMI, begging their mother not to allow George to enrol there because his lack of intellect would disgrace the family name. 'Well, that made more impression on me than all the instructors, parental pressure, or anything else,' Marshall recollected. 'The urgency to succeed came from hearing that conversation; it had a psychological effect on my career.'[8] Sure enough, he became first captain of the Corps of Cadets, played All-Southern football, and graduated high in the class of 1901.

Although it had ended thirty-two years before Marshall arrived at VMI in 1897, the American Civil War still dominated the ethos of the Institute. The building itself had five or six cannonballs from the conflict still sticking out of its walls. Marshall's hero and role model was the Confederate leader Robert E. Lee; watching Stonewall Jackson's widow at a memorial anniversary of the battle of New Market, and seeing the graves of its young dead, made a profound impression on him.

The Spanish–American War broke out in the spring after Marshall joined VMI, and as he told the cadets there fifty-three years later, on what was by then called Marshall Day, 'For the first time the United States stepped into the international picture. At that period, there was not a single ambassador accredited to the United States. We were recognized in the world largely as a country of Indians and buffalo, crude and remarkable manners, and the sudden wealth of a few.'[9] By the time Marshall himself became secretary of state of the United States in 1947, it was indisputably the most powerful country in the world, partly because of what it had achieved during his time as Army chief of staff.

On leaving VMI, and having personally lobbied President McKinley in the White House for the right to sit his lieutenant's examination early – not the action of an overly modest lad – Marshall married his sweetheart, the belle of Lexington, Lily Carter Coles. He had been courting her ever since his last year at the Institute, where he had risked expulsion in order to meet her in the evenings. 'I was much in love,' was his explanation for the risks taken with his nascent military career. They married on 11 February 1902 and he managed to extend his honeymoon from two days to one week before reporting for duty in the Philippines.

Although America's instantaneous victory over Spain meant that Second Lieutenant Marshall served in the Philippines only in peacetime, his career was meteoric after his return in 1903. As senior honor graduate of the Infantry–Cavalry School at Fort Leavenworth, Marshall won promotion to first lieutenant in 1907 and became an instructor there. Fort Leavenworth was then, and was to remain, a centre of advanced military thinking in the Army, and it was there that Marshall formed many of his assumptions about strategy and tactics. During another tour of the Philippines in 1913–16 he organized, as chief of staff for a US field force, a defence of the Bataan Peninsula and Corregidor against a mock Japanese invasion.

As a captain assigned to the General Staff, Marshall sailed to France in 1917, in the first convoy of troops to go there, and was reputedly the first man to alight from the first boat.[10] He found a conflict of deadlock and attrition, very different from the war of movement seen in the last few months of 1914, and then again in the last three months of 1918. Marshall participated in the first entry of US troops into the Allied line, in the Luneville Sector, and – as a Staff officer – in the victory at Cantigny on 28 May 1918, the first American offensive of the war.

After the repulse of the German offensive of June 1918, Marshall was detailed to the Operations Section of US General Headquarters at Chaumont, and in August was attached to the Staff of the First American Army, of which he became chief of operations before the Armistice. General John 'Black Jack' Pershing, the Commander-in-Chief of American forces in Europe, eventually promoted him to colonel. Crucially, in May 1919 Marshall became aide-de-camp to Pershing, under whom he served for the next four years. Although he had not seen action in the field, therefore, Marshall was held to have had an extremely good war. He had witnessed the mutual slaughter of 1917 give way, in the late summer and autumn of 1918, to the open war of manoeuvre that the Allies won. It was to have a profound effect on his strategic thinking.

Alan Francis Brooke was born on 23 July 1883 at Bagnères-de-Bigorre near the French Pyrenees, a fashionable area around Pau where his parents went for the hunting – it was known as 'the Leicestershire of France' – and for the fine climate. He was the seventh and much the youngest child of Sir Victor Brooke, who had inherited, aged eleven, the title of third baronet and the estate of Colebrooke Park in Brookeborough, County Fermanagh in Northern Ireland. Alan's mother was Alice Bellingham, the daughter of another Irish baronet.

On both sides of Brooke's family lay deep roots in Ireland's Protestant

Ascendancy. Nicknamed 'the Fighting Brookes of Colebrooke', they had been soldiers of the Crown for centuries. One had defended Donegal Castle during the English Civil War, another took over Lambert's Brigade to hold the centre of Wellington's line at the battle of Waterloo. No fewer than twenty-six members of the family served in the First World War, and then twenty-seven in the Second, of whom twelve died in action. Yet it was to be the sensitive youngest sibling Alan who was to become by far the greatest soldier of them all.

It is not hard to see from where Alan Brooke's utter fearlessness was derived. Even if his DNA had not included generations of warriors, his father was a Victorian hero–adventurer, as well as that most unusual of phenomena – a genuinely popular Irish Protestant absentee landlord. Born in 1843, Sir Victor Brooke was named after his godmother Queen Victoria. His dead-eye shooting abilities – he could split a croquet ball thrown in the air with one shot and then split the largest fragment with the second – stood him in good stead hunting in India, where 'his life depended more than once upon making no mistake'.[11]

From floor to ceiling at Colebrooke, in halls and passages and many of the rooms, there were heads of every variety, including two tigers and a black panther, and vast elephant tusks were piled up under the billiard table. Handsome, fair-haired, 6 foot tall and 45 inches around the chest, Sir Victor resembled a John Buchan hero. Along with strength of character, an 'open-hearted Irish nature' and immense charm, he was an assured public speaker and universally popular. At the London Fencing Club, he once jumped 5 feet 10 inches in the high jump, and could lift enormous weights. Hearing that a local policeman had won a reputation as an undefeated wrestler, he issued a challenge and duly beat him. He then outran a Canadian champion hurdler. His sporting feats were well known in Ulster, and having such an extraordinary father must have had an effect on his youngest son. When Alan Brooke showed great moral courage at various moments of his military career, it should be recalled that his father had tracked tigers, wolves and bears, and had crossed jungles and deserts in order to do so. He was also a noted biologist with intellectual attainments to match his physical ones. Sir Victor died aged only forty-eight, from fatigue induced by tracking ibex across an Egyptian desert when he was supposed to be convalescing from a lung that he had punctured while hunting in France. Alan was eight years old.

As a child, Alan Brooke lived a self-contained life, close to nature and to his mother.[12] Growing up for most of the year in the Pyrenees, he spoke French (with a heavy Gascon accent) before he learnt English, and spoke both languages very fast, something that some Americans were to come to dislike and mistrust later on, fearing that a fast-talking Limey was trying to get something over on them. Educated at a day school at Pau, Brooke was never sent to an English boarding school, further removing him from the then prevailing Spartan culture of heartiness, but also from interaction with contemporaries of his own age, nationality and social background. In contrast to Marshall's success at football, Brooke did not play team games. Quite how little of a team-player he would turn out to be later in life had yet to make itself known.

For all that he later seemed to others to be cold, restrained, tough and on occasion heartless, Brooke was in fact an emotional man. Churchill's secretary Elizabeth Nel wrote that he 'always seemed to me something of an enigma; he seemed so calm and well controlled, and yet the expression of his face sometimes betokened that he had strong feelings beneath the surface.'[13] He did indeed; Brooke was a loner who had all the self-assurance of the British upper classes of the day. From an early age he knew where he came from, what he liked, what he wanted and how to get it. Class was a vital factor for late Victorians such as Churchill and Brooke. Churchill's aristocratic credentials as the scion of a dukedom created in 1702 impressed and sometimes overawed his contemporaries, though not Brooke, whose ancestors had served the Crown for a similar length of time.

After a short period at a crammer, Brooke entered the Royal Military Academy at Woolwich, but only just, coming sixty-fifth out of seventy-two in the entrance exam (he passed out seventeenth). Had he done any better he would have qualified for a commission in the Royal Engineers, and he would probably not have wound up on the General Staff after the Great War. Lack of success at a crucial moment in life can sometimes prove invaluable later on, however frustrating it might seem at the time. As well as being fluent in French and German, Brooke was soon expert in gunnery. After four years in Ireland with the Royal Field Artillery, he served in India for six years after 1906, showing an aptitude for military life and a natural propensity to command. The outbreak of the Great War found him on honeymoon, having married 'the beautiful, affectionate, vague, happy-go-lucky' Janey Richardson, to whom he had been engaged – secretly, due to lack of money – for six years.

Brooke began the Great War as a lieutenant in command of an ammunition column of the Royal Horse Artillery on the Western Front, and ended it as a lieutenant-colonel. He fought on the Somme and was afterwards appointed to serve in Major-General Sir Ivor Maxse's 18th Division, then as chief artillery Staff officer to the Canadian Corps, where he co-invented the 'creeping barrage', the method by which enemy machine-gun posts were bombarded just as troops attacked them, with the process moving steadily forward as further ground was gained. It was said that fewer casualties were suffered in those units to which Brooke was attached than in similar engagements.[14] Like Marshall, Brooke had had a good war, and he was selected for the very first post-war course at the Staff College at Camberley.

Winston Churchill was fascinated by strategy, tactics and soldiering all his life. When he wasn't actually fighting wars, he was generally thinking and writing about them. He had played with toy soldiers as a child, joined the Army Class at Harrow aged fourteen, and entered the Royal Military College at Sandhurst (on his third attempt) at nineteen. Five hours a day there were devoted to the subjects of Fortifications, Tactics, Topography, Military Law and Military Administration. This involved studying the theoretical and practical side of military engineering, explosives, field guns and ammunition, the penetration of projectiles against defensive structures, the construction of obstacles and stockades, fields of fire, the tactical use of defensive positions, bivouacking, water purification, the importance of terrain in determining tactics, the optimum combination of artillery, cavalry and infantry, the measurement of slopes and embankments, cartography, recruitment, pay and allowances, quartermastering, and the movement of men, horses and equipment.[15]

Yet this was not enough for the young Winston, who recalled in his autobiography *My Early Life* that no sooner had Lord Randolph Churchill instructed his bookseller to send his son any books he might require for his studies than the cadet ordered Lieutenant-General Sir Edward Hamley's *Operations of War*, Prince Kraft zu Hohenlohe-Ingelfingen's *Letters on Infantry, Cavalry and Artillery*, and *Infantry Fire Tactics* by an author named Mayne, 'together with a number of histories dealing with the American Civil, Franco-German and Russo-Turkish wars, which were then our latest and best specimens of wars. I soon had a small military library which invested the regular instruction with some kind of background.'[16] When invited to dinner at the Staff College at Camberley, Churchill was able to talk to the top military experts in Britain about 'divisions, army corps and even whole armies; of bases, supplies, and lines of communication and railway strategy. This was thrilling.'

His early studies imparted to Churchill a thrill that never left him, not as a war correspondent in Cuba, nor during his time with the Malakand Field Force on the North-west Frontier of India, especially not during the Sudanese campaign in 1898. He continued to read widely and voraciously on the subject of grand strategy, and wrote about Marlborough's wars, the American Civil War, the River War in the Sudan and several other conflicts with the self-assurance of an expert military historian. Long before the Great War broke out in 1914, in which he was to have a leading role in the creation of British grand strategy, Churchill had immersed himself in the subject, and even the staggering reverse represented by the Dardanelles disaster in 1915 failed to dent his ardour for it. During the inter-war period, his 'wilderness years', Churchill stayed avidly abreast of all the new technological and intellectual developments regarding military equipment and strategic thinking. By the time Britain declared war on Germany on 3 September 1939 – he was appointed first lord of the Admiralty that same day – Churchill was supremely confident of his ability to discuss grand strategy with the General Staff as much more than an interested and occasionally inspired amateur: he saw himself as their equal.

Once asked which department he disliked more, the Foreign Office or the Treasury, Churchill replied: 'The War Office.'[17] It had been the River War that had left him convinced that the Army's bureaucracy was inefficient and also that the Army General Staff were incompetent, something that was regularly confirmed for him by contact with both in the Boer War and subsequent conflicts. In his book about the Great War, *The World Crisis*, Churchill indicted the General Staff for having narrow vision and rigid minds. He was angered by how long technical innovations, such as the tank, took to gain acceptance, and described Staff officers as men 'whose nerves were much stronger than their imaginations' and whose sang-froid in the face of catastrophe was 'almost indistinguishable from insensitivity'. During the Second World War, Churchill also believed the War Office to be generally 'hidebound, devoid of imagination, extravagant of manpower and slow'.[18] The scene was thus set for titanic clashes with its senior serving officer, Sir Alan Brooke, who was infuriated by his criticisms and sought to refute them at every opportunity.

Lord Halifax, who sat in several Cabinets with Churchill, found the Prime Minister's working methods 'exhausting for anybody who doesn't happen to work that way; discursive discussions, jumping like a water bird from stone to stone where the current takes you'. He blamed Churchill's 'overwhelming self-centredness, which with all his gifts of imagination make him quite impervious to other people's feelings'.[19] Although this certainly had an element of truth to it, Colonel Aubertin Mallaby, the Deputy Director of Military Operations at the War Office, pointed out that with the Prime Minister:

every single thing in the life of each day was an integral part of a work pattern. There was no question of times on duty and times off, no curtain coming down and dividing work from leisure. There was fun and talk and food and drink and films but all these fitted naturally into the very long *working* day. The only real respite from work was a few hours' sleep.[20]

*

By complete contrast with Churchill, Marshall and Brooke, Franklin Roosevelt did not seem to have any strongly held or closely thought-out views on grand strategy when the United States entered the Second World War, except the understanding that his country needed a vastly larger army, navy and air force as soon as possible. Apart from a profound love and knowledge of the US Navy that he contracted while its assistant secretary from 1913 to 1920, military affairs had not affected his career. It was perhaps this very *absence* of any overarching theory of grand strategy that made it possible for him to hold the ring so effectively during the hard-fought contests between the other three principals of this book.

An eighth-generation American of Dutch origin, Franklin Roosevelt was – like his fifth cousin President Teddy Roosevelt – 'of impeccable New York stock, with many generations of prosperity behind them. Insofar as there is an American aristocracy . . . both Roosevelts clearly belonged to it.'[21] After qualifying as a barrister in 1907, Franklin became a New York state senator from 1910 to 1913 before being appointed assistant secretary of the Navy by Woodrow Wilson. He had been impatient for America's involvement in the Great War long before her declaration in April 1917, and since the Navy Secretary, Josephus Daniels, was a 'good-natured, paunchy, puritanical, languid North Carolina newspaper publisher with no maritime background but pacifist and internationalist leanings', it was largely left to Roosevelt to prepare the department for war, which he did with gusto, and somehow without alienating Daniels.[22] He enjoyed reminding people that his cousin Teddy – who had also been assistant secretary of the Navy – was the man who had ordered Admiral Dewey to attack Manila during the Spanish–American War and was the father of modern American maritime power.

From the age of sixteen Franklin Roosevelt was an admirer of the works of the American historian and geo-strategist Admiral Alfred Thayer Mahan, with whom he corresponded until Mahan's death in 1914, and insofar as he can be said to have had views on grand strategy they derived from Mahan's belief in the overwhelming influence of sea power on world history. Mahan had also been a friend and teacher of Teddy Roosevelt, but influenced his cousin almost as profoundly. (Nonetheless, Franklin was convinced that the development of air power meant that Mahan was wrong to claim that the Philippines could not be defended from Japanese attack. In the event the dead admiral would be proved right and the living President wrong.)[23]

A talented sailor who loved the sea, was never seasick, knew how to rig and change sail in all conditions, Franklin Roosevelt had to be persuaded by his father to attend Harvard rather than the Naval Academy at Annapolis and by President Wilson not to leave his Administration to join the Navy in 1917. 'No American president', writes his biographer, 'came to office with as much knowledge of ships, the sea, and sea power and strategy as did FDR. In his first two terms as president he spent an average of forty-five days per year at sea, his preferred escape from the political hothouse of Washington.'[24] Yet the Commander-in-Chief's fascination with the Navy did not develop into a similar interest in America's Army and Air Force or, *per se*, in military strategy during a war that, after all, turned out not to be decided by sea power in the Mahan tradition.

Where Roosevelt did have an acute strategic sense that was to serve his country well in the Second World War was in his appreciation that air power was going to be far more important than it had been in any previous conflict. At the time of the Munich Agreement in 1938 he had instinctively understood the need massively to increase the United States Army Air Force (USAAF), instituting a plan to build 15,000 planes a year. At the time of the fall of France twenty months later, he announced that this should be increased to 50,000, a proposal which Hitler greeted with incredulity but which ultimately, and especially after Pearl Harbor, the United States massively exceeded. (In the Willow Run factory in Detroit alone, Ford built more than eight thousand B-24 Liberator bombers in the last sixteen months of the war.)

Roosevelt's appreciation of the central importance of air power to future operations came at the right moment. Many of the hardest-fought engagements of the war were finally decided by which side had superiority in the air, and Operation Overlord could not have been launched without complete domination of the skies. (As we shall see, whereas the Allies launched more than 13,000 sorties over the invasion areas on D-Day, the Luftwaffe managed only 319.) Although Roosevelt's contribution to the planning of individual campaigns was minimal, his political sense of when it was right for the Allies to return to France was pitch-perfect, and his insistence on a greatly expanded American air force proved invaluable.

Roosevelt was the Democratic Party's candidate for vice-president in 1920, running on the ticket of Governor James M. Cox of Ohio. It was no fault of his that they lost by sixteen million votes to nine million; Woodrow Wilson's brand of liberalism and League of Nations internationalism was by then no longer popular. It was the following August that Roosevelt was stricken by poliomyelitis, leaving him paralysed from the waist down for life. The next thirty-five months were spent in semi-recovery, before he established himself firmly as a coming man at the Democratic convention of 1924, with a scintillating speech nominating Governor Al Smith for presidential candidate, albeit unsuccessfully. He later served as governor of New York between 1928 and 1932, and defeated Herbert Hoover in the presidential elections of November 1932.

Concentrating on the economic, political and legal aspects of his self-proclaimed New Deal to ameliorate the still-debilitating impact of the Great Depression, Roosevelt had little time to consider grand strategy, and since in the 1930s the United States was under no conceivable military threat, there was no reason for him to. Roosevelt had to face the rise of Hitler in his first Administration, although this did not require him to think too deeply about grand strategy either, because

Nazi Germany still posed little direct threat to the United States. The Japanese had already invaded China by the time Roosevelt arrived at the White House, yet beyond criticizing their presence there it was eight years before his Administration took effective action against Japan, by imposing oil and other embargoes. It was not America's duty to act as the world's policeman in the 1930s, a role only thrust upon it in the following decade, and there seemed to be no need for him to master grand strategy or keep abreast of military developments as Churchill, Marshall and Brooke did – at least until 7 December 1941.

Just as the crisis of Churchill's life had come in 1915 over the Gallipoli débâcle, and Roosevelt's when he was incapacitated by polio in 1921, so the crisis in Brooke's came in April 1925 when his adored wife Janey was killed in a car crash while he was at the wheel of their Bentley. Swerving on a wet road to avoid a bicyclist who had turned in front of him unexpectedly, the open-topped car skidded and overturned. Brooke broke his leg and several ribs, but Janey snapped her vertebra and died a few days later, having contracted pneumonia after an operation to save her from paralysis. Their young daughter and son were left mother-less, and two years later Brooke – who used to drive too fast and blamed himself for the accident – wrote: 'I very much wish I could have finished myself off at the same time.'[25]

Several diverse people in a position to know, such as Brooke's biographer General Sir David Fraser, his subordinate General Sir Bernard Paget, Lord Mountbatten and the historian Nigel Nicolson, have seen in the death of Janey the moment when Brooke developed, as Paget put it, 'two distinctive personalities'. One was Brooke the soldier: 'ruthless, decisive, short-tempered to the point of rudeness, remote and in his military life, lonely'. Then there was Brooke the man: 'emotional to the point of sentiment, a lover of nature (especially birds), a family man with deep roots in the past and a sense of responsibility for the future, an easy comradeship with all those who share in his loves and beliefs'.[26] Mountbatten believed that because of his sorrow Brooke 'never let drop the façade which he had created and behind which he hid his kind-heartedness and sensitiveness – perhaps deeming them weaknesses'.

Brooke's emotional defence mechanism was 'to immerse myself as soon as possible in work, and to let absorption in my profession smother pangs of memory'. Whether it worked emotionally is doubtful – Brooke became withdrawn and distant, and scarcely smiled for four years – but it certainly worked professionally. After instructing at Camberley from 1923 to 1926, where he met men such as the sixth Viscount Gort, John Dill and Bernard Freyberg, whose fates were to intertwine closely with his for good and ill, he became one of the first students at the prestigious Imperial Defence College (now the Royal College of Defence Studies), where he later returned for two years as an instructor. Dill was Army instructor there from 1926 to 1928. It was an elite organization intended for the senior officers of all three services as well as a few civil servants, and completion of the year-long course allowed one to put 'idc' after one's name in the service lists. Among other students were Claude Auchinleck, Admiral Tovey, Canada's General McNaughton and Air Chief Marshal Peirse.[27] Alumni were both conscious of their exclusive status and loyal to Dill, their 'headmaster'.

From 1929 to 1932 Brooke commanded the Royal School of Artillery at Larkhill in Wiltshire and in 1934 he took over an infantry brigade. He became a major-general in 1935, after which he was appointed director of military training and shortly thereafter the commander of the British Mobile (that is, armoured) Division. This varied peacetime military experience on top of his wartime success implied that he was being groomed for the top. Away from work, Brooke managed to indulge his passions for ornithology and angling – as solitary occupations as it is possible to have – and he was to become one of the greatest non-professional authorities on birds of all kinds. 'The indefatigable ornithologist is ready to spend hours motionless in a hide,' wrote the *Times* reviewer of Brooke's biography in 1982, 'and is possessed to a high degree of the gift of identifying an object precisely and then never losing sight of it.'[28] Brooke's zeal for bird-watching was all-encompassing: in 1944 he persuaded the RAF to reprieve an island off the Norfolk coast as a bomb-testing area because the roseate tern nested there, and close to D-Day he broke off a conversation with a member of his staff about landing preparations to talk about a photograph he had taken of a marsh tit. At the end of a long meeting at the War Office in August 1943, Brooke asked his director of military operations to stay behind. After everyone had left, he shut the door, opened a drawer in his desk and took out a book, saying: 'Have you read this? It is most remarkable.' It was Edgar Percival Chance's *The Truth about the Cuckoo*.[29] (After the war, the historian Kenneth Rose asked Brooke whether he had ever been tempted to take Churchill bird-watching with him. 'God forbid!' the field marshal replied. As Lady Soames has pointed out, 'Can you imagine Papa ever wanting to go bird-watching?') Brooke's ability to relax – through ornithology, bird-photography and fishing – was, according to his deputy CIGS Sir Ronald Weeks, 'his saving, for he was always highly strung'.

Brooke's grief and sense of guilt over Janey's death were also partly assuaged by his marriage in December 1929 to Benita, Lady Lees, the daughter of one Dorset baronet and widow of another who had died of wounds received in the Dardanelles. We are fortunate that Brooke's second marriage was blissfully happy, since it was partly to inform and amuse Benita that her husband wrote his daily diary throughout the Second World War. (Benita, by then Brooke's widow for five years, also died as a result of a car crash, in 1968.) There are any number of reasons why one might wish to keep a daily record of one's life, which must include narcissism, historical interest, self-justification, financial recompense, to assuage the curiosity of one's children, to amuse oneself in one's dotage, and doubtless many other, darker psychological impulses. It would be naive to believe that none of these (or others) actuated Brooke, but he was certainly also writing for Benita. Brooke's diary acted as a powerful emotional safety-valve too, allowing him to make remarks about colleagues that he might well otherwise have made to their faces, to potentially devastating effect. 'Whatever doubts or fears Brooke may have had,' recalled the politician David Margesson about Brooke's wartime poker-face, 'he kept them from his colleagues.' Projecting confidence in victory was a vital attribute of any Chief of the Imperial General Staff, and confiding his fears to his journal allowed Brooke the more easily to hide them from his colleagues, whose morale was sustained by the sight of a consistently sanguine commander.

Brooke seems to have taken a strangely inconsistent attitude towards security; he would severely admonish anyone giving classified information over non-scrambler telephones, yet he posted his diaries to his wife by Royal Mail.[30] Whether the many journals kept by senior British officials would have helped the Third Reich much had it successfully invaded Britain might be doubted, but they undoubtedly help historians. When the American historian Forrest C. Pogue was researching for his official biography of Marshall, no fewer than four British officers allowed him to use material from their diaries, each on the condition that he never revealed the fact that they had kept them.[31]

In 1942, a Dr Freeman wrote to Marshall to encourage him to 'keep a memorandum of momentous daily happenings', but the general replied that his policy was not to do this. 'Such a practice tends to cultivate a state of mind unduly concerned with possible investigations,' he replied, 'rather than a complete concentration on the business of victory.' He also suspected that diaries might lead 'subconsciously to self-deception or hesitations in reaching decisions', and he reacted 'explosively' when he discovered his subordinates were keeping them.[32]

Having succeeded his friend Archibald Wavell as commander-in-chief of Southern Command in August 1939, the outbreak of war the following month saw Brooke appointed to command II Corps of the British Expeditionary Force (BEF) that was being sent to France under his fellow Irishman Lord Gort in anticipation of a German attack in the west. Brooke chose Bernard Montgomery and Harold Alexander as his divisional commanders, both Ulstermen like him.

At that stage Brooke had not yet met or spoken to Winston Churchill, although he had followed his political career with interest. Two of his elder brothers, Ronald and Victor, had served with Churchill, but such was the multiplicity of Fighting Brookes in the British Army that that was almost a statistical likelihood. Ronald had fought in the River War and on the North-west Frontier in the late 1890s, and was wounded in the Boer War with the 7th Hussars, before commanding the 11th Hussars in the Great War. At the battle of Spion Kop in January 1900, he accompanied Churchill on a dangerous observation mission. 'We crawled forward a short way on to the plateau,' Churchill recalled in *My Early Life*, 'but the fire was much too hot for mere sight-seeing.'[33] The next month three shrapnel shells burst directly above them, killing or wounding nineteen men but leaving them unscathed. Alan's other brother Victor was also wounded in the Boer War, serving with the 9th Lancers, and was killed in action only fifteen days after the Great War broke out. Eighteen years later Alan's second son, who was to become the third Lord Alanbrooke, was christened Victor after his grandfather and uncle.

Having left Pershing's staff in the summer of 1924, George Marshall served for the next three years with the US Infantry at Tientsin in China.

On his return in May 1927, his wife Lily, who was afflicted with health so bad that the couple could not have children, was diagnosed with a goitrous thyroid that was found to be strangling her windpipe. After a thyroidectomy in late August she seemed to recover, but then on 15 September she died suddenly of a heart attack while composing a letter to her mother, the last word of which was 'George'. She was only fifty-three.

Writing to his mentor General Pershing – who had himself lost his wife and three daughters in an hotel fire in San Francisco in 1915 – Marshall admitted that his twenty-six years of intimate companionship with Lily, 'ever since I was a mere boy, leave me lost in my best efforts to adjust myself to future prospects in life. If I had been given to club life or other intimacies with men outside of athletic diversions, or if there was a campaign on or other pressing duty demanding a concentrated effort, then I think I could do better. However, I will find a way.'[34]

It was Marshall's Army superiors who found the way to concentrate his formidable capacities, by appointing him assistant commandant and head of the infantry school at Fort Benning in Georgia for five years between 1927 and 1932. It was there that Marshall showed his capacities as a reformer. His experience of the later stages of the Great War had convinced him that, in any future conflict, officers would not be able to wait for perfect orders written out over four pages of single-spaced foolscap sheets, such as the ones GHQ had provided then, especially with the unreliable intelligence reports that might be expected from a fast-moving battlefield. He therefore took his officers for long morning rides over many miles, and then at lunchtime required them to draw maps of where they'd been. Since no fewer than two hundred of the twelve hundred generals who served in the US Army during the Second World War attended Fort Benning, including Dwight D. Eisenhower, Marshall was able to assess the abilities of many of America's future military leaders for himself. It was reputed that he kept a black book listing the best and worst, which he later drew upon extensively as US Army chief of staff.

On 15 October 1930, three years after Lily's death, Marshall married Katherine Tupper Brown, of Baltimore, with Pershing standing best man. The daughter of a Baptist minister, she graduated from Hollins College in Virginia and moved to New York in order to become an actress. Working for Sir Frank Benson's English Shakespearean Company, she dropped her Southern accent to take roles as important as Ophelia, Portia, Juliet and Viola. In 1911 she had married a Baltimore lawyer, Clifton Stevenson Brown, who died in 1928 (shot by a client, so it was rumoured). Just as Brooke had lost his wife tragically and subsequently remarried in his forties, having flung himself into his military career during the period of maximum grief, and found profound happiness with his second wife, so too did Marshall.

During the Great Depression, Marshall embraced Roosevelt's New Deal, of which many of his brother officers heartily disapproved and considered near-revolutionary. He devoted himself to the work of the Civilian Conservation Corps, training tens of thousands of young men to plant trees, cut firebreaks, clean beaches and rivers, build reservoirs and generally improve America's infrastructure.[35] In the course of this work, which helped him to understand the mentality of American youth and gave him useful insights into how to motivate them – which was to become invaluable when he needed to train eight million of them a decade later – he was finally raised to a substantive colonelcy. Marshall did not win his general's star until 1 October 1936, however, when he assumed command of the 5th Infantry Brigade at Vancouver Barracks, Washington.

Marshall was still only a one-star general in 1938. His career had seemed to plateau, and he readied himself for the disappointment of seeing younger men outstrip him in promotion. Yet owing to an extraordinary confluence of domestic and international circumstances, his own strength of character, Pershing's support and the President's acute judgement of personality, within three years he had become a four-star general and Army chief of staff. He was also helped by the fact that Douglas MacArthur, one of the most prominent and decorated soldiers in America, was widely thought too vain, ambitious and difficult a person to return to the post of Army chief of staff which he had held from 1930 to 1935, and was probably too politically conservative to get on successfully with the President.

In July 1938, having successfully commanded the 'Red' Forces in the Fourth Army manoeuvres at American Lake, Washington State, Marshall was ordered to Washington DC to become assistant chief of staff in the War Plans Division of the War Department. This was a key position, overseeing all the future offensive operations of the United States. Three months later, and a fortnight after the Munich Agreement, he was appointed deputy chief of staff. It was in this post that he attended a conference at the White House on 14 November 1938 to discuss the President's plans to build fifteen thousand warplanes. Others attending included some of the most senior officials in Washington, such as the President's friend and close confidant Harry Hopkins, the Treasury Secretary Henry Morgenthau, the Assistant Secretary for War Louis Johnson, the head of the USAAF General Henry H. 'Hap' Arnold, and Marshall's own boss, the Army Chief of Staff General Malin Craig.

If Marshall was going to make a good impression on the President, here was his perfect opportunity. Marshall had met Roosevelt for the first time in 1928 at Fort Benning, which was close to the polio convalescent clinic of Warm Springs, Georgia, and five years later Marshall had been present at his first inauguration. They had spoken briefly in 1937 on the President's visit to Oregon, but otherwise they were strangers.

According to Arnold's notes of the White House meeting, the President did most of the talking, emphasizing that ideally he would have liked to build twenty thousand warplanes and create an annual capacity for twenty-four thousand, but acknowledging that this would be cut in half by Congress. He also argued that a large air force would be a greater deterrent to would-be enemy powers than a large army. Marshall was unhappy with this reasoning and the way that Roosevelt was concentrating on having more aircraft instead of more soldiers, ammunition and military equipment, especially since the planes seemed mostly destined to be sent overseas. Against Germany's ninety field divisions, Japan's fifty and Italy's forty-five at the time, the USA had a total of only nine, of which not a single one was at full operational strength.[36]

As Marshall recalled of the meeting years later, most of the aides and advisers present 'entirely agreed' with the President, 'had very little to say and were very soothing'. Yet when Roosevelt finally came round to Marshall, saying of his own opening remarks, 'Don't you think so, George?', he replied: 'I am sorry, Mr President, but I don't agree with you at all.' The President gave Marshall 'a startled look' as he outlined his objections. As they left the meeting, the other officials chaffed the Deputy Chief of Staff, saying that they thought his tour in Washington was as good as over.[37] They were probably only half joking. In fact Marshall's calculated risk was perfectly justified. He disagreed with the President's view that a large ground army was not vital, but he must have also reasoned that big men – and FDR was undoubtedly such –

surrounded by yes-men can sometimes appreciate an honest foil. It was also part of his job to argue for a large army, and that would have been understood too. Few people outside Marshall's immediate circle ever called him by his Christian name, and he called his associates and subordinates by their ranks or surnames, in the formal Army manner. He disliked being called 'George', even by the President, later recalling: 'I don't think he ever did it again . . . I wasn't very enthusiastic over such a misrepresentation of our intimacy.'[38]

Marshall well understood Roosevelt's way of suborning people in this way, and refused to be drawn into it. As chief of staff he did not visit Roosevelt's country estate at Hyde Park, saying that he 'found informal conversation with the President would get you into trouble. He would talk over something informally at the dinner table and you had trouble disagreeing without personal embarrassment. So I never went.' Surprisingly, there are also no known photos of FDR and Marshall on their own together. General Thomas Handy described how his boss 'very definitely' and deliberately observed a formality with Roosevelt 'so that he wouldn't be manipulated as "one of the boys" '.[39] He did not want to be drawn into the vortex of Roosevelt's charm, and didn't feel it incumbent on him to laugh at the President's jokes in the way that the press corps and some Cabinet ministers did, yet neither was he stand-offish. (It was also suspected in the Churchill family that Marshall disapproved on moral grounds of the President's affair with Lucy Mercer Rutherfurd.)

Whatever Marshall might have privately thought about Roosevelt at this time, he was chosen by the President to follow Craig as Army chief of staff – the professional head of the nation's military establishment and commander of its field forces. Although Marshall stood no higher than thirty-fourth in Army seniority at the time, with no fewer than twenty-one major-generals and eleven brigadier-generals outranking him, there was an unwritten rule that a chief should be able to serve a four-year term before the age of sixty-four, which made him the fifth-ranking soldier eligible for promotion for the top post. Of those five, Marshall was the President's personal choice. He put his selection down to Roosevelt knowing he 'would tell him what was what, straight from the shoulder, and he knew I was not mixed up with any political clique or other group'.[40]

Marshall also attributed the President's decision to the advocacy of

Harry Hopkins, with whom he had worked closely over the issue of aircraft procurement since Christmas Eve 1938 – when Hopkins became commerce secretary. Because he was not seen as a front-runner he had few enemies, but he did have some very powerful supporters: besides Hopkins, they included Malin Craig, Louis Johnson and especially General Pershing, recognized as the greatest living American soldier. Handsome, just shy of 6 foot tall, grey-haired with fine blue eyes, Marshall certainly looked the part.

When Marshall became chief of staff, the forces under his command stood at only two hundred thousand strong; America's was the seventeenth largest army in the world. When Otto von Bismarck was asked what he would do if the British Army ever landed an expeditionary force on the north German coast, he joked that he would send the police to arrest it. Hitler would have been justified in making such a quip about the US Army of 1939. Within six years, however, Marshall had turned it into a fighting force of more than eight million.

In one of those coincidences of which history is replete, Marshall became Army chief on the very same morning that Adolf Hitler unleashed the Second World War. At 3 o'clock on the morning of his swearing-in, 1 September 1939, Marshall was telephoned with the news that German dive-bombers were attacking Poland. 'Well, it's come,' he told Katherine, and put on his uniform. After the swearing-in ceremony, Marshall went to the White House to brief the President. Except for Hap Arnold, Marshall was the only member of the American and British higher directorate of the war to serve in the same post from Hitler's invasion of Poland all the way through to the surrender of Japan.

Marshall soon established a reputation as a straight-talking Army chief. Despite being, in the words of one of Roosevelt's biographers, 'a courtly and reserved Pennsylvanian', he could be exceedingly blunt when necessary.[41] To a politician who rang up asking for a certain officer to be promoted, he replied: 'Mr Senator, the best service that you can do for your friend is to avoid any mention of his name to me.' Yet when the wife of Teddy Roosevelt Jr asked Marshall to put her husband back into a combat unit after he was hospitalized, but apologized for using her position to get what they wanted, Marshall replied that it was 'always alright to pull strings and favors if what you wanted was a more dangerous job than the one you had'. (Teddy Jr was in the first wave to alight on Utah Beach in June 1944 and the only general to see action that day; he died of a heart attack a month later.)

General John Edwin 'Ed' Hull, of the Operations Division of the War Department, recalled how Marshall worked. When his staff came to him with a problem, they would also have to bring him the various alternative solutions, and their own recommendation. 'He never nodded his head one way or shook it to indicate he agreed with what they were saying until they had finished. Then he'd say yes or no and that was it.'[42] It must have been a nerve-wracking way to work. Hull added, 'When you went into his office he expected you to walk in, sit down in the chair directly across from his desk and sit there while he finished reading whatever he had in his hand . . . and he didn't want you to open your trap about anything until he was finished.' When he looked up he expected his interlocutor to start speaking and he would give a definite decision before the visitor left the office. 'There were never two ways of interpreting his instructions, there was only one.' Hull believed Marshall had an almost photographic memory, and his mind worked fast; he dictated at 150 words a minute. He also had a volcanic temper.[43] Nevertheless, he was constitutionally unpompous: even as a four-star general he drove himself into the Pentagon, stopping to give workers lifts, and when he mislaid his spectacles – which he did often – he bought batches of replacements at dime stores.

In 1940 Marshall bought Dodona Manor, a modest four-bedroom house set in 4 acres on the outskirts of Leesburg, Virginia, in the foothills of the Blue Ridge Mountains and 35 miles from Washington DC. Built in 1786 by a nephew of George Washington, it was a charming, though almost Spartan dwelling, seating a maximum of eight around the dining-room table. Marshall slept in a single bed, with his boots and bright mauve dressing gown – his sole Churchillian affectation – in one small closet, and sharing a tiny bathroom with his wife. Pictures of his heroes Robert E. Lee and George Washington adorned the walls of the house, as they do today. In his retirement Marshall added photographs of Pershing, Dill, Bradley, Mountbatten, Churchill and Roosevelt, but the only one featuring Brooke was a group shot taken at the Quebec Conference.

Because no president had served more than two terms, and in September 1939 Roosevelt apparently had only sixteen months left in office, it would have been inadvisable for Marshall to have become too closely associated with him. The job of Army chief of staff demanded political and diplomatic antennae at least as much as military skill, and in Marshall it found someone preternaturally endowed with them. He understood that the best way of dealing with Roosevelt was through a good-natured but not over-cordial formality, which chimed in naturally with his own personality. 'I often saw the President and Marshall together and was left with the impression that FDR held Marshall in something like awe,' Churchill's confidant Brendan Bracken told Philip Graham, the owner of the *Washington Post*. 'In George Marshall's presence wisecracking and other flippancies were as much out of place as they would be at a solemn service in Washington Cathedral.' (For his part, as late as 1949 Marshall couldn't remember Bracken's name, calling him 'that tousle-headed Information minister', and complaining that he had once congratulated him in front of an elevator operator on getting the Overlord command.)

Marshall's preferred form of contact with Roosevelt was by letter and memorandum. Subjects on which they corresponded during the war were, as one would expect, immensely varied, and included the use of hotels as military hospitals, press leaks from the staff of the Free French leader General Charles de Gaulle, the policy of bombing Germany during daylight, British demands for the recall of General Joseph 'Vinegar Joe' Stilwell from Burma, civil disorder during the Puerto Rican elections, parachute release harnesses, protection against jungle-scrub typhus, the discontinuation of Lend–Lease after the German surrender, senior promotions (which were always agreed to by the President), the frontal armour on German Tiger tanks, relations with the Chinese Generalissimo Chiang Kai-shek, length of tours of duty in Iceland, Turkish neutrality, German reprisals against American pilots, the defence of the Panama Canal from Commando raids, Congressional appropriations, the best way to contact Archduke Karl-Ludwig von Habsburg, whether Marshall should accept the Soviet Order of Suvorov (he did), presidential visits to Army camps, manpower bills, the morale division of the Office of Civil Defense, British pilot training in the US, the financing of Pan-American Airlines to build Latin American airfields, and very much else besides (although Marshall did not pass reports of Lieutenant Joseph W. Alsop Jr's long and very painful-sounding history of syphilis on to the President, as Alsop's mother was a niece of Theodore Roosevelt).[44] Despite Marshall's dislike of the use of his first name in professional situations, Roosevelt did sometimes write 'Dear George'. More often a typed 'Memorandum for General Marshall' would have 'FDR' jotted at the end. Marshall would reply to 'My dear Mr President', although usually it started simply 'Memorandum for the President' and was signed 'G. Marshall' above the designation 'Chief of Staff'. On occasion a proposal was approved with the simple note 'GCM – OK go ahead – FDR'.

Despite deliberately keeping a personal distance from his commander-in-chief, Marshall acknowledged the dangers of being professionally remote. Writing to Harry Hopkins in November 1942, he contrasted the British system – where Churchill saw his chiefs of staff almost daily – with the way that the President saw the Joint Chiefs separately, 'and then the problem is, who summarizes what has occurred and provides a check to see the necessary instructions are sent around. I have often done this on my own initiative and later found out that someone else had been similarly active.' Potentially worse were the 'troubles we get into when we are not aware of what has happened between the President and the Prime Minister', because 'the British here are immediately informed of every detail'. Marshall also worried about 'not knowing the nature of the President's revisions of the drafts of messages we submit to him. All of these things may easily lead to tragic consequences.'[45] Roosevelt conducted his own discussions on military strategy, sometimes without reference to Marshall, who resented it but who saw that the President – charming, organizationally haphazard, brilliant and extremely wily – needed very careful handling.

An early insight into the way that Marshall did this is afforded by a private letter he wrote on 22 November 1939 to Major-General Asa Singleton, Commandant of the Infantry School, with suggestions for how to manage the President's forthcoming visit there. 'Whatever arrangement is made,' Marshall counselled, 'no one press him to see this or that or understand this or that; whatever is furnished him in the way of data [should] be on one sheet of paper, with all high-sounding language eliminated, and with very pertinent paragraphed underlined headings.' If anything needed to be explained, 'a little sketch of ordinary page size is probably the most effective method, as he is quickly bored by papers, lengthy discussions, and by anything short of a few pungent sentences of description. You have to intrigue his interest, and then it knows no limit.'[46] It was the formula Marshall himself stuck to, even though it seems to apply more to a child with attention deficit disorder than to the chief executive of the United States.

It certainly helped Marshall that Roosevelt's first love and primary interest was the Navy, indeed on one occasion he remonstrated jokily: 'At least, Mr President, stop speaking of the Army as "they" and the Navy as "we".'[47] FDR was far more willing to defer to Marshall than to his admirals, recognizing the limits of his own military competence. On the rare, but always significant, occasions that Roosevelt actually overruled Marshall, the reason was always political.

The first thing that the new US Army Chief of Staff needed as the Wehrmacht blitzkrieged its way across western Poland was a US Army. One week into his new job, on 8 September, Marshall drafted a letter to the President arguing that in order to maintain 'peace and neutrality in the midst of our troubled world' the Regular Army had to be increased to 227,000 men and the National Guard reservists to 235,000 by immediate executive order. He warned that the Army's first four infantry divisions were one-quarter under complement, and the remainder mere 'skeleton organisations'. Furthermore, 'Essential Corps troops are essentially non-existent.' The National Guard was at half its regulation peacetime strength.[48] Roosevelt was very receptive to Marshall's demands, but hamstrung by a Congress that was still largely isolationist in temperament.

If the occasion demanded it, especially after Hitler had attacked France and Belgium, by which time he had been in the job for eight months, Marshall was willing to take risks with the President. One such occurred on 11 May 1940, after Congress had decided to cut $10 million out of a $28 million appropriation budget for equipment to detect Japanese aircraft off the western coast of the United States. Marshall visited Henry Morgenthau to apprise him of the supreme importance of getting the full amount and, as he later recalled, 'We went to see the President who, it was quite evident, was not desirous of seeing us.' FDR gave Morgenthau some 'rather drastic handling', which Marshall assumed the President was laying on for his benefit, 'because they [Roosevelt and Morgenthau] were old friends and neighbours'. When finally Morgenthau asked the President whether Marshall could put his case, Roosevelt replied: 'Well, I know exactly what he would say. There is no necessity for me hearing him at all.'

'Well, it was a desperate situation,' remembered Marshall.

I felt that he might be president, but I had certain knowledge which I was sure he didn't possess or didn't grasp. I thought the whole thing was catastrophic in its possibilities and this last cut just emphasized that point. So, recalling that a man has a great advantage, psychologically, when he stands looking down on a fellow, I took advantage, in a sense, of the President's condition.

Marshall walked over to Roosevelt's desk and stood looking down at him, saying, 'Mr President, may I have three minutes? . . . I don't quite know how to express myself about this to the President of the United States, but I will say this, that you have got to do something and you've got to do it today.'[49]

Of course these types of anecdotes have only one outcome, otherwise they would not be told by their heroes any more than by historians: Marshall got all he wanted and more. Equally obviously – in the light of what happened the following year – equipment for detecting incoming Japanese aircraft in the Pacific was about as prescient a spending priority as it was possible for Marshall to promote at the time. His direct method nonetheless shows his confidence by this point, as well as an element of ruthlessness in consciously taking advantage of the President's disability. It was a tactic he was also to employ in a future encounter against an ill Winston Churchill.

The case *against* Marshall, insofar as there is one, was neatly put by Colonel Ian Jacob, the Military Assistant Secretary to the British War Cabinet, who told the Australian historian and broadcaster Chester Wilmot in 1948:

Marshall had been spotted as a bright boy when he served on Pershing's Staff in World War One, but he was essentially a Staff officer rather than a commander, an organizer rather than a director of operations. He had little sense of strategy and no 'feel' of operations. He was a man of great integrity, high character and firmness of purpose. He automatically commands the respect of everyone. His modesty made him reserved and it was rather difficult to penetrate through this reserve.[50]

Whether Marshall had a 'feel' for operations and a sense of strategy is a central question that this book will seek to answer.

From June 1940, the US secretary of war was Henry L. Stimson, with whom Marshall built a close and strong working relationship. Stimson was seventy-seven and a strong advocate of military aid to Britain and of American military preparedness. As he was a leading member of the New York Republican establishment, a former secretary of war under Taft and secretary of state under Hoover, it was a clever bipartisan appointment by the President. In November 1942, by which time it was clear that Marshall completely dominated strategy-making at the War Department, 'one of the less tactful hangers-on of the Administration' asked Stimson how he liked being relegated to the position of 'house-keeper' for the Army.[51] Stimson retorted that 'the question was a foolish one, betraying a fundamental ignorance of the functions of a secretary of war', but in fact it was fair, if a touch cruel. Marshall had effectively removed from the War Secretary the role that the incumbent had enjoyed since before Lincoln, that of being the president's principal military adviser and a central contributor to strategic decision-making.

The same month that Marshall was writing to General Singleton – November 1939 – Franklin Roosevelt inaugurated a correspondence with Winston Churchill that was to have world-historical significance, and in the finest passages of its three published volumes attained far greater significance than those 'pertinent paragraphs' and 'few pungent sentences' that Marshall had recommended to the Infantry School commandant. Writing while Neville Chamberlain was still prime minister and Churchill was first lord of the Admiralty, Roosevelt said: 'What I want you and the PM to know is that I shall at all times welcome it, if you will keep me in touch personally with anything you want me to know about.'[52] Grasping the opportunity with fervour, and signing himself 'Former Naval Person', Churchill responded with the first of 944 letters and telegrams over the next five and a half years. In all, Roosevelt sent 743.

The publication the previous month of a new edition of Churchill's book *Great Contemporaries* would have confirmed to Roosevelt that he was writing to an avowed admirer. In 1937, Churchill had originally published the collection of twenty-one witty and concise potted biographies of famous people, most of whom he had known personally, including figures as diverse as Kaiser Wilhelm II, George Bernard Shaw, Lawrence of Arabia, Marshal Foch, Clemenceau, Adolf Hitler (whom he twice nearly but never actually met), King George V and Lords Rosebery, Asquith, Birkenhead, Haig, Balfour and Curzon. For the October 1939 reissue, however, Churchill added four more essays, the last of which was entitled 'Roosevelt from Afar'. 'A single man whom accident, destiny or Providence has placed at the head of one hundred and twenty millions', wrote Churchill, 'has set out upon this momentous expedition.' He prophesied that Roosevelt's 'success could not fail to lift the whole world forward into the sunlight of an easier and more genial age'.[53] (Had Churchill genuinely recalled meeting Roosevelt at Gray's Inn, he would surely have mentioned it in this essay, and possibly not used the words 'from afar' in the title.)

Though it was more a gushing fan letter than an objective analysis of the New Deal, the piece did nonetheless contain very occasional barbs. In one sentence Churchill suggested that 'the policies of President Roosevelt are conceived in many respects from a narrow view of American self-interest.' As so often in Churchill's writing there was also a detectable (and delectable) element of self-reference, especially in his description of Roosevelt as 'trained to public affairs, connected with . . . history by a famous name . . . he contested elections: he harangued the multitude . . . He sought, gained and discharged offices of the utmost labour and of the highest consequence.'

What Churchill admired above all in Roosevelt was his courage, the attribute that he exalted above all the others. At thirty-six, wrote Churchill, Roosevelt had been 'struck down with infantile paralysis. His lower limbs refused their office. Crutches or assistance were needed for the smallest movement from place to place.' Churchill also had a high regard for luck, and claimed – wrongly in fact – that at one moment in his 1932 race for the Democratic nomination, FDR's victory had turned 'upon as little as the spin of a coin'. This led Churchill to reach for hyperbole: 'Fortune came along, not only as a friend or even as a lover, but as an idolator.' Fortune and Churchill both, it seemed. Within a month of publication, this very public lauding of Roosevelt had paid off superbly with the arrival of Roosevelt's first letter at the Admiralty.

In another of history's regular but nonetheless remarkable coincidences, Churchill became prime minister on the same day that Hitler unleashed his blitzkrieg in the west, Friday 10 May 1940. As Air Marshal Sir Charles Portal, the head of Bomber Command since April, later recalled, the effect of having Churchill at No. 10 was instantaneous: 'He put a bomb under Whitehall. From then till the end of the war he was constantly urging, driving, probing, restless in his search for new ways for getting at the enemy.' Churchill would ring Portal up at all times of the day or night 'and you had to be continually on your toes, always searching into your own mind for the means of improving the job you were set to do.' Portal added that Chamberlain had had one telephone at Chequers, the prime ministerial country house in Buckinghamshire, and that was to be found in the kitchen, whereas Churchill 'at once installed a whole battery on his desk and had them in constant use'.[54]

On the same day that Downing Street braced itself for its new resident, Brooke was facing the whirlwind attack that hit his II Corps on the Franco-Belgian border. Although he had had eight months to prepare his largely raw and under-equipped divisions, and did so as well as possible, he was profoundly sceptical about the proposed strategy – codenamed Plan D – which required the Allied left wing, including the BEF, to advance into Belgium in an attempt to extend the Maginot Line of defence northwards along the River Meuse right up to the sea, to protect Antwerp and the Channel ports. What to do should the Germans attack around the western flank of the Maginot Line, and wheel through Holland and Belgium into France? Brooke thought that Gort (the plan's originator) had been wildly over-promoted and regretted that the job had not gone to the commander of I Corps, his friend and mentor Sir John Dill (yet another Ulsterman). Brooke had confidence in his divisional commanders, Montgomery and Alexander, but not in the leadership, doctrine or morale of the French Army, an organization his upbringing allowed him to understand intimately.

In his diary – which of course he ought not to have been keeping – Gort's chief of staff Lieutenant-General Henry Pownall reported that his boss was 'a bit depressed about [the] Corps Commanders, especially Brooke who has got a very defeatist frame of mind. I fancy he needs a rest, having done so much work in so many different capacities in the last four years.' Pownall thought that Brooke was 'always looking over his shoulder now and shows no confidence that he can withstand attack, especially by tanks'.[55] The word 'defeatist' is a harsh one for one senior officer to use about another in wartime, but Brooke was certainly very doubtful about Plan D. 'From the first Brooke disliked the concept of moving from prepared positions and meeting the German army in open warfare,' records his biographer General Sir David Fraser, 'for which he believed neither the Allied left wing's equipment nor its tactical expertise to be adequate.'[56]

The story of the May–June 1940 campaign is too well known to be rehearsed at length here, and the confusion over it is perhaps best illustrated by the fact that two of the best books on the Dunkirk evacuation are entitled *Strange Defeat* and *Strange Victory*. On one aspect of it, however, there are no two views: Brooke performed superbly. The collapse in the French sector around Sedan by 15 May and the extraordinarily rapid thrusts of the Wehrmacht panzer divisions, combined with the sudden capitulation of the Belgians on 28 May, left Brooke's corps in serious danger of wholesale capture, from which he – through what Fraser describes as 'a series of hazardous manoeuvres of great ingenuity and boldness' – managed to extricate it. In order to cover the gap left by the Belgians, he extended his left flank north of Ypres by sending Montgomery's 3rd Division from south to north in darkness along minor roads close to the front. It got there just in time. He then defended the shrinking perimeter around Dunkirk, handing over his command to Montgomery – tears streaming down his cheeks – only when ordered to return across the Channel before the rest of II Corps, even though he 'felt like a deserter not remaining with it till the last'. London needed him too much to risk his capture. Brooke was taken off the beach at Dunkirk on 30 May, along with 53,820 men that day, and over 338,000 in total.

In 1959, Lieutenant-General Sir Brian Horrocks recalled that the surrender of the Belgians on 28 May had suddenly opened up a 20-mile gap on the British left flank, and that 'If the Germans had got into it, there might have been no evacuation from Dunkirk. It was thanks entirely to Alan Brooke that the gap was closed. He was more responsible than anyone else for the BEF getting back successfully.' The Secretary of State for War between 1942 and 1945, Sir James Grigg, agreed, telling the *Sunday Times* in 1946: 'By almost universal testimony it was due largely to [Brooke's] skill and resolution that, not only his own Corps, but the whole . . . BEF escaped destruction in the retreat to Dunkirk.'[57] Even Pownall admitted in June 1940 that Brooke 'came out trumps'. As we shall see, the experience of the campaign taught Brooke a number of important lessons about how he believed the rest of the war should be fought, lessons that diverged sharply from the ones that Marshall had learnt at Fort Leavenworth, Chaumont and Fort Benning.

On 6 June 1940, only three days after the last troops returned from Dunkirk, Churchill asked the War Office Planners for 'proposals for transporting and landing tanks on the beach', and a fortnight later wrote to suggest 'a Corps of at least five thousand parachute troops'.[58] A fortnight later he set up the Special Operations Executive, whose object, as well as general sabotage in Occupied countries, was to assist future invasion forces. It was an astonishing set of priorities for the leader of a country whose army had only days earlier been flung ignominiously off the Continent, and which must shortly itself face the threat of invasion, but was a sign of Churchill's invincible optimism.

In that same spirit, no sooner had Brooke returned to Britain than he was sent off to command a new BEF which would operate further south on the west coast of France, in Normandy and Brittany, alongside the French Army under General Maxime Weygand. The former CIGS, then Commander-in-Chief Home Forces, General Ironside, wrote in his diary, which he should not have been keeping, 'It has been decided to send to France two Territorial divisions to add to the 51st [Highland Division] already there with the Armoured division. Brooke . . . seemed very distraught over the thought and considered that the Terrier divisions would never stand up to the bombing.' The central question that worried Brooke was: 'Will France stand up long enough to allow us to get them out?'[59] Soon after landing in Cherbourg late on the night of 12 June 1940, he took command of this Second BEF west of the Seine – a total of one hundred thousand line-of-communication troops stretched between Normandy and the Loire, plus the crack 52nd Lowland Division – and made contact with Weygand. 'Refugees again swarming everywhere,' he wrote to Benita, 'and heartbreaking to find oneself back amongst them.'[60]

It soon became clear that the French had lost all will to fight. Indeed at breakfast two days later the seventy-three-year-old Weygand, looking 'very wizened and tired' and nursing a stiff neck from a car crash the previous night, told Brooke: 'he would speak very frankly . . . the French army had ceased to be able to offer organized resistance and was disintegrating into disconnected groups . . . Paris had been given up and . . . he had no reserves whatever left.'[61] Brooke concluded that his own army needed to get back to Brest and Cherbourg for embarkation to Britain as soon as possible. Grigg recalled that Brooke 'had been appalled and distressed at being ordered back to France', believing the mission impossible, but once there he was going to make the best of bringing home as many troops as possible, just as he had at Dunkirk.

This was to be the occasion on which Brooke first came into contact with Churchill. It was also still officially the policy of the Anglo-French Inter-Allied Council to keep an Allied bridgehead in Brittany. Although both Weygand and his second-in-command, General Alphonse Georges, whom Brooke likened to 'a great pink jelly fish – absolutely finished', agreed with him that 'the Brittany Defence Scheme was quite impossible owing to lack of troops', Churchill did not accept this, and a very difficult telephone call resulted. It was the worst possible way for Brooke to be introduced to the Prime Minister's sense of strategy and tactics.

On the evening of Friday 14 June, Brooke told Sir John Dill, who had recently taken over as CIGS, that he had given orders at 4 p.m. that the 52nd Division must 'proceed as soon as possible to Cherbourg'. At 8 p.m. Dill called from Downing Street on a very bad telephone line to ask him about the dispositions of the 52nd. After Brooke repeated what he had agreed with him four hours earlier, Dill said, 'The Prime Minister does not want you to do that.' 'What the hell does he want?' asked Brooke. 'He wants to speak to you,' said Dill, handing over the receiver. Brooke later recalled of Churchill: 'I had never met him, I had never talked to him, but I had heard a good deal about him!'[62] In this, of course, Brooke was no different from any other sentient Briton over the previous four decades; but, as we have seen, Churchill had also fought alongside two of Brooke's brothers, and Benita's first husband had died of wounds sustained at Gallipoli.

Churchill told Brooke that he had sent him to France 'to make the French feel that we were supporting them', and so the 52nd Division must not be evacuated. 'It was impossible to make a corpse feel,' Brooke replied, 'and . . . the French army was, to all intents and purposes, dead, and certainly incapable of registering what had been done for it.' Both men were confirmed lifelong Francophiles, but the facts of the situation were immediately clear to Brooke on the ground, and ought to have been to Churchill in London also. The argument went on for nearly half an hour, with Churchill seeming to imply that Brooke was 'suffering from "cold feet"' because he refused to comply with his wishes. 'This was so infuriating that I was repeatedly on the verge of losing my temper,' Brooke noted afterwards. The idea that one of the Fighting Brookes of Colebrooke was even implicitly being accused of having 'cold feet' must indeed have been fabulously galling. Standing by a window at his headquarters in Le Mans, Brooke looked out and saw two senior officers of the 52nd Division, James Drew and John Kennedy, sitting in the sunshine on a garden seat under a tree, waiting for his decision. The sight of these men 'acted as a continual reminder of the human element of the 52nd Division', and stiffened his resolve not to 'sacrifice them with no attainable object in view'. He was in 'an exhausted condition' by the end of the conversation, when finally Churchill said: 'All right, I agree with you.'[63] Although it had nothing to do with Brooke, the 51st Highland Division were captured virtually en masse at Saint-Valéry that same day, although the 52nd Division were evacuated successfully. Brooke sailed back from Saint-Nazaire to Plymouth on the morning of 18 June aboard the trawler *Cambridgeshire*.

In the second volume of his war memoirs, Churchill wrote (quite wrongly) that Brooke had rung him up to 'press' the evacuation view upon him, and that 'after ten minutes I was convinced that he was right and we must go'. Brooke later commented that, although Churchill was largely ignorant of the prevailing conditions and was attempting to interfere with the judgement of the commander in the field, 'The strength of his power of persuasion had to be experienced to realize the strength that was required to counter it!' In Brooke, however, as we will see repeatedly throughout this book, Churchill's unstoppable force had met its immovable object.

Ten years later, while unveiling a portrait of Churchill at the Junior Carlton Club in London in December 1950, Brooke recalled that 'I had never met Churchill at that time, but even at that distance and through this faulty line, I was at once aware of his dynamic personality and of his dominating influence. It was a useful experience as it gave me an insight into the influence that his magnetic personality might exercise on commanders at a distance.'[64] The intervening decade had lent some enchantment to his views. 'Winston never had the slightest doubt that he had inherited all the military genius from his great ancestor Marlborough!' was a regular Brooke reprise to his diary at the time. 'His military plans and ideas varied from the most brilliant conceptions at one end to the wildest and most dangerous ideas at the other.'[65] Yet against that must be set the equally powerful sentiment he expressed at the Junior Carlton Club, which he believed with equal conviction: 'I shall, till my dying day, thank God for the great privilege of having been associated with him during those momentous war years.'[66]

2

Collecting Allies:
'The finger of God is with us'
June 1940–December 1941

*The distinction between politics and strategy diminishes as the point of view is
raised. At the summit, true politics and strategy are one.*

Winston Churchill, *The World Crisis*

After the war, Sir James Grigg said he 'felt it would be wrong to portray
Brooke as selfless and unambitious, he had a natural and healthy ambition
of the successful professional soldier'.[1] Two days after Churchill visited
Brooke's V Corps near Gosport on 17 July 1940, Brooke was appointed
commander-in-chief Home Forces. His contretemps with Churchill at
Le Mans had clearly not damaged his career prospects, and he was now
faced with the tough job of training an army that had abandoned most of
its heavy equipment in France. He flung himself into the task. Writing to
Wavell some months later, he opined that 'We are not anything like as
tough as we were in the last war. There has been far too much luxury,
Safety First, etc, in this country. Our own idea is to look after our comforts
and avoid being hurt in any way.' That was why, during his time
commanding the Home Forces, Brooke insisted on protracted exercises
lasting several days, in all weathers, with 30-mile marches and longer.

The post also had the advantage, for an ambitious general, of regular
contact with the Prime Minister, including occasional visits to Chequers.
He was there on the afternoon of 3 October when he joined the Chiefs
of Staff to discuss Operation Ajax, a plan to attack Trondheim in
Norway, which had been evacuated by the British only the previous
May. On 16 November 1940, he also stayed for the night at Ditchley,
the country house that Churchill repaired to at weekends when the
moon was high and the Luftwaffe bombers were believed to be able to
pinpoint Chequers.

Despite these visits, however, Churchill did not socialize with Brooke outside their working relationship. The visitors' book at Chartwell, the Prime Minister's home in Kent, which includes the names of all the 780 people who stayed there after 1922, does not feature Brooke's name at all. By contrast, Professor Lindemann (later Lord Cherwell) appears eighty-six times, Montgomery forty-six, Bracken thirty-one, and there are also entries for Alexander, Ismay, Ironside and scores of others.[2] After the war Brooke was elected to the Other Club, founded by Churchill and F. E. Smith in 1911, although he did not often attend. This club was far more widely based than Churchill's cronies, and included several people who profoundly disagreed with him politically, but he had veto rights over its candidates, and so Brooke's membership can be taken as a guarantee that they were not enemies.

Brooke almost never fought Churchill out of pride or pugnacity or perversity, but did so because of the effect of their decisions on the services. With his powerful moral conscience, Brooke was always, figuratively speaking, looking out of that Le Mans window and seeing the young men sitting under the tree in the sunshine possibly being condemned to capture or worse if he settled for an easier life. Like Dill, Brooke was a devout Ulster Protestant, with a deep religious faith that supported him, especially during the tragic time of heartbreak and guilt after the death of his first wife. During the Phoney War, the eight months before fighting began in earnest, when Brooke was in France, he and Benita arranged to read identical Bible extracts every day at the same time. By contrast, Churchill had been convinced since his subaltern days, when he read *The Martyrdom of Man* by William Winwood Reade, that Jesus Christ had been a charismatic and inspired prophet and a profoundly holy man, but was not the Son of God. As Moran put it: 'King and Country, in that order, that's about the only religion Winston has.'[3] Churchill's private secretary Jock Colville believed that the first time the Prime Minister attended a church service during the war, other than a funeral, was in Scotland in March 1945.

On 11 October 1940, staying at Chequers for the weekend, Churchill and Brooke disagreed over the use being made of the eccentric but occasionally brilliant Major-General Percy Hobart (pronounced 'Hubbard'), who was then languishing as a lance-corporal in the Home Guard due to the War Office's extreme disinclination to employ him. 'Brooke said he was too wild,' recorded Colville, 'but Winston reminded him of Wolfe standing on a chair in front of Chatham brandishing his sword.

"You cannot expect", he said, "to have the genius type with a conventional copy-book style." ' That exchange could almost be taken as a template for their future relationship, with Brooke warning against wildness, and Churchill defending it as genius. (Over Hobart, Churchill was right, as the ingenious inventions he deployed at D-Day were later to prove.)

That evening's discussions did not break up until 2.15 a.m., giving Brooke another foretaste of what lay in store for him when he became CIGS. The Chiefs of Staff nicknamed these late-night meetings 'the Midnight Follies', after the famous 1920s hotel cabaret act, *Midnight Follies at the Metropole*. The First Sea Lord, Admiral of the Fleet Sir Dudley Pound, had a subtle way of pointing out how late Churchill was keeping them up; when offered a whisky and soda by Churchill at 2 a.m. he would say: 'I never drink spirits in the morning. I'll have a glass of port.'[4] Churchill seemed addicted to these late nights, but they were to infuriate Brooke, since unlike the Prime Minister he could not take an afternoon nap at the War Office. One historian has pointed out that 'It says something about their code of politeness, duty and respect that none of those who suffered under these afflictions, not even the arch-sufferer Sir Alan Brooke, is known ever to have protested to his face.'[5] Perhaps they were conscious of being present when history was being made, and did not wish to seem petty.

Created in 1924 to provide the Government with expert co-ordinated service advice, the Chiefs of Staff became a permanent committee of the War Cabinet in 1939. When Churchill became prime minister and assumed the title of minister of defence in May 1940, he frequently attended their meetings, which when he was present were called Staff Conferences. Churchill was very conscious that it had been the absence of a constitutionally established, military inter-service authority, working directly for the prime minister, that had led to the dangerous dissensions between the 'Frocks [frock coats]' (politicians) and the 'Brass Hats' (soldiers) during the Great War, the Masters and Commanders of their day. He avoided a repetition of that by working on all strategic problems with and through the Chiefs of Staff, however frustrating they all found it at times.

The Chairman of the Chiefs of Staff Committee when Brooke joined was Admiral of the Fleet Sir Dudley Pound. Like Churchill and Harold Macmillan, Pound had an American mother. Elizabeth Pickman Rogers came from a Massachusetts seafaring family, but her son had few fond memories of her since her kleptomania and profligacy destroyed his parents' marriage.[6] He began his naval career at thirteen, excelled at exams and was appointed naval assistant to the First Sea Lord, Sir John 'Jackie' Fisher, during the Great War. He thus had first-hand knowledge of the combustible relationship between Fisher and Churchill, and learnt lessons about how to deal with Churchill that he later put to good use as chief of the Naval Staff.

In 1916 Pound was given his first independent command, the battleship HMS *Colossus*, which took part in the sinking of the cruiser *Wiesbaden* and a destroyer at the battle of Jutland. During the severe shelling, the range-taker standing next to Pound on the bridge had his arm blown off. In the inter-war years, Pound served first as chief of staff to Sir Roger Keyes, then as second sea lord, and then as commander-in-chief of the Mediterranean Fleet, the Navy's best active command. Partly due to a series of illnesses at the Admiralty, he became first sea lord in July 1939, despite arthritis of the hip that forced him to walk with a cane. It was probably he who ordered the famous 'Winston is back!' signal to be telegraphed to all the ships of the Navy in September 1939, although wags implied that it was as much a warning to the captains as an invocation to their crews.

Drawing on his experiences under Fisher, Pound evolved a way of dealing with the Prime Minister that he vouchsafed to a deputy: 'Never say a direct "No" to Churchill at a meeting. You can argue against it, and as long as you don't exaggerate your case the PM will always let you have your say.' Churchill rarely spoke of friendship, but he did with regard to Pound, often telling his naval aide Commander C. R. 'Tommy' Thompson that the First Sea Lord was one of the three men whose companionship meant most to him, the others being Lord Beaverbrook and the South African premier Jan Christian Smuts. Colville recalled that Churchill 'bullied' Pound over the telephone when he thought the Admiralty was being unimaginative, but the admiral's 'serious wrinkled face would flicker with pleasure and amusement when Churchill teased him'. One summer night at Chequers, walking in the rose garden after dinner, Pound, 'who was lame but unquestionably sober', fell down the steps and lay flat on his back. 'Try to remember', said Churchill, as he and Colville helped him to his feet, 'that you are an Admiral of the Fleet and not a midshipman.' A slow smile spread across Pound's face.[7]

*

Pound's American counterpart, the US Chief of Naval Operations, Admiral Harold R. 'Betty' Stark, drafted 'Plan Dog' in Washington in October 1940, his view of how a global war should be fought were the United States and Great Britain to become active allies against the Axis powers. It was then refined by other Staff officers and subsequently sent to Marshall who also approved it. In essence it set out what came to be known as the 'Germany First' policy, stating that 'If Britain wins decisively against Germany, we could win everywhere; but ... if she loses, the problem confronting us would be very great; and while we might not lose everywhere, we might, possibly, not win anywhere.'[8]

The adoption of the memorandum, first by Marshall and then by Roosevelt – though not in writing – and then by the US Joint Planning Committee, meant that the United States had an outline plan to use during the secret, arm's-length Anglo-American Staff talks, codenamed ABC-1, which were about to start. No such talks could be organized before Roosevelt's third inauguration on 20 January 1941, because during the election campaign he had promised American parents that 'Your boys are not going to be sent into any foreign wars.'

Churchill told Jock Colville emphatically that Roosevelt 'would win the election by a far greater majority than was supposed and he said he thought America would come into the war. He praised the instinctive intelligence of the British press in showing no sign of the eagerness with which we desired a Roosevelt victory.' Four days after that prediction, Roosevelt did indeed win re-election over the Republican candidate Wendell Willkie by 449 electoral votes to 82. Churchill went on to say, his ruminations punctuated with bursts of the song 'Under the Spreading Chestnut Tree', that 'he quite understood the exasperation which so many English people feel with the American attitude of criticism combined with ineffective assistance; but we must be patient and we must conceal our irritation.'[9] His own private irritation was evident from his complaint the next month that 'We have not had anything from the United States that we have not paid for, and what we have had has not played an essential part in our resistance.'[10]

In 1941, some 84 per cent of the munitions used by British and Commonwealth forces originated in Britain. The system called Lend–Lease, whereby American arms were sold to Britain on generally favourable borrowing terms, accounted for only 1 per cent at the time, and the British paid cash for a further 7 per cent under pre-Lend–Lease contracts.

So in the last nine months of 1941 Britain received 2,400 aircraft and 951 tanks from the USA, or the equivalent of six weeks' output from British factories.[11] It was useful, obviously, but not so much as to make a great difference militarily. Where the $14 billion of Lend–Lease aid by the time of Pearl Harbor did help, however, was in Britain's overall financial and food situation. On New Year's Day 1941, Colville, listening to Churchill composing a 'forceful' telegram to Roosevelt on Britain's financial predicament, thought that the Prime Minister 'obviously fears that the Americans' love of doing good business may lead them to denude us of all our reasonable resources before they show any inclination to be the Good Samaritan'.

From his appointment as commander-in-chief Home Defence, Brooke attended meetings of the War Cabinet Defence Committee, a combination of service Chiefs and their political ministers, at least on matters that impinged on his brief. On 10 January 1941, for example, he had discussed there Italian operations in Africa and German naval operations in the North Sea, ending with a long list of the manpower and matériel deficiencies he faced.[12] Even if the memory of their Le Mans conversation had dimmed, Churchill was thus well aware of Brooke's direct manner, his habit of speaking very fast, and his strength of character. He was not always impressed, however, writing to the Secretary for War David Margesson and Dill the next day to say that Brooke's contribution 'did not seem to be at the level of the discussion', and was 'not very illuminating'. He complained that, instead of talking about his strategy for using twenty-five divisions and two thousand guns to counter-attack the expected German invasion, Brooke had merely delivered a list of equipment shortfalls, of which Churchill was 'well aware'. Dill was due to retire as CIGS on Christmas Day 1941: had he done so at the beginning rather than the end of the year, it is safe to assume that Brooke would not have got his job.

Warned by Dill of this threat to his advancement, Brooke acted quickly. A fortnight later he was writing to Churchill about Operation Victor, an anti-invasion exercise in which Neasden power station and the Metropolitan Water Works were captured by two 'German' brigades and thirty light tanks that were landed in London by parachute, but were then fully engaged by his Home Forces. This was much more Churchillian fare, and prompted the Prime Minister to inform Harry

Hopkins and Dill that when the invasion came his broadcast to the nation would end with the words: 'The time has come: Kill the Hun!' Soon afterwards Brooke was again invited to stay at Chequers, where he brought an epidiascope or magic lantern to give a lecture on Operation Victor to Clementine Churchill, the Foreign Secretary Anthony Eden and the Labour leader Clement Attlee, about which the Prime Minister was 'very flattering'.

Brooke would not put his commitment to his career over his obligations to the High Command, however. In the late summer of 1940, Churchill tried to use him to outmanoeuvre the Chiefs of Staff Committee – Dill, Pound and Air Chief Marshal Sir Charles Portal – over Operation Ajax. Having failed to convince the Chiefs of its merits, which he needed to do before it became strategic policy, the Prime Minister sent for Brooke to come to Chequers. 'Then in front of the Chiefs of Staff', Brooke recalled years later, 'he ordered me to prepare an expeditionary force out of my resources for the capture of Trondheim,' and gave him a week in which to do it. Brooke said that he would need the Commanders-in-Chief of the Home Fleet and Bomber Command, the Minister of Transport and several other high officials to help him in the planning, all of whom Churchill promised would be put at his disposal.

The British and French had captured Trondheim and Narvik under Churchill's orders in April 1940, only to be forced to evacuate them by superior German forces. Luftwaffe superiority had cost the Royal Navy dear in that campaign, although it had been the parliamentary debate over that disaster that had, paradoxically enough, brought Chamberlain down and Churchill into the premiership. The idea of returning to Norway only a matter of months later, without air superiority and with the Germans in complete control of the entire coastline, was anathema to War Office Planners.

After seven days Brooke came to the same conclusions that the Chiefs of Staff had, that the operation was unfeasible, mainly because of the lack of aircraft carriers to provide the necessary support. When he reported this to Churchill, he 'received a very unpleasant welcome!' The Prime Minister later tried to persuade the Canadians to undertake the operation, but failed in this too. As well as apprising Brooke of the fundamental impracticality of Ajax – later codenamed Operation Jupiter – the experience also alerted him to the readiness of Churchill to try to bypass the service Chiefs to get his way. In retrospect, far from being the wasted week that it must have seemed at the time, it was one well spent.

Relations between Churchill and Brooke improved in the spring of 1941. In a radio broadcast of 9 February, in which he quoted Longfellow's lines from 'The Building of the Ship!', Churchill declared that he had 'the greatest confidence in our Commander-in-Chief, General Brooke, and in the generals of proven ability who, under him, guard the different corners of our land'. The next day he toured Brooke's headquarters in the reinforced-concrete basement of the Office of Works near the Cabinet War Rooms in Whitehall, and then invited Brooke up to the No. 10 Annexe, the rooms on the ground floor of the same building almost directly upstairs where Churchill lived for much of the war instead of in Downing Street. He showed Brooke his drawing room and dining room, Mrs Churchill's bedroom and bathroom, his own bathroom, and even the kitchen and scullery. After the war, Brooke recalled that, fitted out with elaborate anti-bomb devices, special ventilators, telephones, message conveyors and map rooms, his 'was in every way an excellent battle headquarters, with only one fault, namely its proximity to Winston!'[13]

Brooke later remembered his stay at Chequers on the night of Sunday 9 March 1941 as 'one of the first occasions on which I had seen Winston in one of his really lighthearted moods'. Churchill's friend and scientific adviser Lord Cherwell was also staying, and Brooke recorded that there was much 'flippant' conversation about metaphysics, solipsists and higher mathematics, not subjects that naturally lend themselves to flippancy. After dinner Churchill played martial tunes on a gramophone while giving his guests a display of arms drill with an elephant gun in the Great Hall. His simulated bayonet practice left Brooke convulsed with laughter; after the war he wondered to himself what Hitler would have thought of it all. Churchill – who was wearing a light-blue siren suit, which Brooke thought looked like 'a child's "rompersuit"' – had bronchitis, and so went to bed at the record early hour of 11.30 p.m. The ice was broken, and it is well to remember, when Brooke's relations with Churchill later became stormy and exasperated, that there had been pleasant moments too.

Later that month, in discussions about whether reinforcements should be sent to the Middle East, Churchill put Brooke's demands not to let troops out of the country down to the natural desire 'that every General should try to keep as many troops as possible in his own hands', but 'We must not get too invasion-minded.'[14] Churchill knew from intercepted German messages in late March 1941 that an invasion was definitely off the Wehrmacht's agenda, and he had been impressed by Brooke's willingness to allow large numbers of troops and tanks to leave Britain in the summer of 1940 to protect the Nile Valley. 'When real risks arise in other quarters, risks will be run with courage here,' wrote Churchill, 'as they have been in the past.' If he saw Brooke as willing to run them – as he had shown when the invasion threat was far higher – it could only have redounded to Brooke's credit.

Brooke also learnt how to stay silent at opportune moments. At a dinner at Chequers in late April 1941, attended by Margesson, Ismay, Cherwell and Churchill's daughter-in-law Pamela (née Digby, later Harriman), the Prime Minister kept everyone up till 3.30 a.m. and got into a heated argument with Major-General John Kennedy, the Director of Military Operations at the War Office (and one of the men who had sat outside the window at Le Mans). Kennedy had intimated that there would be worse things strategically than the loss of Egypt, which Churchill took as unacceptably defeatist, saying that he should be made an example of, and citing Admiral Byng's execution by firing squad in 1757.

In his diary, kept surreptitiously like all the many others, Kennedy complained that Brooke completely failed to intervene on his behalf, 'although I knew I had said nothing with which he did not agree'. Had Kennedy the opportunity to read Brooke's own diary entry, where the discussion was dismissed as 'a rather pompous discourse on strategy', he might have expected less, but his superior officer also had two other reasons for remaining silent.[15] The first was that he had a large number of important matters to discuss with Churchill the next day – including pressing manpower and tank shortages – and the second was of course that he did not want to be perceived by Churchill as defeatist himself, which might have been fatal to his chances of promotion to CIGS in due course. Afterwards Brooke wrote effusive thanks to Churchill for his 'great kindness for giving him the opportunity of discussing the problems and putting some of the difficulties' to him, adding, 'These informal talks are of the greatest help to me.'

*

In late January 1941, Marshall's view of the likelihood of 'American boys being sent into foreign wars' became startlingly evident when the press discovered that the War Department had placed an order with a Cleveland metallurgy firm for four million small discs on which soldiers' names and enlistment numbers would be stencilled, known as 'death tags' or 'dog tags'. Marshall tried to explain that they needed two million because each soldier would be required to have two, but in an army of just over a quarter of a million men that argument seemed wanting. 'The whole procedure is routine and the number involved is not large considering the constant use being made of these tags,' Marshall told the President, who would nonetheless have been deeply embarrassed by such a revelation had it emerged before the election.[16] What it did show was that Marshall had high ambitions for the future size of the US Army, ambitions that in the event were to be massively exceeded.

In March, the ABC-1 Staff talks were satisfactorily concluded. After fourteen sessions in Washington over two months, American and British Planners agreed the strategy that would be adopted in the event of the United States entering the war. Germany would be defeated first, Allied interests in the Mediterranean would be maintained and the Pacific theatre would stay on the defensive until victory was secured in the west. This was to form the kernel of Allied grand strategy, once the attack on Pearl Harbor at the end of 1941 catapulted the USA into the Second World War.

Meanwhile, the damaged British aircraft carrier HMS *Illustrious* was repaired in a US shipyard despite America's strict neutrality laws. At that time no fewer than eight thousand RAF pilots were being trained in the United States, and by October 1941 the Joint Staff Mission in Washington numbered two hundred military personnel, their duty to interpret the views of the British Chiefs of Staff to the US Chiefs of Staff and to keep Anglo-American planning up to date. By the end of the war these tasks required no fewer than three thousand people. In their role of keeping in constant touch with the Plans, Operations, Intelligence and Communications branches of the American service departments, the Joint Staff Mission grew to be huge, though not unwieldy.[17]

Although Roosevelt publicly opposed the employment of convoys, which might involve American vessels firing at German ones, he was in favour of 'patrols' that could protect American against aggressor vessels, which often amounted to much the same thing. One such patrol of

10 April 1941 entailed the destroyer USS *Niblack*, while rescuing the Dutch survivors of a merchantman, dropping three depth charges on the U-boat that had torpedoed it. The following day the President told Churchill that he was extending the patrol area to 25 degrees west longitude, a position midway between the western bulge of Africa and the eastern bulge of Brazil. The Royal Navy therefore effectively no longer had to worry about patrolling the western Atlantic.

This strained US naval resources and Marshall strongly advised the President to move some of the Pacific Fleet based at Pearl Harbor on the Hawaian island of Oahu into the Atlantic instead. Marshall was certain that, with heavy bombers and new pursuit planes, American forces were such that the Japanese would not attack. This was, of course, terrible advice, which Roosevelt anyhow turned down since American relations with Japan were at a critical stage, and the removal of ships would be seen by Tokyo as a sign of weakness. The following month, Marshall made an equally dire prediction when he informed the President that 'The island of Oahu, due to its fortifications, its garrison, and its physical characteristics, is believed to be the strongest fortress in the world. With this force available a major attack against Oahu is considered impracticable.'[18]

Churchill described the German invasion of Russia on the night of Saturday 21 June 1941 as 'the fourth climacteric' of the war, the others being the fall of France, the battle of Britain and the passing of the Lend–Lease Act. 'Trust him to find a word no one else had ever heard of,' commented one of his lady-typists.[19] Hitherto the only British grand strategy worthy of the name depended upon blockading Germany, aerial bombardment and attempting to foment revolt in Europe, none of which held out realistic hope of victory over the Nazis. Now Operation Barbarossa so altered the geo-strategic situation that it made it imperative for Roosevelt, Churchill and Marshall – Brooke was still only commander-in-chief of Home Forces – to meet face to face to co-ordinate plans to assist Russia against Germany and deter Japan from attacking in the Far East. After Brooke had estimated to the War Cabinet that Hitler's invasion of Russia could cost Germany as many as two million casualties, Churchill commented: 'It came from God – we did nothing about it.' The Prime Minister added that the 'War can't end in 1942 but optimistically in 1943.'[20] Far too optimistically, as it turned out.

On 24 July at Downing Street, first Harry Hopkins and then Churchill spoke to Roosevelt on the telephone, but the President forgot he was not using the scrambler device, and said 'some things about a certain rendezvous which he afterwards bitterly regretted'. That was Placentia Bay on the southern coast of Newfoundland. The following day Churchill telegraphed Roosevelt to say that Europe would be liberated 'when the opportunity is ripe' by an imposing quantity of tanks being landed 'direct onto beaches' by specially adapted ships. 'It ought not to be difficult for you to make the necessary adaptation in some of the vast numbers of merchant vessels you are building so as to fit them for tank-landing fast ships,' he added. Even three years before D-Day, therefore, the methods of victory were being contemplated, confounding later accusations that Churchill 'never' wanted to invade Normandy.

In a nine-page handwritten letter on 4 August to his cousin and confidante Margaret 'Daisy' Suckley, who lived close to him in Dutchess County, New York, Roosevelt described how he had been secretly transferred from his presidential yacht the *Potomac* on to the heavy cruiser USS *Augusta*, and, with another cruiser and five destroyers as escort, had made his way to Newfoundland. The *Potomac* had continued to fly his presidential flag once he'd left her, in order to maintain the deception: 'Even at my ripe old age I feel a thrill in making a getaway, especially from the American Press.'[21]

The presidential flotilla arrived at Placentia Bay at 6 a.m. on Thursday 7 August 1941. The next day Averell Harriman, Roosevelt's special representative to Britain, and the Under-Secretary of State Sumner Welles arrived by plane and the Americans staged what Roosevelt called 'a dress rehearsal conference' before the British arrived, which included General Marshall, Hap Arnold of the USAAF, the two most senior American admirals, Harold Stark and Ernest J. King, and seven others. 'All set for the Big Day tomorrow,' Roosevelt told his cousin about his meeting with Churchill.

Unbeknown to the Americans, the British were also running a dress rehearsal for what was codenamed the Riviera Conference, with the permanent under-secretary of the Foreign Office, Sir Alexander 'Alec' Cadogan, playing the role of Roosevelt, as he and Churchill strode along the cold and blustery deck of their 35,000-ton battleship, HMS *Prince of Wales*.[22] These rehearsals allowed Roosevelt and Churchill to explore avenues, practise arguments, work out which démarches might be

profitable and which unprofitable, and generally make a verbal reconnaissance of the various combinations and permutations that any future
conversation could take. It focused their minds and lessened the danger
of surprises. Both were to repeat this sensible practice before almost all
of the great wartime conferences.

The general good humour of the Placentia Bay meeting was protected
by both sides staying off subjects – such as Britain's Indian empire – that
they knew would produce discord. There were also moments of humour:
after Roosevelt had said that he couldn't understand the British aristocracy's concept of primogeniture and was going to divide his estate
equally between his five children, Churchill explained that such a distribution was nicknamed 'the Spanish Curse' by the British upper classes:
'We give everything to the eldest and the others strive to duplicate it and
found empires. While the oldest, having it all, marries for beauty. Which
accounts, Mr President, for my good looks.'[23] Since Churchill's father
was a younger son, there was more modesty than mock-vanity in that
remark than the President probably realized.

After meeting Churchill on the USS *Augusta* on Saturday 9 August,
Roosevelt reported to Suckley: 'He is a tremendously vital person and
in many ways is an English Mayor LaGuardia! Don't say I said so! I
like him – and lunching alone broke the ice both ways.'[24] Fiorello H.
LaGuardia was the short, squat but hugely energetic Republican mayor
of New York between 1934 and 1945 who had supported the New
Deal and saw his city through the worst of the Depression. A half-
Italian, half-Jewish dynamo and gangster-buster, he was then serving as
Roosevelt's first director of the Office of Civilian Defense.

The discussions, which confirmed the general outlines of the ABC-1
Staff talks, and drew up some high-sounding war aims, were a great
success. The Americans pledged to assist Russia 'on a gigantic scale' in
co-ordination with Britain, to provide a capital ship and five-destroyer
escort on north Atlantic convoys, to deliver bombers for use by Britain,
and to take over anti-submarine patrols east of Iceland. On 11 August
an exuberant Roosevelt reported to Daisy, 'A day of very poor weather
but good talks.' At dinner with Churchill that night, 'We talked of
everything except the war! and he said it was the nicest evening he had
had!' This prompted Roosevelt to ruminate, 'How easy it is to do big
things if you can get an hour off! The various officers came after dinner
and we are satisfied that they understand each other and that any future

needs or conversations will meet with less crossed wires.' This was unquestionably the case with Sir John Dill and his opposite number Marshall, who struck up a genuine friendship that was to prove invaluable to Anglo-American relations.

On Tuesday 12 August, Roosevelt and Churchill signed the Joint Declaration, later known as the Atlantic Charter, defining the two countries' ideals in the widest sense. These stated that the USA and Great Britain desired no territorial changes that did not accord with the freely expressed wishes of the people concerned, respected the right of all peoples to choose the form of government under which they lived, guaranteed equal access to trade, and so on. After what the President told Suckley was 'a moving scene as they received full honours going over the side', Churchill and his party bade farewell and the *Prince of Wales* left the Bay at 5 p.m.

On his way back to Washington, Roosevelt sailed through Canadian seas that were feared to contain German U-boats. He also faced political risks once he arrived. The House of Representatives had only passed the bill to extend the Selective Service Act by 203 votes to 202, showing that isolationism and the America First movement were still powerful forces in American politics. At the next Cabinet meeting, after commending Churchill's ability as a negotiator, Roosevelt joked: 'But of course, you know Grandpa's pretty good at trading too.' 'You want to look out, Mr President,' someone around the table replied, 'Churchill may be pulling your leg by letting you win the first round.'[25]

Adolf Hitler said that Operation Barbarossa would make the world 'hold its breath', and he was right. The Wehrmacht had completed a land blockade of Leningrad by 8 September, and eleven days later took Kiev. By 1 October it was driving from Smolensk to Moscow, and a fortnight after that was only 26 miles from the Russian capital. Just as King Richard I is said to have gazed upon Jerusalem from afar during the Third Crusade, it was the closest they were to get.

The immense Soviet defeats spawned a powerful movement in Britain for a 'Second Front', an Allied return to the Continent that would draw German troops off the USSR. By late September this had spread far wider than simply among members of the British Communist Party. In a debate in the House of Commons on 30 September, Churchill attempted to counter it: 'If I were to throw out dark hints of some great design, no one would have any advantage save the enemy. If, on the other hand, I were to assemble the many cogent reasons which could be ranged on the other side, I should be giving altogether gratuitous reassurance to Hitler.'[26]

The Sandhurst-educated Conservative MP for Eccles, Richard Cary, nonetheless insisted on a Second Front 'now', and was supported by the Independent Labour MP Colonel Josiah Wedgwood DSO and the Labour MP John Tinker, who said, 'I hope to goodness we do not let it get into the minds of the Russian people that we are prepared to fight to the last Russian before risking any of our own people.' The accusation – that the British had no intention of returning to the Continent and bearding the Nazi beast in its lair – was one that adherents of Marshall were privately to make against Churchill and Brooke before very much longer, and lasts to this day. With the Germans taking Kursk on 3 November, and mounting a second offensive against Moscow later in the month, Churchill's subtle argument was not going to satisfy supporters of the Second Front for ever. At the time, Britain had no longer-term strategy than mere survival.

As early as 28 September 1941, Churchill began considering replacing Field Marshal Sir John Dill, who had been CIGS only since Dunkirk. 'He now has got his knife right into Dill and frequently disparages him,' recorded Colville. 'He says he has an alternative CIGS in mind: Sir Alan Brooke.' Churchill knew that Dill had not actually done anything identifiably wrong as CIGS, and during the Dunkirk campaign he was considered to have done well in command of I Corps. Yet his cruel private nickname for him, 'Dilly-Dally', illustrates the lack of fire that Churchill blamed him for, and there was no personal empathy between the two men. Dill meanwhile regarded Churchill as an arch-meddler whose interventions had to be borne with as much patience as he could muster.

In his published memoirs, Churchill gave no reason for not reappointing Dill beyond his sixtieth birthday, but an earlier draft of them mentioned the CIGS's support for defending Singapore over Cairo during a row over grand strategy in May 1941. At another point that year, Churchill had said of the British High Command in the Middle East, 'What you need out there is a court martial and a firing squad.'[27] Dill only thought up his (hardly crushing) rejoinder – 'Whom would you wish to shoot?' – long after the meeting was over. Against the whip-like wit of Winston Churchill, such mild *esprit de l'escalier* was not enough. Brooke would have had a far sharper retort.

Years after the war, David Margesson, who was secretary of state for war until February 1942, told Brooke that Eden had said to him, 'Brooke will never get on with Winston,' to which Margesson had replied that Churchill needed 'a man who would present the military point of view without fear or favour'. The discussions preceding Brooke's appointment were lengthy, before Churchill finally said: 'Well, David, I'll take him, after all you are secretary of state, but I warn you you may regret it, for I don't think we'll get on.' Margesson recalled that, some years afterwards, Churchill had admitted: 'You were quite right, David, we owe you a lot. Brooke was the right man – the only man.'[28] In a sense, though, they were both right.

With Dill's sixtieth birthday falling on Christmas Day, all Churchill needed to do was tell him that his appointment would not be renewed, as Dill had every expectation that it would be under the special conditions of wartime. On the evening of 17 November Churchill broke this news, offering to make Dill governor of Bombay and rather absurdly emphasizing that he would have 'a bodyguard of lancers' that would follow him everywhere, something that might have thrilled Churchill but meant nothing to Dill.

After helping Dill draft a press statement to the Ministry of Information, John Kennedy noted that the CIGS seemed 'very disturbed but I think not really unhappy and is glad that Brooke is taking over'. Kennedy thought that:

the politicians do not quite realise what they have taken on in Brooke. So far as I know him he is rough and tough and rather impatient. It may be a change for the better in that respect. If he can cut down the time we spend in useless debate with the PM it will be good for the proper conduct of the war ... We may be thankful that Brooke has been chosen to succeed him – it might well have been someone quite unsuitable.[29]

Brooke achieved many things as CIGS, but right to the end of his time with Churchill he was still complaining of the hours wasted in 'useless debate'.

Brigadier Ronald Weeks was on Brooke's special train when the telegram appointing him CIGS arrived at breakfast time on 18 November. 'This is a frightful thing,' said Brooke, 'I don't know how to tackle it.'

Weeks remembered that Brooke 'disliked the idea of "pushing out Dill", his great friend, and was nervous as to how to handle Winston Churchill'. There was no question that both duty and ambition compelled Brooke to accept the Army's most senior post, of course, and he didn't need to have any fears with regard to Dill. It was a publicly smooth transition with the minimum of press comment, although some newspapers ruefully commented that Dill turning sixty was a strange reason for his retirement, considering that the Chairman of the Chiefs of Staff Committee, Dudley Pound, was sixty-four and Churchill himself sixty-seven. In reply to Brooke's letter accepting the appointment, Churchill wrote: 'I did not expect that you would be grateful or over-joyed at the hard anxious task to which I summoned you. But I feel that my old friendship for Ronnie and Victor, the companions of gay subaltern days and early wars, is a personal bond between us, to which will soon be added the comradeship of action in fateful events.'[30]

Sir James Grigg, permanent under-secretary at the War Office, who was 'puzzled at Brooke's great emotion at certain times', told a post-war interviewer that Brooke 'had tears in his eyes' when saying farewell to Major-General Bernard Paget and his fellow officers at Home Forces HQ, and had 'rushed from the room unable to finish his farewell'. This was all the more strange because Brooke had a rather low opinion of Paget and was responsible for replacing him with Montgomery as commander of 21st Army Group before D-Day. It nonetheless does testify to Brooke being a far more emotional man than he allowed himself to seem from the outside.

In a sense, all of Brooke's past life had been but a preparation for this hour and this trial. He arrived at the War Office for his first day as CIGS on Monday 1 December, and Kennedy found him 'very quick and decided'. It was not the last time he was to apply that pair of adjectives to his new boss. Brooke settled into his new role immediately, especially in its most important aspect: the formulation of grand strategy. In his diary for 3 December he wrote: 'I am positive that our policy for the conduct of the war should be to direct both our military and our political efforts towards the early conquest of North Africa. From there we shall be able to reopen the Mediterranean and to stage offensive operations against Italy.' In a mood of distinct self-congratulation after the war, he commented that it was 'interesting to note' that already on his third day as CIGS 'I had a clear cut idea as to what our policy should be . . . It is some gratification to look back now, knowing that this policy was carried out, but only after many struggles and much opposition from many quarters.'[31]

Yet although Brooke did indeed adopt this strategy, which incidentally was already Churchill's, it is also true that, as the distinguished military historian Professor Sir Michael Howard put it, 'None of the British leaders, including Churchill and Brooke, were yet prepared to recommend where and how the decision should be forced – if indeed it had to be forced at all.'[32] The phrase 'against Italy' certainly does not demonstrate that as early as 1941 Brooke envisaged a strategy in which Allied armies fought all the way up the Italian peninsula as far north as the River Po. If this was his scheme, he did not mention it again, even to his diary, for over a year.

At first neither Admiral Pound nor Air Chief Marshal Portal much liked the idea of Brooke joining them as their Army colleague on the Chiefs of Staff Committee, since they thought him 'too abrupt, over-forceful and tactless'.[33] Occasionally described as 'hawk-faced' and 'stoop-shouldered', Brooke would look Churchill in the eye at meetings, say 'I flatly disagree,' and go on to give his reasons.[34] It is true that Brooke was indeed, as Field Marshal Lord Bramall puts it, 'impatient to a fault, even outdoing Churchill in this respect', but very soon his colleagues on the Committee saw his qualities, and especially his readiness to stand up to Churchill.[35] During arguments with the Prime Minister he would sometimes break a pencil in half, a surprisingly forceful – even almost threatening – gesture when closer than 4 feet from Churchill across the narrow green-baize table of the Cabinet Room. In this scion of the Fighting Brookes, the son of the intrepid Sir Victor, Churchill had at last found a CIGS with a determination to match his own.

If he was tough on those above him, Brooke could be tough on those below too. 'He was ruthless where he found anyone at fault,' recalled Weeks, 'and had no use for anyone who had fallen below his standards and had failed him.' For example, in August 1942 he was sent a report on airborne forces which he found sub-standard and which he returned covered in the traditional green ink of the CIGS. Its hapless author found no fewer than thirteen paragraphs of Brooke's corrections, the last, numbered '13', heavily underlined and against it written: 'A most suitable ending to a really lamentable effort.'[36]

Brooke's mere presence at the War Office galvanized those around him. In *The Military Philosophers*, the ninth volume of Anthony Powell's sequence of novels *A Dance to the Music of Time*, the hero Nicholas Jenkins walks down Sackville Street off Piccadilly in London, where his attention was:

unequivocally demanded by the hurricane-like appearance of a thickset general, obviously of high rank, wearing enormous horn-rimmed spectacles. He had just burst from a flagged staff-car almost before it had drawn up by the kerb. Now he tore up the steps of the building at the charge, exploding through the inner door into the hall. An extraordinary current of physical energy, almost of electricity, suddenly pervaded the place. I could feel it stabbing through me. This was the CIGS.

Sackville Street was where Brooke sometimes used to go to relax at lunchtimes, to study ornithology books and prints at the antiquarian booksellers Sotheran's, but Powell's paragraph gives a sense of the pulse that Brooke's presence used to impart to those around him the moment he entered the portals of the War Office on Whitehall.

Because this book is about grand strategy, one should not assume that its four principals spent all, or even most, of their time considering it, even though it was the most important aspect of their duties. Roosevelt was head of state and had simultaneously to carry out those multifarious practical and ceremonial tasks connected with that role. He also spent much of his time overseeing the many government agencies driving the production revolution that turned America into, in his phrase, 'the arsenal of Democracy'. Domestic politics did not end with the attack on Pearl Harbor, and Congress continued to send up bills for his approval. With sustained American economic prosperity now a war-winning weapon, the President had to concern himself with financial questions as much as at any time during the Great Depression.

Nor could Churchill concentrate entirely on grand strategy, having similarly important domestic political calls on his attention. Surprisingly large amounts of time had to be spent honing his speeches, attending Commons debates, lunching with newspaper editors, visiting bomb-sites, factories and military encampments, watching weapons-testing, briefing the King, soothing the Tory Party (which he always feared secretly hated him), meeting ambassadors and foreign leaders, and – to an extent that tended to infuriate Brooke – involving himself in military tactics at a far lower level than grand strategy.

The Cabinet which Churchill chaired furthermore regularly considered subjects far removed from the war, such as whether Noël Coward
should be awarded a knighthood, or whether child allowances should
be paid to the father or the mother. (Of the latter issue, 'The Prime
Minister, in his puckish mood, said it must be left to a free vote and he
would not vote at all lest he lose the votes of the fathers or mothers.')[37]
The most aggressive Cabinet row of the entire Second World War was
not over a military subject at all, but over the £775 million surplus
sterling balances that India had been allowed to build up by November
1944. Churchill was furious with the Viceroy, Lord Wavell, and
demanded to know who 'was responsible for piling up this vast debt
against Great Britain', calling it a 'scandalous intrigue against this
country' and the 'greatest financial disaster in our history'. The Chancellor of the Exchequer, Sir John Anderson, explained that it had been
due to 'our current expenditure in India', whereupon Churchill asked if
there had been 'No effort made to relieve us of a danger worse than the
American debt'. He added that the situation was 'terrible – how can you
get out of a just debt? The day will come when whole position will be
disastrous.'[38] He blamed the situation on Wavell's desire to be popular
with the Indians, whereupon Leo Amery, the Secretary of State for India,
'told him not to talk damned nonsense'. According to a spectator, 'This
shook the PM considerably and there was no end of a row. Amery
withdrew the actual words but not the sense of what he had said.'[39]
After that Anderson threatened to resign, saying, 'I can't go on if there
is feeling in your mind that things are being mismanaged.' Churchill
then himself threatened to resign – 'The PM said he was ready to go etc
etc' – as Admiral Cunningham, the First Sea Lord, put it.[40] Of course
neither man did resign, and their threats were not recorded in the
anodyne report of the discussions in the official Cabinet minutes, but it
was an indication of the non-military political crises that Churchill had
to find time for on top of his consideration of grand strategy.[41]

Similarly, George Marshall was kept busy on very many other matters
than grand strategy, although according to his meticulously kept engagement diary for 1943, the Army Chief of Staff attended no fewer than
fifty-six Joint Chiefs of Staff meetings that year – usually a 1 p.m. lunch
followed by a 2.15 p.m. meeting – and forty-two Combined Chiefs
of Staff meetings in Washington as well. There had been plenty more
meetings, of course, during the total of ninety days that he spent abroad

that year. He visited the White House only thirty-two times in 1943, and not always for military meetings with the President, but also for occasions such as lunches, dinners and receptions for dignitaries including Madame Chiang, the President of Haiti and the Foreign Minister of Brazil. He also gave seven off-the-record press conferences and one on the record.

Although he took four days off with a cold, Marshall went home to Leesburg for only nineteen days in 1943, and officials would occasionally motor out to see him there. Most of the rest of his time was taken up with meetings with generals – who usually got half an hour each – and a large array of senators, admirals, ambassadors and Cabinet members. There were also memorial services, the Foreign Relations Committee, speeches, radio addresses, congressional groups and of course visits to military bases around the country. He would watch propaganda films, such as Frank Capra's *Battle of Tunisia*, before they were released, and lunch with important figures such as the British Ambassador Lord Halifax, Harry Hopkins (four times), General Pershing (thrice) and the Secretary of War Henry Stimson. The Duke of Windsor for some reason got one-and-three-quarter hours of his time on one of his visits from the Bahamas, otherwise he tended to grant people interview slots of fifteen minutes each.[42] The reason why Marshall tended to minimize social functions during the war was well illustrated after a relatively small White House dinner in honour of Anthony Eden in the spring of 1943. On the way back to Leesburg, he showed Katherine eight place-cards, on the back of each of which he had noted requests made by various guests. The subsequent correspondence complying with these favours required thirty-two letters and several telegrams.[43]

The man who had the most time to think about Allied grand strategy was Brooke, who, although he was responsible for running the entire British Army, was expert at delegation. The duties of the CIGS ranged across the entire gamut of war-making, but foremost among them were evolution of grand strategy, appointment of field commanders, allocation of manpower, the equipment and deployment of the Free French, Poles, Dutch, Belgians and Czechs, and the organization of tactical air forces in support of land operations, as well as the protection of sources of raw materials. By far the most important, however, was the first, the provision of an overall global blueprint for how to win the war.

So time-consuming was this primary task that Brooke left the financial, administrative and organizational aspects of the Army to the Secretary of State for War, David Margesson up to February 1942 and thereafter Sir James Grigg, elevated from civil servant to Cabinet minister. Since Grigg got on well with Brooke, but also with his vice-CIGS, Archibald Nye, and the new deputy CIGS, Ronald Weeks, the traditional distrust that had long existed between the military and political sides of the War Office largely disappeared. Brooke could concentrate on strategy and the all-important task of advising Churchill. He was respected in the War Office for being excellent at delegation and almost never got caught up in the details of day-to-day military operations. Like other talented and hard-working individuals at the top of their professions, he only did what only he could do.

On Thursday 4 December 1941 Brooke had his first indication of what life was going to be like as CIGS, when, during a Staff Conference that started at 10 p.m., Churchill's naval plans for raiding Italy were turned down because of the Chiefs of Staff's preoccupations with Japan. 'At midnight Winston banged his papers on the table and walked out,' Brooke told Kennedy the next morning, 'complaining the Chiefs frustrated him in all his offensive projects.' Brooke, who had what Kennedy called 'a delicious talent for mimicry', gave his staff 'a most amusing' account of Churchill's surprise exit.[44] The Chiefs were right to be preoccupied by Japan, however, as events that Sunday confirmed with the attack on Pearl Harbor.

The defeat of Italian forces in Libya in early 1941 had persuaded Hitler to despatch General Erwin Rommel and the Afrika Korps there in the spring, and by the end of the year General Sir Claude Auchinleck had seemingly been out-fought by the 'Desert Fox', whose troops had better tanks and a far better anti-tank gun, the dual-purpose 88mm. Despite Auchinleck having an enlarged army of two corps, his tendency to divide his forces and commit them piecemeal compounded what were at times heavy losses. This in turn meant that the Prime Minister was often in moods alternating between mere ill-temper and what he himself called 'black dog' depression. 'This Libyan "fiasco" is the immediate problem,' Kennedy wrote at this difficult time. 'Winston is very depressed. He had built so many hopes on this offensive.'[45]

The British were right to hold their nerve, as Rommel's bold outflanking movement towards Egypt did not come off, and Auchinleck regained the initiative and relieved the first siege of Tobruk, prior to pushing the Germans back across Cyrenaica all the way to El Agheila on the Gulf of Sirte, from where their offensive had originally been launched in March. Visiting Churchill to discuss the Libyan situation on 5 December, Kennedy found him 'looking pale and rather unwholesome' seated alone at the Cabinet table with his back to the fire in his air-force-blue siren suit, an extinct cigar in the ashtray with half an inch of ash attached. 'But he was in fairly good heart, having apparently got over his fit of depression.' Of the Far East, Churchill told Kennedy that 'The Japs were fools if they come in. Hong Kong will be gone soon, I suppose, he added mournfully.' (It fell on Christmas Day.) The relief of Tobruk by Auchinleck on Sunday 7 December 1941 was not destined to be remembered by history, however, because on that 'day of infamy' all attention was riveted by events half a world away.

Before Pearl Harbor, Churchill had been hoping for a maritime incident similar to the sinking of the *Lusitania* in 1915 to bring America into the war, especially once the US Navy was acting more and more proactively in protection of Lend–Lease ships all the way from Greenland to the Azores. As for Japan, Kennedy recorded that 'Winston always felt Japan would be unlikely to come in and if she did we could leave her to America.'[46]

The day after the attack on Pearl Harbor, Roosevelt went to the rostrum of the House of Representatives to ask for a declaration of war, denouncing the 'unprovoked and dastardly attack by Japan'. The vote took thirty-three minutes, with only the Montanan Republican pacifist Jeannette Rankin dissenting. Simultaneously both Houses of the British parliament voted for war on Japan also. At the United Service Club that evening, Kennedy was unimpressed when he heard Churchill broadcast on the subject: 'He was either very tired or not quite sober. He spoke badly. I wish we had someone in sight in case he breaks up. It is frightful to be so dependent upon a man who is so old and of such luxurious habits.'[47]

On 10 December 1941, just before Germany declared war on America, Churchill told the War Cabinet:

We must address a substantively new situation to that which existed last week. Germany is about to declare war against the US. Japan has attacked Great Britain & the US and placed the right battle group at the right spot, but the US has not lost all her ships, although there has been a disaster in the Pacific. Pearl Harbor was taken by surprise, maltreated ... Japan is in complete control from Cape Horn to Vancouver. In a sense they can ... land at any particular island or place – a situation no one supposed would occur. We will have to put up with a lot of punishment till the situation can be brought around.[48]

Yet he was able to discern that, for all the short-term dangers, in the long run the attack would redound to Britain's benefit:

Looking past the first phase, the real situation is vastly improved, nothing can compare to the US in warfare and now she has to fight for her life. So far as Russia is concerned, Hitler has suffered a colossal defeat which may be turned into colossal disaster, from Leningrad to [indecipherable] German armies are in a frightful condition: mechanised units frozen, prisoners taken in rags, armies trying to stabilise, Russian air superiority. Germany is busted as far as knocking out Russia is concerned. The tide has turned and the phase which now begins will have gathering results. What to face in the Indian & Pacific Oceans we cannot tell. Hong Kong will fight to the death. But we must devise a different kind of warfare and get more ships. It is a bad time but that must not daunt us or inflict doubt upon us in any vital way. They may attack Indian coastal towns, we have lost command of these waters. We and the US will take some time to regain it ... I must meet the President quite soon. There should be no anxiety about the eventual outcome of the war. The finger of God is with us. We must keep our word to Roosevelt, in spite of the fact we don't get enough help from the US.

Asked whether Britain should make an appeal to the Russian dictator Marshal Joseph Stalin to declare war against Japan, Churchill answered, 'He's holding off so much I'm not going to ask him to come into this war. He would not be able to bring back his Siberian army. The battle on the Russian front is going to break the heart of Germany. It would be a mistake to bring them in at this moment. It suits us, it is true, but nothing can compare to the way it suits us in smashing of German armies.' When someone suggested that Churchill should visit Marshall, or that Pound should 'go and concert a policy', the Prime Minister agreed that 'the two Staffs must meet' soon. He added that he didn't 'think you will get a large Japanese force landed in Australia', but considered raids possible against Canadian islands and the Californian coast.

Later that evening the news came through of the loss of HMS *Prince of Wales* (which had conveyed Churchill to Placentia Bay only four months before) and HMS *Repulse*, sunk by Japanese dive-bombers off

Malaya, further underlining how important it was for the Western Allies to co-ordinate their strategy because an American aircraft carrier might have saved them. With Japan's entry into the war, noted John Kennedy, there would be fewer supplies to send to the Middle East for the next spring offensive, and convoys round the Cape of Good Hope would become more precarious. 'The importance of North Africa is therefore greater than ever,' he concluded. 'If we could get the whole North African shore and supply the Middle East through the Mediterranean we should be much happier.'[49] Yet could the United States be encouraged to help in an operation so very far from where they had been attacked in the Pacific?

Lunching in late March at the Carlton Grill restaurant, where Kennedy was outraged that the bill for two with a bottle of burgundy came to £5, Ismay said that Churchill was 'tired and irritable and difficult', and Kennedy complained of the way that 'Winston seems to suck the vitality out of his entourage like a leech,' and how 'very difficult and dictatorial' Churchill had been about the detailed movements of troops, especially the sending of the 18th Division to reinforce India. 'Brooke had great difficulty in stopping him sending it to Burma to carry out a sweep behind the Japanese into Siam. He would not realise that the country in Burma, with its mountains and jungles, was quite impassable for a British mechanised division and furthermore no administrative facilities have been built up for British troops.'[50] It was to be one of very many clashes between Brooke and Churchill where the Prime Minister was convinced that War Office Planners such as Kennedy were deliberately stymieing his vision by their lack of imagination.

Of course Hitler's suicidal declaration of war on the United States – literally so, as it turned out – suddenly and radically altered the relationship between Britain and America, whose leaders immediately realized that they needed to confer again as soon as practicable, if possible before the month was out. No longer was Britain an imploring supplicant to a neutral power. Now, she and America were in a common Manichean struggle. Brooke enjoyed relating that, hearing Admiral Pound still speaking of the United States with the careful phraseology of the pre-Pearl Harbor days, Churchill had turned to him and 'with a wicked leer in his eye' had said, 'Oh! That is the way we talked to her while we were wooing her, now that she is in the harem we talk to her quite differently!'[51]

3

Egos in Arcadia: 'The tremendous hold the Limeys have on Our Boy'

December 1941–February 1942

From the very start of our alliance after Pearl Harbour the President and General Marshall, rising superior to powerful tides of public opinion, saw in Hitler the prime and major foe.　　　　　　Winston Churchill, *The Hinge of Fate*[1]

On Saturday 13 December 1941, the brand-new battleship HMS *Duke of York* left the River Clyde to sail to America for a conference code-named Arcadia. Aboard were Churchill, Pound and Portal, as well as Brigadier Leslie 'Joe' Hollis and Colonel Ian Jacob of the War Cabinet secretariat. Dill also went along, even though he was no longer CIGS, because of his intimate knowledge of strategy. Yet Brooke was left behind 'to mind the shop', as the phrase then went, and also to 'read himself into' his new job, with Ismay staying to assist him. 'How I hated being left behind!' Ismay recalled. In an early draft of his third volume of war memoirs, *The Grand Alliance*, Churchill originally wrote: 'I was anxious that Brooke should remain in London in order to grip the tremendous problems that awaited him.'[2] Although this was removed by the time of publication in 1950, it seems to have been the real reason. (Eisenhower claimed in his book *Crusade in Europe* that Brooke had attended the Arcadia Conference, but that was not the case.)[3]

Although it was indeed important for Brooke to be in London, reporting to the War Cabinet about the terrifyingly rapid Japanese advance across south-east Asia, his absence from Washington cut him out of the vital decisions that were being made there. He approved of some, such as the Germany First strategy for the defeat of Germany before concentrating on Japan, but vehemently opposed others, such as the plan to impose 'unity of command' in all theatres, whereby a single supreme commander would be given ultimate operational control over all Allied forces in each geographical region. Being absent and thus unable to argue his case, or to stiffen the opinions of Pound and Portal, meant that Brooke was reduced to complaining furiously and impotently after they got home. While in London, Brooke gave orders for a scorched-earth policy to be carried out in those areas of the Far East – principally Penang, Borneo and Sarawak – through which the Japanese were marching, seemingly unstoppably.[4]

One of the major lessons that Marshall learnt from Pearl Harbor was, in Stimson's words, 'the importance of unity of command; all the armed forces in any one area must have a single commander.' Stimson felt ashamed that it should have taken such a catastrophe to teach the War Department (and principally himself) something so obvious, considering what had already befallen British arms in North Africa, France, Greece and Crete. 'It was only by the force and tact of General Marshall', he recalled in 1949, 'that unity of command was established in all outposts,' which in the Far East necessitated compromises with the British, Australians, New Zealanders and Dutch.[5]

To have one single supreme commander in each theatre of the war, who had overall control over the land, sea and air forces of all the Allied powers, might seem obvious in retrospect, but at the time it involved each nation giving up long-established and much-prized autonomy of action. Marshall considered it a necessity, however, and first sold the idea to Roosevelt before persuading an initially reluctant Churchill at Arcadia, employing the 'big bribe' that General Wavell should be the supreme commander for the whole of the Far East theatre. Brooke was never persuaded of its merits and was not even present to put forward his objections, but was forced to go along with it because he was outnumbered three to one by Roosevelt, Churchill and Marshall.

Unity of command was to be a vital issue, once the US contribution to the war effort out-compassed that of the British Commonwealth, as both Marshall and Brooke had already calculated that it eventually would. (By late 1943, US warplane production would be equal to that of Germany and Russia combined, and over three times that of the UK.) These key commands – in particular the future supreme Allied commands in North Africa, the Pacific and north-west Europe – would in the long run have to go to Americans. It might be too cynical to ascribe such Machiavellian thinking to Marshall less than three weeks after Pearl Harbor, but he did champion a system that was eventually to work hugely in the United States' favour.

The grand strategy Arcadia agreed was summarized in a document written by Churchill entitled WW1, which was to represent the Allied overall position until superseded by another document, CCS 94, in August 1942. This enshrined the concept of Germany First. Crossing the Atlantic in the *Duke of York*, Churchill dictated four great memoranda in only five days, 'The Atlantic Front' on 16 December, 'The Pacific Front' on 17 December, '1943' on 18 December and 'Notes on the Pacific' on 20 December. Once accepted in principle by the Chiefs of Staff – Pound had the next-door cabin – this immense feat of intellectual effort and foresight made up WW1.

In 'The Atlantic Front', Churchill emphasized the importance of supplying Russia 'without fail and punctually', of an Anglo-American expeditionary force landing in North-west Africa in 1942, and of the movement of American troops and bomber squadrons to Northern Ireland 'as a powerful additional deterrent against an attempt at invasion by Germany'. His next memorandum, 'The Pacific Front', was somewhat over-optimistic, expressing the hope that Singapore might hold out 'for at least six months', while admitting the possibility that it might not. In '1943', he envisaged that the whole of the northern African littoral and the eastern Mediterranean 'would be in Anglo-American hands', after which a 'footing' might be established in Sicily and mainland Italy, with the hope that Italy might be knocked out of the war. Ultimate victory, however, could come only with 'the defeat in Europe of the German armies', aided by 'internal convulsions' in the Reich produced by heavy aerial bombing, 'economic privations' and a collapse in morale.

Overall, it was necessary 'to prepare for the liberation of the captive countries of Western and Southern Europe by the landing, at suitable points, of British and American armies strong enough to enable the conquered populations to revolt'. This he timed for the summer of 1943, with victory coming at the end of 1943 or in 1944. Two days later, in 'Notes on the Pacific', Churchill foresaw 'the burning of Japanese cities by incendiary bombs' as the means of persuading the people of Japan to sue for peace. The danger would be if the United States were to concentrate on defeating Japan before Germany, which would absorb American manpower and supplies – he predicted a ten-million-man US

Army – and preclude the early reconquest of Europe. In order to forestall this, the Americans should be encouraged to fight a naval, but not land-based, struggle against Japan 'to regain their naval power in the Pacific', while their armies were committed to defeating Germany.[6]

Portal recalled in 1948 that 'On the way over on the ship we worked out our approach to the Americans and, thanks to the President, we managed to sell them the "Germany First" idea right away.'[7] This was unfair to Marshall and Roosevelt, since they had quite independently come to favour that strategy and did not need to have it 'sold' to them at all. In the event of a war on two fronts, such as the one the Axis had forced upon them, the American service Chiefs and Commander-in-Chief had already decided to defeat Germany first.

As Dwight D. Eisenhower, whom Marshall had promoted to brigadier-general and brought in as head of the War Plans Division of the General Staff, was to explain to Pogue in 1947: 'In any consideration of global strategy the one dominant factor was that until Germany was defeated a large proportion of British land, sea and air forces would be tied down to the British Isles. For this reason it was necessary to defeat the Axis in Europe in order to release British forces for action.' Eisenhower agreed with Admiral Stark's original assessment in 'Plan Dog' that the defeat of Germany would make the defeat of Japan a matter of time, whereas the defeat of Japan would not materially weaken Germany.

It is true that Churchill's plans for defeating the Axis – WW1 – amounted to one of the great state papers of the war, and have prompted some historians to observe that the Prime Minister's 'conceptual reach, at best, far surpassed that of his professional advisers, including Brooke'.[8] But the memorandum produced by Marshall and Stimson for Roosevelt's use during Arcadia also explicitly recognized the Germany First policy, stating that notwithstanding the attack in the Pacific, 'Our joint war plans have recognised the North Atlantic as our principal theater of operations should America become involved in the war'. The first thing was to ensure 'the preservation of our communications across the North Atlantic with our fortress in the British Isles covering the British fleet'.[9]

In general, Roosevelt accepted Churchill's 'peripheral' view of how to win the war – by blockading and bombing Germany, while containing it with small raids and an attack on North Africa – before finally returning to the Continent for the knockout blow once it had been weakened in Russia. FDR appreciated the advantages of such a strategy, as well as the political dangers of any setback resulting from a military reverse on the Continent. Marshall, Stimson, Eisenhower, Hull and Handy, however, preferred the 'Ulysses S. Grant' view that it should be done with a full frontal assault on Germany via France as early as possible. (The commander of the Federal forces in the American Civil War was thought of as personifying the direct as opposed to peripheral military strategy because of his plans for the invasion of the South.) Yet the President could be swayed, or it seemed as if he could be. As he told Henry Morgenthau: 'Nothing could be worse than to have the Russians collapse. I would rather lose Australia, New Zealand or anything else than have the Russians collapse.'[10] Even though he could never have said so openly – any more than Roosevelt could – Churchill agreed. Yet, unlike the Americans, Churchill and Brooke did not think that Russia could necessarily be saved by an early Second Front anyway.

Marshall drew up another memorandum for Roosevelt that made it clear that he was ready to undertake an attack on North Africa at very short notice, on the assumption that Vichy France would 'invite the United States and Great Britain jointly to occupy and defend North Africa'. One Marine division of eleven thousand men could be ready to embark for Casablanca at ten days' notice by 15 January 1942, along with 160 fighter planes and 114 bombers, with another infantry division of twenty-one thousand men and one armoured division of thirteen thousand to be ready at ten days' notice after 15 February. Anti-aircraft units could be supplied by the British. 'The US Army is prepared to reinforce the foregoing initial contingent with air and ground forces to the extent necessary to maintain its position in North Africa against probable Axis attack,' the general told the President.[11]

Marshall was thus prepared to undertake an operation on the Atlantic seaboard of North-west Africa in early 1942, always provided that the French would not oppose such a landing. Of course once it became clear by mid-1942 that Vichy would indeed fight back everything changed, but this was nonetheless indicative that Marshall was not always diametrically and philosophically opposed to the concept of large numbers of Americans ever arriving in North Africa, as is sometimes suggested.

Churchill arrived in Washington on Monday 22 December, and was given offices in the White House across the corridor from the President. They saw each other constantly. 'He is the only head of state whom

I have ever received in the nude,' the Prime Minister later told Lord Halifax. With Churchill's WW1 paper accepted and the outlines of grand strategy already agreed at Placentia Bay – at least in the widest possible sense and in the short term – the conference tackled the much thornier question of which organization should be ultimately responsible for carrying it out. A body was needed to ensure the most intimate and smoothest co-operation between the British and American High Commands. Marshall appreciated that the ultimate decision-making body – to be named the Combined Chiefs of Staff Committee – needed to consist of those who would be responsible for putting those decisions into effect. In Stimson's words, Marshall 'insisted that the Combined Chiefs should in fact be chiefs, and not merely elders of the council'.[12]

Hitherto, American strategy was not made by the commanders of each service working together in the same committee. Part of the reason Marshall wanted to institute a Joint Chiefs of Staff Committee to mirror the British Chiefs of Staff Committee was his recognition that the way the British High Command had been organized since the early 1920s gave them an undoubted edge in military planning, a lead that he was determined to minimize as time went on. Brigadier-General Thomas Handy, who like Marshall had graduated from VMI, served in the War Plans Division from 1936 to 1940 and returned there in 1941. The next year he was promoted to major-general and became assistant chief of staff in the new, revamped Operations Division (OPD) at the War Department. Working closely with Marshall as a senior Planner throughout the war, he ended up as a four-star deputy chief of staff. 'After Pearl Harbor the Prime Minister descended on Washington with a whole gang of people,' Handy recalled of Arcadia. 'We were more or less babes in the wood on this planning and joint business with the British. They'd been doing it for years. They were experts at it and we were just starting. They'd found a way to get along between the services.'[13] Marshall essentially tried to copy that successful formula.

Certainly, Colonel Ian Jacob was shocked by the haphazard nature of the American system before Marshall created the Joint Chiefs of Staff. 'To our eyes, the American machine of Government seems hopelessly disorganized,' he recorded of a meeting in the White House Cabinet Room on the first day of Arcadia. 'The President, to start with, has no proper private office. He has no real private secretary, and no secretariat for Cabinet or military business.' A stickler for efficient staff procedure, indeed something of a martinet, Jacob was no less astonished by how relaxed the American style of government was. 'The President sat at his desk, with the Prime Minister in a chair on his left hand side, and the rest of the company perched on chairs and sofas in a rough semi-circle facing him,' he recalled of one Oval Office meeting. 'No tables or anything, and very awkward looking at maps or taking down notes.' The meeting was also attended by Fala, the President's Aberdeen terrier, 'who suddenly started barking, and had to be ejected, just as the PM was in the middle of an oration'. It was far more of a country-house-weekend way of doing things than the crisp, clipped formal efficiency of the War Cabinet Office, and it could not survive the rigours of global war. Marshall was right to adopt the British Staff system in its place. 'The Americans are like we were in the days of Jackie Fisher and Kitchener,' concluded Jacob, harking back to the naval and army chiefs of the Great War. 'Personalities each pushing their own ideas, and no real co-operation . . . They will get all right, but they have a hell of a lot to learn.'[14]

At one meeting held at the Federal Reserve building on Christmas Eve 1941, there was no agenda, and the first thing that Admiral Stark, who as the senior American officer present was in the chair, did was to run through the notes he had made at the previous day's meeting. In Britain these would all have been circulated beforehand. 'General Marshall had also dictated, on his return to his office, his idea of what had happened,' wrote Jacob, surprised that this had not been done by a junior secretary. 'We of course had our minutes prepared, and it was a complete waste of everyone's time to go all over the ground again. Not to mention the waste of effort on the part of these Chiefs of Staff to have themselves to go back and put down an account of the meeting.'[15]

The upshot of all this was the realization by the British that the way to achieve results with the Americans was, in Jacob's words, 'to give up all idea of proceeding in an orderly way in accordance with our own machinery, and to deal directly with individuals'. One man going directly to Marshall or Arnold, he believed, would 'achieve much more than any discussion with their Chiefs of Staff in session'. What Jacob did not know was that Marshall was fully aware of the organizational shortcomings of this system and was busy fashioning one based on that of the British. Before it was in place, however, there was a farcical situation by which 'The Director of Plans of their Army, after attending a meeting of the Joint Planning Committee, had to go back and tell General Marshall about it; then he had to attend a meeting of the Chiefs of Staff at which the JPC paper was taken, and on return to his office he had to dictate a note on what took place at that meeting.'[16]

The senior US Army member of the Joint Secretariat of the Combined Chiefs of Staff was Colonel Paul Robinett. Born and educated in Missouri, he had volunteered for the Army in 1917 but was rejected as underweight. He nonetheless persisted, and was finally commissioned into the 1st Cavalry and assigned to the Mexican frontier. An outstanding horseman, Robinett had been in the US equestrian team at the 1924 Paris Olympics. It was his assignment as an ADC to General Malin Craig that put him on the fast track for promotion, involving a course at the Command and General Staff College at Fort Leavenworth in 1932–3. Spells at Harvard, in the Military Intelligence (G-2) section of the General Staff and finally in Marshall's secretariat followed, so by the time of the Arcadia Conference he was an established high-flier. He also kept a (hitherto unpublished) daily diary, which was not against US regulations.

Robinett agreed that the first Arcadia meeting had been badly organized by the Americans, and blamed Lieutenant-Colonel Lawrence Higgins of the foreign liaison section of Military Intelligence, who had not even ascertained beforehand how many people were likely to attend. Instead of initially choosing the imposing board room of the Federal Reserve, Higgins allotted the meeting to Rooms 2064–2066, which proved too small and had to be changed at Robinett's insistence. 'It must have been embarrassing to General Marshall when he saw what had been done,' wrote Robinett. 'But Higgins was supposed to know something about conferences, for he had been a member of the foreign service.'[17] (For someone who had worked in Washington, Robinett retained a touching faith in the efficiency of the State Department.)

'The British sat on the south side of the table and the Americans sat on the north side,' recalled Tom Handy of these meetings at the Federal Reserve. 'The Secretaries sat at the end of the table. After the administrative people had been posted on the outside, the conference settled down to business.' Handy thought the British Chiefs 'not very impressive. First Sea Lord Admiral Pound is lame ... Field Marshal Dill is a fine appearing man of average physique but looks old and tired. Air Chief Marshal Portal is swarthy, eagle-beaked and young. His keen eyes are alert to everything about him.' These were not unfair assessments.

Robinett later wrote that the British secretariat, 'unlike the American, was a working team' under Joe Hollis. Handy thought the ablest member of the Chiefs of Staff was Portal, but as for Pound, 'He went through a couple of those conferences and never opened his mouth.'[18]

One of Marshall's major problems at Arcadia was Roosevelt's habit of getting into almost *ad hoc* meetings with the British without any American minders being present, which wound up discussing important policy issues. Robinett recalled that on one occasion the President 'was wheeled in' to a room at the White House and, 'without his advisers, participated in a meeting with the British. The upshot of the conference was a directive to the Combined Chiefs of Staff for a meeting at 4pm, signed by Brigadier Hollis, to consider the diversion of reinforcements now en route to the Philippines in case MacArthur was unable to receive them.'[19] Whereas the American Chiefs of Staff had had no time to study the directive, the British had heard all the discussions that led up to it and had copies of the directive before they met. Brooke was fortunate that the mirror situation – in which Churchill invited himself into a meeting of the US Joint Chiefs and blithely agreed to call a Combined Chiefs of Staff meeting to approve a policy the British Chiefs had not approved – was utterly unthinkable.

Now, the British system of having the executive officers in charge of each of the services also being the chief policy advisers to the Government was adopted, and Marshall also insisted on the staff for the new Anglo-American body producing a continuous record of consideration and decision and directive. This made the body 'an executive committee for the prosecution of a global war', rather than just an *ad hoc* body that met whenever the British and American service Chiefs happened to be in the same city at the same time.

Marshall believed that the new body needed, for security and political reasons, to be based in Washington. This was why he devised the concept of a Combined Chiefs of Staff Committee which would consist of the British Chiefs of Staff Committee and the American Chiefs, themselves brought together in a new body named the Joint Chiefs of Staff. When the three British members of the Chiefs of Staff were in London, as of course they would be for the day-to-day running of the war, they would each be represented by very senior British officers in Washington, but also collectively by the Chief of the British Joint Staff Mission, who would be Field Marshal Sir John Dill.

This was utterly revolutionary and flung the British into a ferment. For the first time in history the overall decision-making body for the British armed forces, with ultimate powers over when, where and how Britons would fight, would be based outside the United Kingdom. It required careful persuasion of Churchill by Roosevelt and Marshall, and Marshall first had to persuade Roosevelt and Stimson, who both suspected that they were being required to give up powers to Marshall.

Marshall recognized that the existing system, whereby the US Army and Air Force on one side and the US Navy on the other individually made policy that they then tried to persuade the British to adopt, was grossly inefficient and potentially damaging. There were many pitfalls along the way, but the Combined Chiefs of Staff system was an inspired idea, and made a significant contribution to victory. The Combined Chiefs would have a Joint Secretariat and a Joint Planning Staff which would co-ordinate the activities of the Planning Staffs of the Chiefs of Staff and Joint Chiefs of Staff in London and Washington respectively.[20] It looked complicated on paper, but it worked in practice because there was the will on both sides that it should do so – and the prospect of failure was too dire to contemplate.

Yet, even after Churchill had been persuaded, Brooke remained fundamentally opposed to the whole concept. If a body 'to coordinate supply matters, overseas military movements, and broad military strategy' were based in Washington, he assumed, the United Kingdom would lose powers of initiation and perhaps eventually even of veto, whatever the paper guarantees of equality. Brooke instead wanted a mirror committee in London, but the other Chiefs understandably feared that these two committees would be bound either to duplicate bureaucratically or to clash disastrously. Churchill decided to try Marshall's idea out for a month. It was a classic Whitehall manoeuvre, as it is notoriously more difficult to disinvent an idea once it is already in operation than to mount opposition to an untried concept, and the issue never arose again.[21] Brooke's physical distance from the discussions, and being so new to his position, left him too weak to impose a veto. Furthermore his fellow Chiefs of Staff, including their chairman Dudley Pound, were willing to try it out.

The Combined Chiefs of Staff system involved all institutions and players ceding some autonomy for the common good, which everyone – other than Brooke, who was never wholly reconciled – was willing to do. Perhaps those who sacrificed the most were those smaller Allied nations, especially of the British Commonwealth, who made large relative contributions in terms of troops and financial commitments yet had no seats on the Committee. When soon after its inception the Free French leader Charles de Gaulle demanded that France should have a seat on the Combined Chiefs of Staff, Ismay replied that if Canada, Australia, New Zealand, Holland, Belgium, Denmark and Norway also had the right to be represented, 'the only place we could have a meeting is the Albert Hall'.[22]

From January 1942 until his untimely death in November 1944, Sir John Dill was Churchill's personal representative in Washington, but also that of the Chiefs of Staff as head of the British Joint Staff Mission. He was thus also the *de facto* British ambassador for all matters military, while the real ambassador, Lord Halifax, concentrated on the diplomatic and non-military aspects of Anglo-American relations. The British Chiefs of Staff and American Joint Chiefs of Staff were to coalesce as the Combined Chiefs of Staff at ten Roosevelt–Churchill summits and on three other occasions during the war, but otherwise Dill stood in for the British Chiefs of Staff at all CCS meetings, and was thus a key figure in the higher direction of the war.

Dill did not initiate military policy, but was invaluable in guiding it. By suggesting that he stay in Washington after Arcadia, Brooke might have been merely attempting to soften the blow of Dill's recent demotion, and in accepting this advice Churchill may have been acting from similar motives. Whatever the reasons, nothing could have been more effective in creating good relations between the American and British military staffs there. In his sincerity, modesty, frankness, integrity and self-discipline, Dill in many ways resembled his friend George Marshall.[23] This amity between Marshall and Dill transcended simply that of comrades thrown together by force of circumstance. The Marshalls invited the Dills to intimate dinners of family and neighbours at Dodona, such as Thanksgiving, went on holiday and attended church together.

In this respect, it was fortunate that Churchill had left Brooke to 'mind the shop', because it allowed Dill to establish close and friendly relations with Marshall that lasted the rest of his life. This closeness, however, sometimes worried Brooke who, like others in the Chiefs of Staff, feared that his friend and mentor might have 'gone native' in

America. But it is just as likely that Britain benefited as much from Marshall's willingness to trust Dill as vice versa. Brooke acknowledged in 1958 that Dill had 'acted as a great link between Marshall and myself'. Link is one word: another might be shock-absorber. The creation of a buffer between the volatile British CIGS and his American counterpart was later to be invaluable, because Dill proved expert at engineering ways out of seeming deadlocks that did not damage the *amour propre* of either man. What Ismay did in smoothing the often rocky relations between Brooke and Churchill, therefore, Dill managed for relations between Brooke and Marshall.

Marshall's institution of the Joint Chiefs of Staff could not, however, wholly alter the disorganization in the American system which Jacob had commented upon so tartly. Roosevelt's desire to retain power closely in his own hands, and to keep Administration officials competing for his favour, led him to adopt methods that seem incredibly byzantine, even administratively dysfunctional, to modern eyes. 'Mr Roosevelt was always very sensitive about the reports on the conduct of his own affairs,' Marshall recalled after the war. 'He didn't want a record of cabinet meetings. He didn't give us the messages he was sending half the time. He would communicate with Churchill . . . and I would be wholly unaware of it, though it directly affected the affairs of the army and the air and maybe the navy.' The way that Marshall found out what Roosevelt was telling Churchill was often through Churchill's sending copies of the President's telegrams to Brooke who sent them on to Dill, who, unbeknown to Churchill and Brooke, then showed them to Marshall. It was an absurdly roundabout way of keeping tabs on one's own commander-in-chief. 'Dill would come over to my office,' Marshall reminisced, 'and I would get Mr Roosevelt's message . . . Otherwise, I wouldn't know what it was. I had to be careful that nobody knew this . . . because Dill would be destroyed in a minute if this was discovered.'[24]

Knowing the dangers involved, Dill would not physically hand over to Marshall a hard copy of the messages, but would sit opposite him at his desk and read them over to him instead. 'This was quite a risky thing for Dill,' Marshall pointed out, 'but he realized that we just had to have the information. Why should the British Chiefs of Staff have it – it was from our President – and the American Chiefs of Staff not have it? It was just Mr Roosevelt's desire for secrecy.' However irregular, even underhand, this system might have been, it worked both ways, for as

Marshall went on to explain: 'Dill would frequently get messages from Mr Churchill and ask him to ascertain General Marshall's possible view of this. Dill would come over and read me Mr Churchill's communication. Then he and I would make up the reply.' Marshall readily acknowledged that it was 'a rather curious set up', but that it was also 'a very effectual one in this business, because these were all strong men – Mr Roosevelt and Mr Churchill – and the coordination of these matters was of vital importance.'[25] In this and many other ways, Dill would prove a good deal more useful to transatlantic relations than if he had been packed off to Bombay, with or without his bodyguard of lancers.

The distinguished historian of the Dill–Marshall relationship and co-editor of Brooke's wartime diaries, Professor Alex Danchev, has discovered that the extent of Dill's and Marshall's willingness to share secret information went further than Marshall was willing to admit, even to his trusted biographer. 'Dill showed Marshall virtually all the Chiefs of Staff telegrams he received, including those "For His Own Information",' we now know, as well as many of the Joint Staff Mission messages and his personal replies sent to London, in addition to private telegrams and letters from his regular correspondents, for example Brooke and Wavell. 'Churchill's "hot ones" were immediately discussed à deux, and in Dill's absence simply taken to Marshall's office by the senior secretary of the JSM, rather as if the US Chief of Staff were on the regular British distribution list.'[26] If Brooke had suspected this was happening, his fear that Dill had 'gone native' would have been understandable.

On occasion Dill would warn Marshall of issues that were in early stages of development; one telegram had the message appended: 'The question dealt with in the attached is not ripe to put to you officially but I always like you to know the shape of possible things to come.' Dill believed that the earlier that Marshall knew about something, the less likely it was that he would cause problems when the issue had sufficiently 'ripened' into prospective action. Meanwhile, Marshall kept Dill, who intimately understood the intricacies of his relations with Roosevelt, Hopkins and Stimson, in close touch with what he was doing. Neither man liked surprises.

Harry Hopkins was a key figure in Roosevelt's entourage, especially during the earlier part of the war, as the President's confidant, sounding-board, scout and emissary. Ian Jacob described Hopkins as:

a frail anaemic man of great honesty and courage, who lives permanently in the White House and is the President's constant companion. Dumbie [Brigadier Vivian Dykes of the Joint Staff Mission] calls him 'the disused prawn' and the name certainly describes his lanky figure, bent back ... and rather fish-like expression. He has no official appointment, though he has been nominated to various functions connected to Lend–Lease administration.

Jacob believed Hopkins to be 'a real friend of Great Britain, and it is a mercy that we have such a man as the President's chief familiar'. In a Tudor or Stuart court, Hopkins would have been termed a favourite.

On 27 December, Marshall presented to Pound and Portal his plan for a single unified command covering a vast area in the Pacific, stretching across India, Burma, the Philippines and Australasia. He formally pro-posed Wavell as supreme commander, accepting that 'With Australia, New Zealand, Singapore, and British and Dutch islands at stake, it would be difficult for Great Britain to accept an American for this post.' Even Admiral Ernest J. King, the Chief of the US Naval Staff, who prided himself on his defence of American autonomy, supported the idea.

In Stimson's phrase, the Royal Navy 'kicked like bay steers' against having a unified command imposed upon them, let alone the equally horrendous concept of having their ships ultimately being commanded by Wavell, a soldier. Yet the meeting ended with the Combined Chiefs of Staff agreeing to present a joint proposal to Roosevelt and Churchill in favour of the idea. 'The chief of the naval planners rushed to the door to shake hands with me and put his arm around me, which surprised me,' Marshall told Robinett fifteen years later. 'And Dill followed and threw his arms around me, and still another one acted explosively.'[27] For all this obvious – and rather unBritish – enthusiasm, Churchill and Brooke had yet to be convinced.

Churchill discussed Marshall's plan with Roosevelt at the White House that evening, saying that it was probably too geographically wide an area of operations for one man to control successfully, and contrasting it to the responsibilities of the last supreme commander, Marshal Foch, who only had to concentrate on the area between the Vosges and the Channel when he had taken over in March 1918. The analogy was imperfect, since Foch had to relay all orders through the national commands, which was not what Marshall was proposing at all.

It is unclear whether the key meeting between Churchill and Marshall took place on the morning of 27 or 28 December – the sources conflict, and frankly it is immaterial – but whichever it was, the Prime Minister was in bed shortly before noon, suffering from an attack of angina pectoris, deficient oxygenation of the heart muscles. Marshall found him 'propped up in bed with his work resting against his knees' with 'the ever-present cigar in his mouth or swung like a baton to emphasise his points'.[28]

Marshall used the opportunity to impose himself on Churchill, just as he had once before with the President. 'Aware from his talks with Roosevelt that a man on his feet had an advantage in an argument,' Pogue wrote of the encounter, 'the General walked up and down as he talked.' (Anyone capable of taking advantage of presidential polio would not baulk at prime ministerial angina, but of course there is no certainty that this tactic had the slightest effect on men as stratospherically self-confident as FDR and Churchill, whatever Marshall might have thought.)

Churchill took refuge on this occasion, as on others, in Elizabethan historical analogy, but he was clearly given short shrift. 'I told him I was not interested in Drake and Frobisher,' recalled Marshall, 'but I was interested in having a united front against Japan, an enemy which was fighting furiously. I said if we didn't do something right away we were finished in the war.' Churchill went off for a bath, where much important prime ministerial thinking was done, only to return with the delphic remark that Marshall would have 'to take the worst with the best'.[29]

Marshall readily admitted to enjoying listening to Churchill, and to learning much from him. The Prime Minister would regularly deliver mini-lectures on the historic things that had happened in places they visited, as he was to do in Downing Street in a dinner à deux with Marshall later that year, but he soon found that Marshall was never bowled over by his rhetoric as others sometimes could be. (Admiral Stark was similarly suspicious of Churchillian oratory; whenever he heard it, he said, he kept his hand on his watch.)[30] At Arcadia, Marshall instead started to exercise a fascination over the Prime Minister. 'There were few people who could mesmerise Churchill,' Colville later recalled; 'Marshall was one of those few who came close to doing so.' Hopkins went further, alleging that Marshall was 'the only general in the world whom Churchill is afraid of'.[31]

Churchill telegraphed to the Cabinet on 28 December that Roosevelt had 'urged' the idea of unity of command and that Marshall had 'pleaded case with great conviction', and furthermore 'it is certain that a new far-reaching arrangement will have to be made ... Marshall has evidently gone far into detailed scheme.' He ended by saying that he would receive Pound's and Portal's views and then give the Cabinet their conclusions. Yet, instead of waiting for London's response to this revolutionary démarche, Churchill had another conversation with Roosevelt and cabled again later that same day: 'I have agreed with President, subject to Cabinet approval, that we should accept his proposals, most strongly endorsed by General Marshall.' He insisted that the War Cabinet decide before 1 January, urging acceptance of 'this broadminded and selfless American proposal, of merits of which as a warwinner I have become convinced'.

Back in London, however, Brooke was profoundly sceptical, describing the scheme as 'wild and half-baked'. He suspected that Churchill had been outmanoeuvred by Roosevelt and Marshall, and that Wavell was – in Dill's words – merely being set up to be 'responsible for the disasters that are coming to the Americans as well as ourselves'. If Brooke had travelled to Arcadia rather than being left to mind the shop, Marshall would have found it much harder selling the American–British–Dutch–Australasian (ABDA) unified command structure for the Pacific theatre, let alone the idea of a Washington-based Combined Chiefs of Staff.

Once Roosevelt and Churchill had approved the ABDA scheme, recalled Jacob, 'It did not take long for the British Chiefs of Staff and the American Navy to come into line, though the former urged strongly that the Commander should be American. They foresaw disasters in the Far East.'[32] These fears were overruled, however, and the principle was accepted and strongly recommended to London, where the War Cabinet – which then consisted of Eden, the Lord President of the Council Sir John Anderson, the Lord Privy Seal and Leader of the Labour Party Clement Attlee, the Minister of Supply Lord Beaverbrook, the Minister of State Oliver Lyttelton, the Home Secretary Herbert Morrison and the Minister without Portfolio Arthur Greenwood – could do nothing but accept the Prime Minister's fait accompli. Although Anthony Eden had resigned from the Chamberlain Government over appeasement in 1938, had been secretary for war in Churchill's first ministry and was foreign secretary by December 1940, as well as Churchill's heir apparent, it is surprising what little influence he had on strategic decisions during the Second World War. His was the main voice in Churchill's inner circle arguing for close connections with the Russians, but otherwise he seems to have played next to no part in the day-to-day creation of British grand strategy.

'It is, of course, an unequal contest,' observed Moran, who since he was not a government official had the right to keep a diary, although in his case its publication of intimate doctor–patient details possibly contravened his Hippocratic Oath. 'Our Chiefs of Staff miss Brooke, whom we have left in London picking up the threads. The peace-loving Dill is no substitute. What he lacks is the he-man stuff.' Over the issue of having the Combined Chiefs of Staff based in Washington, Moran noted that 'Marshall remains key to the situation. The PM has a feeling that in his quiet, unprovocative way he means business, and that if we are too obstinate he might take a strong line. And neither the PM nor the President can contemplate going ahead without Marshall.'[33]

As Planning discussions about the attacks on the Mediterranean and Atlantic coasts of North-west Africa, now codenamed Gymnast, progressed, General Joseph Stilwell – a protégé of Marshall's and an expert on China – began to notice Marshall's support for it wavering. On New Year's Day 1942, Stilwell, who was originally slated to command the American part of the assault, recorded in his diary that at the outset Marshall had seemed so much in favour that he had 'brushed aside' the argument that U-boats might wreck the operation, arguing that the attack convoy would be so well guarded by air and surface defences that 'subs can't get in', and citing the inability of US submarines to sink the 'Jap troopships all around the Philippines'. Stilwell's conclusion was characteristically scatological: 'The War Department is just like the alimentary canal. You feed it at one end, and nothing comes out the other but crap.'[34]

Yet at a three-and-a-half-hour 'pow-wow' on 3 January, only forty-eight hours later, Stilwell found that Marshall, Arnold and the senior Planning Staff officers, including the tall, gangling Mark Clark, the Chief of the War Plans Division Leonard Gerow and the Assistant

Chief of Staff for Supplies Brehon Somervell, were all now opposed to Gymnast because of the fear that the Germans would march through Spain and cut off the American expedition at the Straits of Gibraltar. Meanwhile the profoundly Anglophobic Stilwell sneered that the 'Limeys claim that Spain would "bitterly oppose" Germans. What rot. The Boches *own* the country. Franco must pay the bill for his [Spanish Civil] war.'

Portal's assurances that an American air attack area around Casablanca could be carried out with relatively few squadrons were denounced by the ultra-suspicious Stilwell as 'too transparent for words, but Our Big Boy [Roosevelt] has swallowed it . . . I also asked George [Marshall] what the basic US objective was in going over there. He said to protect the Mediterranean sea lane, which, if it could be used for convoys, would "quadruple the available British shipping." I don't see it.'[35] Stilwell summed up the American Planners' fears about Gymnast in a meeting with Stimson: 'All agree that the means are meagre, the transport uncertain, the complications numerous, the main facts unknown, the consequences serious.' It was true that an opposed attack from the United States across the Atlantic on to the North-west African coastline would always involve great danger, and the American Planners were right to view it as a risky undertaking.

Having been shown what he called 'an amazing document' – the final report of the Arcadia Conference summarizing the conclusions reached – Stilwell wrote:

It demonstrated the tremendous hold the Limeys have on Our Boy. They shout off their faces as if they were *our* delegates and not theirs. So and so simply must be done. The Magnet Plan [to base US troops in Northern Ireland] will relieve several British divisions, which can now go home to jolly old England, thank you. No hint that they might help elsewhere. And we must keep up the Lend–Lease torrent to our British cousins, even though our people go without . . . And by God the Limeys now say it is impossible for Great Britain to produce even the munitions she needs for herself, and we must keep up our stream of offerings or else. I don't know what 'or else' means, but I would like to ask them. And then tell them what they could do.[36]

Small wonder that Stilwell's nickname was 'Vinegar Joe', or that he was later considered unsuitable to command inter-Allied operations. His account dripped with sarcasm and Anglophobia, and was an unfair summary of the conference, yet it did represent the sincere view of a number of senior OPD officers towards Britain.

'We did look at Great Britain with suspicion at the time,' Marshall admitted to Pogue in 1956, mentioning how once, after the British had proposed something – he could not recall what – his senior Planners Albert C. Wedemeyer and Ed Hull had persuaded him to oppose it because they feared British 'ulterior motives'. Marshall presented their rebuttal to the British Chiefs of Staff, only to learn from Portal that the proposal had originally been taken from an American Planners' memorandum.[37]

The suspicion that Marshall and the Planners, especially Stilwell, Wedemeyer and Hull, felt over Gymnast was that Britain needed the operation not so much to deliver a blow against Rommel as to 'pull the British Empire's chestnuts out of the fire' in North Africa after the fall of Tobruk. The defence of British Middle Eastern interests, in places like Egypt, Iraq and Palestine, was always held by the OPD to be the primary reason why Churchill and Brooke wanted to drive the Afrika Korps out of Africa. As with all really enduring but ultimately wrong conspiracy theories, it had just enough of a leavening of truth to give it life. Of course the British wanted to protect Egypt, but ultimately it was also in America's best interests that the Germans did not capture the Middle Eastern oil fields, which would have been the very next step for the Afrika Korps.

Over at the White House on New Year's Day 1942, Halifax recorded that Roosevelt and Churchill were consulting in Churchill's bedroom, while he and Beaverbrook were sitting on a box in the corridor and Harry Hopkins was 'floating past in a dressing-gown'. The Ambassador can be forgiven for thinking it 'the oddest ménage anybody has ever seen'. Churchill was keen not to overstay his welcome in Washington, and so decided to take some days off. He was not yet ready to return to England, however. The detailed conversations still going on between the military Staffs about how to concert military action, which, after ABC-2 and ABC-3, had got up to ABC-4, were too important, but according to the preternaturally well-informed Ian Jacob he felt that his continued presence in the White House might be irksome and 'liable to cause suspicion' in the minds of American servicemen and Cabinet members, 'who might think he was trying to establish too intimate a connection with the President'. He therefore went, with his assistant private secretary John

Martin, his naval aide Commander Tommy Thompson and Moran, to stay incognito in a bungalow near Miami for five days from 5 January.

Marshall made his own plane available to fly the party to Florida, and FDR's principal bodyguard was assigned to protect Churchill there. On his return, the man related how the Prime Minister would bathe naked and was once 'rolled by rough seas. He had then got up and shaken his fist at the sea, and been rolled again, and reduced to a state of great indignation.'[38] It was nothing compared to the indignation he was to feel once Operation Gymnast seemed to slip from his grasp.

On 9 January 1942, the day that the Eighth Army recaptured the Cyrenaican port of Bardia, Marshall sent Roosevelt a memorandum on French North Africa that breathed freezing cold over Gymnast. It stated that the proposals to occupy Madeira and Tangier and land a large force at Casablanca might indeed have their advantages in protecting the south Atlantic sea lanes and air routes and 'preventing the extension of Axis influence to the West and South', but they also involved serious disadvantages.

Attacking Madeira would sacrifice the element of surprise, would be opposed by Franco, and might prompt a counter-attack by the Axis from the Canaries. Furthermore Spain had 150,000 men in Spanish Morocco, many of them veterans of the Spanish Civil War, and Tangier was vulnerable to air bases in Spain and North Africa. It was also doubtful what support could be expected from 'the natives of North Africa and the opportunist French . . . Self-preservation will undoubtedly be their controlling motive. For planning purposes, it must be assumed that both French and natives will adopt the line of action which involves the least danger to themselves.' Marshall concluded that the only poss-ible plan was one that combined not attacking Spanish territory and being invited by France to occupy French Morocco.[39] The first part was possible, the second was not.

Two days earlier, Roosevelt had proposed to Congress a truly massive increase in the defence budget, designed to produce 125,000 aircraft, 75,000 tanks and no fewer than 8 million tons of shipping by the end of 1943. Marshall had already doubled the size of the US Army to two hundred thousand by the time of Pearl Harbor, but was further to increase it forty-fold by Victory in Japan Day. The total number of Americans mobilized in the Second World War – 14.9 million – was

more than that of Britain (at 6.2 million) and France (6 million) combined, twice that of Japan (7.4 million), more than Nazi Germany
(12.5 million) or China (8 million), and surpassed only by the awe-
inspiring 25 million mobilized by the Soviet Union (whose troops were
however often grossly under-equipped).[40] It was this huge expansion
of America's war-making power that underlay Marshall's ability to
'mesmerise' Churchill, who understandably stood in awe of the sheer
productive capacity of the United States.

The drawback, at least for Britain, of Roosevelt's massive increase in
armaments was Marshall's growing desire that as much of them as
possible should be kept in the United States until it was certain where
they would be needed. Marshall wrote to the President via Hopkins to
warn against Lend–Lease allocations being changed, especially if the US
Army was to be built up to 3.6 million men by the end of 1942. It had
thirty-five divisions in mid-January, and it was hoped that ten of them
would be fully equipped, the remainder averaging 50 per cent by the
end of March.[41] On 25 March the creation of a further thirty-six new
divisions would be undertaken, which would be 50 per cent equipped
during the calendar year 1942, always supposing that all the desired
production schedules were met.

Marshall warned the President that the ammunition-production rates
were 'seriously inadequate' for the number of troops in training, arguing
that any increases in Lend–Lease 'must involve the cutting down on
equipment for units that we may be called upon to commit to active
theatres once we embark on any particular operation'. Basic arithmetic
showed that of the fourteen-and-a-half new divisions – only two-thirds
of which would even exist by the end of the first quarter of 1942 – only
nine were earmarked for Europe and Africa, and the rest to the Pacific
and Latin America. Although this meant that the Germany First policy
still prevailed, it effectively precluded any serious hope of the United
States conducting any significant operations on the European mainland
in 1942.

The last meeting of Arcadia took place on Wednesday 14 January 1942
at the Federal Reserve, after which it was adjourned to the dining room
by Admiral Stark, who presided over cocktails and a lunch that was
repeatedly interrupted by photographers' flashlights. Robinett listed the
conference's achievements as: confirming the Germany First priority;

establishing the Combined Chiefs of Staff mechanism for conducting combined US–UK operations; agreeing on the principle of unity of command in each theatre; prescribing the limits of Pacific reinforcement; drawing up common measures to keep China in the war; and the co-ordinating of shipping.[42] These were impressive, and stand as a tribute to Roosevelt, Churchill, Marshall, Portal and Pound. When the British party left Washington that evening, ninety-two of them fitted into three Boeing Clipper flying-boats; they took 18,000 pounds of luggage, including silk stockings, lipstick, hams, underclothes, soap and oranges, as well as what Jacob called 'other acceptable produce to take to expectant families'.[43]

On his return from Washington, Churchill reported to the War Cabinet, and started off with a soon-to-be-famous story:

The President and his wife were kind and hospitable. I lived in intimacy with them, just up the corridor from the Map Room . . . The first day after I arrived FDR came in and I had time to grab a towel . . . The last thing he said when he came to see me off was 'To the bitter end, trust me.' We are suffering heavy blows . . . but the US is setting about the war with great vigour . . . they have jumped right into it. There is a sense of resolve to fight it out. They have tactical ideas of war, Hitler is the enemy, they will do what can re: Japan. But . . . nothing will get in the way of defeating Hitler. [He then spoke of the United States] occupying the North & West African Coast. If they could win it, it would be a vital factor for our Mediterranean shipping – sixty Infantry Divisions and ten Armoured Divisions is what they are aiming for – they should be enough . . . The Americans are anxious to get into combat with the enemy. There is Olympian calm at the White House.[44]

Churchill was clearly relieved by the reiteration of the Germany First policy; furthermore from this report it seems that Gymnast had been agreed with Roosevelt even while Marshall was still investigating its potential, and was having increasing doubts over it. Churchill reported that Harry Hopkins was 'our great friend' and he seemed surprised by the level of access he had had with Roosevelt: 'Lived in the closest intimacy, lunched nearly every day alone or with Hopkins, their ministers asked me what happened – went to Florida for five days.'[45] The juxtaposition of this last piece of information tends to support Jacob's surmise that Churchill left Washington because he feared he was getting too close to Roosevelt for the comfort of American officials, who had to come to ask him for reports on the President's thinking on various points.

It also seems that the Administration had succumbed to a bout of hubris, although it is unclear how far Churchill shared it: 'All in all their Cabinet feel that have got over the hump of the war. If they do it well they will do it in 1943, if clumsily 1944 or 1945. Supplies of matériel and manpower are overwhelming.' Over the Arcadia agreements, Churchill said: 'Marshall, Hopkins & the President wanted unity of command in the Far East – I was at first against it (I was impressed with Marshall who is not narrow). He proposed Wavell . . . Anyone who has seen what has happened since will see that it shows how necessary it was.'

Churchill said he had also 'proposed to Marshall to take brigades at a time for guard duty on the English beaches – they would like that – troops should be used that want training – not their best troops as they need to keep them for North Africa.' Here was a further indication that Churchill hoped and expected American forces, indeed the 'best' of them, to see action in North Africa long before any cross-Channel invasion. Moreover, he seems to have mentioned it directly to Marshall, who had only agreed, however, to US brigades guarding the invasion beaches of southern England.

Later in his briefing, Churchill told the War Cabinet that he had been 'Impressed with calibre of the 3 American Chiefs of Staff', and that since Admiral Stark's 'number's up', because of Pearl Harbor, 'Admiral King is the man with whom we shall have to deal.' Roosevelt had therefore not hidden from Churchill the fact that the man held ultimately responsible for the Pearl Harbor débâcle would be replaced as soon as the immediate crisis was over, as he duly was in March. Stark's replacement, Ernest J. King, was to provide the British Chiefs of Staff with more problems than any other American in the higher direction of the war.

Pausing only to reminisce about his journey back – 'I drove the plane for a bit . . . the engines purred like happy kittens' – Churchill then gave a global overview of strategy.

According to the classic rules of war, you stave off one and crush the other. Well we've done that; they have only ⅓ of North Africa left and when pushed out of Cyrenaica the picture will then be all altered. If we had squandered our strength we would be thin and dissipated before a war that had not begun. We would have been guilty of grossest error of policy. We should not have got through but for Russia . . . There is no use supposing we can keep a lot of a/c [aircraft] and tanks in Britain. All must be disposed of to best ability. Then the Japanese will feel the . . . power of the US – they'll never forgive Pearl Harbour – they will find her and disarm her . . . They feel over the hump.[46]

It was Brooke who made the worst prediction of the meeting. When asked about the protection of transport on the way to Singapore, he said, 'If we can go on putting stuff in it ought to be all right.' The next day the Japanese invaded Burma, and two days after that Rommel launched a new desert offensive and began driving the British back to the Gazala Line defending Tobruk. 'It was apparent that we could not consider Singapore a fortress, for it seemed that no proper landward defences had been prepared,' Churchill informed the Defence Committee on 21 January. 'Taking the widest view, Burma was more important than Singapore. It was the terminus of our communications with China which it was essential to keep open. The Americans had laid the greatest stress on the importance of keeping the Chinese fighting on our side.' As for reinforcing Singapore: 'We did not wish to throw good men after bad.'[47] The fact that Churchill had been the Chancellor of the Exchequer who had financed the building of the Singapore fortress, with its lack of 'proper landward defences', was understandably not raised by anyone. Even as late as 2 February, Burgis was recording at the War Cabinet that, after Brooke had reported that Commonwealth forces had withdrawn inside the city, where there was four months' food supply and 'satisfactory' water supplies, Churchill said that thirty-seven thousand men had already been sent to defend Singapore and it was the 'Will of the Cabinet to defend it to the last'.[48]

After the meeting on 2 February, Anthony Eden recorded in his Rymans Scribbling Diary that Churchill was tired and depressed and had a cold: 'He is inclined to be fatalistic about the House [of Commons], maintained that the bulk of Tories hated him, that he has done all he could and would be only too happy to yield to another,' adding that the complaints over Singapore, the Australian Government's 'intransigent' demands for the return of two divisions from Libya to defend their homeland, and constant 'nagging' from the Commons 'was more than any man could be expected to endure'.[49] Churchill might well have been suffering a bout of 'black dog' depression, or merely complaining melodramatically, as it is inconceivable that he would really have been willing to hand over the premiership to Eden or anyone else. Equally it is often overlooked how much the Chamberlainite backbench Tories still distrusted Churchill, even after 1941.

That same day, Brooke wrote to Auchinleck, the Commander-in-Chief in the Middle East, about Auchinleck's recent suggestion for 'carrying the war into Germany from Africa' in a letter that affords us a glimpse into the CIGS's strategic thinking at the time. 'I am afraid it does not take account of the shipping situation,' Brooke wrote. 'Shipping is exercising a stranglehold on strategy which is likely to be increased by recent events ... What I am certain of is that North Africa would provide an excellent base of attack on Italy.' This was not something, however, that Brooke had any intention of letting on about to the Americans, who were wary about committing themselves even as far west as Algiers at the time. As late as mid-January 1943, he was deliberately leaving it open whether any North African campaign should be followed up on mainland Italy. If the Americans had suspected that the British wanted to attack Italy after North Africa, rather than land in France, they would not have looked favourably on attacking North Africa in the first place.

As for future operations against Tripoli, Brooke wrote: 'Against Italians all would be well, but against Germans, and with some six weeks to obtain reinforcements, the odds may well be stacked heavily against us.' In the Far East there was no prospect of any offensive in the near future; indeed 'the difficulty is to find adequate resources to cork up the holes ... And yet, if we don't withdraw from the Middle East to reinforce the Far East, we may well lose control of communications in the Indian Ocean to such an extent as to seriously endanger our communications with the Middle East, thus rendering it difficult to reinforce you at all.' The true horrors of the situation, which were only just becoming apparent with the possibility of Singapore's fall, were to blight the next eight months.

In a Defence Committee meeting at 10 p.m. on 21 January, with Churchill wearing a red dressing gown adorned with dragons, the Chiefs argued that it would be better to evacuate Singapore and fight on further south in Malaya, explaining that the Singapore channel was narrow, that British aerodromes could be dominated by Japanese artillery fire from Johore, that mangrove swamps tended to impede fields of fire, and so on, all factors that made Singapore a bad defensive position. 'But Winston', noted John Kennedy, 'thinks the island should be fought to the last man.' After a good deal of discussion, a message from the Australian Prime Minister John Curtin was brought in disagreeing with British proposals for defending Australia, which Curtin thought needed to be done by Australian troops presently fighting in North Africa. 'Winston was so angry at this that the meeting broke up and they were all able to get to bed at a decent hour after all.'[50]

This description of the Defence Committee meeting illustrates the superiority of Kennedy's diary account over the officialese of the minutes, not least because it reveals that Churchill's original position regarding the defence of Singapore was very far from the almost deliberately misleading War Cabinet record that 'We did not wish to throw good men after bad.' Kennedy was impressed with Brooke's first two months in the job, observing that 'Since Winston came back from America of course the strain has been greater,' largely because he 'cannot resist interference in details and a stream of rather nagging telegrams has already been directed upon Wavell. Brooke has a great sense of humour and his descriptions of meetings with Winston are very good fun.'

The first full meeting of the Combined Chiefs of Staff took place at its Washington headquarters in the old Public Health Building on Constitution and 19th Street – today the Department of the Interior, South – on Friday 23 January 1942, with Dill representing the British Chiefs of Staff. Although Admiral Pound, who had returned to London, was fully in favour of the Combined Chiefs of Staff system, he characterized the elaborate command and control system they had established with the Americans, when compared to the still relatively small US forces, as 'plenty of harness but no horse'. Roosevelt and Marshall would soon be providing a stampede, however. Sometimes, though not on this occasion, Dill, Marshall and King would stay behind after the official session was over and once 'the cloud of witnesses' had departed to get down to the real, behind-the-scenes business.[51]

On 18 January 1942, in a memorandum to Roosevelt, Marshall identified what was for Brooke also a key aspect of the war, and one that the British believed justified the Gymnast operation. 'The future effort of the Army is dependent on shipping,' he wrote. 'More shipping than is now in sight is essential if the national war effort is not be neutralised to a serious extent.' Marshall estimated that by December 1942 there would be 1.8 million American troops ready for overseas service, and by the end of 1943 about 3.5 million. The Army therefore needed eighteen extra cargo ships per month solely for military use, to permit an overseas force of three-quarters of a million by the end of 1942, which was still less than half of those potentially available.

The 1943 naval construction programme, announced by the President in February and subsequently increased, was for 10.7 million tons of shipping, which would allow an overseas fighting force of 1.5 million men. 'Immediate steps are urged to increase the tempo of the shipbuilding program to a much higher figure,' wrote Marshall. 'The maximum possibilities in this regard should be exploited.' Throughout the war, the issue of shipping stayed at the very top of the strategists' agenda, especially during the battle of the Atlantic. Without the necessary shipping capacity, the Allies simply could not undertake a cross-Channel attack, however much political will was there.

Yet on 1 February 1942 the German Navy unexpectedly introduced a fourth rotor wheel to their U-boats' Enigma enciphering machines, a seemingly minor addition but one which was to defeat the best efforts of the cryptologists at Bletchley Park to crack the Kriegsmarine codes for the next ten months. The code had first been cracked in May 1941, but now the Royal Navy was once again largely sailing blind in the battle of the Atlantic. Hitherto, Ultra – the codename given by the British to information gained from the decipherment of German radio traffic – had revealed where and when the U-boats would be meeting their supply submarines, allowing the Royal Navy to rendezvous there in force. The increased tonnage and lives lost as a result of being deprived of Ultra suddenly threatened Britain's chances of survival all over again. March 1942 was the worst month of the war so far in terms of Allied shipping losses, with no fewer than 273 merchant ships sunk, totalling 834,184 tons, up from less than four hundred thousand tons in December.

Even supposing that a successful Anglo-American landing in northern France had been possible militarily in 1942 or 1943, it would have demanded a vast flow of men and matériel to exploit any successes all the way to Berlin, which could not have been guaranteed while the battle of the Atlantic was still in the balance. Had German submarines prevented the reinforcement of the beachheads or future campaigns deep within France, it would have been disastrous for the Allies: as long as the battle of the Atlantic was undecided, an invasion could not be safely undertaken. Even after the German naval code was cracked again in December 1942, it took several months for the battle to turn in the Allies' favour; indeed it was not until July 1943 that there was a full calendar month in which a greater tonnage of Allied shipping was launched than was sunk.[52] Apart from a short spat between Marshall and King early in the war over the value of convoys, the politicians and Chiefs of Staff of both Britain and America were united over the best ways to win the vital battle of the Atlantic, which was for a strong protective cordon of warships surrounding the merchantmen, and as much air cover as land-based aerodromes could supply.

The decisions taken by commanders on the ground as a result of Ultra information were enormously improved tactically, but Allied grand strategy was also enhanced when it could be ascertained from decrypts that Hitler intended to defend Tunisia and Italy to the last man. An indication of what the Second World War would have been like had British scientists and mathematicians not early on built a special deciphering machine to unravel Enigma messages came during the Ardennes counteroffensive (the battle of the Bulge), which the Germans launched with complete radio silence and during which, in January 1945, they very nearly reached the River Meuse and split the Allied forces in two.

Far less sophisticated than Enigma was the code adopted in February 1942 between the British Staff officers Brigadier Vivian Dykes of the Joint Staff Mission secretariat in Washington and his friend Colonel Ian Jacob of the War Cabinet secretariat in London, as they began their weekly conversations by transatlantic radio-telephone. These kept the War Cabinet secretariat in London in touch with what the British were doing in Washington, beyond what could be communicated by telegraph. Despite being intended to protect anonymity, the names they gave to their principals showed a certain psychological insight into personality. Brooke, for example, was codenamed 'Colonel Shrapnel' and Marshall was 'Tom Mix', after the actor who had played a cowboy in 325 silent movies between 1910 and 1935. Mix was a Pennsylvanian who had served in the US Army during the Spanish–American War, but his name was probably chosen as code for Marshall because his screen persona was of the clean-cut decent all-American who always saved the day; it might also have reflected Marshall's passion for cowboy fiction. The admirals got far less admiring codenames: King was 'Captain Kettle', Stark was 'Tugboat Annie' and Pound was 'The Whale'.[53] (The jokes didn't all go one way though; behind his back his secretaries nicknamed Jacob 'Ironpants'.)

'Dumbie' Dykes (nicknamed after a Sir Walter Scott character, the misanthropic Laird of Dumbiedykes, with whom he had nothing in common) was an attractive character who, as Jacob recalled, 'had great fun over the CIGS, whose birdlike aspect and fast clipped speech lent themselves to caricature. I have never met a man who so tumbles over himself in speaking ... All this, together with his constant habit of shooting his tongue out and round his lips with the speed of a chameleon, made him an easy prey to Dumbie's imitative wit.' A charming and popular Staff officer, and yet another diarist, Dykes was educated at Dulwich College, the Royal Military Academy at Woolwich and Caius College, Cambridge, and had been temporarily blinded by mustard gas in the Royal Engineers during the Great War, after which he served in Ireland and India. He had suffered a tragedy when his infant son choked to death on a toy.

It was at the Staff College at Camberley between 1932 and 1934, whose commandant was then Sir John Dill, that Dykes discovered his aptitude for Staff work, becoming secretary of the Overseas Defence Committee from 1935 to 1938 and showing a 'wonderful knack of spreading good humour and willing co-operation'. He was also expert at keeping minutes, and master of the way that the language of official committee minute-taking could be an exercise as much in obfuscation as in elucidation. He would write of 'fixing up' and 'cleaning up' the Combined Chiefs of Staff minutes.[54]

Dykes became director of plans at the War Office under Dill in 1940, escorted Colonel William J. 'Wild Bill' Donovan, the future head of the undercover Office of Strategic Services, on a tour of ten countries of the Mediterranean, and after Pearl Harbor joined the American Brigadier-General Walter 'Beetle' Bedell Smith as one of the two senior secretaries of the Combined Chiefs of Staff. They became friends and would drop into each other's offices unannounced, asking questions like 'What's the matter with Uncle Ernie today?' of the famously irascible Admiral King. Their relationship therefore mirrored that of their bosses Dill and Marshall.

Bedell Smith had been a Marshall acolyte since entering the Infantry School at Fort Benning in the early 1930s. 'The ulcerous "Beetle" Smith was an intimidating character, not given to any declaration of emotion other than chronic bad temper,' is the estimation of one historian. 'Tough, profane, intolerant, large, self-educated – reputed to have started his career selling newspapers in the streets – but also loyal and discreet, he was a formidable operator.'[55]

In the early 1970s the US Army instituted a splendidly comprehensive oral-history project in which military historians were sent to the homes of retired admirals and generals to interview them about their wartime experiences. The verbatim transcripts of the Senior Officers Oral History Program (SOOHP) are lodged at the Military History Institute in Carlisle, Pennsylvania, and constitute a treasure trove of anecdote and reflection. Of course retired soldiers tend to remember their old war stories with advantages, which needs constantly to be kept in mind, but nonetheless this gigantic historical resource is full of plums. In 1974 Tom Handy described to a SOOHP interviewer a conversation he had heard between Dykes and Bedell Smith over the draft minutes of one Combined Chiefs of Staff meeting:

BEDELL SMITH: 'This sounds just fine; in fact it's wonderful. There isn't but one thing wrong with it.'

DYKES: 'What's that?'

BEDELL SMITH: 'Hell, this isn't what he said!'

DYKES: 'I know damn well it isn't, but it's what he should have said. You see, these fellows won't object if somebody writes down what he should have said.'[56]

For all that it might have been a sensible approach for the politicians and Staff, especially in the protection of their reputations, such exchanges make historians wary of relying too heavily on official accounts.

Dykes came from Dill's entourage, and Dill knew that Dykes would defend him loyally. There were groups in the armies of both countries during the Second World War that almost approximated to the loose eighteenth-century concept of political parties, whereby loyal supporters and friends grouped around a more senior officer under whom they had served or studied at Staff college. Brooke, Montgomery, Auchinleck, Wavell and Dykes were very much of Dill's clique; Marshall's men included Bedell Smith, Eisenhower, Stilwell, Handy, Bradley, Hull and Matthew Ridgway, several of whom had trained under him at Fort

Benning. This might have approximated to a cabal – and there were certainly accusations of favouritism – but it also guaranteed loyalty. (This closeness was to continue after death: George and Katherine Marshall's grave in Arlington Cemetery is to be found under a tree in plot seven on Roosevelt Drive, not far from that of Dill, and on the next hill from Robert E. Lee's Arlington House. The graves of Hull, Ridgway and Bedell Smith are in the same group, only a matter of yards from each other.)

Marshall had to work hard to persuade Roosevelt of the benefits of setting up a new Joint Chiefs of Staff Committee. The President naturally feared that an overall war-planning body might impinge on his constitutional rights and duties as commander-in-chief. Divide and rule between the Army and Navy had worked well for chief executives before, whereas a Joint Chiefs of Staff system might unite and solidify service opinion in a manner that could become hard to overrule. In spite of Stimson's and Marshall's urgings, Roosevelt had hesitated for some time before accepting the need to appoint a Chief of Staff, but was brought around.

On Monday 9 February 1942 the American Joint Chiefs of Staff met for the first time. From April onwards they were chaired by Admiral William D. Leahy, who had been chief of naval operations from 1937 to 1939 before being brought out of retirement in 1940 to become US ambassador to Vichy France. Roosevelt had known 'Bill' Leahy from his days as assistant naval secretary in the Great War, and as well as chairing the Joint Chiefs of Staff he was chief of staff to the Commander-in-Chief (a post he held until 1949). Marshall had proposed Leahy knowing both that the President liked him personally and that a naval chairman would reduce opposition to the whole scheme from Admiral King, especially as the fourth member of the Joint Chiefs of Staff would be Arnold, who was technically Marshall's subordinate (the Air Force remained part of the US Army until 1947).

Marshall found Leahy an impressive office, large secretariat and new quarters in the Public Health Building, close to where the Chiefs convened. 'I told him where to sit when we next met,' recalled Marshall. 'King was furious when Leahy came in and sat down as presiding officer.' But crucially he did not contest the issue. Marshall told Pogue that 'Leahy was neutral enough for my purpose,' but he was irritated when he acted too much as a presidential aide, going to political meetings with Roosevelt and not reporting back fully to him.[57]

Once inaugurated in early 1942, this four-man Joint Chiefs of Staff – comprising the Chief of Staff (Leahy), plus the chiefs of the Army (Marshall), Navy (Admiral Ernest J. King) and Army Air Force (Hap Arnold) – became Roosevelt's principal military advisers and, in Stimson's view, soon began to exercise 'a most salutary effect on the President's weakness for snap decisions; it thus offset a characteristic which might otherwise have been a serious handicap to his basically sound strategic instincts.'[58] It also helped that each member stayed *in situ* for the rest of the war.

As part of the Senior Officers Oral History Program, General Paul Caraway, who served in the OPD between 1942 and 1944, was asked how Marshall had established personal dominance over the Joint Chiefs of Staff, despite not chairing it nor even enjoying an in-built voting majority over the two Navy men. 'General Marshall by sheer force of character established a moral ascendancy over three very difficult old men, Leahy, Arnold and King,' answered Caraway.

King wouldn't give anyone the time of day. He'd sell you the time of day, but he wouldn't give it to you. He was a strong man, and anybody who could establish any kind of ascendancy over him had to be good, and Marshall could get an agreement out of King. Marshall could get an agreement out of Arnold, although Hap was fighting everyone to show he was the biggest dog in the fight. Leahy went with Marshall almost all of the time.[59]

It helped that at meetings the President used to ask Marshall what he thought before asking Leahy or anyone else.

General Handy, who attended many Combined Chiefs of Staff and Joint Chiefs of Staff meetings as a senior Planner, agreed with Caraway about Marshall's power inside the Joint Chiefs Committee: 'He was the dominant character, there is no question about it. Now that doesn't mean that King wasn't a very strong character; he was.' Marshall rarely used to resort to 'pounding the table', as King sometimes would, recalled Caraway, but he established his dominance nonetheless. King, one of whose daughters thought him 'always angry', on occasion went too far. Once, Marshall did indeed have to pound the table and tell the admiral, who was abusing Douglas MacArthur, 'I will not have any meeting carried on with this hatred.'[60]

The British also saw Ernest King as immensely tough-minded, and recognized the strong degree of Anglophobia in his complicated psychological make-up. 'A hard drinker, to the verge of alcoholism; a womaniser, despite having a happy family life, with six daughters and a son; tactless, petty and parochial; and a hot-tempered and rigid disciplinarian,' is the uncompromising judgement of Field Marshal Lord Bramall.[61] Set against that must be the fact that he led the US Navy in America's victory over Japan. King's drinking was mentioned by Brooke alongside that of the Australian General Blamey and Marshal Voroshilov.[62]

A drunk could not have held down the job King did and have fought the inter-departmental battles he fought for so long. Early on, Jacob spotted him as 'a very dominating personality, who looks as if he would be the man to inspire the American fleet with a strong and offensive spirit. He would be a difficult man to manage, and I should say might be easily roused to obstinacy and pig-headedness.' Colville recalled that King was always 'suspicious and resentful of Churchill's undoubted influence' with Roosevelt and Marshall.[63] To make the atmosphere on the Joint Chiefs of Staff worse, King also resented Leahy, who tended to lean towards Marshall in strategic matters rather than his fellow sailor. Yet King had an ally in Roosevelt, whose secretary Grace Tully remarked that he was 'the type of sea dog whom the President really liked and understood. He loved the stories of King's toughness.' After hearing that Senator Arthur Walsh of New Jersey, an old friend of King's, had given the admiral the present of a blowtorch, Roosevelt wrote to King saying that 'he understood the Admiral was so tough he cut his toenails with a torpedo net-cutter.'[64]

In 1957 Marshall explained to Pogue the contrasting ways that he had worked with King and Arnold on the Joint Chiefs of Staff Committee. Soon after Pearl Harbor, when a turf war had soured relations, Marshall had visited King in his office in the Munitions Building and told him to his face: 'If you and I start fighting at the very start of the war, what in the world will the public say about us? They won't accept it for a minute.' King thought so too, for as Marshall commented: 'We did get along. We had one or two pretty mean fights, but anyone has that.' In the USAAF Commander's case, Marshall recalled that he 'tried to give Arnold all the power I could. I tried to make him as nearly as I could chief of staff of the air without any restraint, though he was my subordinate. And he was very appreciative of this.'⁶⁵ They were two very different approaches, and of course Marshall could not have adopted the Arnold approach with King, who was technically his equal in rank. Yet hard-fought compromise on the one hand and generous devolution of power and responsibility on the other produced the right results in each case for Marshall, a talented man-manager.

On the same day that the Joint Chiefs of Staff met for the first time, Brooke drove back from his country house, Ferney Close in Hampshire, with 'soft rain freezing on the surface' of the roads, to discuss with the Chiefs of Staff the parameters for the Combined Chiefs of Staff's powers, which were to be very wide indeed. After a long meeting Brooke recorded that, ever since Portal and Pound had returned, he had 'told them that they "sold the birthright for a plate of porridge" while in Washington. They have, up to now, denied it flatly. However this morning they were at last beginning to realize that the Americans are rapidly snatching more and more power with the ultimate intention of running the war in Washington! However, I now have them on my side.'⁶⁶ Although the biblical quotation was actually 'mess of pottage', the point was made. Brooke never truly felt that the United States should be allowed to, as he put it once in a letter to Dill, 'butt in' on areas of the British Empire – such as Burma – that he felt did not concern them.

For Brooke the setting up of the Combined Chiefs of Staff in Washington, with its very wide powers, marked a tilt in the balance of power between the British Commonwealth and the American Republic which, with an ultimate US mobilization of 14.9 million people, could only go in one direction. 'Seldom has the passing of power been so concretely defined,' writes one historian of the locating of the Combined Chiefs, noting also that 'Brooke's was the lone voice of protest.'⁶⁷ Nevertheless, within the British Chiefs of Staff Committee itself, Brooke would soon be in a position to assume control of the making of grand strategy, and would then turn the Committee into his personal fiefdom.

On 5 February 1942 Lord Halifax asked Mrs Cary Langhorne of Virginia, at a dinner at the British Embassy in Washington, whether Americans disliked Britain. She replied that 'that was putting it a bit too strong', but volunteered that, both through history and from what they thought to be 'English habits of thought', many Americans had 'a certain inferiority complex'. This begs the question of whether the British were haughty towards Americans, and Brooke was certainly accused by members of the US War Department of adopting an aloof attitude towards them.

There is little in Brooke's diaries that criticizes Americans *per se*, as opposed to the strategic views of Marshall and the Joint Chiefs, although a snobbish anti-Americanism was indeed rife among the British upper classes at that time. Americans were widely thought to be vulgar and plutocratic, although that could emphatically not have been held against the ones that Brooke dealt with during the war, especially Roosevelt, Marshall, Arnold and Eisenhower, who were models of the well-mannered gentleman.

One of the most emotional letters Brooke ever wrote was to an American, the US Ambassador to London Gil Winant, who had sent him a set of Audubon bird books on his retirement as CIGS in January 1946. Writing from the War Office in his customary light green ink, Brooke used unaccustomed superlatives such as 'wonderful', 'deep feelings', 'great joy', 'beyond my wildest dreams', and said that he could not 'remember having been made happier by a present'. To get under the tough exterior of Alan Brooke, one needed ornithology.

Although Brooke decried the practice of commanders in the field writing to Churchill letters that were not copied to himself, he did not mind receiving information from them that he would not have considered passing on to the Prime Minister, even though Churchill also held the post of minister of defence. On 6 February Auchinleck wrote to 'My dear Alan' to say that he was sending out a long official telegram about 'the possibility of our resuming the offensive at an early date', adding that 'I purposely refrained from giving the detailed calculations of our prospective tank strength on which we have reached our conclusions,' because he had 'learnt from previous experience the impossibility of reconciling such figures with those arrived at by calculations carried out at home, particularly by the Defence Minister'. He nonetheless had sent under separate cover his calculations in full, 'as I thought you would like to have them for your own personal information, and that of your advisers'.[68]

The implication was obvious: Auchinleck was going behind Churchill's back to Brooke, providing him with important information that Churchill ought to have had as minister of defence. In this way he hoped that he might limit Churchill's capacity for interference with his command. The CIGS should not have gone along with this underhand activity, but he did. Indeed, as Kennedy recorded, 'Brooke found it an invaluable rule never to tell Churchill more than was absolutely necessary,' recalling him once scoring out nine-tenths of a draft minute to the Prime Minister, saying as he did so: 'The more you tell that man about the war, the more you hinder the winning of it.'[69]

4

Brooke and Marshall Establish Dominance: 'He was very difficult and could be pig-headed'

February–March 1942

It is only by building up the authority of the Combined Chiefs of Staff that we can do anything to curb the tendency of the American Chiefs of Staff to take unilateral action.

Field Marshal Sir John Dill to General Sir Alan Brooke, October 1942[1]

It was almost certainly the fiasco over the escape of the *Scharnhorst*, *Gneisenau* and *Prinz Eugen* on 12 February 1942, blithely steaming up the English Channel and evading every effort of the Royal Navy and RAF to stop them, that persuaded Churchill to appoint Brooke in Pound's place as chairman of the Chiefs of Staff Committee.[2] 'I'm afraid, sir,' Pound had told the Prime Minister on the telephone, 'I must report that the enemy battle cruisers should by now have reached the safety of their home waters.' Churchill had gone quiet for a moment, then said 'Why?' but put the phone down before his old friend had a chance to go through the litany of blunders that had taken place. Unlike Brooke and Portal, Pound was also an operational officer as first sea lord, and Churchill felt that to chair the Chiefs of Staff on top of these other responsibilities was too much for him.

'Pound is necessary to me,' he had told Robin Barrington-Ward of *The Times* the previous December. 'His slow unimpressive look is deceptive!' In fact it was not at all deceptive, but rather the effect of a narcoleptic medical condition that is still undiagnosed today. A month beforehand, Brooke had noted how, in a Chiefs of Staff meeting on shipping shortages, 'During most of the discussion the First Sea Lord went sound to sleep, and looked like an old parrot asleep on his perch!' Although there is still doubt about what ailed the sixty-five-year-old admiral, there is none that he ought to have been retired altogether by Churchill rather than merely relieved of his Chiefs of Staff chairmanship.

Pound's biographer argues that although he did catnap and also had a habit of closing his eyes when concentrating, he also had the ability to perk up whenever words like 'cruiser' or 'destroyer' cropped up in the conversation.[3] Nonetheless, Brooke's testimony on more than one occasion is unambiguous and such an affliction should have automatically removed him altogether from the higher direction of a vital service during a world war.

'You should lighten your load,' Churchill wrote to Pound when relieving him of the chairmanship. 'If therefore you represented this to me,' he continued, very courteously, 'I could arrange for Brooke, as it is the Army's turn, to preside on the Chiefs of Staff Committee.'[4] It is very unlikely that in wartime the principle of Buggins' Turn really existed for the chairmanship of the Chiefs of Staff, as in peacetime; Brooke won it because he was acknowledged to have the best strategic brain, and possibly because at forty-eight Charles Portal was ten years younger than him. Churchill also asked that the Director of Combined Operations, Lord Louis Mountbatten, should join the Chiefs of Staff as an equal member. Churchill admired Mountbatten and wanted Combined Operations – between the Army, Navy and Air Force – to be given a greater role in strategy-making. This greatly irritated Brooke, who later complained that Mountbatten 'frequently wasted both his time and ours'. Pound also opposed Mountbatten's appointment, but they were overruled.

On the morning that Brooke was due to take over as chairman, Monday 9 March 1942, Pound thoughtfully turned up early and took the junior seat that Brooke had formerly occupied, saving his colleague from the embarrassment of physically seeming to usurp him. 'Went off all right,' noted Brooke, 'and both Portal and Pound played up very well.' The phrase – echoing the Henry Newbolt line 'Play up, play up and play the game' – was actually high praise from someone of Brooke's background.

Brooke's RAF colleague on the Committee, Sir Charles Portal, the Chief of the Air Staff (CAS), was born in Hungerford in May 1893. A descendant of a Huguenot family that had arrived in 1695, he played in the Winchester College cricket XI before going up to Christ Church, Oxford. After having won a motorcycle race in 1914, Portal joined the Army as a despatch rider in the Royal Engineers, two days after the outbreak of the Great War, and was mentioned in the very first batch of wartime despatches written by Sir John French, Commander-in-Chief of the BEF. The following month a shell landed near Portal, killing five people; he escaped injury because he was blown through a doorway. He later transferred to the Royal Flying Corps and won the Distinguished Service Order and the Military Cross. He was an authority on falconry, owning a famous game falcon called Sibella, a first-rate shot and a keen fisherman, and was reputed to be able to pilot any type of aircraft then in existence. As a squadron commander in the late 1920s he also won the Minot trophy for bomb-aiming.

Portal's *Times* obituary in April 1971 recorded an apology he once made to Churchill after a Chiefs of Staff meeting: 'I'm sorry if I seemed a bit over-assertive or hot under the collar, Prime Minister,' for which he received the characteristic reply: 'In war, my boy, you don't have to be sorry; you only have to be right!'[5] Colville recalled Portal as a 'quiet, unemotional and unassuming Wykehamist . . . who only spoke when he had something to say, listened intently and neither made promises unless he could fulfil them nor allowed himself to be the victim of undue optimism or pessimism'. Many Americans thought him the most impressive of the three British service Chiefs.

Portal tended to give the CIGS better support against the Prime Minister than did Pound. 'Sir Charles Portal handled Churchill extremely well,' wrote Jacob. This handling of the Prime Minister – which could at times resemble the handling of a Mills grenade – was a vital part of the duties of all the Chiefs of Staff. Fortunately there were only four of them during the whole course of the war after December 1941, and each had his own way of doing it. Brooke was forthright, Pound charming, Portal logical and Admiral Sir Andrew Cunningham – who replaced Pound in the autumn of 1943 – uncompromising.

The Chiefs of Staff system – which Ismay believed to be 'the most perfect machine for the higher organization of the war' – was in theory a triumvirate that made suggestions as to military policy and then had the responsibility for initiating, explaining, expounding and defending them. 'In practice, however', recalled Archie Nye, the responsibility rested primarily on Brooke's shoulders 'and much less on those of the CAS and the CNS [Chief of the Naval Staff]'. This Nye put down partly to the fact that Brooke was the chairman, and 'partly to his overwhelming personality, his exceptional clear brain and his forceful character'.[6]

While still commander-in-chief of Home Forces, Brooke had attended a Chiefs of Staff meeting in autumn 1940 that he had afterwards likened to the Mad Hatter's tea party. Under his own chairmanship, business was despatched promptly and efficiently. The Committee's greatest strength lay in the collegiate nature of its decision-making, but from March 1942 he was acknowledged to be *primus inter pares*, in the same way that Marshall was becoming on the Joint Chiefs of Staff Committee. Brooke despised committee chairmen who allowed meetings to go on too long, or who pursued red herrings. In the month that he took over the Chiefs of Staff Committee, he railed to Kennedy about the way Churchill ran the Defence Committee meetings, 'Here we have been arguing two hours over a simple issue which any one of us could have decided in a few minutes and we have got no answer yet.'[7] In a sense it was no different from many complaints made by subordinates about their bosses the world over.

Pound and Portal were hardly ever steamrollered by Brooke, but were unmistakably guided by him, and until Cunningham joined the Chiefs of Staff after Pound's retirement there are only a very few examples of Brooke failing to get his way, even against the combined opposition of the other two Chiefs. The *quid pro quo* of collegiality was rock-solid devotion to the Committee's decisions once taken. During the whole war there was not so much as a single leak from anyone on the Chiefs of Staff Committee about anything discussed there, however seemingly trivial. Nor were there any examples during the war of Marshall or Churchill managing to turn a member of the Chiefs of Staff against the others once a decision was taken, even if that Chief had vigorously opposed it. It was the reason that Churchill never once overruled – as he was constitutionally permitted to do – any military decisions of the Chiefs of Staff Committee during the Second World War. Jacob wrote of Brooke, Portal and Pound: 'The Prime Minister developed a very strong liking for these three, and a real respect for their judgement and professional attainments.'[8] Colville went even further, claiming that Churchill 'might argue with the Chiefs of Staff, and bark at them fiercely; but he loved them deeply.'[9] Whether that was really true of Brooke will be for the reader to decide.

'They stood in awe of him,' Colville wrote of the relationship between Churchill and the Chiefs, 'but they seldom failed to stand up for their convictions, nor would he have respected them if they had been a tamer body. He complained of their obstinacy and would grumble to their combined faces that he was expected to wage modern war with antiquated weapons; but I never remember him denigrating them as men.' It was hardly likely that Churchill would have denigrated men who had sunk the *Wiesbaden* at the battle of Jutland (Pound), closed the Belgian gap on the retreat to Dunkirk (Brooke), been awarded a Military Cross in the Great War (Portal) and won the naval victories of Taranto and Cape Matapan (Cunningham).

'I cannot say that we never differed among ourselves . . .' Churchill admitted in his war memoirs, 'but a kind of understanding grew up between me and the British Chiefs of Staff that we should convince and persuade rather than try to overrule each other. This was of course helped by the fact that we spoke the same technical language, and possessed a large common body of military doctrine and war experience.'[10] The military historian Captain Basil Liddell Hart believed that Churchill had not overruled the Chiefs of Staff on any issue, 'even when his views were mostly clearly right', because of the Dardanelles disaster, after which he had spent two years in the political wilderness 'as a penalty for putting himself in opposition to the weight of official opinion'. It might well have summoned up occasional ghosts for Churchill, but that was unlikely to have been the main reason. Instead, as Portal told the broadcaster Chester Wilmot, Churchill:

was always the democrat. For all his ardent advocacy of his own point of view he was at heart always a compromiser and he was most thoroughgoing in his search for advice and expert opinion . . . He wanted good hard stones on which to sharpen the knife of his ideas . . . He knew his own weaknesses, and knew that he needed to have around him men who from their experience and their expert training could keep his imagination in check.[11]

Churchill would not have chosen the good hard stone of Alan Brooke as CIGS, and then also as chairman of the Chiefs of Staff three months later, if he had wanted someone he could bully. He knew of the reputation of the Fighting Brookes, and would certainly have remembered the argument during his telephone call to Le Mans in June 1940. It was part of the self-confident bigness of Winston Churchill that he appointed such a foil for his own genius, the best possible person to 'keep his imagination in check' when many another, lesser politician would have opted for a yes-man in that post.

As well as being a master strategist, Brooke was a successful departmental tactician. The Deputy CIGS Sir Ronald Weeks recalled how good he was at choosing the best battlegrounds to fight Churchill, telling an interviewer about the constant stream of requests and demands from the Prime Minister that he passed straight on to Weeks, which were 'frequently annoying to the recipient, and often difficult to answer – sometimes they were trivial.' Yet, as Weeks spotted, Brooke tried never to fight Churchill over small issues and would often refuse to sanction Weeks' combative replies to minutes from No. 10, hoping to let them 'die a natural death'. On big issues, however, Weeks noted that 'Brooke would not give way, and would fight Churchill to the last ditch – Churchill rarely in the end would go against him, but there was many a prolonged fight. The secret of Brooke's success with Churchill lay in the fact that he only fought him on big matters.' Brooke never sought out confrontation with Churchill, any more than Churchill did with him; it was simply in the nature of their jobs that clashes arose, and a consequence of the profoundly serious way that both men viewed their responsibilities.

In the spring of 1942 a confidential guide was drawn up in the US War Department to the personalities of the British Chiefs of Staff. For 'Sir Allan [sic] Brooke', it read: 'Suave, intelligent, politico.' Sir Charles Portal had 'Lots of ability, imbued with offensive spirit, however, primary interest in creating tremendous, and essentially British, air power.' Pound was written off as 'A tired old gentleman, straightforward, quixotic, an eminently successful naval officer of the old school.' Lord Louis Mountbatten they believed to be 'an outstanding naval officer' but one whose 'enthusiasm might result in hastily and ill-considered judgments'. From a distance, and taking into account that it misspelt Brooke's Christian name and could not decide whether Pound was straightforward or quixotic, this estimation was surprisingly accurate, if rather overgenerous about Mountbatten's abilities as a sailor.

The Chiefs of Staff met every day except Sunday, and very often had another meeting in the evening. The number of meetings held after the outbreak of war is testament to the Chiefs' diligence, with 117 meetings in 1939; 441 in 1940; 463 in 1941; 420 in 1942; 372 in 1943; 414 in 1944, and 198 in 1945 before V-J Day, a total of 2,425 meetings, of which 1,329, the majority, were chaired by Brooke.[12] There were also eighty Combined Chiefs of Staff meetings at which Marshall and Brooke were present, and a further thirteen plenary sessions at the various conferences attended by Churchill and Roosevelt as well. By the end of the war, these men all knew each other very well.

Every Monday evening the three British Chiefs attended the War Cabinet to give an account of the global struggle over the previous week. This was the supreme policy-making body of the nation and unless there was urgent news to impart or a foreign guest to welcome they started with long analyses from the CAS, then the CNS and then the CIGS, with only the Prime Minister commenting and interjecting. After they had summed up the news from all the fronts, Churchill began discussions and invited comments, but usually only from those ministers directly involved.

As well as these, at 10.30 a.m. every Tuesday the Chiefs of Staff would meet the Joint Intelligence Committee in Great George Street. 'I had to lead this little choir,' recalled its chairman, Victor Cavendish-Bentinck. 'I gave an aperçu of the position as we saw it. Then they questioned us.' He formed a high opinion of Portal – 'the best, the most intelligent and the calmest' – but disliked Brooke, whom he thought less able than Dill. Bill Bentinck, who later became the ninth and last Duke of Portland, nonetheless recalled how Brooke completed the agenda in half the time that Pound had taken.[13] 'Brooke was a powerful personality,' he recalled. 'He used to gobble like an irate turkey. He was very difficult and could be pig-headed.'

Not all War Cabinet meetings were held at Downing Street, and their location could be changed at the last minute. As well as in the Cabinet War Rooms underneath Whitehall, they were held in the GPO Research Station at Dollis Hill (once), the Rotunda in Horseferry Road, the disused Down Street tube station and Church House beside Westminster Abbey (during the Blitz). In the Cabinet War Rooms seating accommodation was so constrained that Burgis recalled that 'an extra table was put inside the square exactly opposite the . . . mouth holding the cigar'. Pound christened it 'The Dock'. The room was air-conditioned, although on one occasion, squeezing past the very fat Ernest Bevin, Burgis unwittingly switched the air-conditioning off, after which, 'with the PM's cigar, Attlee's pipe and Bevin's cigarettes, the atmosphere soon became unbearable'.[14]

A body Churchill never came to like was the Joint Planning Staff, which provided information and ideas for the Chiefs of Staff and which he called 'the whole machinery of negation', since, as Jacob recalled, all too often they produced papers 'which proved conclusively that what he wanted to do was out of the question'. Churchill told Brooke that 'Those damned Planners of yours plan nothing but difficulties' and on another occasion described them as 'psalm-singing defeatists'.[15] Given that Brooke saw it as an important part of his job to defend the Staff from prime ministerial ire, this provided another fruitful area of contention between them.

Yet, as ever with Churchill, humour infused criticism. Seen in the cold light of six decades later, the word defeatist has an almost sinister ring, but if Churchill said it to Brooke with a twinkle in the eye, it might have been meant humorously, especially when attached to the adjective 'psalm-singing' which summoned up the image of joyless Puritan Planners stymieing the imaginative plans of the Cavalier premier. Without wishing to subject everything Churchill said to the fetters of structuralist analysis, it is important to try to see his many jibes and gags in proper context. There is sadly no special font for ironic humour, since many of Churchill's remarks, which in print simply look downright rude, would undoubtedly merit it. Equally, there is no doubt that occasionally Churchill was often genuinely furious with the Joint Planners, as when they opposed – or more accurately provided the grounds for the Chiefs of Staff to oppose – his plans for operations against northern Norway, Singapore and northern Sumatra.

The Defence Committee was formally a sub-committee of the War Cabinet, which Churchill chaired in his capacity as minister of defence and which included all three service ministers – the Secretary of State for War, the Secretary of State for Air and the First Lord of the Admiralty – as well as the Chiefs of Staff. This met less frequently – only twenty times in 1942, for example – and sometimes went months without convening, but it did consider important issues. The items on the agenda for its meetings in Great George Street covered – to take a typical sample from late 1942 – convoys to Russia, Operation Breastplate (an attack on Tunisia from Malta), the catch-all item 'Future Strategy', a draft telegram to Stalin, the build-up of American forces in Britain, aircraft carriers for the Pacific, the situation in Tunisia, and several more.[16] Non-service ministers attended as and when required, and the secretariat comprised Hollis and Jacob, neither of whom helpfully secreted their verbatim notes like Burgis. From Brooke's diaries it is clear that decisions of the Defence Committee meetings often seem to have produced the most furious rows between him and Churchill.

Eden wanted to abolish the Defence Committee altogether, but recorded that Churchill was 'obstinate about it, and maintains that it is one place where service ministers have a show'.[17] Eden thought that since it effected little and tended to attract criticism in parliament, it ought to go, but Churchill spotted that it would be better for an impotent committee to attract criticism than the real power-house of the war, which were the Staff Conferences – that is, those Chiefs of Staff committees that he attended too. As the war progressed, Defence Committee meetings got fewer, while the number of Staff Conferences increased. By contrast, full-scale War Cabinets – though important in home policy and theoretically the ultimate arbiter – rarely interfered in major issues of grand strategy, to Brooke's intense relief.

A vital cog in each of these bodies was Lieutenant-General Hastings Ismay, who combined the official posts of military secretary to the War Cabinet and chief Staff officer to the Minister of Defence; he was also a member of the Chiefs of Staff Committee, although he didn't sign their reports. This collection of posts emphasizes how much of a hybrid he was, while always being Churchill's faithful 'Pug'. He personified the formal and informal links between civil and military authorities, Churchill and Brooke. Able to interpret each man's views to the other, he tried to ward off some controversies before they blew up and softened others once they had. 'Ismay was the oil-can that greased the relationship between Churchill and Brookie,' says General Fraser[18] – much as Dill oiled that between Brooke and Marshall.

Pug was the son of Sir Stanley Ismay, a judge in India and author of the 1885 work *Rules for the Superintendence and Management of Jails in the Central Provinces*. Born in 1887 and educated at Charterhouse and Sandhurst, Pug had had a varied Army career, having been a cavalryman on the North-west Frontier, served in Somaliland during the Great War, attended the Staff College at Quetta as well as the RAF Staff College in Andover, been military secretary to the Viceroy of India Lord Willingdon, and then assistant secretary to the Committee of Imperial Defence. He was recognized by everyone as a 'good chap' and, despite being a professional soldier, his ultimate loyalty lay with Churchill. Brooke liked him as much as Ismay himself admired Brooke, and so the wheels between Prime Minister and CIGS, which could otherwise have ground against one another on a myriad of occasions, were kept well oiled. In an enclosed world that sometimes witnessed malice and back-stabbing, no one seems ever to have had a bad word to say of Ismay.

Dining with John Kennedy at the Savoy Grill on 4 June 1942, Ismay said that Churchill 'needs someone to use as a whipping boy on whom to blow off steam' and he was 'quite frank in admitting this as his chief function'. He added that 'If someone with sounder and stronger judgment could hold his job it would doubtless be better, but the chances are that such a person would soon be thrown out.' Kennedy concluded that he would not have had Ismay's job 'for anything in the world'. Churchill could be fantastically rude to Ismay: he would occasionally shout 'Appeaser!' at him, or say, 'You have grown fat in honours from your country, and now you betray her,' or 'All you want is to draw your pay, eat your rations and sleep,' or 'Very well, if you don't care about winning the war, go to sleep,' but this was said in mock-anger, and Ismay knew not to take it seriously.[19]

One of Churchill's unfairest remarks was the one he made – 'with frightful sibilant emphasis' – about the Chiefs of Staff Committee to the Minister Resident in North-west Africa, Harold Macmillan: 'Why, you may take the most gallant sailor, the most intrepid airman, or the most audacious soldier, put them at a table together – what do you get? *The sum total of their fears.*'[20] It is sometimes asked why, if Churchill was so difficult to work with, people nevertheless stuck at it, rather than simply resigning, as everyone was always free to do. As Sir George Mallaby of the War Cabinet military secretariat, the brother of Colonel Aubertin Mallaby, wrote:

Anybody who served anywhere near him was devoted to him. It is hard to say why. He was not kind or considerate. He bothered nothing about us. He knew the names only of those very close to him and would hardly let anyone else come into his presence. He was free with abuse and complaint. He was exacting beyond reason and ruthlessly critical.

Yet he concluded that 'Not only did he get away with it but nobody wanted him otherwise. He was unusual, unpredictable, exciting, origi-nal, stimulating, provocative, outrageous, uniquely experienced, abund-antly talented, humorous, entertaining . . . a great man.'[21] Moreover it is given to few to live in the limelight of history, or even near the penumbra of the limelight that was always trained on Churchill during the war. It further afforded anecdotes for friends, children and grand-children. Above all, it was work that mattered.

In May 1943 Brooke wrote to Wavell, who had threatened to resign the command of the Indian Army over Churchill's rudeness, to say that 'If I were to take offence when abused by Winston and given to under-stand that he had no confidence in me, I should have to resign at least once every day!' But Brooke never felt that it would be 'likely to have the least effect in reforming Winston's wicked ways!' It was anyhow unpatriotic to resign in wartime; personal issues needed to be put in the overall perspective of duty rather than one's sense of pique. Brooke considered it his duty to remain CIGS regardless of Churchill's behaviour, and he also knew he was making history. Churchill's occasional bouts of ill-temper were after all a small price to pay in order to be able to say: 'I was there.' (Of course that is not always a commendable reason to continue to serve; it was also partly why Traudl Junge and his other secretaries stayed with their – personally rather considerate – Führer till the end.)

Churchill himself constantly emphasized that Britain was writing a new and glorious chapter in her history, equating the struggle with the days of the Spanish Armada and Napoleonic Wars. Anyone leaving the central stage of world history prematurely would have seemed a small figure indeed, and for all Brooke's strictures against Churchill in his diaries it is hard to find an expression of a sincere intention to resign. He desperately wanted the war won so that he could escape the pressure and Churchill, but he recognized his central place in the struggle and never genuinely considered quitting it prematurely, except once in 1944 when the entire Chiefs of Staff were ready to leave office sooner than permit an attack on northern Sumatra. At a dinner at the Ritz with Kennedy in mid-March 1942, Margesson suggested that 'One of Brooke's great gifts was being able to shake himself like a dog coming out of water after unpleasant interviews with Winston, and another his power of debate (and his rasping voice).'[22]

Churchill's tendency to micromanage could be infuriating. Only three days before Brooke took over the Chiefs of Staff chairmanship, the Prime Minister wrote to him that he had noticed in a press report that a British regiment had 'failed to silence' some machine-guns in the desert and asked the CIGS for an explanation, adding that he understood that the way to silence machine-guns was by artillery fire. 'This is typical of Winston's futility,' thought Kennedy. To inundate the CIGS with minor operational matters down to regimental level, and to write on such a subject to the man who invented the very concept of the creeping barrage, indeed either shows Churchill's 'futility' or alternatively his genius at leaving no stone unturned, or perhaps both simultaneously.

Although Moran's diaries are a somewhat tainted source, as he wrote them up from notes long after the events they described, there is no indication that he invented, and he was an intelligent and perceptive man who was indeed present at many key moments in the war. His view of the relationship is therefore valuable:

The PM got his own way with everyone else: only Alan Brooke would not budge. If he sensed Winston's dislike of criticism he paid no attention to it. He could indeed be brutally frank in pulling to pieces Winston's pet projects. He would even venture to stand up to him when summoned to appear before the Cabinet, and if necessary to answer him back. In short, he kept Winston on the rails in the conduct of the war. That is his epitaph.[23]

It is important not to caricature Churchill's relationship with Brooke as one of constant friction and mutual irritation. Brooke's diaries tend to concentrate on the rows precisely because he used his journals as a way of venting his spleen. Nonetheless in the Burgis and Norman Brook verbatim accounts of the War Cabinet meetings there are astonishingly few accounts of such clashes, which suggests that they could not have been a very regular occurrence. No soldier enjoys interference from politicians, and Brooke was no exception, but he often acknowledged how inspirational Churchill was to the general populace, and he understood how important that was.

There could also be perfectly innocent misunderstandings between the two men, as in this comical account by Churchill's secretary Elizabeth Nel of an incident at which she was the only spectator, other than the Prime Minister's notoriously ill-behaved fluffy grey Persian cat, Smokey:

Mr Churchill sat in bed and Smokey sat on the blankets watching him. The PM's telephone conversation with the CIGS was long and anxious; his thoughts were far away; his toes wiggled under the blankets. I saw Smokey's tail switch as he watched, and wondered what was going to happen. Suddenly he pounced on the toes and bit hard. It must have hurt, for Mr Churchill started, kicked him right into the corner of the room shouting, 'Get off, you fool' into the telephone. Then he remembered. 'Oh,' he said, 'I didn't mean you,' and then seeing Smokey looking somewhat dazed in the corner, 'Poor little thing.' Confusion was complete, the CIGS hung up hastily and telephoned the private secretary to find out what was happening. It took a long time to get it all sorted out, and Sir Alan Brooke assured that it was not his fault.[24]

*

In another of history's significant coincidences, on the same day that Alan Brooke took up the reins of the Chiefs of Staff – Monday 9 March 1942 – George Marshall instituted a very wide-ranging reform of the War Department. With the Joint Chiefs of Staff system now firmly in place he wanted to sweep away the pre-Pearl Harbor deadwood and create a far tighter, smaller, more efficient staff structure with clearer and more direct lines of reporting. Having seen his General Staff grow imperceptibly from 122 to nearly 700, Marshall massively slimmed down the operation, with all the changes coming into effect immediately.

Major-General Joseph T. McNarney, the chairman of Marshall's War Department Reorganization Committee, recommended instituting a tripartite structure for the US Army, and Marshall forced it through. By splitting the US Army into three separate commands – Army Ground Forces, Army Air Forces and Army Service Forces – Marshall re-established central control over the machinery. Within two months, no fewer than six hundred officers had come off the Staff, which became manageable again. Dozens of generals over the age of fifty were retired (although not Eisenhower or Patton). Many senior officers never forgave Marshall, but that was the price of leadership. One of the senior Planners who survived the mass cull, Lawrence 'Abe' Lincoln, was to call it 'a matter of evolution, or perhaps almost revolution, effected by necessity'.[25]

The War Plans Division, which had been in existence since 1921, was thrown by Marshall into an alphabet soup of other forces – including the AGF, A-AF and SOS – to create the all-powerful Operations Division (OPD). It had a Theater Group, Strategy and Policy Group, Operations Group, Troop Control Group, Logistics Group, Current Affairs Group, and so on, which then broke down into sections covering different aspects of Army affairs, and Eisenhower was at the head of it. In addition to its plans and policy role, OPD became Marshall's command post.

'OPD ran the war,' Lincoln recalled. 'Its planners made the plans which under General Marshall's influence guided the strategy around the globe.' Marshall was helped in his massive reorganization by the fact that the President was relatively uninterested in the Army. As he told Pogue, 'While I picked theater commanders without him knowing the people – he never saw Ike before he was appointed – he even intervened in selection of heads of department in the Navy.'[26]

By March 1942, therefore, Alan Brooke in London and George Marshall in Washington had both established dominance over their administrative and bureaucratic hinterlands, and over the military side of the creation of grand strategy. Their clashes with Churchill and Roosevelt – and with each other – could only be conducted once they were entrenched as the two guiding forces on the Chiefs of Staff and the Joint Chiefs of Staff Committees. That achieved, each man now tried to promote his vision for victory.

5

Gymnast Falls, Bolero Retuned: 'It would be the most colossal gamble in history'
February–April 1942

Mr Michael Stewart bets Mr John Lawrence £5 that the Continent of Australia is invaded by the Japanese within six months from today.

Brooks's Club wager book, 27 January 1942[1]

On 14 February 1942 – St Valentine's Day – Bomber Command issued its Area Bombing Directive, aimed at damaging 'the morale of the enemy civilian population and, in particular, of the industrial workers'. Soon afterwards, heavy attacks took place on cities such as Essen, Cologne and Lübeck. Agreed in the ABC Staff talks in Washington as a major part of Britain's contribution to the 'softening up' of Germany before eventual invasion, as well as dislocating German industry and supporting the Russians, this policy was initially supported wholeheartedly by Churchill, although Brooke was less convinced. The CIGS's doubts stemmed not from any humanitarian anxiety for German civilians, so much as from concern about the campaign's effectiveness and that too many raids were being directed against Germany rather than more immediately important targets in North Africa, where General Erwin Rommel and the Afrika Korps would be disembarking later that month. He also saw vast and increasing sums of money being spent on producing bombers that he felt might more fruitfully be spent on tanks and ships.

The news of the surrender of Singapore to the Japanese by Lieutenant-General Arthur Percival arrived at the War Office at 4 p.m. on Sunday 15 February 1942. Five days earlier Churchill had warned the War Cabinet that Britain was 'In for a rough time. There will be smashing blows but we will not come out bust. There must be no gloom or disheartenment. We must send what force we can to Burma. We have to screw down rations and not eat into reserves of food. The Army at home must brace themselves.'² In lightning attacks, the Japanese had crossed over on to Singapore island itself and captured the reservoirs without which the city – already low on food and military supplies – could not survive.

Kennedy considered the wider strategic implications of the fall of the British Empire's strongest fortress in the Far East to attacking forces only one-third the size of the garrison, and noted that 'India is naked.' The main fleet based in Ceylon was 'very bare', fighting was taking place near Rangoon, and Darwin in northern Australia was 'comparatively defenceless'. The Chiefs of Staff's response was to try to take one of the three Australian divisions from the Middle East to protect Burma or Ceylon, and bring back one or two of the Indian divisions from Iraq, even though in Kennedy's opinion the forces in Iraq were not strong enough to meet a German attack, 'and never will be', and were only just enough for internal security. Kennedy recommended suspending the bombing of Germany and instead using the planes 'for essential air reinforcement' of Ceylon, Burma, Australia, New Zealand, India and the eastern Mediterranean.

Like Brooke, Kennedy considered the bombing campaign against Germany 'ineffective' and 'beyond our means'. He repeated to his diary the views he had injudiciously blurted out at Chequers the previous year, that, if it came to the worst, 'It is certainly more important to hold India and Ceylon than to hang on in Egypt. We are getting very little for our effort in the Middle East and certainly not enough to compensate for serious losses of positions in the Indian Ocean.'³ After hearing Churchill's views on Singapore, Kennedy reiterated: 'It is wrong to depend so much on one man who is so temperamental, so lacking in strategical knowledge and in judgment, despite his other great qualities.' This summed up the view of Churchill that was held almost universally among senior British Planners, and especially by Brooke, though none failed to praise those 'other great qualities', principally the fillip he gave national morale.

Waiting for Auchinleck's offensive in the Western Desert, which Brooke was also 'very anxious' for, left Churchill 'often in a very nasty mood these days', as Brooke told Kennedy. In particular he often brought up the subject of replacing Auchinleck with General Harold Alexander, who was considered to have done well conducting the 600-mile-long retreat through Burma. In a diary entry that never made

it into his memoirs, Kennedy concluded that Churchill was 'pretty vindictive sometimes and has a very nasty streak in him . . . What a queer
mixture he is. He is such a bad judge and such a terrific advocate.'[4]

Brooke briefed the War Cabinet about Singapore on 16 February,
saying that Percival had been short of food and ammunition and had
given orders to destroy the giant 15-inch guns there. He then went on
to discuss Burma, where 'the Japs were closing in on the frontier using
elephants'. Sumatra, Libya and Russia were also mentioned, but Singapore was undoubtedly the focus of the tragedy. Churchill summed up,
declaring that in 'spite of difficulties' he had 'confidence that the alliance
would break the enemy'.

The grievous twin humiliations of the fall of Singapore and the simultaneous daylight escape of the German battle cruiser *Gneisenau*, her
sister ship *Scharnhorst* and the heavy cruiser *Prinz Eugen* from Brest to
Wilhelmshaven on 12 February also encouraged Churchill to reshuffle
his coalition government. The Labour leader Clement Attlee became
deputy prime minister, Sir Stafford Cripps entered the Cabinet as leader
of the House of Commons, Sir James Grigg replaced Margesson as
secretary for war, and Lord Beaverbrook resigned from the government
altogether, ostensibly to pursue his campaign for a Second Front.

Beaverbrook was in a powerful position to influence policy: he owned
Britain's highest-circulation newspapers, knew everyone in power on
both sides of the Atlantic – Roosevelt liked him, for example – and was
widely credited as minister of aircraft production between May 1940
and May 1941 with having built the Hurricanes and Spitfires that had
won the battle of Britain, although they were in fact for the most
part produced during the eleven months of peace bought by Neville
Chamberlain at Munich. When he went on political crusades, Beaverbrook was a formidable opponent and to Brooke's profound irritation
and Marshall's delight his next one was to be in favour of an immediate
Second Front in Europe to aid the hard-pressed Soviets. Of course in
fact there were any number of fronts against the Germans already,
including those of the Atlantic, Africa, Murmansk convoys and strategic
bombing, but by then the phrase had stuck. Campaigning for a sixth
front would not have had quite the same ring.

'Brooke had a lively distrust of Lord Beaverbrook,' Margesson told
an interviewer after the war, although it would be fairer to say that he
distrusted all civilians who thought they knew more about grand strategy

than the General Staff. He generally disliked and distrusted many of Churchill's lifelong friends and cronies, but especially Beaverbrook, Brendan Bracken, Cherwell and Admiral Sir Roger Keyes, although he was perfectly willing to acknowledge their qualities as amusing dining companions. This disapproval was based not simply on personality, but on important strategic grounds. Brooke believed that Beaverbrook 'had over-committed British resources without thought of consequence' in his negotiations with the Soviets in Moscow in October 1941 as minister of supply.[5]

Unfortunately for Brooke, the call for a Second Front came not just from the strange alliance of the capitalist tycoon Beaverbrook and the British Communist Party, but also from a British public that was intensely admiring of the courage and resilience of the Red Army. Although public meetings were often sponsored by Communist front organizations, tens of thousands of entirely non-political Britons attended them. The Russian Ambassador Ivan Maisky was cheered in the street and begged for his autograph.

Brooke's adamant opposition to an early Second Front alienated plenty of liberal intellectuals such as C. P. Snow, who believed that Marshall's 'judgment was ultimately better than Churchill's and far ahead of General Brooke . . . whose judgment, particularly about Russia, was abysmal'.[6] In fact Brooke had a far more hard-headed attitude towards the Russians, who had until very recently been allies of the Nazis and had been supplying them with grain and oil right up to the night that Barbarossa was launched. When Alec Cadogan went to visit Brooke in April 1942, he 'found him rather impatient with our attitude of giving *everything* Russians ask and getting *nothing* in return. Of course the Russians are fighting – but for themselves and not for us.'[7]

Far more exasperating for Brooke to deal with than intellectuals like Snow was Churchill himself, but, as Kennedy recorded on 20 February, Brooke was 'standing up well to the strain and keeps his sense of humour' three months into the job. The CIGS said that 'the best way to deal with Winston when he begins to declaim is "to put an umbrella up". He just sits silent and next morning generally finds that Winston has become more reasonable.' Viewing the Far Eastern situation three days later, covering Malaya, Java, Siam and the landings of Japanese troops in Bali and Timor, Brooke predicted that Tokyo's next move would be to march on India itself. As for Singapore, he reported to the War Cabinet that seventy-three thousand Commonwealth troops had been captured there so far, but since Churchill and Brooke had agreed not to reinforce the city in the latter stages there was no recrimination over the defeat – at least none is apparent from the War Cabinet minutes or from the notes on which they were based.

One by-product of the collapse in the Far East was the dissolution of the ABDA Command for which Marshall had lobbied so hard two months earlier. Under it Wavell, the Commander-in-Chief in India, had set up his HQ at Lembang in Java on 15 January 1942, in order to oversee all Allied forces in the Far East, taking his orders direct from the Combined Chiefs of Staff in Washington. By the time it was dissolved – a mere six weeks later – the Japanese were in Sumatra, Singapore, Borneo and Bali, and three days later they landed in Java as well.

Since Pearl Harbor, Marshall had despatched 129,772 American troops abroad, along with 190 planes and 1.12 million tons of cargo. On 16 March he sent the President a report about 'the extreme hazards to which our overseas troops movements are exposed', warning that if the luxury liner *Queen Mary*, now requisitioned as a troop ship, were sunk with nine thousand American soldiers on board, it would inevitably 'produce heavy political repercussions'.[8] Since he was not actually suggesting that the ship should not be used to transport such numbers, this message seems to have been largely an exercise in back-covering, something in which senior soldiers occasionally have to indulge as well as politicians.

It was not down to shipping constraints so much as fear of failure that plans for the Anglo-American landings in Morocco were indefinitely postponed by the Combined Chiefs of Staff on 3 March, much to Marshall's satisfaction. Henceforth he directed his Planning Staff, now under the overall command of Major-General Dwight D. Eisenhower, to begin detailed studies for a direct attack on Germany via north-west France, at least once the Japanese had been stalled. This change of emphasis marked the opening of Marshall's two-year argument with Churchill and Brooke. For the moment, however, the British accepted that the North Africa operation had to be postponed because of the Far East situation and they were not kept fully informed about how advanced and detailed Eisenhower's cross-Channel plans were becoming.

As part of the Senior Officers Oral History Program in 1973, Ed Hull related that he had reported for duty at the War Department on the Sunday of Pearl Harbor, an inauspicious start for what nonetheless turned into a very successful military career. As head of the eight-man Future Plans section of OPD, Hull was invited by Tom Handy to draw up plans for an invasion of Europe. Before then, Hull recalled, although the War Department had detailed plans for invading places like Brazil, Dakar and the Canary Islands, 'There wasn't a plan for putting a force ashore in any place in Europe or Africa that really was of any value.'

Soon after the Japanese attack, Hull and his team 'investigated and studied the approaches to Europe from Norway to Dakar in Africa'. They analysed port facilities and beaches along the entire Atlantic seaboard 'in order to ascertain, as a matter of record, whether or not they were suitable for landing operations'. His section concluded that the best place to land was on the north-west coast of France, after a major build-up of forces in Britain. It was feasible 'under two situations', he recalled thirty years later. Under the first, 'Russia was getting pretty bad blows from the German armies. Nobody knew whether she could withstand it or not.' Because the US desperately needed to keep the USSR in the war – 'otherwise we would have to land against the resistance offered by the entire German military strength' – Hull devised an emergency plan to help Russia by landing a single corps on a beachhead on the Cherbourg peninsula hoping that the British and Americans could 'gradually reinforce that resistance to hold whatever was facing us'.[9] This was later codenamed Operation Sledgehammer.

Under the second scenario there would be 'a build-up of air and ground forces on the island of England across the channel from Normandy where we would base aircraft that could support a landing on the Normandy coast'. Although later in the war there were planes that could fly considerably further, at that time the range of fighter aircraft was limited to about 150 miles. The plan for this build-up became known as Operation Bolero. (In claiming authorship of both Sledgehammer and Bolero, Hull might seem to sound like what the Americans then called a 'glory-hog', not least because that credit was also claimed by Wedemeyer, Handy and of course Eisenhower. In fact all these plans were essentially OPD team efforts.) On 27 March 1942, Marshall set Eisenhower, Handy and Hull to work on a redraft of the Sledgehammer and Bolero plans, which he said had to be ready by 1 April at the latest.

*

On 7 March a bombshell telegram arrived in London from Roosevelt and Marshall, which seemed to put the whole Germany First policy itself at risk. 'We have been in constant conference,' wrote the President, 'to ensure that nothing is left unexplored which can in any way improve our present prospects.' In a detailed global overview of the war, he concluded that the United States 'agrees that the Pacific situation is now very grave, and, if it is to be stabilized, requires an immediate, concerted, and vigorous effort.' This would require 'some of our amphibious forces, and the use of all our combat loaded transports . . . and thus seriously reduces present possibilities of offensive action in other regions'. The difference between an indefinite postponement and a cancellation was then made clear. 'Gymnast cannot be undertaken,' wrote the President, referring to the planned invasion of Morocco, and the movement of US troops to the British Isles must 'be limited', and thus 'any American contribution to land operations on the continent of Europe in 1942 will be materially reduced.'

The shipping available to the US would lift a total of about 130,000 men and, the telegram said, increases from ship conversions during 1942 were estimated at only an extra thirty-five thousand men. By June 1943 new construction would give an additional forty thousand, by December 1943 an additional one hundred thousand, and by June 1944 a further ninety-five thousand. Thus, assuming no losses, 'the total troop carrying capacity of US vessels by June 1944 will be four hundred thousand men.'[10] That June 1944 was to be the very month of D-Day could not have been predicted by anyone at that time, of course, but the figures were a stark warning to the British. Since it was assumed that any early Second Front worthy of the name – one which might take significant pressure off Russia – would require at least one million men, Roosevelt and Marshall were effectively admitting that they would be incapable of providing the necessary manpower for years. Furthermore, once warplanes were allocated to Alaska, Hawaii, the north Pacific islands, Australia, the south-west Pacific, the Caribbean, the China–India–Burma theatre and other 'outposts on lines of communications', the USAAF planes available for offensives against Germany were estimated to total only four hundred by July 1942, no more than 560 by October 1942 and 1,040 by January 1943. As Churchill's official biographer Sir Martin Gilbert has summed up: 'For a landing in September 1942, as desired by Stalin, America could provide only 40% of the landing craft and 700 of the 5,700 combat aircraft needed.'[11] The rest, by implication, had to be provided by the Commonwealth.

'In confiding thus fully and personally to you the details of our military arrangements I do not mean that they should be withheld from your close military advisers,' wrote Roosevelt. 'I request, however, that further circulation be drastically reduced.' Brooke and the other Chiefs could be told, therefore, but otherwise the severe limits of US military capacity in the western theatre should be kept as secret as possible. Roosevelt obviously appreciated the effect that this news might have on Churchill, ending his telegram with the underlined sentence: *This may be a critical period but remember always it is not as bad as some you have survived so well before.*'[12]

A month later, Clementine Churchill described her husband as 'bearing not only the burden of his own country but for the moment of an unprepared America', for as Gilbert concludes: 'American lack of preparedness was the decisive factor ... in the inability of the Allies to mount an amphibious attack against northern Europe in 1942.'[13] Why was it, then, that Roosevelt, only one month after transmitting this cable, sent Marshall and Hopkins to London to try to persuade the British to mount an early Second Front? The answer seems to be that the Americans intended it to be an overwhelmingly British operation.[14]

On the day that Roosevelt despatched the bombshell telegram, Stimson had told a meeting at the White House that Churchill was hoping to disperse American forces to Africa, south-eastern Europe and the Far East, whereas the US should be 'sending an overwhelming force to the British Isles and threatening an attack on the Germans in France; that is the proper and orthodox line of our help in the war.' Marshall also wanted to force Germany to fight on two major fronts as soon as possible, and the following day Stimson found this view fully confirmed by the OPD, by then under the control of Eisenhower. Work was under way for what was to be called the Marshall Memorandum, which was shortly going to offer not one but two separate strategies to attack Germany via France. Roosevelt's cable of 7 March therefore seems to have been an attempt to shock Churchill and Brooke into looking positively at an early cross-Channel attack, while warning them that the British contingent for Sledgehammer would have to be very much larger than the American.

Churchill did not have time to give his considered reply to the telegram before another arrived from Roosevelt, stating his 'purely personal view so that you may know how my thoughts are developing relative to organization'. Rather like the fifteenth-century Borgia Pope Alexander VI dividing the globe into two spheres of influence between Spain and Portugal in 1494, Roosevelt proposed splitting half the world up into British and American areas of responsibility, what cynics might see as 'zones of control'.

'Responsibility for the Pacific Area will rest with the United States,' wrote Roosevelt, whereas Singapore, India and the Indian Ocean, the Persian Gulf, the Red Sea, Libya and the Mediterranean 'would fall directly under British responsibility', although it had to be 'understood that this presupposes the temporary shelving of Gymnast'. (The word temporary in this context was itself temporary; Marshall soon wanted it to be permanent.) A 'third area', comprising the whole of the Atlantic Ocean as well as 'a new front on the European Continent', would be the joint responsibility of Britain and the United States.[15]

London's response to Roosevelt's proposal was not, as might be expected, to protest that the British Dominions of Australia and New Zealand fell within the American sphere, but to fear that the Americans might tend to regard the Pacific war, as Kennedy put it, 'more and more as their main responsibility leaving us too great a share of the war against Germany'. Dill meanwhile wanted Brooke's permission to warn Marshall that unless the United States 'reinforce[s] the Middle East strongly before it is too late', the British might be defeated there, setting back victory 'for years'. He had no trouble in getting it.

On 10 March the Japanese took Rangoon. 'He has not an equable temperament,' Kennedy complained of Brooke in his diary that day. 'He does not laugh so much as during his first weeks, which is a pity.' This was hardly surprising, considering the issues with which the CIGS had to deal, including weighing the relative importance of Ceylon and north-east India; fleet adjustments for Operation Ironclad (to capture the Vichy French port of Diego Suárez on Madagascar); the protection of Australia and New Zealand, and the air reinforcement of India and the Middle East, to which Roosevelt had 'responded to our requests for assistance most nobly'. All that and laughter too was a tall order.

Brooke told Kennedy that, as far as finding talented generals was concerned, 'the real difficulty was that the best men were killed in the last war'. Kennedy disagreed with this analysis, while pointing out to his journal that 'The Germans had heavier casualties than we had and yet they produce lots of good leaders.'[16] (By which he meant military ones.) Brooke, Kennedy and possibly Churchill all feared that the British fighting services were simply not as good as they had been in the Great War. 'The Army has become too soft and is not fighting well,' Kennedy wrote. 'This I believe is because the nation has not the right spirit. We are fighting largely for negatives. The Atlantic Charter, etc, are not enough.' He complained of criticisms of Churchill in the press, which he hyperbolically described as a 'fifth column', but also pointed out that, while it was hard to desert or surrender in ships and aircraft, in the Army it was easier for 'a junior officer not imbued with the right spirit to persuade himself he was doing the best for his men if he surrendered instead of fighting it out'. (This rather skated over the fact that Percival was a lieutenant-general.) The mood of gloom in the War Office extended to pessimism about leadership in the British officer corps, as Brooke 'held forth about the low quality of our people, the lack of ideals, the sloppy thinking encouraged at the universities, the general softness and pleasure-seeking of pre-war years'.[17]

Mid-March 1942 saw a great deal of discussion on both sides of the Atlantic about the possibility of a cross-Channel attack in order to relieve the hard-pressed Russians. These laid the ground for Marshall's visit to London in early April, and for the tough arguments that were to rage between Americans and Britons over the next two years. On 14 March, Hopkins sent Roosevelt a memorandum saying that any bridgehead the Anglo-American forces might set up in France 'does not need to be established unless air superiority is complete. I doubt if any single thing is as important as getting some sort of a front this summer against Germany. This has got to be worked out very carefully between you and Marshall in the first instance, and you and Churchill in the second. I don't think there is any time to be lost, because if we are going to do it plans need to be made at once.'[18] They did indeed, because early September was considered the last time that the weather permitted the English Channel to be safely passable for an assault. Eisenhower's initial plans at that stage involved a landing between Calais and Le Havre east of the Seine, not in Normandy. Beachheads were then to be extended eastwards beyond Dunkirk to Ostend, Zeebrugge and the Belgian coast.[19]

On 16 March Brooke informed the War Cabinet of these American plans for 'spearhead' forces of some twenty divisions plus ten armoured divisions, most of which would have to be provided by Britain in the initial stages. Afterwards he and Kennedy had a long conversation about possible future operations in France. 'The advocates of the Second Front always miss the point that sooner or later a force landed in France must fight a battle with the German Army. We must not be confused by ideas of the French rising, etc. The battle is the thing. We must wait till we have a chance of winning the battle. That cannot be till the Germans are cracking up.' Kennedy's diary does not differentiate which of the two men made this vital observation, which dominated British military thinking until D-Day, but it is safe to assume that it was the view of them both. For the British, this was always the key, yet they constantly feared that, if they seemed to blow too cold on a cross-Channel operation, the Americans might ditch the policy of Germany First and concentrate on the Pacific instead. This in turn would leave the United Kingdom naked if Russia lost in the east, and Hitler turned westward again, reviving Operation Sealion, his plan for the invasion of Britain.

In a handwritten letter from Washington on 18 March, Roosevelt, who had recently held talks with the Soviet Ambassador to Washington, Maxim Litvinov, unburdened himself very openly to Churchill about the future conduct of the war. 'Here is a thought from this amateur strategist,' he wrote.

There is no use giving a further single thought to Singapore or the Dutch Indies. They are gone. Australia must be held and, as I telegraphed to you, we are willing to undertake that. India must be held and you must do that . . . I do not visualize that they can get enough troops to make more than a few dents on the borders . . . You must hold Egypt, the Canal, Syria, Iran and the route to the Caucasus.

This was all broad brush and indeed taken for granted, but then the President went on: 'I know that you will not mind my being brutally frank when I tell you that I think I can personally handle Stalin better than either your Foreign Office or my State Department. Stalin hates the guts of all your top people. He thinks he likes me better, and I hope he will continue to do so.'[20] Of course Roosevelt was putting in as delicate a way as possible the fact that he thought that Stalin hated Churchill's own guts, which was unsurprising given Churchill's attempt to 'strangle Bolshevism in its cradle' in 1919–21. Yet Stalin 'liked' virtually nobody, least of all the leaders of the capitalist West, all of whom ought to have guessed his vicious ruthlessness from the well-publicized purges of the 1930s. Roosevelt's self-description as an 'amateur strategist' suggests he recognized that, unlike Churchill, Marshall and Brooke, he had had no formal training in military matters.

Marshall meanwhile warned Roosevelt about the vulnerability of the Middle East. 'From a military viewpoint, the region invites attack,' he told him, 'and its loss would permit junction by sea between the Japanese and the Germans,' which would have 'disastrous consequences'. Although the ABC-1 agreements with the British before Pearl Harbor 'always placed the Middle East in the sphere of exclusive British responsibility', Marshall had already told Dill that the US Army would help there 'in every practicable way'. The promise of the three hundred Sherman tanks and one hundred 105mm self-propelled guns after the fall of Tobruk should be seen in this context.

The constant fear of both the American and British High Commands centred on what would happen if the Germans moved south-eastwards into the Caucasus and Iraq at precisely the same time that Japanese naval and air forces managed to close the Gulf of Persia and thus the southern exit of the Suez Canal. Brooke's and Kennedy's diaries return to that scenario constantly, for at that nightmare moment the Axis would effect a pincer movement on the near-defenceless Middle East, where much of the oil was derived that ran Britain's war machinery.[21] Fortunately, and of course entirely unbeknown to the Allies, the Germans and Japanese had not co-ordinated their strategies at all and were not to do so subsequently. The Axis essentially fought two entirely separate wars, to their mutual disadvantage. (A Japanese attack on Russia at the time of Barbarossa would have been invaluable to Hitler, for example.)

At the War Cabinet of 23 March 1942, Churchill said that he had heard from Ivan Maisky that:

Evidently Germany is going to use gas in new Russian offensive. We would treat use of gas against Russia as against us. We would retaliate against Germany. We will make common cause with Russia over that and he considered we could deter Germany by making an announcement. If he wanted us to do so we needed plenty of warning. Go into gas mask situation. Furbish them up and it would be a good thing to use them every day.

To this, Brooke added that the British had to 'Work out carefully what our gas reserves are. Must go 100% out if we start.'[22] Gas was never in fact used militarily on the Eastern Front, but Churchill's reaction if it had been was clear.

At a working lunch held around the Cabinet table on Tuesday 24 March, Roosevelt, Stimson, King, Hopkins and Arnold heard a 'fine' presentation from Marshall of the OPD report for a fifty-division – 60 per cent American – cross-Channel invasion in the spring of 1943. Stimson described the British strategy of fighting in North Africa rather than in the Pas de Calais as 'the stopping up of rat-holes', a powerful image. Afterwards Roosevelt, according to Stimson, 'staggered me by a résumé of what he thought the situation was', because it differed profoundly from his and Marshall's. 'He looked like he was going off on the wildest kind of dispersion debauch,' remembered Stimson, 'toyed awhile with the Middle East and the Mediterranean Basin, which last he seemed quite charmed with,' but later the Secretary of War and Marshall 'edged the discussion into the Atlantic and held him there'.[23]

At the end of the meeting, after the President proposed that the OPD plan should be turned over to the British via the Combined Chiefs of Staff, Hopkins said that 'it would simply be pulled to pieces and emasculated' there. Instead, 'someone' – by whom he meant Marshall – should personally take it over to London and see Churchill, Brooke and the other Chiefs of Staff, 'and get it through them directly'. The President agreed. (Stimson's diary is invaluable, because Roosevelt would not allow note-takers at meetings such as this; on one occasion Marshall brought a Staff officer armed with a large notebook to record decisions, and 'the President blew up'.)[24]

The idea behind Marshall's London visit of April 1942 was therefore deliberately to try to bypass the Joint Planners of the Combined Chiefs of Staff, whose British contingent would have put the proposals through their stringent analytical process before Churchill and the Chiefs had a chance to view them. Indeed, had the British been forewarned of the details of Eisenhower's proposals, they would have had their refutations prepared even before Marshall's plane touched down. That evening Lord Halifax visited the White House for a ninety-minute meeting with

Roosevelt and Hopkins. 'The President was cutting stamps from his voluminous correspondence,' noted Halifax in his diary, 'and showed me a few of the addresses under which letters come to him.' (One read: 'His Majesty Roosevelt, United States.') They did not disclose to him the scheme that Marshall had laid before them.

On 27 March, and with Marshall's 'warm approval', Stimson wrote to the President at Hyde Park forcefully advocating that 'You should lean with all your strength on the ruthless rearrangement of shipping allotments and the preparation of landing gear for the ultimate invasion,' because the lack of what he called 'landing barges' was, in his opinion, 'the only objection to the offensive that, after talks with British critics here, I have heard made'.[25] In fact the British had a large number of other serious objections to the cross-Channel operation, primarily the wildly asymmetrical rate of projected reinforcement between the Allies and the Axis in France, but Stimson's and Marshall's pressure on Roosevelt was clear. Stimson also asked the President to send the Joint Chiefs' plans to Churchill and Brooke 'by a most trusted messenger' as soon as they were ready, a reference to Marshall and/or Hopkins.

April Fool's Day 1942 was an inauspicious date to choose for the White House meeting where Marshall persuaded Roosevelt of the viability of the cross-Channel operation, whose target date would be 1 April 1943, 'the earliest possible moment that the necessary tactical forces can be accumulated'. Under the plan worked out by Eisenhower, Wedemeyer, Handy and Hull, but presented as the Marshall Memorandum, there would be three distinct operations proposed. The first was Bolero, under which the US would attempt to ship thirty divisions – approximately 500,000 men – including six armoured divisions, and 3,250 aircraft, to Britain.

Then, under Operation Roundup, these thirty American plus eighteen British divisions, supported by the American plus 2,550 British aircraft, would be landed somewhere between Boulogne and Le Havre to march on Germany. (Roundup is an American ranching term, and one that Churchill found absurdly optimistic because it sounded as though the Germans would be rounded up like cattle. See Appendix C.)

Thirdly, and quite separately from Roundup, there was to be a much smaller-scale operation, codenamed Sledgehammer, which was designed to establish a bridgehead at Cherbourg with nine divisions, and fight in

Normandy as a method of forcing the Germans to draw off significant forces from the Eastern Front and thereby give the Russians a breathing space. Roundup and Sledgehammer were not intended to be mounted simultaneously. Bolero was the precondition for either.

Roosevelt wrote to Churchill later that day that he had completed his survey of the immediate and long-range problems of the military situations facing the Allies, and had 'come to certain conclusions that are so vital that I want you to know the whole picture and to ask your approval. The whole of it is so dependent on complete co-operation by the UK and US that Harry and Marshall will leave for London in a few days to present first of all to you the salient points.'[26] The terms 'Harry and Marshall' unwittingly denoted how much closer the President was to Hopkins than to his Army Chief. The President lunched with Marshall and Hopkins on 3 April and then went to see how his favourite construction project, the new Naval Hospital at Bethesda, Maryland, was progressing. He and Hopkins dined alone together that night, as they had three times that week, so there can be no doubt that when Hopkins visited London he knew precisely his master's mind on every aspect of grand strategy.

Hopkins and Marshall had first met in December 1938, and by Christmas Eve 1941 Marshall was writing to Hopkins to say that ever since then 'You have been a source of confidence and assurance to me.' He was a key friend to have, but Marshall could be helpful to Hopkins too. In October 1941 he had arranged for the Adjutant-General to have Harry's son Robert transferred from the Fort Dix reception centre to the Signals Corps replacement training centre at Fort Monmouth, New Jersey 'without delay' by priority telegram. 'If you will let me know about two months from now what his interests are at that particular time,' Marshall wrote to Hopkins, 'I will do the rest.'[27] Robert saw active service in Tunisia and survived the war; his eighteen-year-old brother Stephen was killed storming the Kwajalein Atoll in the Marshall Islands in 1944.

Marshall decided to take Wedemeyer and Hull to London, on a mission that was given the rather feeble codename Modicum. Wedemeyer had studied at the Berlin Kriegsakademie between 1936 and 1938, learning about armoured warfare from Heinz Guderian and about geo-politics from Professor Karl Haushofer, and meeting Göring, Bormann and other senior Nazis in the process. Gregarious and 6 foot 6 inches tall, he impressed Marshall with his intellect and writing ability.

It didn't harm his career prospects that he was also the son-in-law of Marshall's friend and adviser General Stanley D. Embick, the former chief of the War Plans Division, who had devised the overall plan for the Pacific campaign, codenamed Orange.

It had been Wedemeyer who, in September 1941, had drawn up the report later nicknamed 'The Victory Program' that stated that the US would need armed forces of nine million men to defeat a Germany that was victorious in Russia. Although this was leaked to the *Washington Times-Herald* on 5 December 1941, to the Administration's profound embarrassment, once the Nazis had declared war on the US six days later it looked very different. Wedemeyer had nothing to do with the leak, and was able to assert that 'General Marshall never doubted me'; soon afterwards he was promoted to lieutenant-colonel with a place on the Joint Staff.[28]

On the same day that Marshall finalized the details of his Memorandum in Washington, 2 April, Eden lunched with Churchill, Harriman and Ismay in London, after which he confided to his diary that Churchill would ignore any political pressure to give up the Ministry of Defence: 'He sees himself in Roosevelt's position as sole director of war.' Eden believed, however, that 'It is not what the country wants, nor does it produce good results.' Five days later he complained: 'There is no day-to-day direction of the war except by Chiefs of Staff and Winston,' and, in his opinion, the Chiefs of Staff 'too readily compromise where issues should be decided and Winston's unchecked judgment is by no means infallible'. Eden considered caballing with Lyttelton and Cripps, but the 'difficulty is Winston is probably constitutionally incapable of working any other way'.[29] Not being present at most Staff Conferences, Eden was unable to see that Brooke was in fact by no means too ready to compromise, but it is a sign of how hermetically the meetings with Churchill were sealed even from those in the upper echelons of British politics.

That evening Kennedy had supper at the United Service Club with the Soviet military attaché, Colonel Skliarov, who it turned out had served in the same sector during the Russian Civil War, albeit on the opposite side. After dinner the colonel 'opened up on the Second Front', arguing that 'all efforts should be concentrated on knocking the Germans out in the spring, and all risks taken.' A vigorous discussion ensued. Reporting it afterwards in his diary, Kennedy ruminated upon the whole issue, a passage which, since he was the War Office's most senior Planner, bears quotation at some length:

If we could be sure that the Germans might be knocked out by a maximum effort this spring we could of course do enough to make them divert considerable forces to France. But the fundamental difficulty is that we cannot be sure. We had to carry on this war for over a year without the Russians. We may have to carry it on again without them. It would be the most colossal gamble in history to stake everything on this spring offensive. It could mean for us the sacrifice of the means of defence of the UK both at sea and on land, the sacrifice of everything in the Indian Ocean and the Middle East. For nothing less would provide the naval, military and shipping resources for a big European effort. In fact the gamble for us could be far bigger than it ever was for Hitler to attempt invasion of this country. We are not prepared to risk everything – and it would be everything – on this one throw.[30]

True to the War Office policy of never allowing others to guess the true depth of British opposition to an early cross-Channel attack, Kennedy noted: 'I did not of course say anything of this to Skliarov.'

So far Brooke and Churchill had not been apprised of the Marshall Memorandum. Yet Washington reckoned without the ingenuity of Brigadier Dykes. On 4 April, he somehow 'glanced at' the seven-page Memorandum 'unofficially' and had passed a précis to the War Cabinet Office by the early hours of 5 April. Before the Modicum Mission even landed in Britain, therefore, Brooke knew what it was going to bring.

We do not know how Dumbie Dykes managed to see the document, but he was clearly a resourceful fellow. Paul Caraway, who was the Joint Chiefs of Staff liaison officer with the British members of the Combined Chiefs, was asked after the war which papers the Americans were and were not allowed to show their British counterparts. 'Certain of our top secret Intelligence information they were not allowed to see, because a lot of it came from undisclosed channels or wherever, much of it through the British without their knowledge, and so on,' he explained. 'For example, when we were preparing for a conference, the American positions were not disclosed, because we knew that the liaison people were instructed to report through their Joint Staff Mission, thus being able to checkmate some of our positions.'[31] With Dykes as one of the senior secretaries of the Joint Staff Mission, that is precisely what happened to the Marshall Memorandum.

That same day, 5 April, Kennedy was already recording in his diary that in the War Office it was 'generally accepted that there can be no question of landing an army in France and holding a front for long'. Norway was 'out of court' owing to the impossibility of providing air cover for the supply ships and landing force. This was held to apply to Cherbourg, too. Kennedy quoted Brooke as saying that the most success-ful result that could be expected in Normandy would be to establish a front of 20 miles across the neck of the Cotentin Peninsula, which 'Compared with the great Russian front . . . would be ridiculous', and the Western Allies would 'be the laughing stock of the world if we established such a front and held it up as a substantial contribution to the Russian war'. To emphasize this point, Brooke had a map drawn up for the War Cabinet that juxtaposed the present 1,500-mile-wide Russian with the proposed 20-mile-wide Cotentin front.

'The Germans could turn on us at their leisure and wipe us out', believed Kennedy. 'We cannot afford to lose the twelve or twenty div-isions required. Then there is the question of shipping which is also acute. It is liable to be forgotten that we are *already* containing very substantial German forces in Europe by the mere fact that this country is still holding out and that we hold a *threat* over the Germans.'[32] It is easy to see the general line of opposition that the Planning Staff began to work on, once they received the main features of the Marshall Memorandum from Dykes.

On their way over to Britain, Marshall's and Hopkins' Pan-American Clipper suddenly lost an engine, forcing the party to stay a night in Bermuda while another plane was flown over from New York. The War Office Planners in London were thus unexpectedly afforded a further twelve hours to draw up their detailed objections, which they put to good use. If Marshall had hoped to have any element of surprise when he presented his cross-Channel invasion plans to Churchill, Brooke and the War Cabinet, he had lost all chance of it by the time he and Hopkins finally reached London.

PART II
Engagement

6

Marshall's Mission to London: 'A momentous proposal' April 1942

Let's have this straight for once and for all, Pim! British submarines sometimes, unfortunately, 'fail to return'. But German U-boats are destroyed.

Winston Churchill to Captain Richard Pim, head of the No. 10 Annexe Map Room[1]

When Marshall, Hopkins, Wedemeyer and Hull arrived in London on Wednesday 8 April 1942, the battle-lines had effectively been drawn between the American High Command, who wanted an early – almost an immediate – cross-Channel attack, and the British, who believed this would be grossly premature and preferred to wait until Germany was weaker before taking the risk of returning to the Continent. Marshall and Brooke were to become the standard-bearers for these two, later bitterly opposed, points of view. It is not true, as some Americans – though not Marshall – were to allege, that the British never wanted the cross-Channel attack to take place at all. They did, but not until certain key criteria had been met, primarily the massive diminution of the Wehrmacht's capacity to respond. Roundup was therefore for Churchill and Brooke rather like treachery had been for Talleyrand, a matter not so much of principle as of dates.

Yet because the British desperately needed very substantial American forces in the British Isles to protect them against a German invasion should the Soviet Union suddenly collapse, as it was feared it might at any time, Brooke and Churchill could not simply heap contumely on the Marshall Memorandum, however misguided they might have thought it. Brooke could and did point out the defects in the proposed operations, but only within the overall context of accepting them, especially Bolero, which proposed a massive build-up of American forces in the British Isles.

For Churchill and Brooke always feared – even as late as the autumn of 1943 – that if Roosevelt and Marshall came to believe that they opposed a cross-Channel operation outright, the Americans would switch their attention to the Pacific, adopting a Japan First policy instead. This would leave Britain in renewed and possibly mortal peril, a return to the cold winds of strategic isolation she had experienced in the twelve months between the evacuation of Dunkirk and Operation Barbarossa. Churchill and Brooke therefore had to undertake a precarious balancing act. Their opposition to the Marshall Memorandum had to be presented in such a way that the Americans nonetheless decided to go ahead with Bolero, and flood Britain with thirty divisions, of which six would be mechanized. There is a lively controversy over whether, as some historians have alleged, there was an element of actual deception about the British welcome of the Memorandum during the Modicum talks, and also whether there was also an element of bluff by the Americans in seeming to threaten to abandon Germany First if Roundup were not executed promptly.

There was certainly a high level of suspicion about British motives within the American High Command, matched only by an equally lofty level of disdain for the Americans' strategic expertise from their British counterparts. Neither of these augured well for Allied co-operation, not least because the Americans interpreted Brooke's attitude towards them as insufferably patronizing. His manner seemed to convey the feeling that the Americans were simplistic novices in the world of grand strategy, rather than, as they saw themselves, Britain's ultimate saviours. This was probably a valid criticism; even if Brooke was not intending to give that impression, his diary makes it clear that that was exactly how he felt.

Brooke's view of Marshall never really changed. 'I saw a great deal of him throughout the rest of the war,' he wrote once it was over,

and the more I saw of him the more clearly I appreciated that his strategic ability was of the poorest. A great man, a great gentleman and great organizer, but definitely not a strategist. I found that his stunted strategic outlook made it very difficult to discuss strategic plans with him, for the good reason that he did not understand them personally but backed the briefs prepared by his staff.[2]

The contempt that Brooke clearly felt for Marshall had to be kept from the Americans if at all possible, and the CIGS was not good at dissimulation. If Marshall sensed the disdain Brooke felt for his abilities, he at least did not show it. Although both Churchill and Brooke were of course personally welcoming to the members of the Modicum Mission, the Prime Minister hailed the Marshall Memorandum with salutations and fanfares while Brooke examined it under an unforgiving microscope but nonetheless eventually accepted it, as British self-interest dictated that he must.

'The ultimate objective of all military operations is the destruction of the enemy's armed forces by battle,' reads the US Army's *Field Service Regulations* of 1923, the bible of American strategy. 'Decisive defeat in battle breaks the enemy's will to war and forces him to sue for peace.' This was Marshall's attitude towards war-fighting, and the message he intended to put over in London. It is the Clausewitzian approach to warfare, by which the enemy is relatively quickly brought to a decisive battle on the most important front. By contrast the British adopted an older concept, pioneered by the Chinese military theorist Sun Tzu, by which the enemy is worn down by peripheral attacks and only fully engaged once fatally weakened.

'Pressured hourly to send men and resources to two divergent theaters,' Pogue wrote of Marshall, 'he saw the Middle East and the Mediterranean as peripheral areas . . . He feared the prospect of starving his forces in the Pacific to build up reserves in Britain that might be swallowed up in enterprises in Norway, North Africa or the Middle East or in small raids along the coast of Europe.'[3] The American naval historian Rear-Admiral Samuel Eliot Morison likened the American and British strategies to 'the massive thrust at the enemy's heart' versus 'successive stabs around the periphery to bleed him to death, like jackals worrying a lion before springing at his throat'.[4] Other military historians have employed the analogy of the picadors who weaken the bull with their lances before the matador finishes him off with banderillas and sword.

Reminiscing on television in 1958 about the arguments he had had with the Americans, Brooke put his own view cogently. 'The arguments mainly with Admiral King were connected with the relative effort to be put into the Pacific as compared with the effort to be put in Europe,' he said, whereas those with General Marshall were about where in Europe that effort should go. 'Was it to be put in a cross-Channel operation early on during the war or at a later stage? Were we justified in going into North Africa and on to Italy? . . . In our minds we felt that going across the Channel before the condition was ripe for it,' continued Brooke, 'before Germany had been ripened all round, ripened by the air action, ripened by forcing her to spread and distribute her forces throughout Europe, ripened by the action of Russia on the far side . . . might have had disastrous effects on the war.'

Brooke's sense of superiority was evident from his suggestion that 'Having been involved in operations in France against the Germans, we were perhaps a little better able to gauge the strength of the Germans at that time and the difficulty of obtaining the necessary victories with partially trained troops against the highly efficient and experienced German forces.'[5] Brooke was in effect claiming that the American attitude was over-optimistic, naive and born of not having yet faced the unbloodied Wehrmacht in battle, as he had. General Sir David Fraser believed that Brooke 'reckoned that the Americans had no knowledge of the modern German soldier and seriously underestimated him'.

Brooke's experiences in France in the two BEF expeditions of 1940 had a deciding influence on the assumptions underlying his formulation of grand strategy in the Second World War, principally in convincing him that the French could not be relied upon and that the Germans were very formidable opponents indeed. Only very rarely is it possible to spot a word of criticism of the German fighting man from Brooke, and then only towards the very end of the war when they were conscripting children and the middle-aged. In some American eyes, this counted against him. 'There were officers on Marshall's staff', recorded Pogue, 'who believed that [Brooke's] service in the costly campaigns of Flanders in World War I and in two evacuations of troops from France in World War II were not conducive to the aggressive strategy that they believed necessary for victory.'[6] Pogue was being diplomatic; they constituted the majority, though this did not include Marshall himself.

The Americans' flying-boat touched down at Lough Erne in Northern Ireland on the morning of 8 April. On the flight Marshall read a 25-cent copy of H. G. Wells' *The Outline of History*, given to him by his wife just prior to departure, which he handed on to Hopkins. They then flew on to Hendon Aerodrome where they were met by Churchill, Brooke, Portal and Pound. On the drive into London, Marshall was shown the bomb damage caused by the Luftwaffe in the Blitz. Brooke had come straight from a Chiefs of Staff meeting attended by General Bernard Paget of Home Command, Mountbatten and the head of Fighter Command, Air Marshal Sholto Douglas, in which they had put up what Brooke called 'a thoroughly bad' plan 'to assist Russia through action in France'. He was perhaps not therefore in the best frame of mind to meet someone the sole purpose of whose journey was to propose precisely the same thing. Meeting the plane was purely a courtesy, because he had other meetings throughout the day and next saw Marshall at dinner that evening.

According to Hopkins' notes of the trip, from 4 p.m. to 6 p.m. Marshall presented the 'broad outlines' of his Memorandum to Churchill, who 'indicated that he had told the Chiefs of Staff that, in spite of all the difficulties, he was prepared to go along'. Churchill repeated the objections that the Chiefs of Staff had put, 'all of which we had heard in Washington before coming to England'. Marshall was more optimistic about the interview than Hopkins, thinking that 'Churchill went a long way and he, Marshall, expected far more resistance than he got'.[7]

In explaining the fall of Singapore the previous month, Churchill simply said it had been 'a mess', which had been handled very badly; he offered 'no explanation of the lack of resistance on the part of the British'. Churchill gave Marshall and Hopkins dinner at Downing Street at 8.30 p.m. that evening, with Brooke, Attlee, Eden and the deputy leader of the Labour Party Arthur Greenwood in attendance. Conversation was 'in the main social', with Churchill discussing the Great War and the American Civil War, and the assembled leaders, as Hopkins recalled, 'never really getting to grips with our main business, although General Brooke got into it enough to indicate that he had a great many misgivings about our proposal'. What Hopkins guessed, but Marshall seems not to have, is that Churchill privately opposed an early Roundup and Sledgehammer just as much as Brooke.

'Brooke made an unfavourable impression on Marshall, who thinks that although he may be a good fighting man, he hasn't got Dill's brains,' was how Hopkins reported their first meeting.[8] The American historian Robert Sherwood, who edited Hopkins' papers, thought that that remark might have been made in unconscious resentment of Brooke for having replaced Marshall's friend Dill as CIGS. Nor was Marshall alone in underestimating Brooke. 'Just between ourselves now,' General Handy told his SOOHP interviewer, 'in our opinion, that is at the working level, Brookie wasn't the smartest of the British ... We didn't think that he was really smart. Now maybe that was our prejudice, but we didn't and down on the working level we rated Brookie right down near the bottom.'[9]

Of this first encounter, Moran wrote:

Brooke and Marshall, who now met, had a good deal in common. They both came of virile stock. However, the acquisitive instinct, common enough among full-blooded men, had no part in their lives. Their one ambition was to lead armies in the field, but they would not lift a finger to bring this about. They were both selfless men with a fine contempt for the pressures of the mob. Even if Brooke was not impressed by Marshall's ability, he could not help liking him; he felt he could trust him – and that went a long way with the CIGS ... With his long upper lip and craggy features, Marshall looked more like a painting by Dobson than a modern staff officer.[10]

(The seventeenth-century portraitist William Dobson painted Charles I, Charles II, Prince Rupert and other grandees of the Stuart court, before being imprisoned for debt.)

Pogue agreed that Marshall and Brooke had something in common:

To most Americans, as to some of his colleagues, Brooke was icy, imperturbable, and condescending. Four years younger than Marshall, he was smaller in stature, delicately boned, and with large dark eyes that had a shining, impenetrable stare. Precise, methodical, abrupt to the point of rudeness, he lacked Dill's charm. Marshall later saw a pleasanter side of Brooke ... Some Americans sneered at his bird watching, but Marshall felt at home with a soldier who liked birds, gardening, and fine horses.[11]

The sole reference that Brooke made to the dinner was that 'Neither Hopkins nor Marshall disclosed their proposed plans for which they have come over. However it was an interesting evening and a good chance to get to know Marshall. But did not get back until 1.30am!!'' Churchill's late hours were to produce a large number of double exclamation marks in Brooke's diary over the coming years, as were subsequent meetings with Marshall.

It was important for Hopkins to come on the mission, because as Marshall later admitted, 'I couldn't get at the President with the frequency he could – nothing like it; nor could I be as frank, nor could I be as understanding.' Marshall was impressed with Hopkins' physical courage in flying to London. A severe anaemic, in the previous two weeks 'he had ten blood transfusions, and he had been found crawling up the back stairs at Hyde Park because he wasn't strong enough to walk up.' (If accurate, this was surprising, as the front stairs were supplemented by a lift for Roosevelt's wheelchair.) Marshall maintained that Hopkins 'supported me strongly where I was in difficulties with Churchill, and where I was in difficulties with the President', and he had a gift of making 'the military position – the strategical graphs and all – plainer to the President than I could possibly have done myself'.[12] Yet, as Marshall was to discover on a subsequent trip to London, Hopkins the court favourite spoke only with his master's voice.

The meeting at which Marshall unveiled his Memorandum to Brooke and his colleagues took place at 10.30 the following morning, Thursday 9 April, and lasted two hours. Brooke had called a Chiefs of Staff meeting for 9 that morning, to consider their response before it was even formally proposed. As well as Brooke and Marshall, the others present at the later gathering were Pound, Portal, Major-General James E. Chaney (the USAAF commander in the British Isles), Wedemeyer, Ismay, Mountbatten and Hollis. Marshall presented his three plans, which comprised Bolero (the build-up of US forces in Britain), Sledge-hammer (the nine-division assault to be launched in the autumn of 1942), and Roundup (forty-eight divisions and 5,800 planes attacking France on 1 April 1943). Marshall stated that Sledgehammer would be 'justified only in case' of an 'imminent collapse of Russian resistance unless pressure is relieved by an attack from the west'. He went so far as to label it, somewhat dishearteningly, as 'a sacrifice in the common good', which was an error later pounced upon by its British critics.[13]

The British minutes record Marshall suggesting that, if the war in the east were to develop 'unfavourably' for the Soviets, 'We might have to stage an "Emergency Operation" on the Continent to help them' (that is, Sledgehammer). Equally, if Germany failed to break Russian resistance in 1942, Marshall thought: 'We ought to be prepared to exploit the consequences.'[14] With Mountbatten worried about not having the element of surprise, and Portal doubtful about being able to give proper air support to the Sledgehammer bridgehead, Marshall failed to win support from the British Chiefs of Staff at this first meeting.

For his part, Brooke strongly doubted Sledgehammer's chances of success without a German collapse, which he did not foresee in the near future. He declared flatly that seven infantry and two armoured divisions were simply not enough to maintain the bridgehead against the forces that the Germans would be able to throw against Sledgehammer. 'Worse still,' he argued, 'the Allies would be unable to extricate the units if the Germans determined to expel them.'[15]

Brooke further pointed out that a shortage of landing craft meant that only about four thousand men could be taken across the Channel at any one time. To deliver the necessary reinforcements for Roundup or Sledgehammer would involve these craft going backwards and forwards across miles of open sea under constant Luftwaffe attack, in order to engage as many of the twenty-five German mobile divisions then estimated to be stationed in France as were sent to attack the landing places. Moreover, seven out of the nine divisions that Sledgehammer required would have to be British, and, in the neat understatement of one recent historian, 'given London's strategic views, this presented an obvious problem of motivation'.[16]

According to Brooke's account of the meeting, Marshall 'gave us a long talk on his views concerning the desirability of starting [the] western front next September and [stated] that the USA forces would take part. However the total force which they could transport by then only consisted of 2½ divisions!! No very great contribution. Furthermore they had not begun to realize what all the implications of their proposed plan were!' In fact Marshall knew very well the problems; the day before he had left Washington, his Planners had concluded that by 15 September 1942 only fifteen-and-a-half Roundup divisions would even be in existence in the USA, let alone in Britain, that fighter cover would be insufficient over the battle zone, that landing craft would severely limit the force size, and that the shipping, cargo and port facilities required for Roundup would reduce what food and other supplies could to be exported to Britain. Brooke consistently underestimated the Staff work done by the Americans on all these issues, regularly resorting in his diary to sarcasm and *reductio ad absurdum* to caricature Marshall's stance. None of these undoubted military and logistics problems, however, deterred Marshall, who wanted an agreement in principle rather than an argument about details. He was probably right in thinking that, once the political will for the operation was established, all the other issues would be surmountable.

After the morning meeting, the British Chiefs gave their American counterparts lunch at the Savoy Hotel. 'I liked what I saw of Marshall,' concluded Brooke that night, 'a pleasant and easy man to get on with, rather over-filled with his own importance. But I should not put him down as a great man.'[17] There is of course a contradiction here, in that people rarely genuinely like those who they feel are over-filled with self-importance, but Brooke had no reason to lie to his diary or to Benita. After the war he had occasion to ruminate with hindsight about this diary entry, and he wrote:

These first impressions of mine about Marshall are interesting and of course incomplete. They were based on the day's discussions, which had made it quite clear that Marshall had up to date only touched the fringe of all the implications of a re-entry into France. In the light of the existing situation his plans for September of 1942 were just fantastic! Marshall had a long way to go at that time before realizing what we were faced with. It will be seen from my diary that during the next few days I was busy sizing up Marshall's character and military ability. It was very evident that we should have to work extremely closely together, and for this a close understanding of each other was essential.[18]

It is obvious that with Marshall doubting Brooke's intelligence and Brooke thinking Marshall pompous, relations had hardly started swimmingly, although of course neither knew of the other's views. Marshall cannot have been in the best spirits either, because on 9 April the news came through that thirty-five thousand US troops had been captured on the Bataan Peninsula in the Philippines, the largest single surrender of American soldiers since Stonewall Jackson captured Harper's Ferry eighty years earlier. (Unlike Singapore, however, the Americans at Bataan had held out for five months.)

Even if Brooke was unconvinced, Marshall certainly impressed Ismay, who thought him 'formidable' and recorded in his memoirs: 'He was a big man in every sense of the word, and utterly selfless. It was impossible to imagine his doing anything petty or mean, or shrinking from any duty, however distasteful. He carried himself with great dignity.' He did find Marshall 'somewhat cold and aloof' at first, and noted that, unlike many other senior officers, the general never used either nicknames or Christian names, so Eisenhower was never Ike and Dill was never Jack. 'But he had a warm heart,' concluded Ismay. Marshall admitted to Ismay that the most painful part of his job was telling officers, some of them his close friends, that they could expect no further employment. 'His integrity was unshakeable,' Ismay believed, 'and anything in the nature of intrigue or special pleading was anathema.'[19]

While Marshall met Brooke and the Chiefs of Staff, Hopkins was with Churchill in the Cabinet Room at 10 Downing Street, where the Prime Minister was left in no doubt about 'the serious weight which the President and Marshall gave to our proposals'. Hopkins made it plain that the two men had made up their minds that a cross-Channel attack was 'by far the most advantageous from a strategic point of view', and so they 'mean business'. Hopkins furthermore told Churchill that the President and Marshall 'were prepared to throw our ground forces in', because Marshall had got the impression that Churchill's advisers felt 'that the ground attack would never be made, at least for nearly a year'. According to Hopkins' notes of the meeting, 'Churchill took this very seriously,' saying he hadn't hitherto fully taken in 'the seriousness of our proposals'.

Might the Americans have misunderstood the negative British reaction to Sledgehammer and an early Roundup – but not to Bolero, which they always favoured – because of Churchill's manner? As Portal told the Australian war correspondent Chester Wilmot of the Chiefs of Staff's own attitude towards the Prime Minister: 'We used to listen to him enthralled with his words, but once we got to understand him we were never taken in. We knew him too well. We knew what to discount, what to accept. We got to know when he was in earnest, when he was only flying a kite.' Portal felt that the Americans never understood this aspect of Churchill, however.

They admired him and respected him but they were doubtful of his strategic judgment and suspected his political motives. They so often heard him raise wildcat ideas and they were never able to tell, as we were, when he was serious and when he was just leading them on. They didn't realize, as we did, that half the time these ideas were just put up because Churchill believed that in the sharp clash of discussion ideas could be thrashed out and developed.[20]

Portal went on to explain that if Churchill was worried about a particular course of action, he would sometimes throw out a suggestion in such a way as to make his listener think he was strongly in favour of it, and then would put forward arguments violently in support, not because he genuinely believed them but because he wanted to draw every possible objection out of his interlocutor. He would then drop the idea completely, satisfied that it did not measure up. 'We were well aware of this technique, this means of clarifying his own thoughts,' explained Portal, 'but the Americans always took him seriously and literally, and so to them he appeared unpredictable, unreal and extravagant.' In a rare bout of anti-Americanism, he added, 'The reason why the Americans distrusted Churchill, in spite of their affection for him,' was because 'their rather pedestrian and matter-of-fact minds couldn't keep track of the voluble outpourings of his fertile imagination.'[21] Yet was it really so pedestrian or matter-of-fact to expect a British Prime Minister, at high-level global-strategy meetings, where lives could be at stake, actually to believe what he was saying?

The Modicum Mission was based at Claridge's, where the Americans took over an entire floor of the hotel, with Marshall booked under his codename Mr C. G. Mell. Wedemeyer's bedroom was next to Marshall's sitting room, and in his 1958 autobiography *Wedemeyer Reports!* he recounted being woken by a valet on the morning of 9 April who gave him bacon that was stringy, mushrooms that were tough and coffee that was 'execrable'. There were no eggs.

'I recall vividly this initial joust with the British concerning definite plans for a cross-Channel operation,' Wedemeyer wrote, juxtaposing the Americans as 'always keeping uppermost in mind the basic idea of concentrating and making a decisive effort against the heartland of the enemy', against the British, who 'kept returning to a concept of scatterization or periphery-pecking, with a view to wearing down the enemy'. According to Wedemeyer, Brooke:

talked in low measured tones and was cautious as he commented upon the American concept as described by Marshall. The British were masters in negotiation – particularly were they adept in the use of phrases or words which were capable of more than one interpretation. Here was . . . all the settings of a classic Machiavellian scene . . . When matters of state are involved, our British opposite numbers had elastic scruples. To skirt the facts for King and Country was justified in the consciences of these British gentlemen . . . There was no expressed opposition to Marshall's ideas at this first meeting, just polite suggestions that there might be difficulties in undertaking this task or that. What I witnessed was the British power of finesse in its finest hour, a power that has been developed over centuries of successful international intrigue, cajolery and tacit compulsions.[22]

Even sixteen years after the conversations, Wedemeyer's bitterness is plainly evident, and there was plenty more overt Anglophobia to come, along with such splendidly mixed metaphors as: 'It is true, I thought, that the sun never sets on the British Empire. But neither does the dove of peace. Moreover, the wings of justice had constantly been clipped as British influence and possessions were increased all over the world. I reflected upon the history of the British as I sat watching these senior military leaders carefully parrying, sidestepping and avoiding a head-on collision at this stage of the scheduled conferences.'[23]

Wedemeyer believed that the British were 'familiar hands at using the intimidation of latent force or resorting to subtle deals, doing anything and everything to protect and extend British interests', whereas by total contrast, 'We Americans, who were adolescents in the international field, had no clear-cut conception of our national interest.' The image of a trusting, naive America, personified by Marshall, being led down the garden path by wily, cynical, perfidious, aristocratic Old World Limeys, as personified by Brooke, was widely accepted among American Planners during the Second World War – indeed the 'garden path' metaphor was explicitly used by them later on in the conflict.

At lunch at the Savoy Hotel that day, Wedemeyer sized up the British as individuals. 'I surmised that Brooke would be a source of future trouble for my chief, General Marshall,' he wrote years later. 'I sensed in Brooke a quick, incisive mind. He was articulate, individual, sensitive; one who would nibble away to gain his ends, meanwhile skilfully avoiding the necessity of coming to grips frontally with a basic issue. Over the following months this initial impression was many times confirmed by many fellow Americans, and even by a few Britishers.' Pound, by contrast with this Machiavellian monster, was a courteous gentleman with 'a twinkle in his clear blue eyes', small in stature and 'rather taciturn'. Portal had a 'large nose and high forehead. He seldom raised his eyes,' and was careful when choosing his words. Poor old Ismay was written off as merely 'a smoothie', 'insincere', a 'Mr Fix-It', a man 'without real convictions and incapable of reaching sound conclusions', although Wedemeyer had a much more positive view of him later on. Not so of Brooke.

*

In March 1948, Portal reminisced to Chester Wilmot about the British Chiefs' reactions to the Marshall Memorandum:

We wouldn't look at it for 1942 and even then we were very doubtful for 1943. The Americans had tremendous confidence in their own troops and by and large their confidence was justified, for they did learn very quickly once they got into action – far more quickly than our chaps did, and once they got the experience they fought extremely well. But this doesn't mean they could have carried out a successful invasion in '42 even if the craft had been available, which they weren't.

He then made an astonishing admission – especially as he was talking to a television journalist – which confirms much of what the Americans were saying privately about the British Army at the time. 'I'm afraid that we never had the same confidence in our troops,' the Chief of Air Staff openly stated. 'It was clear to us that they had been very badly shaken by the early defeats and we were very much afraid of setting them a task beyond their capacity.'[24]

As for the accusation that Churchill had perfidiously and deliberately misled Marshall over seeming to support his Memorandum, as Stimson, Wedemeyer and several American historians were to allege, Portal told Wilmot that 'When Winston described Marshall's suggestion as "a momentous plan", he was certainly not thinking of Sledgehammer but rather of Marshall's idea that the final blow to Germany must be delivered across the Channel. We had learnt not to take Winston too seriously, and no doubt he was being polite to Marshall and was anxious not to discourage his general conception, but at no time did he or we agree to Sledgehammer.' This was because 'None of us believed that we could possibly hold any bridgehead.'[25] Yet that was not at all the impression that Marshall took away with him when he returned home, and began to send troops to Britain under Operation Bolero.

On Friday 10 April, Eden saw Churchill after Cabinet: 'We spoke of [the] American plan. He feared [the British] General Staff would say "Yes" and make this a pretext for doing much less elsewhere.' After a long day's work, 'mainly concerned with trying to save India from the Japs', including two Chiefs of Staff meetings, Brooke drove down to Chequers with Pound and Portal for dinner with Marshall, Hopkins and Churchill. 'We were kept up till 2am doing a world survey, but little useful work,' he recorded. After the war, Brooke recalled being 'amused at Marshall's reactions to Winston's late hours, he was evidently not used to being kept out of his bed till the small hours of the morning and not enjoying it much!' Marshall used to say that no important decision was ever taken after four o'clock in the afternoon, but he soon learned that Churchill thought otherwise.[26] He was also astonished at the level of access that Brooke had with the Prime Minister, saying that he frequently did not see Roosevelt for a month or even six weeks. This prompted Brooke to note with feeling: 'I was fortunate if I did not see Winston for 6 hours.'[27]

The Modicum visit brought Churchill and Marshall into close contact over a long period, allowing the general to observe the Prime Minister's habits closely. He admired Churchill's gift for language, of course, as well as 'his knowledge and sense of history, his splendid contempt for the enemy, his capacity for boldness'. Yet, according to Pogue, Marshall was also 'appalled' by Churchill's 'swift changes of plans, a flexibility that brought chaos in planning, a daring that brushed aside careful details'. Most of all he 'dreaded' Churchill's influence on Roosevelt, who he said 'also delighted in the dramatic and the unexpected, and who was determined, now that he was in the war, to strike a sudden and vital blow at the Nazis'.[28] It was very commendable, but where would it fall?

On Saturday 11 April, three days into the visit, Hopkins cabled Roosevelt to say that discussions with Churchill and the British Chiefs of Staff were progressing well. He complained that 'my underwear is itching like the devil', which was probably more information than the President needed. The Buckinghamshire countryside led him to eulogize about how 'It's only when you see that country in spring that you begin to understand why the English have written the best goddam poetry in the world.' It was a calm weekend, although Mrs Churchill was feeling exhausted and they again stayed up till the early hours.

Unfortunately a telegram from Roosevelt about Stafford Cripps' mission to India arrived at 3 o'clock on Sunday morning. Cripps was attempting to secure Indian political support for the Allied war effort, in return for full autonomy after the war, but the President thought that large measures of Indian self-government should be accorded while the war was still being fought. Churchill told Hopkins that he 'refused to be responsible for a policy which would throw the whole subcontinent of India into utter confusion while the Japanese invader was at its gates'.[29] Churchill deeply resented Roosevelt's attempted interference in imperial matters, and did not mind making that clear. Hopkins' scribbled notes also indicate that Churchill threatened to resign 'if that would do any good in assuaging American public opinion' with regard to India, an absurd notion on so many levels that it can safely be put down to one of Churchill's late-night rodomontades.

Churchill's late nights had tired out Hopkins, who anyway suffered from anaemia. 'Please put Hopkins to bed and keep him there under 24-hour guard by Army or Marine Corps,' Roosevelt joked in a telegram to Marshall later that day. 'Ask the King for additional assistance if required on this job.' Brooke got away from Chequers in time to spend the day pruning roses at home, and playing with his daughter Kathleen, nicknamed 'Pooks'. That afternoon, Marshall telephoned McNarney, his deputy in Washington, to say that Churchill had 'indicated that he had virtually accepted *in toto* the proposals I submitted to him, and that the Defense Cabinet Committee would undoubtedly approve. I regard this as acquiescence in principle. We must now get to business and arrange the details.'[30]

McNarney duly told the President that Marshall's message 'indicates good progress', adding that the general was due to sit with the full Cabinet on Monday the 13th and the Cabinet's Defence Committee the next day, but already 'Naval Person [that is, Churchill] told me he accepts our proposal.'[31] Considering how opposed Brooke had been on the morning of 9 April to two of the three proposals – Roundup and Sledgehammer – it was surprising that Marshall had taken Churchill's word as representing 'acquiescence in principle' to the proposals '*in toto*'. Perhaps there was as much an element of self-delusion by Marshall as there was dissembling by the British.

There were lighter moments at Chequers that Sunday morning also. Hopkins had asked Wedemeyer to have two crates of fresh food put on the Clipper in Bermuda to present to Churchill and the request had duly gone down the command chain. Churchill, accompanied by Hopkins and Marshall, went to watch the gardener carefully prise open the crate and to see 'what precious gift had been brought him'. Some pineapples, perhaps, or paw-paw, or some other delicious, exotic tropical fruit? It turned out to be Brussels sprouts, one of the vegetables in plentiful supply in England. 'Churchill started to laugh, and soon everyone joined in.'

Churchill telegraphed Roosevelt about the Marshall Memorandum that Sunday, to say that he had 'read with earnest attention your masterly document about future of the war and the great operations proposed. I am in entire agreement in principle with all you propose, and so are the Chiefs of Staff.' His sole proviso was that 'We must of course meet day-to-day emergencies in the east and west while preparing for the main stroke.' Although he stated that the whole matter was going to be discussed by the Defence Committee on the evening of Tuesday the 14th, to which 'Harry and Marshall' were of course invited, Churchill concluded: 'I have no doubt that I shall be able to send you our complete agreement. I may say that I thought the proposals made for an interim operation in certain contingencies this year met the difficulties and uncertainties in an absolutely sound manner. If, as our experts believe, we can carry this whole plan through successfully, it will be one of the grand events in all the history of the war.'[32]

This was at best disingenuous, because as the Chiefs of Staff meeting on the morning of 9 April showed, the British experts plainly did not believe that Sledgehammer, the 'interim operation' referred to, could be carried through successfully in 1942. Although Churchill had not been present, having been in conclave with Hopkins at the time, he must have known this, having been in regular and close contact with Brooke since then. Churchill's seeming wholehearted endorsement of the Marshall Memorandum in its entirety stored up great – and largely unnecessary – problems later on.

Monday 13 April brought three meetings of the Chiefs of Staff and Cabinet, each dedicated to 'trying to frame a reply for Marshall'. These took up most of the day, and Brooke only got away after 8 p.m., when he had Marshall to dine with him. 'The more I see of him the more I like him,' he noted at the time. Yet after the war Brooke added this damning assessment of that dinner: 'There was a great charm and dignity about Marshall which could not fail to appeal to one. A big man and a very great gentleman, who inspired trust, but did not impress me by the ability of his brain.'[33] Since this was written several years later, long after Brooke had got to know Marshall very well indeed, it must be taken as his considered appraisal.

Marshall had meanwhile heard the reservations that Brooke and others had about the cross-Channel proposals, telegraphing McNarney that his Memorandum would be accepted in principle, 'but relative to avoidance of future dispersions particularly of planes, such acceptance will have to be considerably and continuously bolstered by firmness of our stand.' He concluded perceptively that 'Virtually everyone agrees with us in principle but many if not most hold reservations regarding this or that.'

Marshall attended the Chiefs of Staff meeting the next morning, Tuesday 14 April, at which Brooke handed him a document giving their detailed reply to his Memorandum. As Marshall had surmised, it was favourable overall but contained strong caveats. They agreed with his 1943 timetable, while stating that any action in 1942 would need to await developments on the Russian front, and that India and the Middle East would have to be made safe before anything else. Marshall then said that 'within the next three or four months we were very likely to find ourselves in the position when we were forced to take action on the Continent,' and this 'might be either because we might not be able to hold back while the Russians were being driven back or borne down, or because a favourable opportunity had presented itself'.[34] At the time everyone assumed that the Manichean struggle between Nazism and Communism must come to a swift and decisive conclusion; it seemed unlikely to anyone that the conflict might still be going on a full three years later.

Brooke concluded that if Britain was 'forced to undertake an operation on the Continent, it could only be on a small scale'.[35] Marshall ought to have taken from this, as well as all the earlier discussions and the conversations over the various lunches and dinners, that there might have been a serious disagreement between Brooke and Churchill over future operations, yet he does not seem to have, or if he did he obviously thought that Churchill's will must prevail. Mountbatten also pointed out the lack of landing craft, which was to become a central issue very soon. 'Marshall, who has made a splendid impression here,' Hopkins cabled Roosevelt, 'has presented our case with moderation but with great force. I believe we will achieve not only agreement in principle but a real meeting of minds.'[36]

Welcoming Marshall and Hopkins to the crucial Defence Committee meeting at Downing Street at 10 p.m. that Tuesday evening were Churchill, Attlee, Eden, the Minister of Production Oliver Lyttelton, the First Lord of the Admiralty A. V. Alexander, the War Minister Sir James

Grigg, the Air Minister Sir Archie Sinclair, Brooke, Pound, Portal, Ismay and Mountbatten, along with Hollis and Jacob from the secretariat. Churchill opened the meeting by saying that Hopkins and Marshall had brought over what he again called 'a momentous proposal', one that had been fully discussed and examined by the Staffs.

For himself, he had no hesitation in cordially adopting the plan. The conception underlying it accorded with the classic principles of war, namely, concentration against the main enemy. One broad reservation must however be made – it was essential to carry on the defence of India and the Middle East. We could not possibly face the loss of an army of six hundred thousand men and the whole manpower of India. Furthermore, Australia and the island bases connecting that country with the United States must not be allowed to fall, as this would inevitably prolong the war. This meant that we could not entirely lay aside everything in furtherance of the main object proposed by General Marshall.[37]

So far, therefore, the British had not made any direct criticisms of Roundup or Sledgehammer, but had only put their imperial self-interest first, as the Americans must have expected they would.

Marshall was the next to speak, saying that 'It was a great relief for him to know that there was a basic agreement on general principles. All were in complete agreement as to what should be done in 1943.' He mentioned a strong bombing offensive and the importance of occasional raids on the French coast, in order to raise morale and provide a battle-hardened nucleus of American troops. 'The availability of troops presented no problem,' he promised. 'The main difficulties would be in providing the requisite tonnage, the landing craft, the aircraft and the naval escorts.'

The general then brought up the 'two points of doubt' which had arisen in his discussions with the British Chiefs of Staff. The first was whether the United States would contribute enough matériel to support Britain in the Middle East and India, and the second was the practicality of making 'a landing on the Continent' in 1942. 'We might be compelled to do this, and we must, in any case, prepare for it,' he said. He thought that air superiority would solve most of these problems, and this would come both from the Allied plane-building programme and from the Luftwaffe's preoccupations in Russia.

He admitted that there had not been much time before he left the United States to study the problem of operations in 1942 and, on the data available, he had concluded that they could not be undertaken before September. 'If they had to be done before then, the United States contribution would be modest,' he admitted, 'but whatever was available in the way of American forces over here at the time could be used to the full. The President had particularly emphasised that he wished his armed forces to share to the greatest extent possible in whatever might be undertaken.'

Marshall then discussed trebling production of landing craft by the United States, the defence of Australia and the south-west Pacific by the US Navy, the garrisoning of Iceland, and other projects that could be undertaken concomitant with Bolero, adding that full provision was being made for holding the Australia–Hawaii–Alaska line in the Pacific. He ended by using a legal analogy he was to return to in the future, stating that it was 'essential that our main project, i.e. operations on the Continent, should not be reduced to the status of a "residual legatee" to whom nothing was left'. It was a powerful appeal, presented with the assurance of a man who had already practised it several times by then.

Brooke spoke next, saying that the British Chiefs of Staff Committee had:

examined very carefully General Marshall's proposals. They were in entire agreement with him on the project for 1943. Operations on the Continent in 1942 were governed by the measure of success achieved by the Germans in their campaign against Russia. If they were successful, we could clearly act less boldly. If, however, the Russians held the Germans, or had an even greater measure of success, our object should be to force the Germans to detach air forces from the Russian front. This could be done either by air operations or by the landing of troops, which would force heavy air battles.

Brooke then stated unequivocally that 'The Chiefs of Staff entirely agreed that Germany was the real enemy. At the same time, it was essential to hold the Japanese and ensure that there was no junction between them and the Germans.' He conjured up the by now familiar lurid scenario in which the Japanese won control of the Indian Ocean, allowing the Middle East to be gravely threatened and oil supplies prevented from going through the Persian Gulf. Under those circumstances, Germany would seize Persia's oil, the southern route to Russia would be cut off, and Turkey would be isolated, destroying any hope of her joining the Allies, while Germany and Japan could exchange any hardware they needed. American assistance was thus vital to prevent Japan taking control of the western Indian Ocean. Churchill unsurprisingly agreed, readily acknowledging that Britain was 'unable to cope unaided' with the Japanese threat there.

This led to a discussion between Churchill, Brooke and Pound about the Indian Ocean, with the Americans as spectators, although that did not prevent Churchill from suggesting moves that involved their fleets, and even individual ships. If the new battleship USS *North Carolina* was to sail to the British naval base at Scapa Flow in the Orkney Islands, he posited, HMS *Duke of York* could be released for duty in the Indian Ocean. When Brooke said he 'welcomed the idea of an offensive in Europe' after measures had been taken 'to prevent a collapse in the Indian Ocean', which would require American assistance, he was of course asking for an open-ended commitment. This prompted Hopkins to warn the Committee that 'If public opinion in America had its way, the weight of American effort would be directed against Japan. Nevertheless, after anxious discussion, the President and American military leaders had decided that it would be right to direct the forces of American arms against Germany.' He explained that the Germany First policy had been adopted because the US High Command wanted to fight on land, at sea and in the air, as well as 'in the most useful place, and in the place where they could attain superiority, and they were desirous above all of joining in an enterprise with the British'.[38] He might have been more honest, if less comradely, if he had added that Roosevelt and Marshall realized how more difficult the task would be if Britain lost to Germany during the time that it took for the United States to defeat Japan.

Attlee, Eden and Mountbatten then all had their say, before Churchill summed up, affirming that although the details of the cross-Channel plans still needed to be worked out, it was clear that there was 'complete unanimity on the framework'. He would request help in the Indian Ocean, 'without which the whole plan would be fatally compromised', and rather grandly he acknowledged that 'It would gradually be known that the English-speaking peoples were resolved on a great campaign for the liberation of Europe.'[39]

Although Churchill had inserted the Indian Ocean, and not the Middle East, into his caveat, the mention of the Indian Ocean only made sense in terms of the Middle East, and he had mentioned the Middle East unequivocally earlier on. Moreover Brooke had already inserted his several caveats about the Roundup operation. The way that matters stood, it did sound as though the British would be able to veto any cross-Channel operation until they were satisfied that both the Indian Ocean and the Middle East were safe from Japanese and German aggression, and the United States was fully committed to bringing about that seemingly Utopian state of affairs.

Marshall expressed confidence that the Indian Ocean theatre could be protected at the same time that Roundup was undertaken and Hopkins emphasized that once the decision had been taken to cross the Channel it would become the United States' major contribution to the war effort and would therefore become irreversible. The Prime Minister then gave him a solemn assurance that the British would support the great enterprise energetically and unreservedly.

Churchill also promised Marshall 'that nothing would be left undone on the part of the British Government and people which could contribute to the success of the great enterprise on which they were about to embark'. Perhaps the very extravagance of Churchill's remarks, especially as no dates were given and as there was no indication of whether he was referring to Sledgehammer, Bolero or Roundup, ought to have given Marshall pause for thought. What he could not fail to spot, however, were the implications that British support for Continental attacks entirely depended upon America protecting British positions in the Indian Ocean and Middle East.[40]

Unsurprisingly, Marshall telegraphed Stimson soon afterwards to say that he'd been successful on all fronts. Yet in his memoirs Churchill stated that he only ever saw Sledgehammer as a feint, always preferring operations in North Africa and Norway. 'I had to work by influence and diplomacy in order to secure agreed and harmonious action with our cherished Ally,' he wrote, 'without whose help nothing but ruin faced the world. I did not therefore open any of these alternatives at our meeting on the 14th.'[41] Crucially, therefore, and in order to secure Bolero, Churchill hid his deep reservations about Sledgehammer and entirely failed to mention his preference for both Ajax and Gymnast, the attacks on northern Norway and North Africa. Marshall had indeed been misled, by Churchill if not by Brooke, and he understandably came to resent it.

Ismay was adamant in his own memoirs that:

Everyone at the meeting was enthusiastic about the prospect of the despatch of a mighty American army to Europe, and of the English-speaking peoples 'marching ahead together in a noble brotherhood of arms' as Churchill put it . . . Everyone seemed to agree with the American proposals in their entirety. No doubts were expressed; no discordant note struck. It is easy to be wise after the event, but perhaps it would have obviated future misunderstandings if the British had expressed their views more frankly . . . I think we should have come clean, much cleaner than we did, and said: 'We are frankly horrified because of what we have been through in our lifetime . . . We are not going to go into this until it is a cast-iron certainty.'[42]

With a classic penchant for understatement, Ismay concluded: 'This misunderstanding was destined to have unfortunate results.' In the controversy over whether the British deliberately misled the Americans during the Modicum Mission, Ismay, who was writing in 1960 – during Churchill's lifetime – confined himself to suggesting that they might have done so unintentionally.

The most 'unfortunate result' was that Marshall became convinced later on that Churchill and Brooke had indeed broken faith with him, and suspected that their opposition to Sledgehammer and an early Roundup in fact extended to any Roundup operation at all. Moran recalled his puzzlement over the way that Churchill had 'agreed with Marshall, almost, as it were, without a fight. It was not like him.' His explanation – and since he was not present at any of the meetings this was mere speculation – was that Churchill 'may have decided that the time has not yet come to take the field as an out-and-out opponent of a Second Front in Europe. Anyway, 1943 seems a long way off, and a good deal may happen in the meanwhile.' Moran added that Churchill told him it was no time for an argument with Roosevelt, who might be driven by domestic public opinion to concentrate on the Far East instead.

In two private interviews with Chester Wilmot in March and April 1948, Ian Jacob also went a good way towards admitting that the British had deliberately misled the Americans six years earlier. After explaining that Marshall wanted to use all American shipping possible so that a cross-Channel attack could be launched in 1942, he said:

We were convinced that it could not but we were reluctant to say so too strongly lest Marshall should pack up his divisions and take them to the Pacific . . . Consequently at the last meeting when the PM was most enthusiastic about the importance of crossing the Channel as soon as possible, Marshall went home thinking that an attack could be launched that year. The basic difference centred on the question of when it would be possible. Marshall thought this year, we thought not before 1943, if then.[43]

Something significant seems to have taken place between the pessimistic morning Chiefs of Staff meeting on 14 April, when objections were raised about air cover, landing craft and so on, and the wildly over-optimistic Defence Committee meeting at 10 o'clock that evening, at which the Chiefs were also present. Had the Prime Minister persuaded Brooke that Britain would be far better defended in the event of a German victory in Russia should there be seven or eight American divisions in the United Kingdom to help her? Brooke was under personal as well as professional pressure on 14 April; between the Chiefs of Staff meeting and another meeting with Lyttelton he had to dash back to his flat where his wife was suffering from an ear infection – mastoids – that his son had earlier only narrowly escaped. The operations to counter it were complicated and painful in those days, and Brooke admitted he found it all 'Most distressing' and 'a source of deep anxiety'.[44]

Brooke's diary entry for that evening stated that there had been 'A momentous meeting at which we accepted their proposals for offensive action in Europe in 1942 perhaps and in 1943 for certain. They have not begun to realize all the implications of this plan and all the difficulties that lie ahead of us! The fear I have is that they should concentrate on this offensive at the expense of all else! We have therefore been pressing on them the importance of providing American assistance in the Indian Ocean and Middle East.' This was a fairly accurate summation of the double-act that he and Churchill had put on for Marshall's and Hopkins' benefit, right down to repeating Churchill's own adjective 'momentous'. His reference to 1943 operations 'for certain' shows that he did not himself feel he was misleading the Americans, because he also believed – or at least hoped – that an attack might be possible by then. There was no reason why Brooke should lie to his diary: in April 1942 he felt himself committed to Roundup some time in 1943.

Sir Michael Howard believes that, as he puts it, the American plan 'did not conflict in any essentials' with such ideas as Churchill and Brooke had entertained, and that the 'evidence suggests that both Mr Churchill and his Chiefs of Staff were, in April 1942, entirely sincere in their acceptance of the Bolero–Roundup plan as expounded by General Marshall, and were determined to put it into effect. There was certainly not, at that time, any alternative and conflicting "Mediterranean strategy".' Although careful studies were conducted in London into the practicability of Operation Sledgehammer, 'about Roundup no doubts were expressed at all'.[45]

This was true, but after the war Brooke noted that with the situation prevailing in April 1942, 'it was not possible to take Marshall's "castles in the air" too seriously! It must be remembered that we were at that time literally hanging on by our eye-lids!' With Australia and India under threat from the Japanese, a temporary loss of control in the Indian Ocean, Germany threatening Iranian and Iraqi oil supplies, Auchinleck hard-pressed in the Western Desert, and the battle of the Atlantic hanging in the balance, 'We were desperately short of shipping and could stage no large scale operations without additional shipping. This shipping could only be obtained by opening the Mediterranean and saving a million tons of shipping through the elimination of the Cape route. To clear the Mediterranean, North Africa must be cleared first.'[46] Of course Brooke had not made this case at the Defence Committee meeting, mentioning neither North Africa nor the Mediterranean, let alone the idea that Marshall was building 'castles in the air'.

That same night, Marshall telegraphed McNarney to report that, although the British Chiefs of Staff had expressed some reservations, he believed there was 'complete agreement as to 1943'. On the following afternoon, Wednesday 15 April, Marshall visited Brooke's room at the War Office in Whitehall for two hours to hear about British military dispositions. 'He is, I should think, a good general at raising armies and providing the necessary links between the military and political worlds,' recorded Brooke. 'But his strategical ability does not impress me at all!! In fact in many respects he is a very dangerous man whilst being a very charming one!' Marshall led Brooke to believe that because Admiral King was 'proving more and more of a drain on his military resources, continually calling for land forces to capture and hold bases' in the Pacific, while simultaneously the Commander of the south-West Pacific Area, General Douglas MacArthur, now based in Australia, was also demanding forces to develop an offensive from there, his advocacy of Roundup was intended to 'counter these moves'. Nor was this a bluff; only three weeks later Marshall talked Roosevelt out of sending more men to MacArthur, since it would detract from the cross-Channel project.[47]

Brooke believed that protection of personnel and resources for Germany First was the real reason that, in his words, 'Marshall has started the European offensive plan and is going 100% all out on it!' He considered that Marshall was making 'a clever move which fits in well with present political opinion and the desire to help Russia', one which was 'also popular with all military men who are fretting for an offensive policy. But, and this is a very large "but", his plan does not go beyond just landing on the far coast!! Whether we are to play baccarat or chemin de fer at Le Touquet, or possibly bathe at Paris Plage is not stipulated! I asked him this afternoon – do we go east, south or west after landing? He had not begun to think of it!!'[48] Paris-Plage was a popular part of the seaside resort of Le Touquet, but Brooke presumably did not in fact ask whether the expedition would move west after landing, because that would take it back into the English Channel. Yet as the official US Army history admits: 'Only the most hurried and superficial investigation of the complex logistic problems involved had been made before the London Conference, and the Conference contributed very little to an understanding of them or to argument about them. Everything remained to be done.'[49]

Brooke's contemporaneous sarcasm was not really softened when he made extensive notes on his diaries years afterwards, in which he stated that his conversation with Marshall that afternoon had been 'an eye-opener! I discovered that he had not studied any of the strategic implications.' Of course this was unfair, as Marshall had been doing little else over the previous fortnight, and the discussions on 15 April centred on the difficulty of the landing compared with what happened afterwards. Brooke believed that 'our real troubles would start after the landing. We should be operating with forces initially weaker than the enemy and in addition his rate of reinforcement would be at least twice as fast as ours. In addition his formations were fully trained and endured [inured] to war whilst ours were raw and inexperienced.'[50]

In May 1957, Samuel Eliot Morison vigorously defended Marshall against Brooke's criticisms, some of which had been published earlier that year in Brooke's memoirs *Turn of the Tide* as ghosted and edited by the historian Sir Arthur Bryant. Several of Brooke's harsh diary entries about Marshall had been quoted in the book, and in a tart rejoinder during a lecture series in Oxford, Morison said that 'General Marshall has adopted a policy of dignified silence about these war controversies, so I cannot quote him; but I am confident that his strategic ideas did not stop at the water-edge; that he had a very definite concept of land strategy, namely the double envelopment of the Ruhr, which was actually carried out in 1945 against the strong objections of Field Marshals Brooke and Montgomery.'[51] While it is actually unlikely that Marshall had planned anything like that far ahead in 1942, it was right to defend Marshall from Brooke's repeated and unfair strictures about his strategic abilities.

Brooke's exasperation about Marshall in his diaries – 'almost impossible to get him to grasp the true concepts of a strategic situation', for example – needs to be put in the overall context of his blistering rudeness about almost everyone else, very often over their supposed lack of strategic understanding. Churchill was described as 'temperamental like a film star' and 'peevish like a spoilt child' with 'no long term strategic vision ... He can never grasp a whole plan'; Lord Gort's 'brain has lately been compared to that of a glorified boy scout' who 'just fails to be able to see the big picture'; Eisenhower 'literally knew nothing of the requirements of a commander in action' and had 'a very very limited brain from a strategic point of view'; Eden was 'dangerous – rather obstinate, featherheaded – and with no strategic sense'; Alexander had 'many fine qualities but no very great strategic vision ... It was very doubtful whether he was fit to command his Army'; Mountbatten was 'quite irresponsible, suffers from the most desperate illogical brain, always producing red herrings'.[52]

A definite pattern thus emerges, of people who might have charm and even some organizing ability, but are not gifted enough to grasp Brooke's overall strategy. Over and again identical criticisms appear in Brooke's journals. Almost the only senior wartime figures to emerge unscathed were Douglas MacArthur, the South African premier Jan Christian Smuts, and Stalin. Yet it simply cannot be the case that no one on the Allied side but Brooke had any grasp of strategy. Far from being strategic amateurs, as Brooke constantly implied about the American Joint Chiefs, Marshall had been a senior Planning officer on General Pershing's staff, Stark had been flag secretary to Admiral Sims in 1917–18, King had been assistant chief of staff to Admiral Mayo, Commander-in-Chief of the Atlantic Fleet in the Great War, and Eisenhower had been chief of staff to the XXII Corps and after that of Walter Krueger's Third Army, hugely successful in the famous 1941 Louisiana manoeuvres. Just because these men disagreed with Brooke, it did not follow that they were all dunderheads over strategy.

'Those who did not know Marshall were apt to think of him as a cold man,' wrote his biographer, 'but compared with the hard, distant, lofty Field Marshal Brooke, he was a ball of fire.'[53] After the publication of *Turn of the Tide*, Pogue asked Marshall for his comment on what the British official historian John Ehrman had written of Brooke:

Possessing a clear and acute mind, great professional integrity, and – a useful attribute on occasions – a strong but controlled temper, his views always commanded the respect of the army, of his naval and air colleagues, and, even when the two men differed, of the prime minister. Insofar as the Chiefs of Staff designed British strategy, that strategy bore his impress.

If Pogue was trying to draw Marshall into criticizing Brooke, he failed, as the gentlemanly general merely answered: 'Of the Chiefs of Staff, I thought Portal was probably the most brilliant, but I had a great respect for Brooke and I think the characterization regarding him is quite correct.' On another occasion Marshall told Pogue that 'Brooke was cold, but did not give the impression he disliked him. Portal had the best mind of the lot and was thus the most difficult to deal with in the sense of putting anything over.' He also recalled being 'Affectionately fond of Pound.' Others leapt to Marshall's defence by criticizing Brooke in 1957, but Marshall did not stoop to it, writing to thank him for his signed copy 'and for your gracious inscription, the sentiments of which I deeply appreciate'.[54]

On 16 April 1942 Brooke attended an important Chiefs of Staff meeting to discuss Bolero, Sledgehammer and Roundup. 'The plans are fraught with the gravest dangers,' he confided to his diary. 'Public opinion is shouting for the formation of a new western front to assist the Russians. But they have no conception of the difficulties and dangers entailed!' One of the advantages of having a CIGS from the nearest British equivalent to the Junker class was that he despised public opinion, and felt it almost a point of honour not to bend to it. All the political campaigns, newspaper editorials and demonstrations in Trafalgar Square in favour of the

Second Front meant absolutely nothing to him, except insofar as they might affect politicians' judgement, despite the suspension of general elections for the duration of the war. 'The prospects of success are small and dependent on a mass of unknowns,' he opined about an early Second Front, 'whilst the chances of disaster are great and dependent on a mass of well established military facts.'[55] Foremost among these, as we have seen, was the proven ability of the German Army, of which he had a healthy – although many Americans thought too robust – respect.

The next morning, on 17 April, after a short Chiefs of Staff meeting, Brooke went over to Downing Street to bid Marshall farewell. 'He was very charming as usual and hoped I would be able to return his visit,' he wrote afterwards. They had been in each other's company – sizing each other up, as Brooke put it – for nine days. Hopkins and Marshall left London later that day, stopping the night in Londonderry. Before they got back, however, Churchill cabled Roosevelt, stating baldly: 'We wholeheartedly agree with your conception of concentration against the main enemy, and we cordially accept your plan with one broad qualification,' that it was 'essential that we should prevent a junction of the Germans and Japanese'. This could have been a qualification so broad that it might have wrecked Roundup, since the hinge between the Middle East and the Indian Ocean was as large as it was vulnerable, covering several countries. Nonetheless, Churchill sought to reassure Roosevelt by adding that 'Marshall felt confident that we could together provide what was necessary for the Indian Ocean and other theatres, and yet go right ahead with your main project.'[56]

The Prime Minister then made a reference to Sledgehammer that, considering his and Brooke's severe private reservations, was at best a red herring, at worst deliberately misleading. After saying of Roundup, 'The campaign of 1943 is straightforward, and we are starting joint plans and preparations at once,' Churchill wrote:

We may, however, feel compelled to act this year. Your plan visualised this, but put mid-September as the earliest date. Things may easily come to a head before then. Marshall explained that you had been reluctant to press for an enterprise that was fraught with such grave risks and dire consequences until you could make a substantial air contribution; but he left us in no doubt that if it were found necessary to act earlier, you, Mr President, would earnestly wish to throw in every available scrap of human and material resources. We are proceeding with plans and arrangements on that basis. Broadly speaking, our agreed programme is a crescendo of activity on the Continent.[57]

This seems like nothing less than an endorsement of Sledgehammer in only five months' time, or even 'before then', but only once United States troops were flooding over under Bolero, or what Churchill was soon confusingly to call 'Super-Bolero'. Yet were the British *really* committed to such a 'concentration against the main enemy' in a cross-Channel attack? The historians Warren Kimball and Norman Rose describe Churchill's message as 'disingenuous'.[58] The writer Leonard Mosley went further, writing of Churchill's 'insincerity'.[59]

Brooke meanwhile told John Kennedy over at the War Office that 'he had backed the first stage – that is movement into the UK – because that suits all eventualities including defence of the UK and possible action overseas elsewhere, as well as the main conception put forward by Marshall,' and that Marshall 'has gone back with substantial agreement in principle, but a better realisation of practical difficulties and a better appreciation of the possible dates'. This seems a clear indication that Brooke had accepted Sledgehammer in principle because he wanted Bolero in practice, and knew that Sledgehammer could be argued out when the time came. Some in the War Office thought the Americans themselves would drop Sledgehammer as soon as their Planners had further analysed its deficiencies.

In Ulster, Marshall inspected the Bolero advance guard, American units under Major-General Russell P. Hartle. While he and Hopkins were there, and then while they were on their way to Port Patrick in western Scotland prior to flying home, a situation developed that led Churchill to suggest that the two men return to London. Roosevelt had heard that Marshal Philippe Pétain, the dictator of Vichy France, was restoring the pro-German former premier Pierre Laval to power, under pressure from Hitler, leading the President to ask Marshall to consider reviving Gymnast, on the ground that the French in North Africa might prefer the Allies to Laval.

Although Roosevelt did not think the situation was fast moving enough to justify Marshall returning to London, his instinctive reaction – that Gymnast be reactivated – was indicative of a continued presidential fondness for the North African operation that the American Chiefs intensely disliked. Marshall can hardly have failed to infer as much.

Often, even when nothing comes of them, such démarches can throw sudden light on a scene, like a split-second flash of nocturnal lightning. Marshall, in Eden's phrase, 'would not be stopped' from returning home, but he now knew that his various memoranda had not killed off Gymnast in Roosevelt's restless mind.[60]

After the war, Albert Wedemeyer told the BBC interviewer Richard Dimbleby that, on their flying-boat going back to Washington, Marshall had turned to him and said: 'I think the British have bought your plan, but I think they did so with tongues in their cheeks.'[61] Wedemeyer also claimed to his SOOHP interviewer that Marshall could tell by the questions that the British had asked him that they preferred to invade Europe through the so-called 'soft underbelly' of the Mediterranean.[62] Was this *ex post facto* rationalization, or did it really happen? Did Marshall really spot Brooke's future foot-dragging over Roundup as early as on his return flight home? If so, it makes his later professed shock at British perfidy rather deceptive itself. Wedemeyer was a senior member of the Joint Planning Staff and was present at strategy conferences before he became US Chief of Staff of the South-East Asian Command in 1943. His testimony – even at three decades' remove – is therefore noteworthy. In the end, however, it is simply not convincing: as Michael Howard notes, the British had no conception of invading Europe from the south in April 1942, with the phrase 'soft underbelly' not being coined by Churchill until the following year.

Ismay essayed a metaphor of seduction when describing the Modicum Conference to Pogue four years later, saying the British had been 'swept off their feet when Marshall and Hopkins visited them on Roundup and that they agreed to it without considering what it entailed'. According to him, Churchill and Brooke 'were swept away by the prospect of going back [to France] in 1943'.[63] This was only partly true, and to continue the analogy, Churchill might have been flirting outrageously, but Brooke was still only mildly teasing. Certainly neither had by then decided to go all the way.

7

The Commanders at Argonaut: 'The easiest road to the centre of our chief enemy's heart'

April–June 1942

There are no laws of war which say that men can die but staff officers mustn't be vexed.   Ronald Lewin[1]

As Marshall and Brooke bade farewell in London their two sets of Staffs got down to putting meat on the bones of the three operations, which in the case of Roundup and Sledgehammer were still positively skeletal. As they did so, a process now known as 'group-think' got under way on both sides of the Atlantic. Knowing that Marshall was committed to an early cross-Channel operation, the Operations Division of the US War Department found arguments and strategies to support one. Similarly, knowing that Brooke was in favour of a later operation, once Germany had been 'ripened', the War Office Planners found arguments to oppose an early crossing. It is surprising, despite their all being intelligent, independent thinkers, how few American Planners supported a later and how few British Planners an earlier cross-Channel operation. Meanwhile, London watched intently for any sign in Washington that the President could be detached from Marshall in his support of Sledgehammer. If the tectonic plates gave any hint of shifting between them, speed in exploiting it would be of the essence.

Foremost among the sceptics of an early Second Front in the British War Office was the man whose duty it now was to begin planning for one, the Director of Military Operations, John Kennedy. He receives relatively little attention as an important figure today, but his testimony from the very heart of the military decision-making process is compelling. 'The slowness of getting ashore and the certainty of ultimate defeat while the Germans are in a good state are the governing factors in knocking a French operation out of court at the moment,' he wrote, neatly encapsulating British thinking in April 1942. 'We cannot afford to lose a big Army detachment so long as there's a chance of the Germans surviving the Russian battle and then attempting invasion.'[2]

Over the next four months, the War Office were planning to send abroad five divisions – four infantry and one armoured – while only receiving a maximum of two American divisions into the United Kingdom, leaving a large net loss of troops from the mainland. Another problem with Roundup was the widespread belief in British military circles that the Free French were unreliable. After General de Gaulle's chief of staff Pierre Billotte came to see Brooke, bringing maps and plans for sabotage in France and for raising the local population, Kennedy commented: 'It is hard to get all of these on the right basis because of the difficulty that de Gaulle cannot be trusted.'[3] Relations between de Gaulle and the British Government had deteriorated soon after his flight to London in June 1940, and worsened when loose talk by Free French officers was blamed by the British for the Allied attack on Dakar that September turning into a débâcle.

On Sunday 19 April, as Lord Halifax drove to the airport in Washington to welcome Marshall home, and Hopkins stopped off at Hyde Park to report on Modicum to the President, Kennedy told his diary: 'I think the Germans will be able to advance a long way when the weather permits but that they will not be able to knock the Russians out and the Russian front will remain and prove a fatal commitment in the end.' That was just the sort of prescience he was hired for. 'We can do a lot by raids and air action to cause a diversion of strength from the Russian front,' he continued, 'although not much to affect the German *Army* dispositions.'

Although a constant American refrain would be that Britain was not doing enough to help the beleaguered USSR, of the two countries it was even more in Britain's interests than America's that Russia should remain fighting. If Germany won in the east she would have been able to move into Iran and Iraq, cutting off Britain's oil supply.[4] Yet even that danger would not have justified Sledgehammer, so Brooke believed, because it would have made no appreciable difference to what happened in the east, while costing the Allies dear if it failed, as he was certain it would.

Marshall had mentioned Sledgehammer as a 'sacrifice' to avert an imminent collapse of Russian resistance, and was to refer to it as such for the rest of his life. He had made it clear in London that this sacrifice would be made mainly by the British, since so few American divisions could be transported for 'immediate' action by the autumn of 1942. By 15 September, the US could provide only 700 of the planes required, for example.[5] Yet by presenting two operations instead of one – that is, Sledgehammer as well as Roundup – Marshall allowed the British authorities to concentrate their arguments on the weaker of the two. In strategic and tactical terms this was undoubtedly Sledgehammer, which the War Office Planners soon came to view as little more than the modern equivalent of the 'forlorn hope' that was sometimes seen in seventeenth-century battles.

When Marshall returned to Dodona he saw all the improvements that Katherine had made in his absence in Washington and London. 'As we drove along the narrow roller-coaster road,' she recalled,

the honeysuckle along the fences filled the air with fragrance, the cows were heading towards the barns for their evening meal, while sheep grazed contentedly ... George gave a sigh of contentment, and then followed the happiest hours of the past three years. He stopped the car, got out and walked around, taking in every detail. Finally he turned and said in a husky voice, 'This is home, a real home after forty-one years of wandering.'

They sat on the lawn until 9 p.m. as he told her of the highlights of his trip to 'war-torn Europe'.

On Thursday 23 April Churchill spoke to a secret session of the House of Commons, the first since the fall of Singapore in February. In one of his longest speeches of the war, he made no attempt to minimize the various disasters of the previous six months. He described how the *Ark Royal*, *Barham*, *Prince of Wales* and *Repulse* had been sunk, and how the battleships *Nelson*, *Valiant* and *Queen Elizabeth* had been crippled, pointing out that in a mere seven weeks one-third of the Navy's battleships and battle cruisers had either been sunk or severely damaged. He related to an understandably sombre House that one hundred thousand Commonwealth troops had surrendered Singapore to a mere thirty thousand Japanese.

Yet he concluded his wide-ranging recitation of these 'ugly realities' with an upbeat assessment of the future, and said that the Germany First policy was 'earnestly and spontaneously shared by the Government and dominant forces in the United States'. He went on: 'The visit of General Marshall and Mr Hopkins was to concert with us the largest and swiftest measures of this offensive character. It will no doubt become common knowledge that the liberation of the Continent by equal numbers of British and American troops is the main war plan of our two nations.'[6]

He then read out a recent telegram from Roosevelt, which stated that the President was 'delighted with the agreement which has been reached between you and your military advisors and Marshall and Hopkins. They have reported to me of the unanimity of opinion relative to the proposal which they carried with them and I appreciate ever so much your personal message confirming this.' In fact the numbers of troops earmarked for the operations were far from equal – seven British divisions to two American for Sledgehammer, or eighteen British against America's thirty divisions for Roundup – but then neither was there genuine 'unanimity of opinion' between Marshall and Brooke over either operation, as was soon to become evident.

At the War Cabinet of 27 April, with Norman Brook taking verbatim notes, Portal gave details of the bombing raids on Cologne and Rostock. Churchill said: 'Don't make too much of this in Press – we're hitting them three times as hard – don't give impression this is quits . . . Tone down, and keep in proportion. Don't give out photos.' (A red-haired Scottish sergeant-pilot from Motherwell who was being honoured at the White House for heroic feats bombing Germany told Halifax of the tremendous fires in Rostock. 'I'm afraid we not only killed them,' he said; 'we cremated them.')[7] Later on in the War Cabinet, discussing the modifications proposed for 'R'-class battleships, Churchill denounced them as 'Floating coffins. Unsafe to face any modern vessels or air attack.' Pound replied that any increase in their armour would produce bulges that would reduce their speed to 15 knots, further reducing their value as ocean-going warships.

'On 13 January last,' Marshall wrote to Roosevelt on 5 May, 'you authorized an increase in the enlisted strength of the Army to 3.6 million by 31 December 1942. Authorization for additional men in 1942 is now essential to our plans.' In the intervening four months the Army had had to garrison the lines of communication to Australia, and rush reinforcements to Hawaii, Alaska and Panama. Anti-sabotage measures and enemy internment sucked in further troops. 'The Bolero plan requires a material increase in special troops, eventually requiring some three hundred thousand men for the ground forces alone.' Marshall therefore asked for another 750,000 men, making a total of 4.35 million, by January 1943.[8]

Although Roosevelt did not reply to that immediately, the next day he sent his own memorandum to Marshall, Stimson, Arnold, King and Hopkins, which was not to be circulated beyond them. 'I always think it well to outline in simple terms and from time to time complex problems which call for overall planning,' he began. 'Therefore I would like you gentlemen to read the following.' Emphasizing that with 'the world situation changing so rapidly' the memorandum could only apply to the present time, it nonetheless set out future strategy from the very first sentence: 'The whole of the Pacific Theater calls, at the present time, fundamentally for a holding operation.' Two bombing offensives – against the Japanese mainland and its lines of communication – were desirable in the furtherance of 'defence of all essential points'. Further west, the President believed 'We can very nearly separate India and Burma into a separate theater of war.' Here, the primary responsibility was British, but with the US assuring air connections with China and helping British armies in India.

Roosevelt characterized the recapture of previously British- and Dutch-owned islands as 'premature'. In the Near East and East African theatres, the responsibility was again British, although America 'must furnish all possible matériel' in Libya, Egypt, Palestine, Syria and the Persian Gulf. Britain and America would split responsibility for the Atlantic, while 'The principal objective is to help Russia,' since 'It must be constantly reiterated that Russian armies are killing more Germans and destroying more Axis matériel than all the twenty-five United Nations put together.' (At that time the phrase 'United Nations' was shorthand for the global anti-Axis coalition, and had nothing to do with the post-war multilateral organization.) The President then brought up the possibility, once they had air superiority, of fast Commando raids and week-long 'super-Commando operations' of up to fifty thousand men on the Continent, and finally 'a permanent front' big enough to ensure that it was not 'pushed into the sea'.[9]

While the Pacific, Indo-Burmese and Near East theatres all required merely to maintain existing positions 'for the next few months', Roosevelt opined, the Atlantic theatre needed 'active operations to be conducted in 1942. I fully realize difficulties in relation to the landing of armed forces under fire,' he wrote. Nonetheless, 'The necessities of the case call for action in 1942 – not 1943.' Roosevelt concluded: 'If we decide that the only large scale offensive operation is to be in the European area, the element of speed becomes the first essential.' This memorandum would have come as no surprise to Marshall – who would probably have been forewarned of it by Hopkins – and it merely reiterated his own Germany First views on grand strategy. However for the US Navy, and especially for Admiral King, it represented a serious reverse to their Pacific ambitions, at least in the medium term.

Nonetheless there were also alarming implications for Marshall in Roosevelt's memorandum. While seeming to reaffirm his own Memorandum, it actually contained a bombshell that threatened it profoundly. The phrase 'action in 1942 – not 1943' effectively meant that the British could ultimately decide where the first military blow against Germany and Italy fell, just as Brooke was starting to consider outright opposition to Sledgehammer and having severe doubts about an early Roundup. Since the British had an effective veto on both operations, which could only be staged from the United Kingdom, Roosevelt's insistence on 'action in 1942' handed over the final say to Churchill and Brooke. Seeing the President's memorandum in terms of his turf war with Admiral King, Marshall seems to have failed – at least for the moment – to spot what for him would soon be a far more dangerous portent.

Meanwhile, in London, Brooke had to consider who should be in charge of any cross-Channel attack were one to take place that September. 'We have to nominate commanders now and make plans for conversion for forces to an offensive basis on a phased programme which at the outset must still take account of the possible revival of the invasion threat,' said Kennedy, ever the professional Planner. 'Staffs have to be reorganized and augmented and arrangements coordinated with the Americans. We have considered whether we could get a Supreme Commander appointed in order to obviate a soviet of Commanders – Army, Navy and Air Force and American and British.'

Brooke was against the idea, 'because he does not think the man could be found. Or if he could it would be somebody like MacArthur who would be more of a nuisance than anything else.' Brooke's distaste for an early operation left him feeling that if anyone should be making the running in terms of organization, it should be Marshall. (He was to alter his view of Douglas MacArthur, who by the end of the war he considered the finest Allied commander of them all, a view perhaps aided by the fact that he was operating more than ten thousand miles away.) Kennedy put the opposing view over the issue, saying that, since plans were obviously taking definite shape over in Washington, any future commander could be handed plans for whose execution he would be responsible even though he had not been party to drawing them up. Kennedy told Brooke that such a problem 'would of course be solved if he himself or one of the other Chiefs of Staff, who had seen the plan grow up, took charge', but Brooke failed to rise to the bait, merely restating his view 'that at the moment we should only get Marshall or MacArthur if we pressed it and that would not help'.[10] This is the only recorded occasion in the war that Brooke was dismissive of Marshall by name in front of someone else.

On 26 May 1942 Churchill asked for the construction of special piers that could unload supplies from ships during an invasion of France. 'They must float up and down with the tide,' he minuted. 'The anchor problem must be mastered. Let me have the best solution worked out.' Planning for the vast 'Mulberry' harbours and the Pipeline Under The Ocean (PLUTO) was thus put in motion a full two years before D-Day, testament to British commitment to an eventual return to the Continent. Nonetheless, two days later Churchill sent Roosevelt a couple of telegrams setting out his doubts about any cross-Channel attack in 1942, and in support of the old Operation Ajax attack on Norway (now rechristened Jupiter), which he hoped would address the President's concerns about helping Russia. In the first he wrote: 'We must never let Gymnast pass from our minds. All other preparations would help if need be towards that.' In the second he went into the three great difficulties that Brooke felt worked against an early Sledgehammer and what Churchill now called 'Super-Roundup', namely the Luftwaffe's command of any paratroop landing sites, the lack of landing craft – only 383 would be operational by August 1942 and 566 the next month – but most importantly the American contribution, which would not be up to strength until 1943.

Churchill reported to Roosevelt that he had told the visiting Soviet Foreign Minister V. M. Molotov that it was 'the earnest resolve of the British Government to see what could be done this year to give the much needed support to the valiant Russian armies', but had significantly added that 'It was unlikely that any move we could make in 1942, even if it were successful, would draw off large numbers of enemy land forces from the Eastern Front.'[11] From this, Roosevelt could have been left under no misapprehension about Churchill's and Brooke's profound doubts about Sledgehammer.

On his return to Washington from Modicum in April, Marshall had appointed Eisenhower to command all American forces in the European Theater of Operations. Eisenhower then flew to London. In his diary for 28 May, he recorded a meeting on Roundup in which Brooke 'outlined the position in detail', stressing the need for agreement on a supreme commander 'at an early date'. As we have seen, this was probably a bluff, as Brooke was in no hurry over anything to do with attacking the Continent. Eisenhower replied that any operation taking place in 1942 would 'be under British command, with our forces attached to the British in suitable capacities'.[12] He went on to say that Marshall might be prepared to accept Mountbatten as the commander, which was unlikely to commend itself to Brooke, who was sceptical about Mountbatten's abilities.

Eisenhower noted, 'It is quite apparent that the question of high command is the one that is bothering the British very much and some agreement, in principle, will have to be reached at an early date in order that they will go ahead wholeheartedly into succeeding steps.' The fact that Eisenhower had written of 'the assault echelon of Bolero' rather than Roundup, just as Churchill had written of 'Super-Roundup', is indicative of the severe nomenclature confusions that both allies were soon to experience over the codenames of these three distinct but almost overlapping operations.

The next day Kennedy met Eisenhower for the first time and found him 'a very pleasant, intelligent and forceful character', as did almost everybody. They discussed the maps in their offices. On Kennedy's, 'America is shown twice in order that one may realize the size of the Pacific, the usual maps being cut through the middle of the Pacific.' Eisenhower said he did the same thing, but as Kennedy, a son of the Empire, observed with astonishment: 'He had had his map cut through India because that was the place in which he was least interested!' Brooke also liked Eisenhower, and was greatly amused when Ike quoted one of his young Staff officers telling him: 'Well, you know sir, there are two things in which, from the beginning of time, amateurs have always considered themselves experts – one is military strategy and the other is prostitution.'[13]

The same day that Eisenhower met Kennedy, Molotov arrived at the White House at 4 p.m. for a series of meetings that lasted until after midnight. The next day he was back to see the President, Hopkins, Marshall and King, with interpretation undertaken by Vladimir Pavlov and Samuel H. Cross, the Professor of Slavic Languages and Literature at Harvard. In answer to the Soviet Foreign Minister's long and impassioned request for a Second Front in 1942 to draw off forty German divisions from the USSR, Roosevelt asked Marshall 'whether developments were clear enough so that we could say to Mr Stalin that we were developing a Second Front'. Marshall replied, 'Yes, Mr President.' Then Roosevelt 'authorized Mr Molotov to inform Mr Stalin that we expect the formation of a Second Front this year'.[14]

There was a good deal of theatre to all this, since the President knew perfectly well what had been agreed the previous month in London. Yet after his reception there, where Churchill and Brooke gave no such cast-iron commitment, Molotov could have been forgiven for assuming that there were divided counsels among the Western Allies, and that it was the Americans who were driving the idea forward. The fact that this was true does not make it any less reprehensible of Roosevelt for promising something that was still in its early developmental stages and which Churchill had recently downgraded in his telegram. For all the kudos that the Americans gained from the Soviets over the British by this play-acting, the President had set the Allies up to disappoint Stalin when no Second Front emerged in France either in 1942 or in 1943. Of course it could not have happened had Churchill and Brooke been more straightforward with Marshall about the genuine likelihood of any operation in the near future.

Marshall outlined to Molotov what seems to have been a hybrid Sledgehammer–Roundup operation, emphasizing that the transport difficulties 'were complicated by the necessity of sending tonnage to Murmansk'. He laid out his strategic vision, which was 'To create as quickly as possible a situation on the Continent under which the Germans would be forced into an all-out air engagement, but they will not engage on this scale without the pressure of the presence of our troops on the ground.' He envisaged shipping enough men across the Channel 'to provoke an all-out battle for the destruction of the German air force'.[15] Under this idea, Cherbourg was intended as a kind of honey-trap, to draw in the Luftwaffe to its destruction by squadrons based in the Channel Islands, the Cotentin Peninsula and mainland Britain.

At the end of Molotov's visit, Roosevelt issued a communiqué to the press which included the fateful words: 'Full understanding was reached with regard to the urgent task of creating a Second Front in Europe in 1942.'[16] This raises the question whether he genuinely believed a Second Front was still possible, or whether he merely wished to placate Molotov. Did he hope to calm public opinion rather than inflame it with such a remark? Or did he hope such a promise might encourage the British to plan more proactively? We do not know. It is one of the many aspects of America's enigmatic commander-in-chief that historians continue to debate. The most likely explanation was that given by Ian Jacob to Chester Wilmot in the spring of 1948: 'Roosevelt went much too far in his assurances to Molotov and did so without reference to us and on the strength of what Marshall told him.'[17] Since Marshall only told him what Churchill had explicitly told *him* at the key Defence Committee on the evening of 14 April, Churchill was effectively deceiving the Russians by proxy, as an unexpected by-product of misleading the Americans about Britain's true enthusiasm for Sledgehammer and an early Roundup.

Molotov flew back to London from Washington. Although Churchill did not want to put his name to a 'full understanding' for 'a Second Front in Europe in 1942', Roosevelt had effectively sold the pass, and just such an Anglo-Soviet communiqué was indeed released on 11 June. At the same time, Churchill gave Molotov an aide-mémoire in front of his colleagues in the Cabinet Room which stated that, although the Allies were making preparations for a landing in August or September 1942, the shortage of landing craft meant that 'We can therefore give no promise in the matter.' Molotov did not care about this private caveat, having secured the all-important public declaration. As he later pointed out, he did not think that the Western powers would create a Second Front in 1942, but their promise to do so had immense political and propaganda potential once they had reneged upon it for two calendar years running. 'This undermined faith in the imperialists,' he wrote.

'All this was very important to us.' For all that Roosevelt would have been outraged at being classed as an imperialist, that was of course how both the Western Allies were regarded by the veteran Bolshevik.

Meanwhile Roosevelt told Churchill that he had 'got on a personal footing of candor and of friendship as well as can be managed through an interpreter'. This was again frankly naive, as Molotov was notoriously lacking in either candour or friendship with foreigners, which could carry with it a death sentence, even in Stalin's closest circle. Roosevelt added: 'I am more anxious than ever that Bolero proceed to definite action beginning in August and continuing so long as the weather holds.' Changes were made to this draft by Hopkins who cut the words after 'beginning' and inserted 'in 1942. We all realize that because of weather conditions the operation cannot be delayed until the end of the year.'[18] While resiling from the August date, therefore, Roosevelt and Marshall still hoped to keep to 1942, or at least they wanted Brooke and Churchill to think that they still supported that date. Since the weather in the English Channel in autumn is notoriously changeable, in effect Hopkins' altered wording mattered little.

Delivering the graduation speech at West Point on Sunday 31 May, Marshall further built up public expectations by announcing that 'American troops are landing in England and will land in France,' at which the Corps of Cadets let out 'a mighty roar' of cheering. He ended his address by saying: 'We are determined that before the sun sets on this terrible struggle, our flag will be recognized throughout the world as a symbol of freedom on the one hand and of overwhelming power on the other.' He gave no timetable for the French operation, obviously, but this speech indelibly thereafter connected his name and reputation to an early cross-Channel operation.

Eisenhower returned from London on 4 June to report back to Marshall. He had met the Chiefs of Staff, Bernard Paget of Home Command, Bernard Montgomery of South-eastern Command and Archibald Nye, as well as all the most important Americans there, and he told his diary, 'It is necessary to get a punch behind the job or we'll never be ready by spring 1943 to attack. We must get going.' That is also what he told Marshall, and a week later, after recording speculation that he might go over to command the US forces in whichever operation was chosen, he wrote in his diary: 'The chief of staff says I'm the guy.' With II Corps earmarked to be shipped over and Mark Clark as its commander, the man universally nicknamed Ike (except by Marshall) commented: 'Now we really go to work.'

In May 1942, the US Army moved into the first half-million square feet of the Pentagon building, even though the $87 million construction project would not be complete until the following January. Previously the General Staff had worked in the Old Munitions Building, a temporary Great War structure on the Mall, so to provide security for the move soldiers were posted along the whole route from there, across the Fourteenth Street bridge, and through the fields to the vast new structure. The Army's files were transported from the old building to the new in armoured cars. At a mile in circumference, the Pentagon was the largest building in the world at the time. With office space for forty thousand people it had been built in a little over a year, albeit with an accident rate four times the average for US building sites.[19] The size of the edifice gave rise to jokes about how easy it was to get lost, such as the one about the pregnant lady who asked a Marine guard to help her get to a maternity hospital, saying it was an emergency as she was in labour. When he said that she shouldn't have gone there in that condition, she replied: 'When I came in here, I wasn't.'

At a War Cabinet on Monday 1 June, Churchill 'congratulated' Portal and Bomber Command on the previous night's raids, when 1,137 bombers had 'left this island and almost as many go tonight', describing the attacks as a 'Great manifestation of air power. US like it very much. Give us bigger action again early next month.' Portal replied that although fifty-one bombers had been lost, photographs taken over Cologne and the Rhine 'show results are good'. In a reference to bombing Germany earlier in the war, Churchill had said that he did not see why 'the disgusting stertorous slumber of the Boche should remain undisturbed,' and on another occasion, urging that the size of bombs dropped on Germany be increased, he complained: 'We might as well drop roasted chestnuts.'[20]

Pound had far less happy news; a convoy from Iceland to Russia had been very heavily mauled – of the thirty-five ships the Cabinet had chosen to send, six were sunk by bombers and one by a U-boat, with the overall loss of 147 tanks, 37 aircraft and 770 motor vehicles. Churchill wanted the next convoy postponed, but Eden pointed out the disastrous effect on Anglo-Russian relations that would result. Brooke, his forearms badly swollen after being stung trying to shake a swarm of bees off a branch that weekend, reported on the fighting taking part in that area of the Western Desert known as 'Knightsbridge'.[21] On 2 June, Lieutenant-General Sir Neil Ritchie, commander of the Eighth Army, was forced to abandon the Gazala Line, a series of fortified positions stretching 40 miles from Gazala on the coast to Bir Hacheim.

On the morning of Sunday 7 June, Churchill telephoned Eden at his seventeenth-century country house, Binderton, near Chichester in Sussex. Eden had stayed in bed to work on the contents of his red despatch boxes as it was too cloudy to enjoy gardening, and instead talked to the Prime Minister about the disappointing reports from Libya, which he admitted 'depressed' them both as Rommel appeared to be retaining the initiative. 'I fear we have not very good generals,' said Churchill. Eden recorded that the Prime Minister 'was also depressed by the Chiefs of Staff's sudden decision to cancel their own previous plans to take [a] certain place in north', a reference to Trondheim. Churchill feared it was because of the 'extreme reluctance' of the Commander-in-Chief of the Home Fleet, Admiral Sir John Tovey, to continue sending convoys to Russia. 'The politicians are much abused, but they get little help or inspiration from their service advisers' was Churchill's comment, to which Eden added, 'It can hardly be denied.'[22] Before this rather self-pitying conversation evokes too much sympathy, it ought to be pointed out that Operation Ajax/Jupiter had been a Churchillian idea that Brooke and all the Chiefs had always opposed equally doggedly.

Churchill was enormously cheered by the news that started to come in on 8 June about the great American naval victory at the battle of Midway, indicating that two Japanese carriers had been sunk and two damaged, prompting him to declare to the War Cabinet:

Losses at sea have produced signs of fear on part of the Japs – the Navy is a political force in Japan, which will perhaps be more inclined to a restrictive and cautious policy. This policy might be in harmony with sending out submarine raiders. If we think of this as having an effect on the Jap situation, I think they will go for China and Chiang Kai Shek conquest. I don't think they'll try India or Australia. This gives us two or three months' breathing-space. We must come to rescue of China – it would be an appalling disaster if China were forced out of the war and a new government set up. The General Staff should think of attacking lines of communication in Burma. If carrier losses are confirmed we can review the consequences of diminution of enemy forces. If Japan adopts conservative course it is a chance for us to get teeth into her tail.[23]

Brooke evidently agreed, as he answered that there was a 'possibility of opening up the Burma Road in the next two or three months to land practical military aid'. Churchill answered: 'Tell Chiang Kai Shek to hold the fort because we are coming next October or November.' Asking for a report from Wavell, Churchill said: 'The further removed from danger and fighting, the more dangers and responsibilities seem to weigh with officers,' which might or might not have been a criticism of Wavell, or even of the Chiefs of Staff.

By mid-June 1942, Churchill desperately needed to know Roosevelt's state of mind over Roundup and Sledgehammer. The man he cannily chose to sound him out was Lord Louis Mountbatten, who as director of combined operations had been promoted by Marshall as a possible commander for the operations, despite being a sceptic of Sledgehammer due to the paucity of landing craft. On 15 June, Mountbatten dined with the President and Harry Hopkins at the White House, but crucially, and to their chagrin, Marshall and King were not present. In post-war interviews, Marshall spoke of how Mountbatten used to go to him behind the backs of the British Chiefs of Staff, telling him privately how he supported various American stances that Brooke opposed. This time, however, he would very effectively undermine Marshall himself.

Mountbatten told his hosts frankly that, since the Germans had twenty-five divisions in France, however hard-pressed Russia might be she would not be aided by a massacre of Allied troops there. Roosevelt replied that he did not want to accumulate one million American troops in Britain if all they were going to do was form a home guard while British troops fought in the Middle East and India.[24] Shoring up the British Empire was not the proper role of the US Army, he implied. Nonetheless, Mountbatten could spot a chink between Roosevelt's and Marshall's views on future operations, and he encouraged Churchill and Brooke to come over to try to widen it.

'Lord Mountbatten was closeted with Roosevelt for five hours,' noted a highly suspicious Wedemeyer. 'I understand that no American officer was present . . . Now we had an extremely articulate Britisher endeavouring to raise bogeys about the hazards of a cross-Channel operation.'[25]

Although Wedemeyer was chronically suspicious of British motives on most occasions – he joked that he would never meet a British officer without a witness being present – he was right to be on this occasion.

Mountbatten's report of his White House meeting, which according to his aide Commodore John Hughes-Hallett had in fact lasted six hours, indicated that Roosevelt had agreed with him that there were not enough landing craft to mount a Sledgehammer sufficiently powerful to force Germany to take pressure off Russia. Yet the President still wanted the British to make some sort of landing, in order to damage German morale. Furthermore, he was 'resolutely opposed to sending a million men to England on the off-chance that Roundup' would be launched. Roosevelt wanted absolute guarantees of an attack by 1 April 1943, but Mountbatten spotted that in Roundup's absence the President was much more sympathetic than his advisers to the idea of landings in North Africa.[26] Mountbatten flew back via Montreal and went straight to Chequers, where he was kept up until 4 a.m. reporting to Churchill. He then made it to the Chiefs of Staff meeting at 9.30 a.m., and worked a full day until 9.30 p.m.

Here was the news for which Churchill had been waiting and hoping. Alerted that there might be a chance of splitting Roosevelt off from Marshall over Sledgehammer and an early Roundup, and guiding him instead towards North Africa – where Rommel was once again on the offensive – Churchill leapt. 'In view of the impossibility of dealing by correspondence with all the many difficult points outstanding,' he telegraphed the President, 'I feel it is my duty to come to see you. I shall hold myself ready to start as weather serves from Thursday 18th onwards, and will advise you later. I shall bring the CIGS, General Brooke, whom you have not yet met, with me, also General Ismay.'

The need for an American attack in North Africa could not have been more pressing. On 10 June Churchill had held a two-hour Staff Conference on Middle Eastern strategy, saying that he doubted Auchinleck's offensive spirit – 'I don't know what we can do for that Army, all our efforts to help them seem to be in vain' – and complaining that the one hundred thousand Commonwealth soldiers defending Egypt 'all come up for their rations but not to fight'. These were terrible libels on the British Army, and Brooke regarded it as his duty to defend his men. Just before he went to bed, he told his friend and aide-de-camp Barney

Charlesworth, who shared his flat in London, 'Well, that is one of the bloodiest days I have had for a long time.'[27]

Three days later, having lost 230 tanks, British forces withdrew to the Sollum–Sidi Omar Line near the Libyan frontier with Egypt. Churchill cabled Auchinleck to say 'he presumed there was no question in any case of giving up Tobruk', for, 'as long as Tobruk was held, no serious enemy advance into Egypt was possible'. Auchinleck replied the following day that he had no intention whatever of doing so. Taking time off from weeding and hoeing the long herbaceous border at his country home, Thatched Cottage, near Northam in Sussex, Alec Cadogan wrote of Rommel's victories: 'I suppose he's a very good general, but I am quite convinced that our own (including CIGS) are blockheads, who cannot learn anything.' It was a rare criticism of Brooke, and probably prompted by frustration at the lack of success in the Western Desert rather than by genuine doubts about his intelligence.

June 1942 was the worst month of the war so far for Allied shipping losses, with 83 per cent of the tonnage lost through U-boat action, and 60 per cent in the Caribbean and Gulf of Mexico. Marshall wrote to King in mid-June, 'The losses by submarines off our Atlantic seaboard and in the Caribbean now threaten our entire war effort,' pointing out that 22 per cent of the bauxite fleet and 3.5 per cent of all tanker tonnage had been lost in that month alone. King replied two days later that the US had little in the way of anti-submarine forces and that convoys would be used as 'the only way that gives any promise of success'. It was far too late in the conflict for him to be discovering that, given how valuable convoys had been in the Great War.

Pound meanwhile had to announce to the War Cabinet that he was not very hopeful of a convoy being able to reach Malta at all. That island had been heavily bombed by Axis planes based in Sicily, yet holding it was vital in order to interdict resupply of the Afrika Korps from Italy. Churchill answered that this was 'Grave; have another shot. Lie [to the press]; say all important stores got in.' In the previous week alone thirty-six ships had been sunk, with the loss of 225,000 tons in eight days, and the losses were getting more widespread. Brooke then went through the 'two distinct phases' of the battle then being fought at Bir Hacheim and Knightsbridge, which had forced the British withdrawal. Churchill said this did not strengthen 'the case for passive defence'. He expressed anxiety about Auchinleck's lack of reserves and said that the Eighth Army seemed to have been 'outmanoeuvred and outfought' and had 'lost the battlefield', estimating Rommel as having only 150 tanks.

Churchill then went into the history of the training of the 10th Armoured Division and instructed the CIGS 'to look into counter offensive at first possible date', accepting that there had been 'Not a lot of change in the actual positions. Total advance of Germans has not been very great after three weeks fighting.' Many of these criticisms accorded with Brooke's own view, and it is hard to escape the conclusion that, even at these great crisis moments of the war, Brooke was confining his fury to his diary rather than openly rowing with the Prime Minister.

In response to the Lidice massacre in Czechoslovakia – where the Germans had shot 181 innocent villagers and taken away forty-nine women to be gassed at Ravensbruck, in revenge for the assassination of Reinhard Heydrich, acting Reich Protector of Bohemia and Moravia – Churchill reported a conversation he had had with the Czech leader, Eduard Beneš, 'about the possibility of reprisals for the savage cruelties now being practised by Germans in Czechoslovakia'. Churchill 'Suggested wiping out German villages (three for one) by air attack', and proposed that one hundred bombers would be required to drop incendiaries from low levels in bright moonlight on three unprotected German villages, with the reason announced afterwards. If it was 'thought worthwhile' by the Cabinet, Churchill said he would give the RAF discretion to carry out such a raid 'to fit it in when they can'. As might be expected, this prompted a lively debate.

Sir Archibald Sinclair, the Secretary for Air, thought it a diversion from proper military objectives, as well as an unnecessary risk to RAF planes and crews. Attlee doubted whether 'it is useful to enter into competition in frightfulness with the Germans'. The Home Secretary, Herbert Morrison, considered the likelihood of reprisals on English villages which had no air-raid sirens or shelters. He feared the public would say: 'Why have you drawn this down on us?' Eden approved of 'the deterrent element' and Bevin argued that 'Germany responds to brute force and nothing else,' whereas Stanley Bruce, the Australian High Commissioner, feared that it might lead to even greater atrocities in Czechoslovakia. The Lord President of the Council, John Anderson, was opposed because 'It costs us something and them nothing.' Brendan Bracken and the Colonial Secretary Lord Cranborne were also opposed.

'My instinct is strongly the other way,' said Churchill. The Secretary for India, Leo Amery, asked: 'Why a village? Why not a residential town?', but the Lord Privy Seal, Sir Stafford Cripps, said that the operational argument against such a reprisal raid was very strong. 'I submit (unwillingly) to the view if the Cabinet are against,' concluded Churchill.[28]

After a report from Brooke on the Levant–Caspian front, where Churchill said 'We're not very strong – we've only two men and a boy there, and Libya has lost us some face again,' the discussion got round to public morale. Churchill told the Cabinet that it was 'Likely the news for next months will cure undue optimism,' and there would be 'No practical change on the Second Front which newspapers think so easy'. Bracken reported that a Second Front was being demanded at public meetings all over the country. 'Can't be helped,' replied Churchill, 'but ministers should be careful – the less said the better.' As for the demonstrations themselves, he felt that the 'Important thing is to make clear that close relations with Russia does *not* involve coddling our Communists.'[29]

In Libya, meanwhile, the South African brigades had retreated from Gazala into Tobruk. The 50th Division were stranded behind the German lines and feared lost, orders for their withdrawal having been sent too late. 'CIGS tells me however that Winston thinks this movement a piece of fine generalship,' wrote Kennedy, whose own view was that it was 'a piece of bad bungling', not least because the Germans were claiming to have captured twelve thousand prisoners. Kennedy put the episode down to bad British and good German generalship, the inferior quality of British to German tanks in both manoeuvrability and firepower (British tanks were mostly armed with a 2-pounder gun against the German 4½-pounder). 'Winston has been especially wrongheaded throughout this battle,' opined Kennedy. 'There has been a stream of telegrams from him, practically none of which should have been sent by a PM. These telegrams have not been the result of calmly considered advice from the Chiefs of Staff.' It was true that Churchill had deluged Auchinleck with questions, such as that of 14 June, 'To what position does Ritchie want to withdraw the Gazala troops?', and exhortations such as 'This is a business not only of armour but of will-power,' but these hardly affected the outcome of the struggle.

The usual procedure was for Churchill to draft telegrams and then get agreement for them over the telephone, which Kennedy considered

1. The Masters and Commanders at the Casablanca Conference, January 1943

Back row from left to right: General Brehon Somervell, General Henry 'Hap' Arnold, Admiral Ernest J. King, Lieutenant-General Sir Hastings 'Pug' Ismay, General George C. Marshall, Admiral Sir Dudley Pound, General Sir Alan Brooke, Air Chief Marshal Sir Charles Portal, Rear-Admiral Lord Louis Mountbatten. Front row: President Franklin D. Roosevelt, Winston Churchill

2. General John J. 'Black Jack' Pershing and his aide-de-camp, Colonel George C. Marshall, in France in 1919

3. Alan Brooke in the uniform of a lieutenant of the Royal Horse Artillery, 1910

4. Without entourage or even a detective, Winston Churchill arrives at Downing Street for a Cabinet meeting on 15 May 1940, two days after his 'blood, toil, tears and sweat' speech

5. President Roosevelt telling Congress about the attack on Pearl Harbor the previous day, 7 December 1941, which he memorably dubbed 'a date which will live in infamy'

6. Churchill and Roosevelt meet for the first time since 1918, on board the USS *Augusta* on 9 August 1941, during the top secret Riviera Conference. Roosevelt is receiving a letter from King George VI. From left to right: Averell Harriman (right hand in pocket, smiling), King (right hand in pocket, head turned), Churchill, unidentified naval officer, Franklin Roosevelt Jr, Sumner Welles, Captain John R. Beardall USN, President Roosevelt and Captain Elliot Roosevelt.

7. 'The tremendous hold the Limeys have on Our Boy': Churchill making a point, which Roosevelt is enjoying, on board HMS *Prince of Wales* at Placentia Bay on 14 August 1941

8. Marshall, Churchill and US Secretary of War Henry L. Stimson watch a mass demonstration of paratroopers at Fort Jackson, South Carolina, on 24 June 1942

9. Alan Brooke's lunch for Marshall at the Savoy Hotel in July 1942. From left to right: Portal, Pound, Marshall and Brooke

10. Harry Hopkins, General Mark Clark, President Roosevelt and General Dwight D. Eisenhower eat lunch from mess kits after inspecting American troops in North Africa on 31 January 1943

11. Eisenhower and Marshall meet in Algiers, 3 June 1943

12. Churchill, recuperating from pneumonia in his blue and gold bedragonned dressing gown on Christmas Day 1943 in Carthage. On the steps behind him, Eisenhower and General Sir Henry 'Jumbo' Maitland Wilson, are, from left to right, Air Marshal Sir Arthur Tedder, unidentified soldier, Admiral Sir John Cunningham, General Sir Harold Alexander, Major-General Sir Humfrey Gale, Brigadier Leslie 'Joe' Hollis and General Walter Bedell Smith

13. Generals George S. Patton (note the pearl-handled revolver), Omar Bradley and Bernard Montgomery smiling together in France. Their intense rivalry suggests the photograph was highly posed

14. The Combined Chiefs of Staff at Casablanca, January 1943. Note Brooke's jabbing finger. Brigadier Vivian 'Dumbie' Dykes, the dark-haired moustachioed man near the top left of the photograph, has only a few more days to live

'dangerous', because 'although he appears to be covered in each step by professional advice, the general trend of strategy is too much influenced by his personality.' If the battle went well, the directives and orders he constantly issued 'would give the impression that he had directed all operations successfully as a sort of Commander-in-Chief'. Yet if they went awry, 'he as PM is bound to be in a very bad position.'

Churchill's plans for the reconquest of Burma Kennedy thought 'quite impractical because the resources . . . are not even in sight. But he bores away at it.' Equally, Operation Jupiter was 'impractical from the naval, air and army points of view . . . out of the question. Yet Winston has been pressing it and wasting our time on it for several weeks.' In Kennedy's view the most dangerous matter of all was Churchill's pressure on Auchinleck to hold Tobruk under any circumstances. He spoke to Brooke about how, back in April 1941, Wavell had taken the decision to defend Tobruk, and that it had come right owing to the unexpected withdrawal of German forces – especially Luftwaffe – to the Russian front. Yet now he felt that for all of Tobruk's obvious importance to the defence of Egypt it could only be held at too high a cost, especially to the Navy, in view of its long desert perimeter. 'If we are to lose it,' he advised the CIGS, 'it is better to lose it without throwing away a big garrison at the same time.' Yet he feared that its political value would weigh too heavily with Churchill to permit evacuation.

Only in one area would Kennedy give Churchill much credit, and this was over Roundup and Sledgehammer, where he said that 'Winston was good on this in his talks with Marshall. We have got it established that nothing big is to happen this year unless the Germans begin to crack. The Americans are still in confusion and the Marshall plan for moving an enormous army over here is being disputed by the other Services in the USA. This will be the next big thing to clear up.'[30] With the news Mountbatten had brought back post-haste from the White House, that was exactly what was about to happen.

Just before leaving for America, Brooke took the day off to go to the Farne Islands off Northumberland to photograph seabirds, but his dinghy overturned alongside the naval launch and his £250 camera was damaged by seawater. 'Although his day was spoilt he could laugh about it,' recorded Kennedy. Warned that it was very hot in Washington in June, Brooke managed to get a lightweight military suit and white denim jacket made in twenty-four hours, although there was no time for his impressive rows of medal ribbons to be sewn on to the suit, because at 11.30 p.m. on Wednesday 17 June he and Churchill flew from Stranraer to Washington in a BOAC flying-boat called the *Bristol*. Brooke later told Kennedy, who had been to school at Stranraer, that Churchill had walked along the pier there singing the Great War ditty 'We're Here Because We're Here'. According to Ismay, 'Our Boeing flying-boat was the acme of comfort – plenty of room, full-length bunks, easy chairs, and delicious food ... We were in the air for twenty-six hours at a stretch, but the time passed quickly and pleasantly.'

At 4.30 p.m. US time, still four hours from Washington, Churchill asked Commander Thompson for dinner to be served. On being told of the time difference, and that he was going to have dinner at the British Embassy that night, he replied: 'I go by tummy time, and I want my dinner.' He put his personal method of avoiding jet-lag more delicately in his war memoirs, stating: 'I adhered to my rule in these flights that meals should be regulated by stomach time. When one wakes up after daylight one should breakfast; five hours after that, luncheon; six hours after luncheon, dinner. Thus one becomes independent of the sun, which otherwise meddles too much in one's affairs and upsets the routine of work.'

Before they left they heard from Auchinleck that he intended to hold Tobruk, covered by fighters from the port of Sollum in Egypt, which pleased Churchill greatly. He also had time to write to the King, saying, 'Sir, In case of my death on this journey I am about to undertake, I avail myself of your Majesty's gracious permission to advise you that you should entrust the formation of a new Government to Mr Anthony Eden.'[31]

While the British party was in the air, a shocked Henry Stimson attended a meeting at the White House at which Roosevelt reopened the whole issue of cross-Channel operations. 'The President sprung on us a proposition which worries me very much,' the Secretary of War wrote afterwards. 'It looked as if he was going to jump the traces after all that we had been doing in regard to Bolero and to imperil our strategy of the whole situation. He wants to take up the case of Gymnast again.' Stimson hoped that the reason he was doing this was 'in his foxy way to forestall trouble that is now on the ocean coming towards us in the shape of a new British visitor', but he could not be sure.

The President's new démarche in favour of Gymnast met with 'robust opposition' from his advisers. Marshall had already prepared a paper against it, as he had 'a premonition of what was coming'. Stimson also 'spoke very vigorously'.[32] During what the War Secretary called 'a disappointing afternoon', the Navy were nothing like so vigorous. Marshall could begin to see he was about to find himself in a minority of one, unless he was able to bring Brooke over to his point of view.

Marshall's paper stated that 'The Gymnast Operation has been re-examined on the assumption of full French co-operation, utilizing British shipping, and the use of US troops only.' Among many problems he identified were the lack of aircraft carriers and naval escorts for the convoys in the Atlantic, delays over reinforcements, and the late availability of US anti-aircraft units (ready in October) and air forces (December). 'The operation is not possible unless Bolero is abandoned,' he concluded, 'or at best indefinitely postponed.'

Furthermore, since the Gymnast operation involved occupying North-west Africa with 220,000 US troops, comprising six divisions and air squadrons protected by the Navy, it was extremely worrying that Casablanca harbour could only accommodate a maximum of forty-three vessels, with no more than twelve capable of being unloaded at any one time. Marshall estimated it would take a month to disembark forty thousand troops with all their equipment and supplies, and so to land the entire force 'would consume at least six months', while of course the enemy would try to 'bomb Casablanca during the early phases'. Furthermore, the use of scarce British shipping 'would seriously affect' the flow of vital reinforcements to the Middle East and to India, and so in the event of a Russian collapse 'the resulting threat to the Middle East would require a maximum of British shipping to block a probable Axis advance into that theater'.

Marshall also believed that, although French co-operation and Spanish neutrality were 'essential', nothing suggested that either would be forthcoming. With so little in its favour, he seemed almost to be understating his case when he concluded: 'It appears that this operation should not be undertaken in the present situation.'[33] Stimson also drew up a strong defence of the Sledgehammer operation for Roosevelt – who had by then gone to Hyde Park – which Marshall copied to Arnold, McNarney and Eisenhower.

At 8.15 p.m. local time on Thursday 18 June, the flying-boat *Bristol* landed at the Anacostia naval airstation in Washington. The party were met at the airport by Halifax, Dill and Marshall and driven to the Lutyens-designed British Embassy on Massachusetts Avenue where they sat down to a second dinner. Brooke later observed of Churchill: 'As I had to share every one of these meals with him and they were all washed down with champagne and brandy, it became a little trying on the constitution.'[34]

When Churchill discovered that Roosevelt was at Hyde Park rather than Washington, he was 'rather put out at the President being away, and inclined to be annoyed that he hadn't been diverted to New York'. Halifax noted that 'He got into a better temper when he had some champagne.' The party was kept up till 1.30 a.m., with Lady Halifax 'much impressed' by the way that Brooke, Ismay, Moran and John Martin 'all took advantage of the darkness on the porch to snatch bits of sleep while Winston talked'.[35]

The Second Washington talks, codenamed Argonaut, started on the morning of Friday 19 June, with Marshall conferring with Churchill at the Embassy. The Prime Minister then flew up to Hyde Park to stay with Roosevelt for the weekend, as Marshall telegraphed the President warning that 'your guest . . . is pessimistic regarding Bolero and interested in August Gymnast and another similar movement in Norway.' This forewarned FDR, although the destination of the second operation should certainly not have been mentioned *en clair*. Marshall meanwhile warned Stimson that Churchill 'was full of discouragement and new proposals for diversions. Therefore the importance of a firm and united stand on our part is very important.' It was not to be.

Roosevelt's great-grandfather had moved to the Hyde Park area in 1818, and half a century later James Roosevelt – the President's father and seventh in a line of Roosevelts who were prominent in New York City – bought the Springwood estate there. They were not ostentatious, and James lived the equine, estate-management and hunting-based life of an English country squire. Franklin Roosevelt had been born at Springwood and had spent his boyhood there, privately tutored until the age of fourteen, before attending Groton, Harvard and (briefly) Columbia Law School.

Although it had been built as a farmhouse around 1800, FDR turned Springwood into a country house in 1915 by adding two substantial wings and extensively redesigning the rest, largely using his mother's money. He adored the place, and returned there no fewer than two hundred times during his presidency. The house is still kept as it was during his last stay in March 1945, complete with the telephone in his bedroom that used to be connected straight through to the White House.

Because it was unchanged from his death, and opened to the public a year later, it is possible today to view the rooms where the two men planned the outlines of Allied grand strategy very much as they were then. Churchill would doubtless have felt at home seeing the large collection of seafaring prints on the hallway walls, at least until he noticed that they mostly commemorate the War of 1812. Elsewhere in the hallway a 1783 Isaac Cruikshank cartoon illustrates *British Valour and Yankee Boasting* and another *The Fall of Washington 1814*.

A beautiful panelled library extends the entire southern width of the house, with a large marble fireplace flanked by two high-backed chairs commemorating Roosevelt's gubernatorial term in New York. Churchill stayed in the Pink Bedroom immediately on the right of the top of the staircase, one of many guest bedrooms in the thirty-five-room house. At one point on this trip, driving Churchill around the estate in his specially adapted Ford, FDR – who was 6 foot 3 inches tall and weighed 190 pounds – invited Churchill to feel his biceps, which were tremendously powerful.

The two men had much to discuss besides the cross-Channel versus North African operations, and reached full agreement for complete co-operation over the project to build and deploy the atomic bomb. While at Hyde Park, Roosevelt received a ten-point letter from Stimson, which had the unanimous endorsement of Marshall and his Staff, and which argued vigorously that an early Second Front in Europe was the best way of 'keeping the Russian Army in the war and thus ultimately defeating Hitler'. Despite the Axis powers controlling every other 'feasible landing spot' on the Continent, 'By fortunate coincidence one of the shortest routes to Europe from America led through the only safe base not yet controlled by our enemies, the British Isles.' Hence Bolero, which Stimson described as 'an essentially American project, brought into the war as the vitalizing contribution of our fresh and unwearied leaders and forces'. He listed its advantages as not requiring carrier-based air cover, and as being launched from a site that did not need time spent on its development and fortification and where 'we could safely develop air superiority'. Furthermore, 'Geographically and historically, Bolero was the easiest road to the centre of our chief enemy's heart . . . Over the Low Countries has run the historic path of armies between Germany and France.'[36]

Stimson argued that the sinking of four Japanese aircraft carriers at the battle of Midway earlier that month had alleviated the danger of raids on American aircraft factories, and (less credibly) that unrest in Occupied Europe was increasing. He believed that a German victory over Russia might lead to a German invasion of Britain, when it would be 'imperative for us to push our forces into Britain at top speed', which would be impossible if American shipping was 'tied up with an expedition to Gymnast'. Moreover, Gymnast's need for large-scale aircraft-carrier support 'could not fail to diminish the superiority over Japan which we now precariously hold in the Pacific'.

As for the consideration of 'other plans' – by which Stimson meant Gymnast – 'When one is engaged in a tug of war, it is highly risky to spit on one's hands even for the purpose of getting a better grip.' The tug of war he was referring to was against Britain, not Germany, as was clear from the next sentence: 'No new plan should even be whispered to friend or enemy,' which through a process of elimination meant Britain, since Roosevelt would hardly whisper American plans to the Nazis. Stimson ended by saying that Gymnast would detract from Bolero, and nothing should be allowed to do that.[37] It was a formidable indictment, and it is easy to detect Marshall's hand in helping to draw it up, not least because many of its points precisely mirrored those from Marshall's own note to the President on the same subject.

Back in Washington on 19 June, Brooke and Marshall met in Room 240 of the Combined Chiefs of Staff Building at 12.30 p.m., along with a large gathering of British and American senior servicemen, including Vice-Admiral Horne (representing Admiral King), Arnold, Ismay, Dill, McNarney, Eisenhower, Clark, Rear-Admiral Charles 'Savvy' Cooke, Colonel Jacob, Lieutenant-General G. N. Macready and Dykes. It was the first time that Brooke had met Arnold, whom Ismay described as 'a veteran airman, with an unlimited belief in air power'. He certainly was a veteran, having first flown with the Wright brothers themselves. Jacob found Arnold a 'cherubic little man, with white hair and humorous blue eyes. Quick on the uptake and rather impatient.' Wedemeyer thought he had 'a good understanding of human beings and the use of air power.

I think he was a great administrator . . . self-effacing, never a strutter.'[38] Marshall was fortunate to have Arnold as his right-hand man on the Joint Chiefs of Staff.

Brooke reviewed the strategic situation, and told the Committee that 'the object of the PM's visit was to clear up a number of points which had been mentioned by the President in a conversation with Lord Mountbatten'. These all related to grand strategy for 1942 and early 1943. He then touched upon some broad areas, indicating the Chiefs of Staff's views on each. Sometimes minutes of Staff meetings, especially those written with an eye to history, or inter-departmental struggles, or minimizing inter-Allied disagreement, or anything other than a strict record of what took place, can be as difficult to decipher as the hieroglyphics of Linear B. Yet at this meeting Brooke rattled through the strategic objectives straightforwardly, which were:

The importance of employing in an active theatre the United States forces which were being sent to England, and in this connection the fact that no operation in France might be possible in the event of a Russian collapse, permitting large German reinforcements to return to Western Europe.

The difficulties which would arise, in the event of the Russians being hard pressed, in establishing a Second Front in Western Europe in 1942 in accordance with our promises.

The possible establishment in late 1942 or the winter of 1942/3 of a bridgehead in the neighbourhood of Cherbourg or the Brest salient as a base for 1943 operations.

The possibility of carrying out some kind of Gymnast operation in 1942.

The undertaking of offensive operations based on Australia against the Japanese.

Brooke stated that 'The crux of the matter was the degree of reliance we could place on the Russian front holding,' about which he was fairly confident, since 'the Russians' showing, both at Sevastopol and in the Kharkov area, was encouraging.' If the Russians held, he argued, 'our chances of a successful offensive on the Continent in 1943 were good, and the Middle East situation would be relieved, as there would be no German threat to the oil fields and the Persian Gulf and therefore our scale of reinforcements could be cut down'.

If the establishment of a Western Front was impossible in France in 1942, Brooke continued, 'then some form of Gymnast should be considered and forces now set up for Bolero might be used.' Consideration needed to be given to training, shipping and whether it 'was to be undertaken entirely by the United States forces or on a combined basis', which would be largely dependent on whether the Vichy French resisted. Unlike Marshall, Brooke believed it improbable that there would be a 'rapid arrival of German forces through Spain'.

Brooke then went through the four options that the British Chiefs of Staff had been considering 'aimed at relieving pressure on the Russians'. These were, 'A landing in the Pas de Calais Area', 'Establishment of a Bridgehead at Cherbourg or the Brest Salient', 'Large Raids' and 'Operations in Northern Norway'. The first, even if undertaken by six divisions, Brooke believed would be 'unlikely to achieve important results', such as the diversion of significant German forces from the Eastern Front. The Cherbourg operation would require fifteen divisions, but was 'worth further careful study'. Large raids were also being considered actively, even those intended to last two or three days.

As for Norway, the opportunity to free the northern convoy route of German aircraft was highly attractive, but it would mean keeping four-and-a-half divisions supplied north of Narvik. If it was going to be done, it had to be undertaken immediately because the most dangerous period was during the extremely short Arctic summer nights when the Luftwaffe could operate virtually around the clock. Brooke was similarly non-committal about Burma, saying that since five or six Japanese divisions were within easy reach of Rangoon, a seaborne attack there could not be undertaken 'except simultaneously with some other offensive against the Japanese', by which he meant ones undertaken in conjunction with the Americans.[39]

Brooke's exposition was short – all over by lunchtime – but it set the agenda for the rest of the Argonaut discussions. Nothing was ruled out, and Gymnast was on the long list of possible operations. In the revised version of the minutes he inserted two other aspects that had not been in the originals, either because he had not in fact mentioned them at the meeting or because of stenographical error, most probably the former. These were: 'The possibility of economizing shipping by dispatching substantial US forces direct to the Middle East rather than by reinforcing the Middle East by British forces from the UK', and the fact that 'It would be necessary during the present visit to give urgent consideration to the question of command arrangements for a Continental offensive' (that is, of who was going to command Roundup).

Dykes privately recorded the American reaction when Brooke said that the President was 'worried because US troops would not be engaged with the enemy on any scale this year'. Because this effectively meant that 'Bolero had dropped back in the batting,' Dykes could see that Marshall and Eisenhower 'were considerably perturbed by this'.[40] After lunch, Marshall, Brooke, Dill, Ismay, Eisenhower and some other members of the US War Department sat down at 2 o'clock for an 'informal meeting' in Marshall's office, one which was nonetheless minuted. The discussions, in the catch-all phrase, 'generally revolved around the basic reasons that led initially to the adoption of Bolero as the principal effort of the United Nations and the possibilities of conducting an offensive operation either in Western Europe or North-west Africa during 1942 as a means of assisting Russia'.[41]

Brooke and Marshall seem to have agreed at that meeting that, although Bolero 'should constitute the basis of our future strategy', nonetheless logistical factors precluded its being mounted before the spring of 1943. Any other offensive operations undertaken in 1942 must 'not materially delay the date at which Bolero can be mounted', they agreed, and 'should contribute directly to the success of Bolero'. Both men used Bolero as shorthand for Roundup and Sledgehammer, which stemmed from it.

As for Gymnast in 1942, Marshall and Brooke agreed that it 'would seriously curtail reinforcements in the Middle East with possibly disastrous consequences in that theatre', that it would 'thin out' naval concentrations in other theatres, especially aircraft carriers and escort vessels, that it was impossible to predict the various psychological factors pertaining in North Africa that would be crucial to its success (that is, the French reaction), that it would have a 'marked effect in slowing up Bolero', and that it would generally 'tend to disperse further our available resources and weaken our effort'. Pressing his theme, Marshall had the conclusions minuted that 'It was the considered opinion of the conferees (a) That Gymnast should not be undertaken under the existing situation. (b) That United States and Great Britain should adhere firmly to the basic decision to push Bolero with all possible speed and energy.' From this is it clear that Brooke was almost as sceptical about Gymnast as Marshall, however much Churchill liked the idea.

Brooke did agree that any 1942 operation – that is, including Gymnast – 'should be undertaken only in case of necessity or if an exceptionally favourable opportunity presented itself'. Anyone present at that meeting on 19 June 1942 would have been forgiven for assuming that there would be no Operation Gymnast at all, but a cross-Channel invasion some time around May 1943. Yet in fact American troops were to go ashore in North Africa that very November, and Operation Bolero – by then conflated with Roundup and renamed Operation Overlord – would not take place until June 1944. Brooke's principal objective for Argonaut was to scotch Sledgehammer and a pre-1943 Roundup, as well as the northern Norway operation, whereas Churchill's principal objective was to secure Roosevelt's support for Gymnast.

Brooke concluded, after the two Combined Chiefs of Staff meetings on 19 June, that 'we made further progress towards defining our policy for 1942 and 1943. Found that we were pretty well of the same accord as to our outlook.' That was a fair summation, but although Marshall and Brooke were agreeing in Washington, so too were Roosevelt and Churchill in Hyde Park – to something quite different. After dining with Dill at his house that night, Brooke noted: 'On the whole made fine progress today, but am a little doubtful as to what PM and President may be brewing up together.'[42]

After the meetings that day, Eisenhower composed a memorandum that neatly encapsulated American strategic thinking about how to win the war against Germany. This read suspiciously like the letter Stimson had written to the President which Marshall had already copied on to Eisenhower. 'To defeat Germany we must operate against her with overwhelming superiority,' it began in classic Clausewitzian style. 'For logistic reasons North-west Europe is the only front on which this superiority can be achieved.' Eisenhower assumed that it would be clear by September 1942 whether Russia would 'crack' or not, and Bolero covered both eventualities. In the event of a Russian collapse, Eisenhower accepted that a cross-Channel attack in 1943 could not take place and 'an alternative front would have to be found,' but overall he concluded that 'Continental operations on a large scale at the earliest possible moment should be the principal offensive effort of the United Nations.'[43]

The memorandum was discussed at a meeting of the Combined Chiefs of Staff at 11 a.m. the next day, Saturday 20 June, at which Brooke met Admiral King for the first time. Brooke – who sweltered in the intense heat as he had put on the wrong uniform – said 'that he had been much encouraged to find that there was complete unanimity of opinion between the US and British Staffs on general strategic policy and the merits of the Bolero plan as a whole'. This was slightly disingenuous, since it was not Bolero but Sledgehammer that had been the bone of contention between them, and was another example of codeword nomenclature being used with advantages.[44]

Ernest King then bluntly stated that he was 'entirely opposed to any idea of carrying out Gymnast in 1942. An entry into North-West Africa would open a ninth front with all the increase in overheads and escort and transportation problems involved therein.' He did not want naval forces withdrawn from the Pacific even for Bolero, let alone for Gymnast. 'He was tough as nails and carried himself stiffly as a poker,' recalled Ismay of King. 'He was blunt and stand-offish, almost to the point of rudeness. At the start he was intolerant and suspicious of all things British, especially the Royal Navy; but he was almost equally intolerant and suspicious of the American Army.' King resented any American resources being used for any other purpose than to attack Japan, and – like Marshall, Leahy and Wedemeyer – he deeply mistrusted Churchill's powers of advocacy and was apprehensive that the Prime Minister was persuading the President to neglect the war in the Pacific.

As a fellow Anglophobe, Wedemeyer predictably admired Ernest King, declaring, 'I thought he was the strongest man on our Joint Chiefs of Staff.' Yet even he described the admiral as 'rather cold – not an attractive man.' He added:

They say in the Navy he kept a tight ship and didn't have many good friends although he was highly respected. I would say that he protected America's interests more than any other member of the Joint Chiefs of Staff. People say that he made the remark, 'What's good for the American Navy is good for the United States.' He was always thinking in terms of conserving US men and materials, and he resisted the British, who were trying, always, to get everything they possibly could from us.[45]

Brooke also met Henry Stimson for the first time on this occasion, characteristically noting that he had 'a limited strategic outlook. He was one of the strong adherents of breaking our heads in too early operations across the Channel. Consequently a strong supporter of Marshall.'

At the 20 June meeting, Marshall added that since a commitment to a cross-Channel invasion in 1943 would concentrate all efforts, any change of plan, by which he meant the adoption of Gymnast, would achieve nothing. 'To defeat the Germans we must have overwhelming power,' he said – paraphrasing the Eisenhower memorandum – 'and North-west Europe was the only front on which this overwhelming superiority was logistically possible. It was, therefore, sound strategy to concentrate on this front and divert minimum forces only to the other fronts. From the military point of view, therefore, there seemed no other logical course than to drive through with the Bolero plan.'⁴⁶ This too was pure Clausewitzian military thought, the concentration of maximum forces on the most important point of the battlefield. It was what Napoleon, Grant and the elder Moltke did whenever possible, and it was what was taught at the US military academies. In arguing for this strategy, Marshall had history as well as classic strategic teaching on his side. But did he have his commander-in-chief?

At lunch at Dodona on the 20th, Marshall and Brooke speculated about what Roosevelt and Churchill, in Brooke's words, 'were brewing up together at Hyde Park ... We fear the worst, and are certain that North Africa and North Norway plans for 1942 will loom large in their proposals, whilst we are convinced that they are not possible!' The 'We' in this sentence definitely includes Marshall and possibly King too.⁴⁷ They were right to worry, for just as the two military Commanders were damning Gymnast in Washington, Churchill was handing Roosevelt a note at Hyde Park that was to alter the whole course of the war.

8

The Masters at Argonaut:
'Please make it before Election Day'
June 1942

Mr Churchill was sure that only by the premature invasion of France could the war be lost. To postpone the evil day, all his arts, all his eloquence, all his great experience were spent.

Lord Moran, *Winston Churchill: The Struggle for Survival*[1]

'We are bound to persevere in the preparation for Bolero if possible in 1942 but certainly in 1943,' conceded Churchill in his 20 June memorandum to his host, Franklin Roosevelt, at Hyde Park. 'The whole of this business is now going on.' He stated that arrangements for Sledgehammer – 'a landing of six or eight Divisions on the coast of Northern France' – were under way, but the British Government were firmly opposed to any operation 'that was certain to lead to disaster for this would not help the Russians whatever their plight'. Furthermore, it would 'compromise and expose to Nazi vengeance the French population involved' and would also seriously delay Roundup. 'We strongly hold to the view that there should be no substantial landing in France this year unless we are going to stay,' he wrote, adding in a phrase that he was to use often in the future: 'No responsible British military authority has so far been able to make a plan for September 1942 which had any chance of success unless the Germans became utterly demoralized, of which there is no likelihood.'[2]

Quite why Churchill felt the need to write an official memorandum to the man whose bedroom was only a few yards up the corridor from his, and whom he was with – often alone – throughout the weekend, is hard to explain, except that he always had an eye to history and found that his brain worked especially well through the written word. Roosevelt would also need it in order to convince Marshall and the other service Chiefs. Whatever the reason, he then posed a barrage of questions that Marshall had to answer supposing the Germans were in no way demoralized by 1943. 'Have the American Staffs a plan?' he asked. 'If so, what is it? What forces could be employed? At what points would they strike? What landing craft and shipping are available? Who is the officer prepared to command the enterprise? What British forces and assistance are required?'³ These were not rhetorical, although they sounded it, and of course Marshall was not present to answer them.

If any of those questions could not be answered satisfactorily, Churchill had more: 'What else are we going to do? Can we afford to stand idle in the Atlantic theatre during the whole of 1942? Ought we not to be preparing within the general structure of Bolero some other operation by which we may gain positions of advantage and also directly or indirectly to take some of the weight off Russia?' Then came his own answers to these questions: 'It is in this setting and on this background that the Operation Gymnast should be studied.' Churchill was straining credibility in putting Gymnast 'within the general structure of Bolero', because it necessitated the severe postponement of Sledgehammer and Roundup, yet he understood that it was necessary to present it that way after his enthusiastic reception of Bolero in London only two months before.

So, just as Marshall, King and Eisenhower were trying to consign Gymnast to a strategic, logistical and even 'logical' grave in Washington – hardly resisted by an almost equally sceptical Brooke – Churchill resurrected it at Hyde Park. In getting Roosevelt on his own there, Churchill had a considerable advantage, as the 'amateur strategist' President tended, at least at this stage in the war, to defer to him on military matters in a way that he would not have done had Marshall been present.

'I must emphasize', admitted Wedemeyer after the war,

that President Roosevelt did not have the knowledge, the military knowledge, the strategic knowledge that Prime Minister Churchill did. He wasn't in close proximity with the leaders of our Army and Navy. Churchill worked with his [Chiefs of Staff] and Planners down in his offices and Map Rooms ... He supported his chiefs on their negotiations. I am sure it wasn't a question of supporting General Marshall and the Joint Chiefs of Staff with Roosevelt. He really didn't know enough about it and didn't pretend to.⁴

Although Churchill's visit to Washington was public, his trip to Hyde Park was not. During those few days the two men held long confidential conversations with only the house staff, the secret service and a very small group on hand, including John Martin, Commander Thompson and FDR's secretary Grace Tully. Tully recalled her boss saying to her, 'I think Winston is terribly worried, Grace, and well he might be.'[5] She had been ordered specifically not to make any appointments or permit any callers whatever; even when Marshall's aide-de-camp Frank McCarthy arrived with an important message he was kept waiting outside the main house.

While Marshall and Brooke considered the risky strategic implications of Operation Gymnast, Roosevelt and Churchill had also to consider the risks of *not* getting American ground troops into direct combat against the Wehrmacht before the mid-term elections in early November. Roosevelt had served in Woodrow Wilson's Administration when it lost control of Congress only days before the end of the Great War, and he had seen the disastrous effect that had on subsequent peace-making attempts. He therefore wanted American troops fighting Germans before the polls opened, if at all possible. If the isolationists and Republicans triumphed, both Roosevelt and Churchill reasoned, the whole Germany First policy might be placed in jeopardy. For all Admiral King's talk of already having eight fronts, neither Churchill nor Roosevelt wished to contemplate a situation where no major offensive was undertaken on the ground against Germany for another ten months or so. After all, Hitler had declared war against the United States back in December, and had yet to feel her wrath anywhere. Roosevelt therefore had Churchill's note telephoned through to Marshall in Washington, and Jacob was ordered to circulate it to Brooke, Dill and Ismay.

Churchill and Roosevelt returned to Washington on the night of Saturday 20 June on the presidential train. En route, Grace Tully was charmed by what she thought 'an amusing manifestation of Victorian shyness' on the part of Churchill, 'this normally uninhibited statesman'. She was sitting with him and Roosevelt in the presidential carriage when Sawyers, Churchill's short, bald valet who always travelled with him, came in carrying his master's initialled slippers, leant down and proceeded to untie the prime ministerial shoelaces: ' "God, no," Churchill spluttered in obvious embarrassment, "not here." Retreating hastily into his own compartment he made the change and came back – slipper-clad.' Tully was surprised that Churchill should have been discomfited by other people witnessing this nightly ritual.[6]

The train arrived back in Washington at 9 o'clock on the morning of Sunday 21 June, and Brooke went to visit Churchill at 11 a.m. Marshall meanwhile saw the President, taking with him the Combined Chiefs of Staff memorandum agreed the previous day that had so deprecated Gymnast. As the Introduction depicts, at noon Roosevelt, Churchill, Brooke and Marshall all met for the first time together, along with Hopkins and Ismay. At lunch an hour later the President recalled Brooke's father and brother.

Brooke was standing with Churchill next to Roosevelt's desk after lunch when the news of the fall of Tobruk was brought in. Tobruk had long been a talisman for Churchill, a totem. A year earlier at Chequers – after a day seeing the 'horror and desolation' of bombed Plymouth, where he had kept repeating, 'I've never seen the like' – he had fallen into one of his 'black dog' depressions, a deep melancholic despair. He had received a long telegram from Roosevelt explaining why the United States could not co-operate in preventing the Germans from seizing the Azores or Cape Verde islands, and had heard that the sinking of one of Mountbatten's destroyers HMS *Jersey* had probably blocked the entrance to the Grand Harbour at Valletta in Malta. It was then that Churchill explained to Harriman that the fall of Tobruk could lead to the eventual triumph of Hitler's 'robot world order'.[7]

After Tobruk's fall, the place selected by Auchinleck to try to hold back Rommel's victorious forces was Mersa Matruh, only 125 miles east of the Egyptian border, and a telegram from the Middle East Defence Committee warned that even if the vast quantities of stores in Tobruk had been successfully destroyed before the surrender – which it later transpired they had not been – the enemy was 'now stronger than we are in all types of troops essential for battle in open country and is well provided with transport'. Furthermore, the political consequences of withdrawing to Mersa Matruh 'may give rise to internal difficulties in Egypt and may change Turkey's attitude towards us'.[8]

The American response to the news about Tobruk was instinctive, and was often later recalled with powerful nostalgia by all the Britons present. 'For a moment or two no one spoke,' recalled Ismay, but then the silence was broken by Roosevelt. 'In six monosyllables he epitomized his sympathy with Churchill, his determination to do his utmost to sustain him, and his recognition that we were all in the same boat: "What can we do to help?" '[9] Brooke also vividly remembered 'being impressed by the tact and real heartfelt sympathy that lay behind these words. There was not one word too much or too little.' Churchill agreed: 'Nothing could exceed the sympathy and chivalry of my two friends,' he wrote. 'There were no reproaches; not an unkind word was spoken. "What can we do to help?" said Roosevelt.'[10] Marshall offered an American armoured division, to be shipped off to the Middle East immediately.

In June 1958, Brooke went to the London office of the American broadcast network NBC in order to record his reflections on Marshall. In his clear, clipped, upper-class voice, he recited the events of sixteen years before, concluding that 'Acts of that nature bound us together. We were bound to have differences and we had many differences during the war, but we were always able, even after the most heated discussions in conference, to walk out arm-in-arm and go to lunch together still exactly the same friends.'[11]

The Oval Office meeting continued until 5 p.m. on 21 June, and then continued throughout the morning of the 22nd, finally coming to much the same conclusions as before on every subject except Gymnast, namely that 'Plans and preparations for the Bolero operation in 1943 on as large a scale as possible are to be pushed forward with all speed and energy.' Over Gymnast, however, whereas the Combined Chiefs of Staff had comprehensively rubbished the operation among themselves, the six men now agreed that 'plans will be completed in all details as soon as possible', with forces to be employed in Gymnast to come from Bolero units that had not yet left the United States.[12] Churchill's attempt to move the entire direction of the Anglo-American war effort from north-west Europe to North-west Africa was well under way, although Marshall and Brooke were still far from convinced.

Since Churchill, who had been in Hyde Park during its composition, had not seen the Combined Chiefs of Staff memorandum on Gymnast, it had to be decided whether or not he should be shown it. Ismay and Jacob felt that in view of the conclusions just reached at the White House, 'no good purpose would be served' by submitting it, and Brooke agreed.[13] Marshall had already given it to FDR, but Churchill did not see it.

Roosevelt and Churchill expressed themselves in favour of operations 'in France or the Low Countries' in 1942, but only if Brooke and Marshall could agree upon what Churchill called 'a sound and sensible plan'.[14] Yet both Roosevelt and Churchill well knew that Brooke would simply not agree a plan for any cross-Channel attacks in 1942 beyond minor raids. This somewhat disingenuous form of words therefore allowed Roosevelt and Churchill to seem to be supporting something that they knew would never happen, in which case Gymnast would be the only alternative.

Ismay's minute of the meeting was a veritable masterpiece of misleading prose. Marshall could have had no cause for complaint until the end of the second paragraph, where the full extent of his having been outmanoeuvred must have become apparent. 'If, on the other hand,' wrote Ismay of the two cross-Channel operations, 'detailed examination shows that, despite all efforts, success is improbable, we must be ready with an alternative.'[15]

That alternative was made all the more pressing by the Tobruk news. 'The possibilities of French North Africa (Operation Gymnast) will be explored carefully and conscientiously,' continued Ismay, 'and plans will be completed in all details as soon as possible. Forces to be employed in Gymnast would in the main be found from Bolero units that have not yet left the United States.'[16] The effect of Gymnast on the cross-Channel operations was thus obvious from the start. Ismay's wording effectively killed off Sledgehammer, pushed back Roundup and resuscitated Gymnast, which had otherwise been moribund and was opposed by Marshall and Brooke.

If Roosevelt hadn't come round to Gymnast on his own by the time that Churchill announced he was visiting Washington – as Stimson suspected that he had – then Churchill converted him at Hyde Park, and the rest of his protestations about supporting Sledgehammer and an early Roundup was essentially window-dressing for Marshall's benefit. As Sir David Fraser puts it, the agreement of 21 June merely 'paid lip service to Sledgehammer'.[17] In the television series The World at War, Mountbatten claimed it had been he who had converted Roosevelt to Gymnast. He said the President had told him: 'My nightmare would be if I was to have a million American soldiers sitting in England, Russia collapses and there'd be no way of getting them ashore.'[18]

There is evidence to suggest that Roosevelt favoured Gymnast long before his supposed conversion by Mountbatten in June 1942. In late October 1941, after representing Britain at the International Labour Organization in New York, Clement Attlee flew to Washington to stay with Lord Halifax, and in the course of his visit he was invited by the President for a short cruise on the presidential yacht. Attlee later recalled that Roosevelt had taken down an atlas and, putting his finger on Algiers, had said, 'That is where I want to have American troops.'[19] The story is reminiscent of the Duke of Richmond being told by Wellington at the famous ball in Brussels that he would stop Napoleon's advance at a small village which the Iron Duke marked with his thumbnail on a map, called Waterloo. Nonetheless, Attlee was no fantasist, and Roosevelt, as he put it, 'had had experience in the Navy Office and his mind took a broad sweep of world strategy'.

In 1956, Marshall was frank about the political imperatives behind Gymnast: 'We fail to see that the leader in a democracy has to keep the people entertained,' he told Pogue. 'That may sound like the wrong word, but it conveys the thought. People demand action. We couldn't wait to be completely ready. Churchill was always getting into side shows . . . But I could see why he had to have something.'[20] Much more was this true of his own President than of the Prime Minister. General elections had been suspended for the duration of the war in Britain, but in America the Democratic Party faced difficult mid-term Congressional ones in November 1942.

As well as a political desire to see American troops fighting German ground forces in 1942, Roosevelt was convinced by reports he was getting from US consular authorities across North Africa, especially Algeria, that the French authorities might actually welcome an American landing. Nonetheless, many Americans preferred to assume that Churchill's rhetoric had somehow seduced Roosevelt; Stilwell adopted characteristically forthright language when he noted in his diary: 'Besides being a rank amateur in all military matters, FDR is apt to act on sudden impulse. On top of that he has been completely mesmerized by the British, who have sold him a bill of goods . . . The Limeys have his ear, and we have the hind tit.'[21]

Marshall told Stimson that there had been a 'good deal of pow-pow and a rumpus up at the White House'. He said that Churchill had been 'particularly disturbed' by some remarks that Roosevelt had made to Mountbatten 'about the possibility of having to make a "sacrifice" cross-Channel landing in 1942 to help the Russians'. According to Marshall, 'Churchill started out with a terrific attack on Bolero as we had expected . . . The President, however, stood pretty firm.' Hopkins told Stimson that 'Marshall made a very powerful argument for Bolero,' managing to dispose of 'all the clouds that had been woven about it by the Mountbatten incident'.[22]

Had Roosevelt really 'stood pretty firm' on 21 June, or had he and Churchill tacitly agreed at Hyde Park that Gymnast would by default have to be the major operation of 1942? Might 'the American Houdini' have been miming support for Roundup and Sledgehammer in order to lull Marshall, Stimson and the Joint Chiefs? The circumstantial evidence suggests that Roosevelt, who had been talking loosely about a diversionary strategy ever since March, had come around to Churchill's way of thinking, or at least had appreciated that since Roundup and Sledgehammer – which in White House parlance had been shortened to Bolero – could not be mounted without Churchill's and Brooke's enthusiastic support, which was clearly not on offer, so another offensive operation must be considered, and of the limited options available Gymnast made the most sense.

For all that Brooke and Churchill were cast in the role of mendicants in the White House on 21 June, receiving welcome munitional charity from Marshall, they were not about to relent on their opposition to an early cross-Channel operation. The spontaneous gesture of the armoured division was gratefully accepted, but the Britons soon got down to undermining the 'momentous proposal' that Churchill had agreed to with such seeming enthusiasm only two months earlier. The official historian of British grand strategy for this period, J. R. M. Butler, described 21 June 1942 as 'The Day of the Dupes'.[23] Churchill visited the USA fifteen times in his life, but this was the most important of all.

At dinner at the White House that night, after the climactic moment in the afternoon, Brooke, Marshall, Churchill, Roosevelt and five others stayed up until 1 a.m. discussing global strategy, finally getting round to the Middle East. Brooke recorded that he and Churchill 'accepted offer of American Armoured Division for Middle East' believing that 'This may lead to a USA front in the Middle Eastern at expense of the European front.'[24] It was exactly what FDR, Churchill and Brooke wanted – but Marshall didn't. The British had effectively used their own vulnerability as a trump card, somehow trading on their very weakness after the fall of Tobruk to get what they wanted.

That long day was not yet over for Marshall. At the end of his fourth and final meeting with the British, Roosevelt asked him to stay on in his office after everyone else had gone off to bed. To Marshall's 'consternation' the President then suggested sending a large American force to take control of the entire region between Teheran and Alexandria, including the whole eastern Mediterranean seaboard and Levant. Marshall told Stimson he was 'terribly taken aback' by this new and very unwelcome development and almost lost his temper with the President. Not trusting himself to hold his tongue if the conversation continued much further, Marshall politely declined to discuss such an important subject so late at night, turned and left the room.[25]

One can imagine the tremendous frustration for the Army Chief of Staff, exhausted after a long day of meetings and negotiations, seeing a strategy, which he had proclaimed publicly and only very recently at West Point, now subtly undermined by the President and the British, despite his immense generosity in offering to help plug the gap between Rommel and Egypt. Then in the early hours he had been asked by his commander-in-chief – seemingly off the cuff – to consider a massive deployment of American troops in a huge and entirely unfamiliar part of the world (there are 1,250 miles between Teheran and Alexandria) very far from Berlin. The wonder is that this 'reserved and courtly Pennsylvanian' did not react more aggressively to his boss's midnight musings. Had MacArthur, Patton or King been the dominant figure among the Joint Chiefs of Staff, Roosevelt would probably have been the one 'terribly taken aback' by their response. 'Roosevelt had a habit of tossing out new operations,' Marshall told Pogue years later. 'I called it his cigarette lighter gesture.' At that, Marshall made an expansive move of his hands as though gesticulating with a cigarette lighter.

The next morning Marshall sent Roosevelt a memorandum entitled 'American Forces in the Middle East'. The President had praised the view of the US military attaché in Cairo, Colonel Bonner F. Fellers, who had repeatedly urged American intervention in the Middle East. Marshall wrote: 'Fellers is a very valuable observer but his responsibilities are not those of a strategist and his views are in opposition to mine and those of the entire Operations Division.'[26] With regard to the United States locating large ground forces in the Levant, 'It is my opinion, and that of the Operations Staff, that we should not undertake such a project. The controlling reasons are logistical, serious confusion of command (further complicated by strong racial and religious prejudices), and the indecisive nature of the operation.'

The points Marshall made were cogent. The south-west Pacific was 8,000 miles away from Washington, the central Pacific 3,000, Alaska 2,000, the Caribbean 1,000, Greenland and Iceland 2,000. To support large forces in the Middle East would 'deny the probability of assembling American forces of decisive power in any theater in this war'. It would, Marshall argued, only be of use in:

gaining a foothold on the southern but indecisive fringe of the European continent. We would still be a long distance from Germany, with extremely difficult natural intervening obstacles. You are familiar with my view that the decisive theater is Western Europe. That is the only place where the concerted effort of our own and the British forces can be brought to bear on the Germans. A large venture in the Middle East would make a decisive American contribution to the campaign in Western Europe out of the question. Therefore, I am opposed to such a project.[27]

*

'All day and half the night they have gone on since the news of Tobruk came through,' Hopkins told Moran, who noted: 'Winston has battled with the Americans; he has not allowed the facts, damaging as they are, to handicap him.' Moran saw Roosevelt as 'The big man on the American side in this dismal time', whose 'brain goes on working as if it were packed in ice'. But the President's 'prop', insofar as he needed one, was Marshall, who had 'seen the British collapse in the Middle East end in the success of the PM's efforts to postpone a Second Front. A smaller man would have turned sour.'[28] Instead it was Stimson, Wedemeyer and later Hopkins who turned very sour over the seemingly indefinite postponement of Marshall's plans.

On 22 June, at another meeting at the White House, Churchill told Stimson that the British Planners all thought a cross-Channel attack impossible in 1942, and that if it went ahead the carnage would be like the Great War. Stimson felt Roosevelt spoke 'with the frivolity and lack of responsibility of a child' during this sombre discussion. Meanwhile Brooke and Marshall went into the implications of sending an American armoured division to serve in the Middle East. Marshall's offer had been made promptly, but the working out of the details took much longer than Churchill later recollected in his memoirs. The command was earmarked for General George S. Patton Jr of the Desert Training Center in California. Only after it was pointed out that the unit couldn't get into action until October was the offer altered – to the relief of Brooke, who foresaw huge organizational difficulties in having an American division in the middle of the Eighth Army – to three hundred Sherman tanks and one hundred self-propelled 105mm gun-howitzers, to be sent by fast convoy.[29]

Marshall was as good as his word, despite the unpopularity of the decision in the US 1st Armored Division, which had only just taken possession of the three hundred brand-new tanks. The tanks and guns were shipped out to the Middle East post-haste, arriving in time to play an important part in the victory of El Alamein four months later. When one of the transport vessels containing seventy tanks was torpedoed, Marshall immediately made up the numbers for the next shipment, without even telling Brooke what he had done. 'Any mistrust which there may have been in those early days between the British and American chiefs of staff did not extend to their political masters,' recalled Ismay, pointing out that under normal circumstances the request for more tanks would have been sent by Auchinleck to Brooke, who would have got on to the US War Department, whose 'probable reaction' would have been to promise the next consignment to come off the production line, before discussions began about the shipping required. In this instance, Roosevelt and Marshall cleared the way within days.

In explaining his actions to his advisers, Marshall discounted generosity as having been a motive. Instead he explained that the tanks and guns had been given in order 'to hold the British to their promise to mount Roundup'.[30] Yet of course great operations perhaps costing thousands of lives are not affected by gratitude for generous gestures, and it would have been naive of Marshall's advisers to believe that they would be. The interesting point was that Marshall's advisers were too hard-hearted, or sceptical of the British, or doubtful of Auchinleck's chances of using the tanks to good effect, or irritated by the American 1st Armored Division being deprived of the Shermans, to respond positively to the offer unless it was cloaked in the motive of advancing American interests.

On the morning of Wednesday 24 June Churchill, Brooke, Marshall, Stimson, Dill and Ismay arrived by air-conditioned train at Fort Jackson in South Carolina, where they reviewed the sixty thousand troops that had been stationed there since the Great War-era military camp had been reopened in September 1941. Marshall wanted to impress on Churchill and Brooke the battle-readiness of the troops earmarked for Roundup. Both men undoubtedly got the message, but as Churchill put it in his memoirs, 'I consistently pressed my view that it takes two years or more to make a soldier.' As with so much of Churchill's six-volume history, this possibly had a degree of hindsight to it, since Overlord did indeed take place two years after the visit, to the very month.

The commander at Fort Jackson, Major-General Robert L. Eichel-berger, recalled in his autobiography, *Jungle Road to Tokyo*, that his assignment had been to put on a corps demonstration of the 8th, 30th and 77th Divisions for his British guests. Generals Marshall, Eisenhower and McNair (commanding Army Ground Forces) had each 'impressed on me, in turn, the importance of making a good military showing; at the moment I did not just know why.'[31] Many years later Marshall told him 'why that successful demonstration was so important. Up until that time Churchill had refused to believe the Americans capable of raising an army of sufficient size or excellence to manage a cross-Channel invasion in the foreseeable future.'

Eichelberger put on demonstrations undertaken by ten thousand men from infantry and paratroop units, as well as manoeuvres by seventy-four tanks, on a day that was 'infernally hot, hot almost beyond belief'. Churchill said it reminded him of his days as a subaltern in India. The Prime Minister particularly enjoyed operating a walkie-talkie for the first time, and the use of live ammunition in the machine-guns excited him. Out of the six hundred parachutists of the 77th Division who jumped, there were only three casualties.

'Nothing was left to chance,' recalled Eichelberger. 'A printing plant in Columbia had clanked all night to provide programs for the demonstration. Chemicals had been applied to the road to keep down the dust. I had been warned that Churchill's health was such that he should be spared the spinal rigors of a jeep ride; and a kindly motor-car dealer in Columbia gave us without charge the use of a convertible coupé.' As it was, Churchill 'hopped in and out of the car and walked at such a tremendous pace that most of the faithfully following entourage had to run to catch up.'

During the day further despatches were brought to Churchill about

Tobruk, prompting him to tell Eichelberger, over a lunch of celery-heart salad and 'several' Martinis, that he wished he had a field commander of Rommel's calibre. In the afternoon, wearing headphones to listen to the air–ground communication for the parachute drop, Churchill heard an impatient commander say over the airwaves, 'Goddammit, I told you what to do!' Stimson was shocked; Churchill roared with laughter. Brooke found something to laugh at too; after Sawyers was invited by Marshall to join the lunch party, he drank too much and would not let Churchill pass him on the plane ride home unless he turned down the brim of his panama hat, which Churchill, 'rather red and looking angry', duly did.[32]

Marshall was soothed by the positive reactions of his British visitors to what had been achieved by the US Army in the six months since Pearl Harbor. Of all the British contingent that day, Ismay was the least impressed, afterwards telling Churchill that 'it would be murder to pit them against continental soldiery', meaning the Wehrmacht. Churchill agreed the US troops 'were still immature' although 'magnificent material who would soon train on'.[33]

Brooke concluded that the American system of individual and elementary training seemed excellent, 'but I am not so certain that their higher training is good enough, or that they have yet realized the standard of training required.'[34] After the war, he added: 'They certainly had not – and had a lot to learn! I next met them in Northern Ireland, and they still had a lot to learn, but seemed to prefer to learn in the hard school of war itself. As a result they learned a great deal more in North Africa!' This reference to the American defeat at Kasserine Pass would have been spiteful if Brooke had not followed it up with the remark that, with Americans, 'in the art of war, as in polo, lawn tennis, golf etc, when they once got down to it they were determined to make a success of it.'[35]

Despite the hospitality of Fort Jackson, Anglo-American relations were not completely back on track. 'One further incident is worthy of record,' wrote Jacob in his diary of the conference, 'as it very nearly wrecked the harmony of the last day.' On 24 June, an Anglo-American Air Agreement, covering the transfer of American planes to Britain under Lend–Lease, had been signed by Arnold and the Assistant Chief of the Air Staff, John Slessor, and also initialled by the President, before it was discovered that the Joint Chiefs of Staff (except of course Arnold) actually opposed it. By lunchtime on 25 June there was 'a feeling abroad' that the Agreement had reached the President through Churchill rather than from the proper source, namely the Joint Chiefs of Staff. Jacob put it bluntly: 'They were accusing us of sharp practice.' The accusation was that Slessor had persuaded the Prime Minister to take the Agreement to the President and that he had initialled it before the Joint Chiefs of Staff had even been able to examine it. The British secretariat of the Combined Chiefs of Staff had been told that Arnold was submitting the Agreement simultaneously to the President, and they had 'naturally assumed that Arnold had squared his colleagues and masters', just as Slessor had the British Chiefs.

When all this was explained to Marshall, Arnold confessed to having given the Agreement to FDR without informing him or Admiral King, but said it was only because Slessor had been 'anxious to get everything finished' by the time the conference broke up the next day. 'This rather lame excuse may or may not have been believed by Marshall and King,' recorded Jacob, 'but in any case our name was cleared of suspicion. The US Chiefs of Staff then found themselves in the position of being unable to do anything but accept what the President had already initialled.' For Jacob the whole incident was 'only one more illustration of the extraordinary lack of coordination in the direction of the US Services'. It beggared belief that Arnold got to the point of signing an important international protocol and getting presidential approval for it without clearing it beforehand with Marshall and King.

The incident had other implications, however. The speed with which the suspicion 'got abroad' among the Joint Chiefs of Staff that Brooke and Churchill had in effect suborned the President did not bode well for the future. Similarly the assumption that Roosevelt could be so easily 'got at' by Churchill. As the British party took off on board the *Bristol* from Anacostia at 11.30 on the evening of 25 June, Marshall, Stimson and King might have been forgiven for thinking that they had left a fifth columnist behind, in the shape of a president who seemed fully wedded to the idea of Gymnast, an operation they thought too risky to undertake, and also certain to delay Roundup.

Stimson – whose attitude towards the Prime Minister is evident from his diary entry: 'Marshall and I had Churchill on our necks for three days' – even suspected that Roosevelt might have been responsible for the whole Argonaut visit from the very start. In his autobiography he stated that although the initiative for the meeting clearly came from Churchill, 'he might well have acted on the basis of an indication that the President was not completely certain about the wisdom of Bolero,' despite its having been 'the brain-child of the US Army'.[36] This was shrewd, for that indication almost certainly came during the Mountbatten meeting, although it was not so much Bolero – the US build-up in the United Kingdom – that Roosevelt had his doubts about, as the two early cross-Channel operations that were intended to follow from it.

Ian Jacob told Chester Wilmot in 1948 that the Argonaut negotiations in Washington had been 'pretty sticky', because they had agreed:

that Bolero should continue full blast until 1 September but we never seriously contemplated using these forces for a cross-Channel invasion that year. Our idea was to get them onto our side of the Atlantic so that they would be committed for the war against Germany, and we hoped that if we had the forces in England we could persuade Roosevelt to use them in Africa . . . Unfortunately once again because we had agreed to Bolero and had accepted cross-Channel in principle, the Americans thought we were willing to go into France that year.[37]

There, in black and white, is Jacob's admission that Churchill and Brooke had deliberately misled Roosevelt and Marshall into thinking that if the United States poured troops into the United Kingdom in 1942 they might be used to attack France that year, when in fact they had no intention of allowing that to happen. Perfidious Albion, good strategy, clever footwork – whatever it was, many key Americans were deceived about British intentions.

Historians have long debated Stimson's accusations – which were indignantly denied by soldiers and statesmen at the time – and Brooke's biographer denies that Marshall had been misled. Yet Wilmot's record of the interview of Jacob lodged in the Liddell Hart Centre for War Studies at King's College London makes it quite clear what Churchill and Brooke were up to, and few were in a better position to know the truth than he.

Gymnast was the codename given to a landing by Americans alone on the Moroccan coast, while Super-Gymnast was a landing on the Algerian and Moroccan coasts by both the British and Americans; it was Super-Gymnast that was eventually to be renamed Operation Torch. Marshall recalled of Roosevelt: 'When I went to him with Torch, he put up his hands' – at which Marshall raised his own hands in an attitude of prayer – 'and said, "Please make it before Election Day."' However, when it became clear that there might not be quite enough time for this, Roosevelt never complained (although his veteran press officer Steve Early certainly did). The Torch landings in fact took place on 8 November 1942, five days after the mid-terms.

General Handy was blunt on why Roosevelt decided on Gymnast in June 1942: 'Of course, we had to do it because the President wanted us to get some place to use the Army. We were building up a sizable army and we didn't really have any place to use it.' General Hull was blunter still: 'That, you see, was an election year.'[38] As for Churchill, Hull suspected that he had wanted Anglo-American landings in North Africa 'from the very beginning. I can't swear to that, but he was having difficulty in Northern Africa.'

In his memoirs Churchill attempted to depict Sledgehammer as little better than a suicide mission. 'Out of many plans the fittest might survive,' he wrote in the fourth volume, *The Hinge of Fate*, effectively claiming that the operation succumbed in a Darwinian struggle against the better plan of Gymnast. 'I did not have to argue against natural selection myself. It fell of its own weakness.'[39] In fact Sledgehammer fell because only Marshall wanted it enough, Roosevelt and Churchill actively preferring Gymnast.

A fabulously ill-informed NBC interviewer who was hoping to make a programme about the 'friendship' between Brooke and Marshall in 1958, naively asked Brooke the moronic question: 'The differences between you and George Marshall were not major, were they?' After an understandably long pause, Brooke replied:

They were major in the early part of the war and insofar as they affected the time at which we should start the liberation of France, as to my mind any attempt at carrying out this operation before we were really ready for it might have resulted in a ghastly catastrophe. I considered in the early stages Marshall had not quite appreciated all the difficulties that would arise and his ideas of this early crossing were not of the best, and that is where our main differences occurred ... As things turned out I think finally we came across exactly at the right time, and I do not think we could have done that crossing earlier in the war.[40]

*

When explaining to themselves Churchill's and Brooke's reasons for not wanting to cross the Channel in 1942 – when after all the Germans had not yet built the 'Atlantic Wall', concrete defensive fortifications all along the western European seaboard – the Americans kept returning to the experiences that they assumed the Prime Minister and CIGS had both undergone between 1914 and 1918. Charles Moran – unhelpfully, even disloyally – went so far as to warn Marshall that Churchill was 'fighting the ghosts of the Somme'. Wedemeyer could be relied upon to be far ruder than Moran, of course, in his diagnosis of what he called Churchill's 'lamentable deficiencies as a strategist'. After 1941, he contended that 'the problem was to restrain the pseudo strategist in Churchill' which he considered 'inherent in his islander's psychology' and 'conditioned by his experience in World War One'.[41]

Writing in 1955 to a would-be biographer of Brooke, Marshall said that the CIGS had:

had a hard schooling in the battles of the Somme in earlier years and had suffered the shock of the highly modernized Nazi Army. This, for a year or more in our earlier negotiations, made it difficult for him to meet our theory and battle inexperience with his practical and rather desperate experience. All this washed away as the war developed, and we came more and more into mutual understanding.[42]

It was a diplomatic way of explaining their clashes, and except for the last sentence was insightful about Brooke's thinking. The Somme had indeed been such a cataclysmic event in recent British experience that it would have been extraordinary if had not affected the outlook of the large number of British Planners – including Brooke and Kennedy – who had fought there.

The Pentagon developed the view that, because of their terrible bloodletting in the Great War, the British would commit only to relatively small-scale diversionary attacks. 'The British are a great people, you know, but they hit around the edges,' explained General Handy to his SOOHP interviewer. 'Some of the British thinking was affected by what happened in World War One. Our losses weren't even in the same class as the British and the French ... We figured that if you were going to lick the Germans the only way to do it was to face up to them and fight them, and the British maybe wanted to do it some other way.'[43] While it is quite true that the Americans suffered 50,500 battle deaths in the First World War, against 908,300 from the British Empire and 1,357,800 from France, the OPD made far too much of that, and eventually assumed that the British secretly opposed a cross-Channel invasion on that basis.

Put at its crudest, the hack-historian Leonard Mosley, an employee of Lord Beaverbrook, wrote that if the cross-Channel operation had to happen, Churchill 'was determined that the more fecund United States should provide the cannon fodder. And that meant postponing any cross-Channel operation until American troops were ready to make up the bulk of the invading force.'[44] It was a disgraceful accusation not supported by the facts; Churchill no more saw American troops as cannon fodder than he did the British. Ismay meanwhile readily accepted that the Americans 'suspected that we were haunted by memories of Passchendaele and the Somme and that we would always shrink from undertaking an assault on Fortress Europe. This suspicion persisted for a very long time, and lay at the root of future misunderstandings.'[45]

Churchill himself fully admitted being haunted by the ghosts of the Somme and in *Closing the Ring* he wrote of Roundup: 'The fearful price we had had to pay in human life and blood for the great offensives of the First World War was graven on my mind. Memories of the Somme and Passchendaele and many lesser frontal attacks on the Germans were not to be blotted out by time or reflection.'[46] Furthermore, since the horrors of those days, Churchill pointed out, 'The firepower of the defence had vastly increased. The development of minefields both on land and in the sea was enormous.' Set against that were the great advantages of air supremacy and paratroopers' ability to disrupt counter-attacks from behind enemy lines, neither of which existed in the Great War.

Yet it wasn't so much returning to Passchendaele and the Somme that worried British strategists in 1942–4 as the Dunkirk and Brest campaigns of the summer of 1940. Rommel's and Guderian's seemingly unstoppable blitzkrieg campaign across France featured more in their fears – especially Brooke's and Dill's – than the mud and blood of Flanders of a quarter of a century before. Even more recently, April 1941 had seen the British make a sacrificial attempt to help Greece and being ignominiously flung off the Continent once again as a result. Memories of these humiliations naturally served to work against Sledgehammer a mere fourteen months later.

Some senior OPD members understood this, of course. 'The British had a bloody nose from Dunkirk,' recalled Hull.

They were leery of amphibious operations against a main force like a landing against Germans in France. Churchill was a great man, but as a strategist, I think, you can criticize him . . . He was responsible for Gallipoli . . . he was always thinking about what he called 'the soft underbelly' of Europe . . . He was responsible for the Greek débâcle when he put troops in there . . . All during the war he was thinking of landing troops in the south.[47]

The problem with such easy explanations is that they imply irrationality on behalf of the British, even at its worst a form of cowardice.

When Marshall was asked by Pogue after the war whether Churchill's insistence on a Mediterranean strategy was due to a desire to vindicate Gallipoli, his answer was indicative of the American misunderstanding of Churchill's true views regarding Roundup. 'No, I think it was not Gallipoli, but the fact that he was opposed to the Channel crossing. He had a horror of bodies floating in the Channel . . . The dominant thought is that they didn't think we were capable of manufacturing the troops.'[48]

A much fairer criticism of Churchill and Brooke was that they were willing to fight not to the last American, but rather to the last Russian. 'If his willingness to allow the Russians to bleed the German Army was cynical,' the distinguished military historian Max Hastings has written, 'it was a great service to his own country.'[49] Hastings is right, and for all the inescapable admiration for the Red Army's sacrifice, there was much less genuine sympathy for it among the British High Command than among the public at large. The senior strategists recalled Stalin's non-aggression pact with Hitler that lasted up until the evening of Barbarossa itself. In September 1945 Grigg and Kennedy agreed that Stalin's 'claim to have won the German war single-handed' was 'silly', because 'Hitler not Stalin was our friend in bringing Russia into the German war.' (Brooke was profoundly anti-Bolshevik and Kennedy had fought against them in the Russian Civil War.)

Yet it was fortunate that the Americans provided a counterpoint to the British strategy, for by 1944 the Allies needed to go further than Churchill's and Brooke's strategy of 'probe, jab, bomb, subvert'. As Samuel Eliot Morison put it: 'The elephant does not like mice, but a thousand mice cannot kill an elephant.'

'It has been a very interesting trip and real good value,' Brooke wrote in his diary while on the flight back between Newfoundland and Stranraer.

'I feel now in much closer touch with Marshall and his staff and know what he is working for and what his difficulties are.' A primary difficulty was clearly Admiral King, who Brooke now appreciated had a Pacific-based outlook quite opposed to Marshall's European one. As for Roosevelt, Brooke acknowledged he had 'wonderful charm' but he was not about to be admitted to that small, elite circle – Smuts, Stalin, himself and later MacArthur – who Brooke thought really understood strategy. 'I do not think that his military sense is on a par with his political sense,' he wrote of the President. 'His conceptions and plans are not based on a full grasp of all the implications.'

Brooke admitted that he had found it difficult in the first few meetings to know how much importance to attach to the President's strategic ideas, and also he 'did not know how Marshall would react' to them. With Roosevelt and Churchill planning grand strategy on their own at Hyde Park, in the Map Room and even seemingly in the bathrooms of the White House, Brooke found it 'difficult at first to carry on business with Marshall. However, I finally got on sufficiently intimate terms with him to discuss freely with him the probable reactions of both President and Prime Minister to the plans we were discussing,' and vice versa. Brooke wrote these comments while sitting on the Clipper only a few feet away from Churchill, who he noticed was mouthing the phrases of the speech he was writing for the forthcoming debate of no confidence in the House of Commons.[50]

After the war, Brooke added: 'The President had no military knowledge and was aware of this fact and consequently relied on Marshall and listened to Marshall's advice. Marshall never seemed to have any difficulty in countering any wildish plans which the President might put forward. My position was very different.' Because Brooke conceived part of his duty as weaning Churchill off impractical schemes, which he sometimes needed to do in other ways than mere outright opposition, Brooke became 'convinced that on many occasions Marshall imagined that I was in agreement with some of Winston's wilder ideas; it was not easy for me to explain how matters stood without disloyalty to Winston. On several occasions I believe that Marshall thought that I was double crossing him.'[51] He did indeed, and understandably so after Brooke's apparent support for Roundup (with reservations) during the Modicum talks and then two months later coming out so strongly against it at the Argonaut Conference.

Lew Douglas, who as deputy director of the War Shipping Administration from 1942 to 1944 and US ambassador to London from 1947 to 1950 was in a good position to know, claimed after the war that Marshall thought Brooke had been dishonest over the date of cross-Channel operations, while Brooke 'thought Marshall was being stupid. Marshall had the full support of Stimson and was bitter against Churchill.'[52] This is a fascinating insight, quite at variance with all Marshall's later encomia about Brooke's integrity and gentlemanly behaviour. After the way that Brooke had seemed to change his tune, it was perhaps natural for Marshall to feel let down, indeed hoodwinked. Bitterness against Brooke, perhaps extending to Churchill too, would have been a reasonable enough reaction, though obviously not one that could have been much alluded to at the time or – for political reasons – for some time afterwards.

Whether or not Marshall was ever bitter, Wedemeyer certainly still was in 1958 when he wrote that 'The insincerity of the British about Bolero–Roundup was ultimately to be exposed long after the war, when Sir Alan Brooke confessed that the promotion of Gymnast was specifically designed to stall the cross-Channel operation scheduled for 1943.'[53] In fact Brooke never confessed any such thing, not least because that was not the rationale. Gymnast was promoted by Churchill and Roosevelt rather than Brooke; it was undertaken because Brooke and Churchill had concluded that both Roundup and Sledgehammer were impossible in 1942, and the politicians wanted definite military action in the calendar year 1942 for perfectly comprehensible political reasons. Brooke's own preference was for Roundup in 1943 if possible, but of course a successful operation that helped finally to win the three-year, pendulum struggle along the length of the North African littoral was very welcome too, especially as late June 1942 saw Rommel forcing the Eighth Army into full retreat to Mersa Matruh, and Auchinleck taking over day-to-day command of the battlefield from Ritchie. Simultaneously, on the Eastern Front, the Germans scored some spectacular early successes in their summer offensive.

Soon after his return to Britain, Brooke sent Marshall a telegram thanking him 'for all your kindness' and expressing 'the conviction that our discussions have gone a long way towards ensuring that close co-operation and understanding so essential between us in the execution of the task we are engaged in'. Marshall replied: 'Greatly relieved that you have made a safe trip and deeply appreciate the gracious message just received from you. If nothing else was accomplished during the visit of the Prime Minister, I feel that the intimate accord and I believe understanding developed between us justified the trip. It was an honour to the Army to have you here and a great privilege to me.' One wonders whether Marshall intended the deep ambiguity of the phrase 'If nothing else was accomplished . . .' in quite the way that it read.

Brooke thought much had been accomplished, and told Kennedy back at the War Office that it was 'a good thing in many ways' that he and Churchill had been in Washington when the 'bad news' about Tobruk arrived, since 'It was easier to explain things to the Americans by word of mouth than by telegram. The Americans showed a fine spirit and great eagerness to get into the war and particularly to help in the Middle East.' Bolero would 'go ahead with the movement over here at the greatest possible speed which is good'. As for Gymnast versus the cross-Channel projects, Brooke merely said that they 'were discussed and put on a proper basis'.[54]

Although Marshall was enough of a soldier–politician to see the auguries clearly, he did not consider the battle for an early cross-Channel operation entirely lost, although he did deeply deprecate Gymnast, and Roosevelt's support for it. Had he been offering to contribute more than around 700 of the 5,800 aircraft it was estimated that such a venture required, he would have been in a much stronger position to insist. Instead, he prepared to implore in London once more, and in person.

9
Torch Reignited: 'A new and rather staggering crisis in our war strategy' July 1942

If Rommel's army were all *Germans, they would beat us.*

Winston Churchill, July 1942[1]

'The criticism of Sledgehammer is that we had so little and that it could virtually have been destroyed,' argued George Marshall after the war. 'This overlooks the fact that the Germans had little in the West and that little was of poor quality.' It was true; the Germans had only twenty-five divisions ready to repel Sledgehammer in 1942, but over twice that number in France and the Low Countries in 1944. Marshall continued: 'I thought I had a firm commitment on Roundup,' but in Washington 'the backing of the President weakened.'[2]

When considering German divisions, there is a severe problem comparing like with like. A full-strength panzer division simply cannot be compared with one just brought out of the line after heavy losses, and an elite division like the Grossdeutschland had little in common with a rear-areas security division; it would be like comparing the Guards Armoured Division with the Home Guard, and tells us next to nothing. Yet divisions are the standard currency of grand strategy, and have to be employed as such. Because German divisions were constantly being moved around the Reich, Russia, Africa and Occupied Europe during the war, and some were of superb fighting quality and well equipped while others were little better than militia, made up towards the end of the war of middle-aged men and adolescents, any statement of what the Allies might or might not have faced had they invaded France in 1942 or 1943, or even of what they did face on 6 June 1944, when there were around fifty-seven German divisions in France and the Low Countries, must be inexact. Although these fifty-seven divisions were clearly many more than the average of around twenty-five estimated to have been in France for much of 1942, they were smaller and weaker from losses sustained in the meantime.[3] Intelligence reports on where German divisions were at any one time were necessarily inexact also.

It is vital, when considering the Allies' attitudes towards cross-Channel operations, to try to remember that they did not know what we do today, that Operation Roundup (rebranded as Overlord) took place on 6 June 1944 and was a success. Amphibious operations generally did not have a happy history in British arms; indeed many had led to humiliating evacuations. Churchill was very conscious of this, telling Eden in July 1941: 'Remember that on my breast are the medals of the Dardanelles, Antwerp, Dakar and Greece.'[4] It was true. Churchill had been directly responsible for four of the worst amphibious operations and evacuations of both world wars, and while in his mood of self-decoration he might also by then have added Norway, Dunkirk and Crete.

Moreover, the number of German divisions stationed in France, and even their quality, was not really the essence, because until the spring of 1944 they always had strong reserves that could be rushed to wherever the coastal blow fell. In December 1961, J. R. M. Butler, the general editor of the Grand Strategy series of the *Official History of the Second World War*, wrote to Brooke to ask him to comment on one of the volumes then in production. In his reply, Brooke complained that one of the chapters failed to expound upon the geographical and communication factors that affected the German position. He explained his overall view of strategy and thus his fundamental difference of view from Marshall. 'Having been forced to fight on two fronts during the 1914–18 War', he began, the Germans 'had further developed their East–West communications with double railway lines and autobahns, to meet the possibility of being again forced to fight on two frontiers. They were capable of moving some six to eight divisions . . . simultaneously from East to West.' Yet the Germans had far less easy manoeuvrability in southern Europe and the Mediterranean, where, he argued, the rail and road communications from northern France to southern Italy and the Mediterranean 'were very poor. Furthermore, command of the sea gave us the power of selecting our point of attack at a suitable point on the outer circumference of German- and Italian-held territory.' Brooke then lamented that he:

could not get either Marshall or Stimson to realize that operations across the Channel in 1942 and '43 were doomed to failure. We should go in with half-trained divisions against a superior number of war-hardened German divisions and the Germans would have the facility of reinforcing that front at a rate of two to three divisions for every one we might put in. Any idea of a cross Channel operation was completely out of the picture during 1942 and '43, except in the event of the German forces beginning to crack up, which was very unlikely.[5]

Samuel Eliot Morison was unconvinced by these arguments, pointing out in his Oxford lectures the various excellent north–south connections in southern Europe and especially Italy, where Field Marshal Kesselring was able to rush troops swiftly and effectively to Salerno and Anzio in 1943–4. There were also some who thought that Brooke was wrong in thinking that there was any great mass of reserves in Germany which could be called upon to repulse Sledgehammer or Roundup. One of these was the man theoretically best placed to know, Victor Cavendish-Bentinck, head of the Joint Intelligence Committee throughout the war, who thought Brooke was always too pessimistic on this score and that the Germans were stretched to the limit on the Eastern Front and 'substantially without reserves' in central Europe.

Yet because each service had its own intelligence section, and each Chief of Staff was reading the Ultra decrypts that arrived from Bletchley Park every day, they were able to form views that were often at variance with those of Cavendish-Bentinck and the Joint Intelligence Committee, whose principal job was to collate all three services' intelligence reports to see if any overall pattern emerged. The Joint Intelligence Committee therefore played a more minor role in the Second World War than its name suggests, and than it has in later conflicts.

Brooke was keen to scotch the assumption of so many American Planners that he was viscerally opposed to any cross-Channel operation ever taking place, something to which even Dwight Eisenhower's war memoirs, *Crusade in Europe*, lent some credence. 'A major factor in all American thinking at that time', Eisenhower wrote, 'was a lively suspicion that the British contemplated the agreed-upon cross-Channel concept with distaste and with considerable mental reservations concerning the practicability of ever conducting a major invasion of north-west Europe.'[6] Just as politicians are advised never to use the word 'never', so the word 'ever' should be absent from grand strategy. (Strictly speaking, Eisenhower was right, however, because he was writing of a general 'lively suspicion' rather than saying it was justified.)

'I was all for assembling US divisions in England to complete their training and in preparation for the ultimate cross-Channel operation,' Brooke told Butler. 'I could never get Marshall to appreciate the fact that the North Africa and Italian operations were all part of one strategy preparing for the final blow. I feel it is most important that our strategy should be looked on as one large whole, and not detached theatres, and one in which we made full use of our historical maritime strategy.' Here, Brooke was using too much hindsight, as invading Italy was nowhere near the agenda on either the Modicum or Argonaut missions.

Until Germany was weakened by bombing and on the Eastern Front, and there was enough shipping available for a massive attack, Brooke was determined not to undertake anything more than large-scale raids, on places like Dieppe and Saint-Nazaire. Over the Second Front, David Fraser writes, 'Churchill's and Brooke's minds were complementary, with the former producing the visionary, the latter the prosaic elements.'[7]

Meanwhile, Roosevelt made clear to American strategists – though not to the British – that he too had concerns about an over-hasty Roundup. 'We were largely trying to get the President to stand pat on what he had previously agreed to,' Marshall later said of this very trying period for him. 'The President shifted, particularly when Churchill got hold of him ... The President was always ready to do any sideshow and Churchill was always prodding him. My job was to hold the President down to what we were doing. It was difficult because the Navy was pulling everything toward the Pacific, and that's where the Marines were, and they got a lot of publicity.'[8] It was to try to break this seeming log-jam over future policy – with the cross-Channel operations stalled by the British and Gymnast opposed by the American military – that Marshall visited London again in July 1942, accompanied once more by Hopkins.

The historian Richard Overy has convincingly argued that because Britain had no large settler communities in the Mediterranean, nor vital economic interests, and took much of her oil from the New World by 1942 – although she did need to deny Germany access to Middle Eastern supplies – 'The chief argument for Allied presence in the Mediterranean was that here they were fighting a corner of the war that they could win in 1942, against weak Italian forces, spiced up with a handful of German divisions and air squadrons. It was not a glamorous alternative, but it was a realistic one.'⁹ This went to the heart of the Anglo-American disagreements about whether to undertake Torch. If it was just a defensive measure to 'close the ring' around Germany while still relying on bombing and blockade for victory, the American OPD would be excused for concentrating strength in the Pacific instead. Yet if it was the vital first stage in clearing the North African littoral, turning the Mediterranean into an Allied lake and thus easing pressure on shipping, press-ganging hitherto neutral Turkey into a war that Italy was simultaneously being forced out of, then threatening the so-called 'soft underbelly' of southern Europe and drawing German divisions away from Russia, Torch could be absolutely integral to Germany First. Preparing for an eventual Roundup by these methods made much more sense to British strategists.

General Hull recalled that although the OPD had discarded plans for invading North Africa by mid-1942, 'The British hadn't though. They still had a problem in Africa and that was the area of principal concern to them. That's the area where Churchill wanted to go. We tried our darnedest not to get into it but Mr Roosevelt decided that we would have to go into North Africa or someplace else and start fighting Germans face to face and that was why we agreed to go into there.'¹⁰ Many Americans, especially the Anglophobes of whom there then seems to have been no shortage in the OPD, assumed that Churchill and Brooke – proud sons of the Empire both – had ulterior imperialist motives for wanting Gymnast, namely the defence of the British Empire in Egypt and its interests in the Middle East.

Though no Anglophobe himself, Marshall astonishingly did not reprimand Albert Wedemeyer for installing a secret tape recorder in his office, one that he could activate with his knee from behind his desk. He later played Marshall a recording in which British officers from the Joint Planning Staff had made 'unreasonable demands, while using big names like Roosevelt and Hopkins to intimidate me or influence my action. Marshall was extremely interested and advised me to record all future discussions, which I gladly did.'¹¹ Wedemeyer claims he later also told Dill, who 'was surprised, but sympathetic, too'. (There is no record of Dill warning any British Planners about this underhand activity.) Perhaps it was true that the British took advantage of the fact that American Planners were not always au fait with presidential intentions over grand strategy, but it was a devious thing for Wedemeyer to have done, and if the British had discovered it before the end of the war it would have wrecked Anglo-American trust, especially if it had been revealed that Marshall had not forbidden such disgraceful behaviour.

On 1 and 2 July 1942, Churchill faced the House of Commons for a two-day debate on the fall of Tobruk. Supporting the motion 'that this House ... has no confidence in the central direction of the war', the Labour MP Aneurin Bevan said that the War Cabinet as a whole should be able to meet the Chiefs of Staff 'instead of the Prime Minister'. At this Churchill interjected: 'They do.' But Bevan continued: 'Instead of the Prime Minister seeing them before the War Cabinet sees them. Because then the Prime Minister goes into the War Cabinet defending his own decisions.' Churchill again interjected: 'That is not true.' In fact it was very substantially the case, because Staff Conferences very often preceded War Cabinets. Much less credibly another critic, Admiral Sir Roger Keyes, complained that Churchill was too easily swayed, and 'could never be induced to override the advice of the Chiefs of Staff Committee, or to undertake any exercise, unless they were prepared to share fully with him in the responsibility'.[12]

Later on in the same debate, Churchill said he would 'meet a complaint which I have noticed that the Minister of Defence [was] in Washington when the disaster of Tobruk occurred'. In fact, Churchill argued, still speaking of himself in the third person, 'Washington was the very place where he should have been. It was there that the most important business of the war was being transacted.' He prayed Brooke in aid: 'When I left this country for the United States on the night of 17 June, the feeling which I had, which was fully shared by the CIGS, was that the struggle in the Western Desert had entered upon a wearing down phase, or a long battle of exhaustion, similar to that which took place in the autumn.'[13]

The debate was not all bitter and hard fought; at one point the Prime Minister had the House roaring with laughter when he said of the prototype A-22 tank: 'It had many defects and teething troubles, and when these became apparent the tank was appropriately christened "the Churchill".' The Government survived the debate easily – by 475 votes to 25 – and the malcontents were well-known troublemakers, extremists or a combination of the bitter, disappointed and marginalized. Any senate of more than six hundred politicians will always be able to cobble together a couple of dozen professionally perverse members, even in wartime. After the debate, Roosevelt cabled Churchill just three words: 'Good for you.'

On the second night of the debate, Churchill and his brother Jack dined with Eden, and the Prime Minister argued repeatedly that Britain had not done as well during the war as she should have, adding, 'I am ashamed.' They then discussed the problems of the Army, 'its Trade Union outlook, paucity of talent, etc'.[14] Churchill suggested 'another journey', probably a reference to his going out to the Middle East to see the situation for himself. When Eden deprecated this, saying he'd just get in the way, Churchill replied: 'You mean like a giant bluebottle buzzing over a huge cowpat!' Eden agreed that that was exactly what he meant.

Churchill employed humour regularly, and his words should not always be taken entirely literally. At a War Cabinet meeting in July 1942, when discussing the issue of what to do with the senior Nazis, Churchill said: 'If Hitler falls into our hands we shall certainly put him to death. He was not a sovereign who could be said to be in hands of ministers, like the Kaiser. The man is the mainspring of evil. The best instrument to use would be the electric chair, which is used in America for gangsters and would no doubt be available on Lease Lend.'[15] Churchill was serious about executing Hitler, of course, but probably not about the means, and certainly not because of the thrift involved.

Nor was it just the Nazis whom Churchill considered executing. 'What led thirty thousand men to surrender at Tobruk?' he asked the War Cabinet on 6 July. 'Orders were given to generals not to surrender without express permission from home. Some order should be given that generals will be put on capital charge after the war if they surrender.' This would include the Commander-in-Chief, although 'some other generals should not be affected. Enquire about the four thousand tons of petrol Tobruk left undamaged.'[16] General Nye answered that there was 'no evidence', but that a secret inquiry was under way in Cairo to discover what had happened.

Although he would hardly have said so to Churchill, Brooke agreed with the problem of paucity of talent in the Army. Writing to Wavell in India – 'My dear Archie', and signing off 'Yours ever, Alan' – he said that Harold Alexander was 'one of the few good commanders we have got. But the shortage of good ones is one of my worst troubles. The last war casualties took the cream of the generation that should be providing division and Corps commanders now.' Brooke also wrote:

I like Marshall and find him an easy person to deal with. There is a good deal of divergence of outlook which will want watching. King, the Admiral, has his eyes riveted on the Pacific whilst Marshall looks out towards Europe, and between them the 'prima donna' MacArthur must maintain the Australian front in the limelight! I think the President as Commander-in-Chief may well find himself puzzled at times![17]

*

On 7 July the Chiefs of Staff pointed out to Churchill that preparations for a cross-Channel operation should be continued because it was important that the Allies should 'Not lead the Russians to think there is no chance of our attacking this year'. The double negative is instructive. Brooke did not want Stalin to know that there was no hope of a Second Front in Europe in 1942; however, he did want the Americans to be told that Sledgehammer was off the agenda 'at once'. He added that there was 'no satisfactory solution' to the problem of how to take and hold the northern tip of Norway.

At a War Cabinet at 5.30 p.m. that day memoranda written by both Churchill and the Chiefs of Staff were read out, because they had only just been finished and could not be circulated in time. Churchill then summarized the discussions he had had about Sledgehammer, Roundup, Gymnast and Jupiter. To be able to 'summarize' was a great advantage for the Prime Minister, who naturally put his side of the case forcefully and the opposing arguments with far less rigour. He was objective, however, in stating of Sledgehammer that 'the conditions which would make this operation practicable in 1942 were now extremely unlikely to arise,' and the War Cabinet agreed 'only to make such preparations for it as would enable us to deceive the enemy, but not at the expense of interference with Roundup'.

Of Gymnast, however, Churchill said that 'the President had always expressed the keenest desire to carry it out.' They hoped an attack on Rommel from the west, and the capture of air bases in French North Africa, might transform the Mediterranean situation. It would consist of six US divisions, with 'a spearhead of British assault troops'. As for Jupiter, Churchill admitted that the Chiefs of Staff found the 'difficulties insurmountable', but he nonetheless thought that a lodgement in northern Norway, possibly in conjunction with the Russians and Norwegians, was worth planning for. 'The War Cabinet fully agreed with this view,' recorded the minutes, which committed them to nothing.

Pound then pointed out, with regard to Jupiter, that in view of the disaster that had recently overtaken Convoy PQ-17 – only eleven of whose thirty-five merchantmen had reached port – 'we would have great difficulty in setting up an air striking force in Murmansk. Furthermore, we should have to gain simultaneous control of all airfields between Petsamo and Narvik,' a potentially vast operation. The naval forces required 'were beyond our present capacity', not least because anti-submarine craft would have to be withdrawn from the Western Approaches, which he said was 'quite unacceptable'. Finally, the Germans had spent two years acclimatizing to Arctic conditions and had five divisions in northern Norway and its environs, with another three capable of being deployed there rapidly, as well as 255 aircraft.

'After further discussion', the minutes relate, the War Cabinet agreed that 'the Americans should be encouraged to proceed with Operation Gymnast, and that we ourselves should undertake Jupiter, if by any means a sound and sensible plan could be devised'.[18] It was an excellent let-out clause. They also decided that Jupiter was 'to be carried out largely by Canadian troops'. The poor Canadians were obviously regarded as the forlorn hope by the Cabinet, though not by Brooke who had been attached to them in the Great War. The following month it was a largely Canadian force that was earmarked for the Dieppe Raid (codenamed Operation Jubilee). The lack of Canadians on the Combined Chiefs of Staff meant that, for all their superb loyalty and generosity, they were often assigned the toughest tasks. In the event, though, Jupiter was considered beyond even their capacity for self-sacrifice.

After that meeting, Eden went back to the Foreign Office and told Cadogan of the 'Very gloomy outlook and Cabinet', complaining that strategically the 'Chiefs of Staff have no ideas and oppose everything.' Churchill had said: 'We'd better put an advertisement in the papers, asking for ideas!'[19] It was one of the few examples of Winston Churchill ever telling an unfunny joke. Meanwhile, across Whitehall back at the War Office, Brooke agreed that the Cabinet had been 'a thoroughly bad one', telling Kennedy that 'Winston was in one of his worst moods and abused the Army and criticised our generalship and our failure to use the masses of men in the Middle East.' When Bevin had asked him whether it could not be right to urge Auchinleck to attack, Brooke replied that a false move would be fatal and the Auk 'might lose Egypt in five minutes if he made a mistake'.[20] Churchill apologized to Brooke after the meeting.

On 6 July Churchill wrote to Roosevelt to try to clarify the terms used for operations. Astonishingly enough – and this cannot but have added to the mood of mistrust between the two Staffs – the British and Americans often used the same term to denote quite different operations. At one stage, for example, Roosevelt had been writing 'One-Third Bolero' to mean Sledgehammer. Churchill suggested a final clarification, but of course could not explicitly name in the cable the places due to be attacked, which made the whole process even more convoluted. He also stated that the Joint Chiefs of Staff had been using the term 'Semi-Gymnast', which in fact they had not.

The US Chiefs regularly used 'Bolero' to mean both the build-up in the UK and one or both of the cross-Channel invasion plans, whereas for the British it just meant the build-up. The distinction was vital for Brooke, who supported Bolero and an eventual Roundup but not Sledgehammer, whereas for Marshall it was less important. In several cases Planners seemed to use whichever term best supported their ideas, so British Staff officers occasionally criticized Roundup for attributes that in fact only pertained to Sledgehammer, causing further confusion in Washington. For a group that prided themselves as much as did Hollis and Jacob on the superiority of their organization and the precision of their language, it was extraordinary that the War Office secretariat could have allowed such ambiguities to pervade their counsels for so long.

Some British Planners had been using 'Roundup' to mean a limited, opportunistic attack across the Channel in the event of a German collapse, which was actually Sledgehammer, whereas for the Joint Chiefs of Staff 'Roundup' was (rightly) taken to mean the vast, forty-eight-divisional attack on France. 'To the befuddlement of both contemporaries and historians, both plans were labelled Roundup,' records Warren Kimball.[21] Churchill anyway disliked the term Roundup, which he thought either over-confident or over-gloomy, depending on who was likely to be rounded up.

As for 'Gymnast', the British originally thought it meant an Anglo-American attack east of Tunis. Yet by mid-1942 it had morphed into an American attack on Morocco near Casablanca, a different operation entirely. The situation was becoming absurd, so in his reply two days later Roosevelt defined the terms clearly, in accord with Churchill's suggestions. Henceforth, Bolero would mean the preparation and movement of American forces into the United Kingdom. Sledgehammer was to be a limited, nine-divisional cross-Channel attack in 1942 in the event of either a German or a Russian collapse. Finally, Roundup was to mean the massive forty-eight-divisional cross-Channel invasion to liberate the Continent, 'to be carried out by combined American and British forces in 1943 or later'. No mention was made of Gymnast, the mythical term Semi-Gymnast (found nowhere else in the records) or Super-Gymnast. Here, however, was an indication – the first in any document – that Roosevelt understood, in stark contrast to what he had told Molotov only the previous month, that the cross-Channel attack would not take place until possibly 'later' than 1943.

The next War Cabinet meeting addressed the question of who should command Roundup. Churchill said that on his recent visit to Washington he had gathered that if the supreme command was offered to Marshall – as the Americans were proposing to employ twenty-seven divisions to Britain's twenty-one – the general 'would be very pleased to accept it', and Churchill added that 'our interests would be best served by the appointment of an American as Supreme Commander'. If Marshall were appointed he would, of course, 'receive loyal and effective aid and support from the Staffs in this country, who might be expected to exercise some influence over his views', and he would also be best placed to obtain the maximum resources and equipment from the United States. If the operation failed, Churchill said, 'there would be no reproach' against 'a British generalissimo'.[22] He added that in so large an operation there would be plenty of British corps commanders. It was a deeply pessimistic stance to adopt, indicative of the Prime Minister's fears.

Dill had written a telegram emphasizing Marshall's claims and the War Cabinet expressed 'general agreement' that Roosevelt should be invited to appoint him, with ultimate responsibility to the Combined Chiefs of Staff. The minutes are quite specific; though they do not record exactly what Brooke said, they do state that 'The three Chiefs of Staff all expressed agreement with the proposal that General Marshall should be invited to be Supreme Commander.' Meanwhile, Dill warned

Churchill and Brooke that Marshall was seriously looking at a Pacific alternative if they continued to oppose Sledgehammer. It seems likely that Marshall himself drafted the telegram in which Dill first alerted London to the likely consequences of rejecting Sledgehammer in favour of Gymnast.[23] But was Marshall bluffing?

Churchill replied to Roosevelt that the War Cabinet had been considering what he called 'maximum Bolero' (meaning Roundup) and asked whether 'General Marshall would undertake this supreme task in 1943. We shall sustain him to the last inch.' A few moments later, he sent another one-sentence telegram to say that he assumed that the appointment of Marshall 'over Bolero 1943 does not prejudice operations of immediate consequence such as Gymnast'. Churchill must have known that Roundup was near impossible in 1943 if the troops earmarked for it were to be used in Gymnast, and certainly Marshall and Brooke knew it, since it was one of the primary reasons why Marshall opposed Gymnast, and one of the only reasons why Brooke could be prevailed upon to support it.

The morning after the telegrams were sent, Thursday 9 July, saw a flurry of almost identical news reports that Marshall was about to be appointed supreme commander of all Allied military forces, with Marshal Foch cited as the precedent. A report came from a US radio despatch quoting 'British and American circles in Washington', and was printed by the *Evening Standard, Daily Mirror, News Chronicle* and *Daily Express* in Britain. The following day saw a categorical denial by the War Department.

Yet Marshall was not to be fobbed off with the command of an operation still far in the future, especially once he had learnt from Dill that the War Cabinet meeting of 7 July had definitely decided against Sledgehammer and in favour of Gymnast. (Dill ought not to have told him; at that Cabinet meeting Brooke had specifically said he did not think the Americans should be informed of the decision too soon.) Once he realized what was happening, Marshall determined to fight back. He first went to Stimson to report what he called 'a new and rather staggering crisis that is coming up in our war strategy'. He explained that the British were 'going back on Bolero and are seeking to revive Gymnast – in other words, they are seeking now to reverse the decision which was so laboriously accomplished when Mr Churchill was here a short time ago.' He cannot have yet digested Roosevelt's telegram on definitions: Churchill supported Bolero, but opposed Sledgehammer.

Marshall complained to Stimson that Gymnast 'would be simply another way of diverting our strength into a channel in which we cannot effectively use it', meaning North Africa. 'I found Marshall very stirred up and emphatic over it,' recorded Stimson. 'He is very naturally tired of these constant decisions that do not stay made. This is the third time this question has been brought up by the persistent British and he proposes a showdown which I cordially endorsed. As the British won't go through with what they agreed to, we will turn our backs on them and take up the war with Japan.' Stimson later wrote in a post-war addition to his diary that it showed the effect of Churchill's 'obstinacy on Marshall and me, for we were both staunch supporters of the "Europe First" plan and argued for the Pacific only because we were desperate at British inaction'.[24]

The same day as the press leaks, Eden's private secretary Oliver Harvey recorded in the diary that he too kept throughout the war in defiance of regulations, that because of the British rejection of Sledge-hammer the US Ambassador to London, John 'Gil' Winant, 'is worried and so is Dill, both think it may cause the Americans to shy off the West and go all out for the Pacific.'[25] Sure enough, the very next day, Friday 10 July, there was a great démarche in Washington. Marshall – with the enthusiastic support of Admiral King – telegraphed Roosevelt in Hyde Park to say that 'If the United States is to engage in any other operation than forceful, unswerving adherence to Bolero plans, we are definitely of the opinion that we should turn to the Pacific and strike decisively against Japan; in other words, assume a defensive attitude against Germany, except for air operations; and use all available means in the Pacific.'[26]

Did Marshall genuinely want to ditch Germany First because the British had turned against Sledgehammer, or did he simply hope the threat of it would persuade the British to change their minds? The fact that Dill took the threat seriously implies that it probably was genuine, unless Marshall was bluffing Dill too, or – much less likely – Dill was in on it and putting his friendship with Marshall before his duty to the British Chiefs of Staff. Also, since the British were not in receipt of Marshall–Roosevelt communications, there would be no point in Marshall sending that telegram to Hyde Park if he was indeed bluffing.

Years later Marshall said he used the Pacific option 'as a club' with which to bludgeon the British whenever necessary: 'In my own case, it was bluff, but King wanted the alternative.'[27] Marshall was therefore claiming to have used King to give the bluff verisimilitude. (It is not known whether King ever knew Marshall was using him in this way.) Equally, that might well have been an *ex post facto* rationalization of Marshall's actions, because they did not come off. King was certain that the British did not have their heart in the cross-Channel operation and said at a Joint Chiefs of Staff meeting on 10 July that they would never invade Europe 'except behind a Scotch bagpipe band'.[28] (In a sense he was right; one of the very first men ashore on Sword beach on D-Day was Piper Bill Millin of Lord Lovat's No. 4 Commando, playing 'The Blue Bonnets' and 'keeping the pipes going as he played the Commandos up the beach'.)

The journalist Leonard Mosley believed that Churchill's and Brooke's volte face over Sledgehammer had left Marshall 'appalled. Later he would become much more sophisticated in his reaction to examples of political expediency. But on this occasion he was genuinely shocked by what he considered Mr Churchill's casual perfidy, and he did something that was quite unusual for him. He hit back.'[29] On this occasion at least, Mosley was right. If it was all a bluff by Marshall, it was a daring and gigantic one, intended to terrify the British into re-engaging meaningfully with Roundup and Sledgehammer and dumping Gymnast. Yet such a scheme could be pulled off only with Roosevelt's full support. There was certainly appetite for a new policy among American decision-makers. Eisenhower had already suggested that, if the British blocked the cross-Channel operations further, America should turn away from the Atlantic war and try to defeat Japan first. Douglas MacArthur continually emphasized the help that the United States could give Russia by drawing off Japan, thus allowing the Soviets to move divisions to their western front, as well as the popularity of a Pacific-centred war among Americans. Stimson was in favour and so too, of course, was Admiral King, who had opposed Germany First from the beginning.

Having digested Marshall's telegram, Roosevelt telephoned from Hyde Park to ask for a full assessment of the Pacific alternative by the afternoon of Sunday 12 July. He wanted to know how many men, planes and ships needed to be withdrawn from the Atlantic, and its likely effects on the situation in the USSR and Middle East. Marshall immediately left Dodona for the War Department to draw up this appreciation. Of course it was impossible to reconfigure the whole of US grand strategy over one weekend, as both FDR and Marshall must have known.

On 14 July Churchill sent Roosevelt a five-sentence telegram that left no doubt about his stance. 'I am most anxious for you to know where I stand myself at the present time,' it began. 'I have found no one who regards Sledgehammer as possible. I should like to see you do Gymnast as soon as possible, and that we in concert with the Russians should try for Jupiter. Meanwhile all preparations for Roundup should proceed at full blast, thus holding the maximum enemy forces opposite England. All this seems to me as clear as noonday.'[30] It was clear as noonday to Roosevelt too – except the Jupiter part – as Marshall, Stimson and King were about to discover in equally forthright terms.

'I have carefully read your estimate of Sunday,' wrote Roosevelt to Marshall and Stimson on Tuesday 14 July, before his return to the White House the next day. 'My first impression is that it [the Pacific option] is exactly what Germany hoped the United States would do following Pearl Harbor. Secondly, it does not in fact provide use of American troops in fighting except in a lot of [Pacific] islands whose occupation will not affect the world situation this year or next. Third: it does not help Russia or the Near East. Therefore it is disapproved of at present.' In order to underline the gravity of this disapproval, as well as to remind Marshall, Stimson and King who was ultimately in charge, he signed his memorandum, very unusually for him, 'Roosevelt C-in-C'.

After Marshall had explained Roosevelt's view to the Joint Chiefs later that day, it was concluded that, although the Pacific would be preferable, 'apparently our political system would require major operations this year in Africa,' and the President wanted Gymnast. Marshall was effectively not blaming Churchill, therefore, but Democracy itself for the decision. The President was under political pressure, Marshall slightly disloyally explained to his Planning Staff, and that took precedence. Although Roosevelt still adhered to the formulation that he supported the concepts of both Roundup and Sledgehammer, Marshall fully appreciated 'that he was leaning towards the North African strategy'.[31] Without firm opposition to Gymnast from the President, Marshall faced defeat.

On the same day that Roosevelt comprehensively rejected the Japan First alternative, Dill told Churchill about a recent conversation he had had with Marshall. He warned that Marshall and King would be seeing

Roosevelt the next day, and 'would recommend turning to the Pacific as, in their opinion, the only practical alternative to action in Europe', something Marshall had confirmed to Dill personally. Marshall's objections to Gymnast, Dill reported, were that it would take carriers from the Pacific which were urgently required for operations already under way there, that it would necessitate a new line of sea communications, and that to strike at Casablanca would not draw off any German troops from Russia. Furthermore, to attack Bizerte or Algiers 'would be too hazardous' should the Germans attack through Spain to cut the Americans off, and finally it would also destroy any possibility of Roundup even in 1943.[32]

These were a formidable list of objections. Indeed Wedemeyer said of them after the war: 'I couldn't have presented the American case better if I had tried. One wonders where Dill got this information, which certainly would be considered classified.' Dill further alerted Churchill that Marshall had warned him that any future concentration on the Pacific 'would reduce the air forces sent to Britain by some two-thirds', adding of Germany First:

There is no doubt that Marshall is true to his first love but he is convinced that there has been no real drive behind the European project. Meetings are held, discussions take place and time slips by. Germany will never again be so occupied in the East as she is today and if we do not take advantage of her present preoccupation we shall find ourselves faced with a Germany so strong in the West that no invasion of the Continent will be possible. We can then go on pummelling each other by air but the possibility of a decision will be gone. Marshall feels, I believe, that if a great businessman were faced with pulling off this coup or going bankrupt he would strain every nerve to pull off the coup and would probably succeed.[33]

This analogy broke down on a number of levels, of course. Great businessmen are virtually never faced with such stark alternatives, and in any case the difference between commercial bankruptcy, from which it is possible to return, and being flung off the Continent and possibly invaded was enormous.

Dill's advice to Churchill was straightforward: 'Unless you can persuade him of your unswerving devotion to Bolero, everything points to a compete reversal of our present agreed strategy and the withdrawal of America to a war of her own in the Pacific leaving us with limited

American assistance to make out as best we can against Germany.' Pogue believes that this letter was almost certainly sent with Marshall's encouragement, which implies that if it was a bluff, Dill was probably in on it. Having read it, Churchill must have known that calling Marshall's bluff risked a return to the dark twelve months when the British Commonwealth stood alone in 1940–41. He would be taking a gamble, with stakes that could hardly have been much higher.

Dill also mentioned another unwelcome fact to Churchill in his telegram, namely that the American Chiefs of Staff were reading Field Marshal Sir William Robertson's two-volume memoir about the grand strategy of the Great War, *Soldiers and Statesmen*, and that Marshall had sent him a copy with the third chapter of the first volume heavily annotated. Churchill would have understood immediately what that meant. Robertson, who had been CIGS from 1915 to 1918, was a Clause-witzian, and volume I chapter III of his book covered the Dardanelles expedition. 'An essential condition of success in war being the concentration of effort on the "decisive front", or place where the main issue will probably be fought out,' it began, 'it follows that the soldiers and statesmen charged with the direction of military operations should be agreed among themselves as to where that front is.' Like Marshall, Robertson believed it to be in north-west France, not – as Marshall suspected Churchill did in both 1915 and 1942 – in the Mediterranean.

Over the Dardanelles, Robertson did not deny that 'it might be desirable to threaten interests which are of importance to the enemy, so as to oblige him to detach for their protection a force larger than the one employed in making the threat, thus rendering him weaker in comparison on the decisive front,' which was to be precisely Churchill's and Brooke's Italian strategy for 1943–4, but Marshall is unlikely to have underlined that for Dill's attention. Much more likely candidates for annotation were Robertson's strictures on 'ministers' – primarily Churchill himself – who were 'indifferent to, or ignorant of, the disadvantages which always attend on changes of plan and the neglect to concentrate on one thing at a time'. Churchill was also criticized by name for having briefed the supreme strategy-making body, the War Council, directly, instead of allowing the Admiralty professionals to do it, 'as was, in fact, done after Mr Churchill left that department'.[34]

Marshall further read that the General Staff had been opposed to the

Gallipoli Expedition from the very beginning, but had been overruled by the politicians, particularly Churchill. There were long quotations from the Dardanelles Commission Report about 'the atmosphere of vagueness and want of precision which seems to have characterised the proceedings of the War Council'. Robertson criticized Churchill for sending messages to the Russians promising a diversionary attack on Turkey that had not previously been cleared by the War Council, and complained that 'even the Prime Minister', Herbert Asquith, 'did not see the reply until after it had been dispatched'. Marshall would have also approved of the BEF commander Sir John French's opposition to the plan to 'draw off troops from the decisive spot, which is Germany itself'. French believed that in the Near East there were 'no theatres . . . in which decisive results can be attained'. Gallipoli therefore precisely mirrored Marshall's criticisms of Gymnast.

Robertson's book was a critique as much of the process of decision-making – 'the whole affair was so irregularly handled', he wrote – as of the Gallipoli decision itself. He criticized the methods by which Churchill had dragged the General Staff into the expedition, by taking advantage of ambiguous phraseology in War Council conclusions and by pressurizing the First Sea Lord, Jackie Fisher, at every opportunity. Robertson also charged the War Council in general and Churchill in particular with gross over-optimism, writing that 'The stress laid upon the unquestionable advantages which would accrue from success was so great that the disadvantages which would arise in the not improbable case of failure were insufficiently considered.'[35] He quoted the report as saying that Churchill had been 'carried away by his sanguine temperament, and his belief in the success of the undertaking which he advocated'. Robertson charged that Churchill's 'unsystematic methods' had led to disaster.

It is easy to see why Marshall and Stimson felt this should be required reading for Roosevelt in July 1942, but what they failed to take into account was that it was precisely *because* of the Dardanelles débâcle that Churchill behaved in so relatively restrained a manner towards his General Staff when he became prime minister. He railed, ranted, wept sometimes, often cajoled and relentlessly pressurized, but the Dardanelles had been the defining crisis of his life so far, and he had genuinely learnt from it.

Ismay believed that, while he was serving on Pershing's staff, Marshall had:

Like almost everyone, both American and British, who fought in the main theatre
. . . regarded our failure at the Dardanelles as well-deserved retribution for an
unjustifiable strategic gamble, and he condemned the expeditions to Salonika
and Palestine, etc, as diversionary debauches. In the Second World War he
used all his influence to prevent the slightest deviation from the principle of
concentration of all available force at the decisive point.[36]

Handy spoke for almost everyone in OPD when he said of Churchill:
'He was a wonderful man, but we never thought that he was too hot as
a military strategist . . . He was one of the main instigators of the
Gallipoli operation in World War One. It had more or less been a
complete disaster.' When Churchill proposed action in the eastern Mediterranean, therefore, Handy and others suspected, 'Maybe he wanted to
prove he had been right then.'[37]

When Churchill learnt from Dill about Marshall's annotations in
Robertson's book he of course spotted the implications, but replied in
surprisingly good humour, saying of Gymnast: 'Soldiers and Statesmen
here are in complete agreement.' Privately, however, he was pained
by Marshall's choice of reading matter. As Brooke told Kennedy,
'Winston's hackles were up,' and the Prime Minister 'indicated that
he would make short work of Marshall if he tried to lay down the law
over *Soldiers and Statesmen*'. Brooke had heard that in Washington
'the supplies of Robertson's book in the libraries have run dry and
Ministers are walking about with Volume One under their arms.' He
went on: 'It may do some good, but I doubt it.' To have brought up the
disastrous Dardanelles campaign directly with its principal author would
have been a brave – or more probably a foolhardy – thing for Marshall
to have done.

Even if the British soldiers and statesmen were not in complete agreement – especially over Jupiter – they were not as split in outlook as the
Americans, for as Brooke commented to his diary that evening: 'Harry
Hopkins is for operating in Africa, Marshall wants to operate in Europe,
and King is determined to strike in the Pacific!' Since Hopkins represented Roosevelt, it is clear what the President wanted, and all that
the British now needed to do was coalesce together under the coming
onslaught from Marshall and the OPD. In his memoirs, Jock Colville
recalled that Hopkins was 'privately supporting Churchill and Alan
Brooke in their determination that the first Anglo-American enterprise

should be in North Africa and not . . . take the form of a premature, ill-judged invasion of Europe'. Without knowing it, Marshall was travelling to London with a cuckoo in the American nest.

On the evening of 14 July, at a meeting at No. 10 of representatives of Allied countries grandly entitled the Pacific War Council, Kennedy recorded: 'Winston in his blue romper suit but with a clean white shirt with cuffs . . . looked well and serene, lit a cigar and proceeded to give a general survey of the war, speaking slowly and without effort.' After asking the New Zealand High Commissioner Sir William Jordan to stop taking notes because it distracted his attention, he talked of shipping losses, the efforts to sustain Russia, and the Eastern Front, and pointed out that the Germans had only seventy-five days before winter fell there. He believed 'The Japs would attack Russia when the moment came – they would stab her in the back . . . But for the moment they were gorged with their prey.'

Speaking in front of the representatives from Australia, South Africa, China, New Zealand and Holland, Churchill said that the loss of half of Japan's aircraft carriers since the war started had made India's and Australia's positions safer. As for taking on Japan across the Bay of Bengal and through the islands of the south Pacific, this could be done only if Auchinleck beat Rommel and the threatened German attack through the Caucasus did not materialize. If it did, 'We would have to devote all our efforts to meeting that onslaught.' During the hour-long meeting, Churchill caught sight of some of his memoranda lying on the Cabinet table. As he spoke he took a paste pot and stuck one of his red labels saying 'Action This Day' on to a file, and rang for a secretary to take it away. Kennedy was impressed by the way that 'He shows no sign of the terrific strain to which he is subjected.'[38]

Alec Cadogan's diary entry of 15 July was similarly understated. After discussing the global situation in a conversation with Eden, he wrote: 'We have made up our minds against a Second Front this year. This, I'm afraid, is right – sad though it might be. We want Americans to do Gymnast, President would probably be willing. But Marshall against. I fear that his idea is that, if Sledgehammer is off, America must turn her attention to the Pacific. This is all rather disquieting.'

Meanwhile, in Washington, Wednesday 15 July became, as Churchill acknowledged in his war memoirs, 'a very tense day in the White House', as the President's willingness to carry out Gymnast came under severe threat from his advisers. When Roosevelt returned there from Hyde Park that morning, Stimson and Marshall visited to try to dissuade him one more time. They took along a copy of *Soldiers and Statesmen*, telling the President that it showed that Churchill had always been addicted to 'half-baked' schemes that drew attention from the vital front in northern France, and they pointedly left the book with the President afterwards, advising him to read it carefully.[39]

The meeting resulted, in Stimson's words, in 'a thumping argument' between Roosevelt and Marshall over American involvement in the Middle East, which Marshall thought even more ill advised than involvement in the North African operation, as it would endanger the war in the Pacific and relegate any significant cross-Channel operation to 1944 at the earliest. Instead, Marshall argued, a Pacific-centric policy might be able to prevent Japan and Germany linking up in the Indian Ocean. The President nonetheless ruled out any major offensive in the Pacific. They did however agree that Marshall, Hopkins and King should go to London immediately, and Roosevelt cabled Churchill to that effect.

The President further told Stimson that he had disliked the Japan First memorandum he had been sent at Hyde Park at the weekend, which struck him as like 'taking up your dishes and going away' – that is, being childish and petulant. Stimson had to admit the truth in that, but said that 'it was absolutely essential to use it as a threat of our sincerity in regard to Bolero if we expected to get through the hides of the British,' which Roosevelt did not deny.[40] Stimson then employed an equestrian analogy to illustrate his point: 'When you are trying to hold a wild horse, the way to do it is to get him by the head and not by the heels [*sic*], and that is the trouble with the British method of trying to hold Hitler in the Mediterranean and the Middle East.'

As far as Stimson was concerned, his and Marshall's Japan First threat was a bluff, but in his autobiography he admitted that 'Mr Roosevelt was not persuaded, and the bluff was never tried. It would not have worked in any case, for there was no real intention of carrying it out,' which Stimson 'supposed that the British knew as well as he did'.[41] In fact, however, the British – especially Dill, unless he was a co-conspirator with Marshall – could never be certain of that, and Lord Halifax was writing in his diary at the time: 'Just because the Americans can't have a massacre in France this year, they want to sulk and bathe in the Pacific!'

After Marshall produced a set of instructions for himself for the London visit, Roosevelt returned it with 'Not approved' written at the bottom right-hand corner, and 'See my substitute' underneath. At the end of these alternative instructions, the President wrote in his own hand the ominous words that, if he failed to persuade the British of the benefits of Sledgehammer, Marshall would have to determine 'upon another place for US troops to fight in 1942'. Their watchwords in London, Roosevelt emphasized, must be 'speed of decision on plans, unity of plans, attack combined with defense, but not defense alone'. Once again he signed the document as 'Commander-in-Chief'.[42]

Stimson's autobiography attests that he 'pushed his disagreement with the President to the limits prescribed by loyalty'. Had he and Marshall by then spotted, as he later claimed, that Gymnast 'destroyed Bolero even for 1943'? It would have taken supreme foresight to have looked eighteen months ahead at that stage of the war, and might be doubted. Old men forget, but old statesmen forget selectively. Stimson did, however, foresee the way the impasse would be resolved, stating that since Churchill and Brooke categorically refused to undertake a cross-Channel attack in 1942, while Roosevelt 'categorically insisted that there must be *some* operation in 1942', the only way that both demands could be satisfied simultaneously was through Gymnast.

Thus before Marshall even landed in Britain he was in a minority of one among the four Masters and Commanders, and he had failed to get presidential authority to try to bluff the British. It is unclear, even at this distance of time, whether King was aware that the memorandum he co-signed with Marshall might have been intended as a bluff. If he had thought it was Marshall's genuine reaction to disappointment over Sledgehammer, he was about to be gravely disappointed.

In *Hinge of Fate*, published in March 1950, Churchill wrote of this 'very tense day in the White House' that, although there were discussions about ending the Germany First policy, 'There is no evidence that either General Marshall or Admiral King harboured such ideas.' Yet in fact he knew from Dill's telegram that that was not the case, and that both King and Marshall did harbour 'such ideas', or claimed to do so. Churchill blamed 'a strong sense of feeling in the powerful second rank of the American staff' for the resurgence of Japan First, adding that 'The President withstood and brushed aside this fatal trend of thought.' According to Dill's telegram it was the Joint Chiefs themselves, not their lieutenants, who considered the Pacific 'the only practical alternative to action in Europe'.[43] Churchill had certainly been reminded of the contents of that telegram in 1950, because he had bowdlerized it on the previous page of his book. He played outrageously fast and loose with the facts in his six volumes of *The Second World War*, especially with regard to people who were still in positions of power at the time of publication, and Marshall had only just ended his post-war stint as US secretary of state.[44]

Churchill edited out Dill's final paragraph in which he had somewhat cheekily informed him that Roosevelt, Marshall and King were reading *Soldiers and Statesmen*. In an early draft of *Hinge of Fate*, Churchill had included that fact, along with the observation: 'Naturally General Marshall could not know how perfect was the harmony of thought between his British comrades, and that the divisions and quarrels of *Soldiers and Statesmen* which had disfigured the previous war belonged to a vanished epoch.' In the event, he cut this somewhat arch sentence before publication.

The news from Russia was bad in mid-July 1942, with the Germans pressing forward along a wide front and the British War Office fearing that there seemed little chance of stopping them 'before they get to the Caucasus and the oil'. It was hoped that by the time winter came, there might still be Russian resistance on the Volga. Auchinleck remained on the defensive in North Africa, with the initiative lying with Rommel. The 'big commitment' to the desert campaign worried British Planners, especially if the Germans got to the Caucasus and threatened Iran and Iraq. 'Altogether, the war is in an anxious phase,' noted Kennedy. This seems to contradict Wedemeyer's criticism that 'The truly depressing thing about [Gymnast] was that, even as it was being accepted, the danger of a junction of the Germans and Japanese in the Indian Ocean had completely evaporated. The fall of Tobruk had been a temporary worry.'

Wedemeyer believed that, after the battle of Midway in June 1942, where the US Navy sank four Japanese aircraft carriers, 'The British could no longer logically play up such a threat as they had in the past in order to obtain resources from American production for their African ventures. The die had been cast, however, before the full implications of Midway had dawned on the authors of [Gymnast] . . . our feet were on the ladder.'[45] Yet even if the Japanese were no longer so serious a threat in the Indian Ocean after Midway, they continued to pose a threat to northern India until the victories at Kohima and Imphal in the spring of 1944, and even had there been no link-up at all, Germany's control of the Iraqi and Iranian oil fields would have been disastrous for the Allies.

On Thursday 16 July, Marshall, Hopkins and King flew from Washington 'to finalize the competing claims' between Gymnast and Sledgehammer (which was still confusingly called Bolero by some Britons and many Americans). Although Churchill knew from Dill that Marshall was considering dropping Germany First, he also understood that it was the Commander-in-Chief who took the ultimate decision, and he had squared him a month before at Hyde Park.

Roosevelt's written instructions to Marshall, Hopkins and King – of which Churchill was unaware – had left the field completely open in the conflicting claims between the two operations. He ordered the three men to 'reach immediate agreement', however, on 'definite plans' for action in 1942, as well as 'tentative plans for the year 1943'. He ordained that 'Absolute co-ordinated use of British and American forces is essential,' that 'It is of the highest importance that US ground troops be brought into action against the enemy in 1942', and that 'If Sledgehammer is finally and definitely out of the picture, I want you to ... take into consideration ... a new operation in Morocco and Algiers designed to drive in against the back door of Rommel's armies.'

Roosevelt then reiterated his commitment to Germany First – 'I am opposed to an American all-out effort in the Pacific' – making the extraordinary prediction that 'Defeat of Germany means the defeat of Japan, probably without firing a shot or losing a life.' His final line was about his 'hope for total agreement within one week of your arrival'.[46] Having thus completely hobbled his delegates, Roosevelt needed only to wait for Churchill and Brooke to turn down a 1942 Sledgehammer for Gymnast to be adopted *faute de mieux*. Roosevelt had not even permitted his team the possibility of stretching out negotiations, let alone coming home empty-handed.

Before Hopkins left for London, he and the President worked out a series of private codenames for their messages to one another. Marshall was 'Plog', named after Roosevelt's mother's superintendent at Hyde

Park, William Plog; Churchill was 'Moses Smith', named after a friend who rented a farm from Roosevelt; Portal was 'Rev. Wilson', the rector of the local church, and Brooke was 'Mr Bee', who was the caretaker of Roosevelt's hilltop village there.[47] The last would perhaps have not been as impenetrable to eavesdropping Germans as some of the others, but the characterization of Marshall, Churchill and Brooke all as tenants or employees of Roosevelt was an unconscious indication of the way that the relationship between the British and Americans was moving.

Marshall's TWA Stratoliner landed at Prestwick in Scotland at 5.19 p.m. on Friday 17 July 1942; his party also included Admiral King, Harry Hopkins (and his doctor), FDR's press secretary Stephen T. Early, Brigadier-General Walter Bedell Smith and Marshall's aide-de-camp Major Frank McCarthy. When they arrived, the weather was too bad to go on to London by plane, and Churchill had sent his train with Commander Thompson to take the senior members of the party direct to Chequers. This led to the first clash of the trip, however, since Marshall wanted to talk to Eisenhower and other senior London-based American Staff officers before seeing the Prime Minister, whereas Churchill understandably wanted to talk to Marshall, Hopkins and King before their views were informed and strengthened by Eisenhower and his Planning team. Marshall therefore instructed the train to go straight to London, where it arrived before breakfast on the morning of Saturday 18 July, Harry Hopkins and Steve Early playing gin rummy on the journey down. It was met at Euston Station by Brooke, Eisenhower and Winant. 'Since Thursday, Ike has been working night and day, preparing reports for General Marshall's use,' recalled Eisenhower's naval aide and family friend Harry C. Butcher in his bestselling diary, *Three Years with Eisenhower*, published in 1946.[48]

The American party took sixteen rooms on the fourth floor at Claridge's Hotel, where Marshall and King conferred with Eisenhower, Mark Clark, the Commander of the Eighth Air Force Carl Spaatz and Admiral Stark. Each of the rooms had a military sentry, with King insisting upon a Marine. Churchill was angry with Hopkins for snubbing his hospitality, and telephoned him at the hotel to upbraid him over the diplomatic protocols broken by such behaviour. 'The Prime Minister threw the British Constitution at me with some vehemence,' Hopkins later joked to Roosevelt. 'As you know, it is an unwritten document, so no serious damage was done.' He nonetheless went off to Chequers, while Marshall and King used the weekend to confer with the Planners.

Brooke was 'delighted' that the Chequers weekend, at which he would have had to have been present, had fallen through, and was 'very amused' by a rumour that Marshall did not want to assume the post of supreme commander, 'so long as he would have to deal with Winston as prime minister'. Brooke and the Chiefs of Staff also prepared for the visit, recognizing from Dill's warning telegram to Churchill the potential danger that it posed to Germany First. 'They have come over as they are not satisfied that we are adhering sufficiently definitely to the plans for invading France in 1943, and if possible 1942,' Brooke noted. 'In my mind 1942 is dead off and without the slightest hope. 1943 must depend on what happens to Russia. If she breaks and is overrun there can be no invasion and we should then be prepared to go into North Africa instead. But Marshall seems to want some rigid form of plan that we are bound to adhere to in any case!' After the war, Brooke wrote of these meetings that he had 'found Marshall's rigid form of strategy very difficult to cope with. He never fully appreciated what operations in France would mean – the different standard of training of German divisions as opposed to the raw American divisions and to most of our new divisions.' Brooke feared that Marshall still 'hankered after direct action in France without appreciating that in the early days such action could only result in the worst of disasters'.[49]

At 4 p.m. on Saturday 18 July, just as Brooke could see the end of the day's work in sight and was thinking of going home, he was informed that all the Chiefs of Staff were 'wanted at Chequers for the night!!' He got there just in time for dinner, finding Pound, Portal, Mountbatten, Ismay and Cherwell already present. After dinner there was a long meeting reviewing the situation until 2 a.m., when they were taken to see a short film, finally getting to bed at 2.45 a.m.

Chequers, a fine Elizabethan country house in Buckinghamshire with some neo-Gothic alterations, had been given to the nation by Lord Lee of Fareham in 1917 as a retreat for prime ministers who had nowhere to live near London. Churchill used it instead of the more distant Chartwell. Although we do not know which room was used for his meetings with the Chiefs of Staff, it might well have been the long and spacious library, where there were large sofas on either side of the fireplace and places for chairs to be drawn up. At their meeting on the Saturday night,

Churchill and Brooke agreed to tell the Americans that the 'only feasible proposition' for 1942 'appeared to be' an American landing in French North Africa, with 'more easterly' support from the British.[50] They could not be certain that they were not about to be presented with an ultimatum and effectively blackmailed into undertaking some sort of cross-Channel operation, of course, but Churchill trusted to whatever deal he had worked out with Roosevelt at Hyde Park the previous month.

With both sides thoroughly prepped for their coming clash, the British assembled at 12.30 p.m. on Monday 20 July. Brooke had originally intended to meet the Americans at 10 a.m. 'for a private talk', but Churchill was 'very suspicious and had informed me at Chequers that Marshall was trying to assume powers of C-in-C of American troops which was President's prerogative!' Churchill knew that Brooke was highly sceptical of Jupiter and had been critical of Gymnast, and so presumably did not want him discussing operations with Marshall beforehand and in his absence. Instead Churchill and the Chiefs of Staff worked over lunch on the details of the British reply to Marshall's proposals.

When the Americans arrived at 3 p.m., Marshall put forward his case for Sledgehammer, stating that 'We would be guilty of a gross military blunder if Germany should be permitted to eliminate an Allied Army of eight million men, when some stroke of ours might have saved the situation.' If Russia and Germany were approaching their crisis, he argued, 'Sledgehammer would constitute the most effective action that the Allies could take on behalf of Russia and, indeed, for the Allied cause in general.' This led to what Brooke described as a 'long argument' with Marshall and King. 'They failed to realize that such an action could only lead to the loss of some six divisions without achieving any results!' the CIGS noted afterwards.

When the Americans next argued for a Sledgehammer-style bridgehead to be established in Normandy in 1942 that would serve as the base for a Roundup-style operation in 1943, Brooke tried to convince them that there was no hope of it surviving the winter. Afterwards Gymnast was discussed, but the Americans said they preferred action in the Pacific to North Africa. The threat was therefore made, even if Roosevelt had already likened it to a child breaking dishes in a tantrum.

Six years later, Jacob recalled some of the arguments used against Sledgehammer in those hard-fought July meetings:

We said that if you do this there is no guarantee that you will be able to carry it out. In the meantime the Germans might stabilize the front in Russia and concentrate one hundred divisions in France. In that case Roundup could not be launched in '43, six months would have been wasted, and we would still have undertaken no offensive action. The essential thing was, we said, to operate in 1942 in some area where we could meet the Germans on reasonable terms regardless of what happened in Russia.[51]

This was an exaggeration: merely stabilizing the Eastern Front could not have freed up one hundred German divisions out of the 185 or so stationed there; only pushing the Soviets back beyond the Urals could have achieved that.

At 11 a.m. the next day, Marshall presented Brooke with another memorandum in favour of Sledgehammer, in which he argued that it needed to be executed before 15 October 1942 and 'regarded as the opening phase of Roundup with a consequent purpose, not only of remaining on the Continent, but of building up ground and air forces and logistic facilities, and expanding our foothold, to the limit of our capabilities'. He wanted a task-force commander appointed immediately, with an initial objective of taking the Channel Islands and Cherbourg. Such a foothold, Marshall argued, 'will afford some relief from the continued inaction we are now enduring' and would 'provide valuable training and experience' as well as tending 'to promote an offensive spirit throughout the entire British and American armies, and peoples'. It would also bolster Russian morale, 'and will either force some material diversions from the Russian front or will allow us to operate in France against the weakest forces we can ever hope to find there'.

Meanwhile, the occupation of Cherbourg would seriously interdict German communications through the English Channel, and could take place before the enemy had an opportunity further to fortify the beach defences. This was a strong point to make, since it was mainly in 1943 and 1944 that the Germans built the formidable coastal fortifications known as the Atlantic Wall. Cherbourg's capture, Marshall argued, would also give 'positive evidence to the people of France and other occupied countries of our will to carry the war to the enemy'.

Marshall used his memorandum to counter some of Brooke's arguments. On the question of lack of air support, he pointed out that, other than the Pas de Calais, Cherbourg was the best place for English-based air operations, which could be concentrated closely there. The 250 German bombers then stationed in France could be made to pay a high price at Cherbourg through fighter attack by day and barrage balloons and anti-aircraft defence by night. Their aerodromes would also be bombed continually. If it were true that Sledgehammer would have 'no effect whatever on the Russian Front', as Brooke alleged, then Marshall suggested that 'our prospects for future success will be enhanced'.[52] It was a curiously circular but logical and undeniably optimistic stance.

It also had absolutely no effect on Brooke, however, who wrote in his diary: 'Disappointing start! Found ourselves much where we started yesterday morning!' They argued for two hours over the Russian front and weather conditions in the Channel, 'during which time King remained with a face like a Sphinx, and only one idea, i.e. to transfer operations to the Pacific'. Yet Brooke's physiognomy was equally unrevealing; Hopkins told Moran that 'Nothing that they said appeared to make the slightest impression on General Brooke's settled convictions . . . he kept looking into the distance.'[53]

At 11 o'clock that night Brooke had to go to report to Churchill, who had returned to Downing Street. Although Eden and Hopkins were also there, he was not allowed to join them, 'for fear that Marshall and King should hear of it and feel that I had been briefed by Hopkins against them according to the President's wishes!!' Instead Churchill came to the Cabinet Room to hear the results of the meetings so far, and Brooke did not get home to his flat at 7 Westminster Gardens until half-past midnight.

It is clear that something significant was going on at Downing Street that night. If he had nothing to reproach himself for, Hopkins would have been able to meet Brooke, as he had on numerous occasions when Marshall was present (including the meeting early that same day). It is therefore likely that Hopkins was passing on to Churchill and – at one remove therefore – to Brooke, Roosevelt's view, which coincided with the British stance but clashed with that of Marshall and King.

The President was thus possibly – via his intimate confidant Hopkins – encouraging Churchill and Brooke to stay firm in their opposition to Sledgehammer, leaving Gymnast as the only alternative, under his written instructions to Marshall. Despite travelling over and staying at Claridge's with them, Hopkins was nonetheless there to undermine Marshall's and King's position if they couldn't get Sledgehammer past Brooke. Not meeting Brooke directly in Marshall's absence would preserve deniability if necessary. Although this seems a lot to infer from a single diary entry, there is further circumstantial evidence to suggest that that was indeed what was going on.

Negotiations began again at 11 a.m. the next day, Wednesday 22 July, when Marshall handed Brooke another memorandum supporting an attack on the Cherbourg salient as the preliminary move for a larger assault in 1943, and Brooke put the case against. Rather than arguing the same issue yet again, the Americans merely stated that they would now have to put the matter to the President, asking to see Churchill beforehand. Brooke therefore fixed a meeting for 3 p.m. that day, and went to Downing Street to explain to the Prime Minister 'how matters stood and to discuss with him most profitable line of action'.[54]

After lunch with his wife – 'it was such a joy and rest from my labours having a couple of hours with you' – Brooke went back to Downing Street for what was clearly going to be the major showdown of the visit. In the Cabinet Room between 3 and 4 p.m., Marshall, King and Hopkins argued for a Cherbourg bridgehead before Churchill and the Chiefs of Staff. 'Without Sledgehammer,' said Marshall, 'we are faced with a defensive attitude in the European theatre.'[55] Hopkins passed a piece of Downing Street paper on which he had written 'I feel damn depressed' to Marshall, who concluded the third meeting with the British by saying that he needed to report the stalemate to the President. Churchill – who Brooke thought was 'very sound on the whole business' – told Marshall 'that no decision was constitutionally valid until confirmed by the War Cabinet, of which Brooke was the senior military adviser', thus effectively warning him that Brooke's decision was final.[56] Marshall cannot have hoped the War Cabinet would endorse his own views over those of the CIGS.

At 5.30 p.m. Brooke reported to the War Cabinet on the negotiations with the Americans and read out the salient points of Marshall's memorandum for Sledgehammer, which he said 'was envisaged by the American Chiefs of Staff as either (a) a desperate venture to assist Russia; or (b) a valuable prelude to the Roundup operation of 1943'. He then told the politicians what he had told the Americans, namely that Cherbourg was on the fringe of the area over which fighters could operate from Britain and thus no continuous fighter protection could be guaranteed. The maximum force that could be maintained there was six divisions plus ancillary units such as anti-aircraft, which he did not believe would 'get very far inland'. To hold a peninsula 16 miles at the top and 65 at the base would require at least ten divisions, he argued, and since the United States could contribute only three by mid-October, 'the rest of the force would have to be provided by this country'.

Brooke conceded that the Luftwaffe had only 250 bombers operating from four French aerodromes, but they could quickly fly in reinforcements from Germany. He reminded the Cabinet how they had reinforced their air forces bombing Malta in late 1941, despite the Russian campaign. The Germans had about twenty-seven divisions in France, he estimated, of which about fifteen were stationed in the coastal areas between Belgium and Brest. A force of some six to ten divisions could therefore quickly be built up and brought to bear against the Sledgehammer forces without weakening the Eastern Front.

Taking no account of the Italians, therefore, the Wehrmacht could amass a considerable force against the Cherbourg peninsula. The Allies would soon be bottled up there, Brooke argued, and a systematic air and land attack would be directed on the troops and port facilities. 'The result might well be that our forces would be driven back into the sea,' he feared, 'while the troops themselves, and their equipment would be lost.'[57] It was a dour assessment wildly at odds with the US memorandum, but it came from the man who had commanded both II Corps at Dunkirk and the Second BEF, much of which had been evacuated from Cherbourg itself.

Portal told the War Cabinet that a large fighter force could be available on the invasion day for about three hours, after which thirty-five fighters could be kept aloft at any one time, but they would be in the vicinity of a German first-line aircraft strength of 324 bombers and 426 fighters, three-quarters of which would be serviceable from the outset, so he could not guarantee 'that we should not suffer severe losses during the assault'. As the campaign progressed, however, things might get worse. Even if two aerodromes in the peninsula and two in the Channel Islands were captured, he estimated, the fighter force would still be 280 or fewer, and would require protection by six hundred anti-aircraft guns.

'Assuming that the Russians were still holding on the Volga, that

Rommel had been driven back, and that Germany was not involved in major operations in Persia, Iraq or Spain,' Portal said, 'Germany might possibly be able to build up against us a force of 700 bombers and 930 fighters,' of which he estimated three-quarters would be serviceable. Within six months, therefore, Cherbourg and its environs would be 'a heap of ruins'. For good measure Admiral Pound gave statistics for how bad the weather could be in the Channel in the autumn. The four consecutive days of good weather needed for the attack happened, as an average over the last ten years, only 4.3 times in September and 2.2 times in October.

Churchill said that Cherbourg was attractive only as part of a much larger plan of simultaneous landings along the French coast, but would not work on its own and would delay Roundup indefinitely. 'On the other hand,' he told the War Cabinet in the Americans' absence, 'we must not show ourselves too ready to raise difficulties.'[58] He was still very conscious of the pressures on Roosevelt and did not want to provoke Marshall's ire unnecessarily, chiefly because of the Pacific alternative. Only Mountbatten supported the Cherbourg operation, as 'the one area of the coast on which we could stage a successful assault this year'. Since it was Mountbatten's criticisms of Sledgehammer during his marathon meeting with Roosevelt the previous month that had precipitated the whole change of plan, it is understandable that Brooke and the other Chiefs of Staff occasionally became exasperated with him. He was also the major progenitor of the calamitous Dieppe Raid the following month, and it is probably fair to assume that an attack on Cherbourg would merely have been a vastly larger version of that.

Churchill then stated that 'If Sledgehammer was abandoned', discussion of Gymnast 'would at once be started' and that 'planning for Roundup in 1943 would, of course, continue.' At this, Brooke told the Cabinet that Gymnast had to be carried out before the end of 1942, therefore 'many months before Roundup', and before the enemy could anticipate an attack in North Africa. While there had been 'no definite discussion' with the Americans on Gymnast, he said, 'they were favourably impressed with the importance of the scheme'.[59] It is hard to see how he could have come to this conclusion, given that the operation had indeed been considered, and dismissed, by Marshall only very recently.

At the end of the meeting, Churchill went around the table asking each member in turn for his views, and got a unanimous verdict against

Sledgehammer and in favour of Gymnast (or actually a hybrid of it codenamed Mohican, which comprised several more attacks along the North African coast). Brooke recorded afterwards of Sledgehammer: 'I had no trouble convincing Cabinet who were unanimously against it.'[60] Ismay was then instructed to inform Marshall that the War Cabinet had turned Sledgehammer down, which Churchill thought would 'open the way for discussion on the alternative operation with the least difficulty'. Eisenhower reacted somewhat melodramatically to the news, telling Butcher that Wednesday 22 July 1942 could well go down as 'the blackest day in history' if Russia was defeated by 'the big Boche drive now so alarmingly under way', and the West had done nothing to save her.[61]

That night the British Chiefs of Staff gave their American counterparts dinner at Claridge's, which must have been a somewhat strained social occasion as they all awaited the President's response to Marshall's gloomy report of the three days' impasse. Nonetheless, Brooke recorded that 'On the whole [it] went well.' The least desirable *placement* for a Briton that evening must have been to be sat next to Ernest King, for as Marshall told Pogue in 1957: 'I had trouble with King because he was always sore at everybody. He was perpetually mean. I made it my business to be on a very warm and understanding basis with the British, and they were appreciative of that. We were more suspicious of them than they were of us. This may not have been a compliment. I think they just thought we didn't know enough.'

Back in Washington, Stimson insisted on seeing the standoff in terms of 'a fatigued and defeatist government which had lost its initiative, blocking the help of a young and vigorous nation whose strength had not yet been tapped'.[62] Brooke of course feared that the US strength would not be 'tapped' so much as wrecked, and Britain's along with it. The experience of these negotiations with Marshall and King must have been rather like reliving his June 1940 conversation with Churchill at Le Mans.

That day John Kennedy was given a full briefing on the negotiations by Brooke, who told him that Roosevelt had 'given instructions to Marshall to the effect that the American Army must get into action somewhere against the Germans and that he was to go and make plans accordingly'.[63] This is so remarkably accurate that Brooke simply must have known at least the gist of the secret instructions that Roosevelt had given Marshall and Hopkins before they left. Had Hopkins leaked them to Churchill, who passed them on at the 11 p.m. meeting at Downing Street? However he came by the information, Brooke knew that if he stayed utterly intransigent over Sledgehammer – if he kept 'looking into the distance' – Marshall was under orders finally to buckle. Thus forewarned, Brooke could hardly fail; it was like playing poker against an opponent whose cards were face up on the table.

Brooke told Kennedy that, after he had expounded on all the defects of Sledgehammer, Marshall had said to him: 'Well, how are you going to win the war? You cannot win it by defensive action.' Brooke replied that 'That was another matter and the question was not so easy to answer.' It was hardly an inspiring response from Britain's senior military strategist, who still thought Gymnast risky and Jupiter suicidal.

Kennedy emphasized to Brooke that, over Sledgehammer, 'much as one might hate constant refusal to take offensive action, there seemed no doubt that the project was impracticable in the near future. What we had to do in this phase of the war was avoid *losing* it. We could only expect a disaster if we landed prematurely.'[64] He estimated that the Germans might concentrate up to thirty divisions against an invasion without taking a single one from the Russian front, since they had forty-seven in central Europe and twenty-seven in France. Moreover, the Allies would only have small aerodromes and few port facilities in the Cherbourg area, the use of which could be denied them by German air action alone. As Churchill also put it to Kennedy: 'We should be eating up the seed corn of later and bigger projects.'[65]

Kennedy therefore believed it 'really a good thing' that Marshall had written his Sledgehammer memorandum, 'for it is easier to point out the fallacies'. He considered it 'most curious that Marshall has no sense of reality in considering the position that would arise if we found ourselves in the same land theatre as the German army with a tiny army against their huge forces'. He loyally concluded of his boss: 'We are indeed fortunate in having Brooke to expound these matters, for he is clear and decisive and most practical in his outlook.'[66]

At a 3 p.m. meeting the next day, Thursday 23 July, Churchill told the Chiefs of Staff that 'Roosevelt had wired back accepting that western front in 1942 was off. Also that he was in favour of attack in North Africa and was influencing his Chiefs in that direction. They were supposed to be working out various aspects with their staff and will probably meet us tomorrow.' Roosevelt, Churchill and Brooke had therefore decided fundamentally to alter the future direction of the war southwards. The phrase 'western front', with its Great War overtones, was instructive; after all, Casablanca is six degrees further west than Cherbourg, and North Africa was never referred to as the 'southern front'. Brooke was certainly not enthusiastic about Gymnast – soon to be rechristened Operation Torch for security reasons – and recorded that Churchill was 'anxious that I should not put Marshall off Africa by referring to Middle East dangers in 1943. Told him I must put whole strategic position in front of Americans. Foresee difficulties ahead of me!!'[67]

That evening Marshall dined at Brooke's flat in Westminster Gardens with Brooke, Grigg, the Quartermaster-General Sir Walter Venning and the Deputy CIGS Lieutenant-General Ronald Weeks. Marshall was 'in very pleasant and friendly mood', which is a fine testament to his strength of character since just about everything he had worked for on the strategic level since Pearl Harbor was collapsing around him, and his bluff had also been called successfully.[68] If anything confirms the universal description of Marshall as a great gentleman it was his good nature when dining with his arch-antagonist (at least in strategic terms) that night.

'We were at a complete stalemate,' Marshall recalled of that period years later when talking to Pogue. 'Churchill was rabid for Africa. Roosevelt for Africa. Positive reaction by both. Both were aware of political necessities. It is something we fail to take into consideration.' The need for an alternative plan was now overwhelming, and the President's instructions were unambiguous, so 'One morning before breakfast, I sat down at Claridge's in my room and began to write. I recognized we couldn't do Sledgehammer and that there was no immediate prospect of Roundup. What was the least harmful diversion? Always bearing in mind that we didn't have much. I started into writing a proposal . . . It called for an expedition into North Africa with operations, limits, nature.'[69] Just as Marshall was finishing the paper, King entered the room. He had obviously come to the same conclusion. 'It is remarkable now,' recalled Marshall fourteen years later, 'but King accepted without a quibble, despite the fact that he usually argued over all our plans'. King's acceptance of the situation illustrates how hopeless it was for the American delegation.

Portal believed that the decisive factor in making the Americans accept Operation Torch was 'our argument that we couldn't amass enough shipping for Roundup until the Mediterranean was open, and we were saved a long [journey] all around the Cape. When we put this argument to Marshall and the others in conjunction with the Torch plan, they argued it out amongst themselves at the Dorchester all one evening, and finally came round to accepting our point of view.'[70] He was wrong, and not just in the name of the Mayfair hotel. In fact it was a very clear order from their commander-in-chief that forced Marshall and King to think again.

Like Portal, Kennedy also thought that superior British arguments rather than presidential diktat had won the day. 'The last week has seen a development in our planning with the Americans that may govern the future course of the war,' he wrote. 'Marshall and King came over with very fixed ideas of what they wanted to do and were convinced that our ideas were sounder.'[71] This was a wholly inaccurate interpretation. Marshall and King had run up against two immovable objects in the shape of Churchill and Brooke, and in the event of not being able to shift them had been ordered by Roosevelt to take the North African route.

Once Marshall had completed his memorandum after breakfast on Friday 24 July, he contacted Brooke to arrange a noon meeting with the Chiefs of Staff. 'I was a bit nervous as to what they might have been brewing up since our last meeting!' Brooke confessed to his diary. 'I wondered what new difficulties and troubles I might have to face!' He needn't have worried; Marshall's paper, he discovered, contained 'almost everything we had asked them to agree to at the start'. This document, officially entitled 'Operations in 1942–3' but known by its designation 'CCS 94' (that is, the 94th memorandum put out by the Combined Chiefs of Staff), was adopted with only minor alterations. In return for an agreement to give up Sledgehammer and an early Roundup, and to prepare for an attack on North Africa, Brooke had to agree to cuts in proposed air allotments in the Far East and to having a US armoured division stationed in Iran. Neither was problematic.

'It having been decided that Sledgehammer is not to be undertaken as a scheduled operation,' CCS 94 began,

we propose the following plans for 1942–3 ... If the situation on the Russian front by 15th September indicates such a collapse or weakening of Russian resistance as to make Roundup appear impracticable of successful execution, the decision should be taken to launch a combined operation against the North and North-west coasts of Africa at the earliest possible date before December 1942.

Marshall's memorandum stated that Sledgehammer would be continued only for the purposes of deception; that an American armoured division would be sent to the Middle East in British ships; that the operation against Casablanca would be wholly American, while others against Algiers and Oran would be British 'but under a United States veneer', the whole to be under the overall command of an American; that 'plans should be made . . . to re-enter Europe as soon as opportunity offered', and that it was hoped that the US would have 3,100 first-line aircraft in Britain by April 1943, but 800 would be diverted to the Pacific and 700 to the North-west African operation, leaving 'a substantial Air Force in this country for bombing Germany or for operations against the Continent in 1943'.

As so often in hard-fought compromises between Staffs, the key detail was to be found towards the end, almost in the small print. Under paragraph C subsection 4 it stated: 'That it be understood that a commitment to [Torch] renders Roundup in all probability impracticable of successful operation in 1943 and therefore that we have definitely accepted a defensive encircling line of action for the Continental European theatre, except as to air operations and blockade.'[72]

That might sound like Brooke's strategy, but there was a catch, one that Michael Howard has even likened to a Faustian compact made between the British Chiefs of Staff and the Americans. CCS 94 seemed to imply that Churchill's original WW1 document from the Arcadia Conference had now been officially superseded, and that instead of Germany First, the phrase 'defensive encircling line of action' meant that the Americans could now also concentrate more on the Pacific. This rogue interpretation seemed to be supported by the specific provision in CCS 94 for the transfer of fifteen US aircraft groups – about 800 planes – 'for the purpose of furthering offensive operations in the Pacific'. (They were to prove invaluable for the conquest of Guadalcanal.)

The US Navy soon professed to believe that CCS 94 now rendered the Germany First policy, as delineated in Churchill's WW1 document, obsolete. On 11 August Dill warned the Chiefs of Staff that CCS 94 'gives to American Naval Staff the extra emphasis on the Pacific theatre they have always wanted and intend to maintain', and that in Washington it was now being 'quoted verbatim as the present "Bible"'. That same day Dill wrote to Marshall about this interpretation of strategic holy writ, saying: 'At present our Chiefs of Staff quote WW1 as *the* Bible whereas some of your people, I think, look upon CCS 94 as the Revised Version.'[73]

Marshall's reply was direct; in his view CCS 94 did indeed revise WW1 by diverting the air groups to the Pacific and by instituting Torch, which would, 'in my opinion, definitely preclude the offensive operations against Germany that were contemplated in WW1'. Meanwhile, Brooke wrote to Dill to say that the reference in CCS 94 to 'a defensive encircling line of action' only meant that a longer prelude was needed before the cross-Channel assault than the one 'that we had in mind when we accepted the Bolero plan'. So whose interpretation of CCS 94 was to prevail? If it was Admiral King's, it would lead to a very different war from that of Sir Alan Brooke.

At the time, however, CCS 94 looked to Brooke very much like victory over Sledgehammer and at 4.30 p.m. he took it to Churchill, who immediately approved it, seeing that it gave the green light to his beloved Torch operation. Yet at the 5 p.m. Cabinet, in Brooke's words: 'From the start things went wrong!' Usually a lapdog over grand strategy, suddenly the Cabinet expressed their doubts, and argued over perceived flaws in the deal. 'I perspired heavily in my attempts to pull things straight,' Brooke recalled, 'and was engaged in heated arguments with Eden and Cripps with most of Cabinet taking sides.' This was an unexpected development, and for Brooke – who had spent seven hours in discussion with Marshall on 22 July – a most unwelcome one. He was infuriated to have politicians such as Attlee, Lyttelton and A. V. Alexander, who had not attended the Combined Chiefs of Staff meetings, threatening to wreck the agreement at the last moment. In the end CCS 94 was passed without a word being altered, after Churchill had thrown his considerable weight behind Brooke. 'Any changes would have been fatal,' Brooke believed, because 'the Americans have gone a long way to meet us, and I should have to ask them for more.'

Nor had Churchill, Brooke or the War Cabinet spotted the problem with CCS 94 of which Dill was later to warn. Instead the discussion centred on the effect of Operation Torch on the timing of Roundup, with emphasis being put on whether planning for both could proceed concurrently. Brooke said there was 'complete unanimity' between the British and American Staffs over CCS 94, and thus, theoretically at least, over the future course of the war. Yet Marshall might have been surprised to hear Brooke claim that 'Both the British and the United States Chiefs of Staffs believed that it was unlikely Roundup would be carried out in 1943, and that ... Operation Torch therefore held the field,' since CCS 94 had a target date for Roundup as 'before July 1943' in its very first paragraph. 'A very trying week,' concluded Brooke after the Cabinet sceptics finally piped down, 'but it is satisfactory to think that we have got just what we wanted out of the USA Chiefs.'[74]

In Washington, Stimson tried to make a last-ditch protest against Torch, but all the President did was good-naturedly to offer to wager him on how the operation would turn out, which the War Secretary accepted. If either Stimson or Marshall had been commander-in-chief, Sledgehammer or Roundup would have been launched in 1942 or 1943. Asked later whether he thought the war could have been ended sooner, Stimson stuck to the formulation that 'if he were faced with the problems of 1942, he would argue again as he had then.' So his answer was yes.

Meanwhile Admiral Sir Andrew Cunningham, Pound's representative on the Combined Chiefs of Staff in Washington, was vociferously in favour of Torch, arguing that 'It would go a long way toward relieving our shipping problem once the short route through the Mediterranean was gained.' He also believed it would 'jeopardize the whole of Rommel's forces and relieve anxiety about Malta. It would shake Italy to the core and rouse the occupied countries.'[75] He was to be proved right on all fronts.

That night the Admiralty gave a dinner for the Americans in the Painted Hall of Wren's baroque architectural masterpiece, the Old Royal Naval College at Greenwich, where Nelson had lain before his state funeral. The chief guests were taken there by launch from Whitehall, possibly in order to show them how badly the City of London had suffered in the Blitz. After dinner, Churchill and Hopkins sang around the piano, which was played by the First Lord of the Admiralty, A. V. Alexander. 'One of the highlights', John Martin told his parents, 'was Admiral Stark singing "Annie Laurie" solo. Even the grim Admiral King thawed.'

One who had not thawed was the American Ambassador Gil Winant, who had seen Eden that day and 'was very critical of us' for abandoning

Sledgehammer. The Foreign Secretary reminded him 'that his people did not suggest anything before October' that could be useful to the Eastern Front. 'He had no arguments, but was obstinate,' recorded Eden. 'I have never seen Winant so put out. He dislikes Gymnast.'[76] Eden was right, however: the Americans had not offered enough divisions or planes to be able to insist on Sledgehammer taking place. With Roundup, it would be different.

The British Commonwealth had already been fighting in North Africa for two-and-a-half years, and as the historian Sebastian Haffner put it, if the Americans 'did not want to wait until they could wage their own war in two or three years' time – and America is an impatient country – they had no choice but to join in Britain's war as it stood and reinforce her with their initially slender but gradually increasing military resources.'[77] Furthermore, as Handy told the SOOHP, 'You've got to remember that the British were our principal ally and as a matter of fact were the only ones putting anything worthwhile into it' (a statement that might have surprised Stalin). 'They had the RAF and that was pretty good and they had some British divisions and they had the Royal Navy and we had to fight the war with them.'[78]

On Saturday 25 July, Churchill and Brooke were back at Chequers, where Marshall, King, Hopkins and Harriman were shown Oliver Cromwell's death mask and a ring that had once belonged to Queen Elizabeth I. The Americans then left by special train for Scotland, from where they flew back to the USA. The Britons meanwhile first watched the movie *The Younger Mr Pitt* (which became one of Churchill's favourites: it was a propaganda film, the great orator Pitt, played by Robert Donat, representing Churchill, and Napoleon, played by Herbert Lom, representing Hitler, with costumes by Cecil Beaton). They then discussed the visit before getting to bed at 2.45 a.m., in Brooke's words 'Dog tired and grateful this week is over.'

Although it had indeed been a successful week for the British in terms of getting what they needed out of Marshall, there was a heavy price to pay in terms of American suspicion, as well as the – possibly deliberate – ambiguity of CCS 94. As late as August, Marshall was complaining that Torch 'represented an abandonment of the strategy agreed in April', and of course he was right.[79] The change of Allied policy from attacking Cherbourg in France to attacking Casablanca in Africa, swivelling the whole focus of grand strategy 1,150 miles to the south, cannot but have rankled with Marshall. Even ten months later, walking to a meeting together in Washington, he told Brooke: 'I find it very hard even now not to look upon your North African strategy with a jaundiced eye!!'[80] Considering that even the US Secretary of War had bet the President that the American invasion of Morocco would fail – something that would surely have forced his resignation if known publicly – there was much ground to be made up.

IO

The Most Perilous Moment of the War: 'I am convinced that man is mad' July–November 1942

It is well that we should avoid unwarranted complacency and remind ourselves that if we did win the last war it was not without moments of extreme peril.
Field Marshal Lord Alanbrooke at a Unionist luncheon in Belfast, 1949[1]

No sooner had one great argument between the British and Americans ended than the next ones began, primarily over the issues of where, when and how to carry out Operation Torch. Marshall wanted to land on the Atlantic coast near Casablanca, and gradually move eastwards along the coast towards Algiers, whereas Brooke wanted to land at Casablanca and Algiers but also further east too, indeed as far to the east as possible, in order swiftly to gain control of the vital channel between Tunisia and Sicily, over which the Afrika Korps was resupplied. The final compromise, which was to attack at eight points along the North and North-west African coast, three near Casablanca, two near Oran and three near Algiers – but nothing further eastwards – came about only once Roosevelt and Churchill intervened.[2]

'No staff officer as far as I know, certainly none in the Operations Division, recommended the North African operation,' recalled General Hull, 'but they supported it completely once the decision had been made.' When it came to departmental unanimity, or group-think, the OPD was even more monolithic than the British Planners. Even thirty years later, speaking to the SOOHP, Generals Hull and Handy had views so similar on almost every aspect of personality and strategy that they might have been Tweedledum and Tweedledee.

On the debate with the British Chiefs of Staff, Hull said the Joint Chiefs of Staff 'insisted on going to the west coast of Africa because we wanted a foot toward the home base so that at least we could get out of there and we couldn't see [ourselves] throwing everything into the

Mediterranean. The Germans could have gone right down to Gibraltar most any time they wanted to ... We were scared to death they would come down there even after we went into North Africa.' It was the reason that the US 3rd Division was held back from the Tunisian campaign until almost the last moment. The Straits of Gibraltar, only 8 miles wide, Handy said, provided 'a focal point for the German subs, too'.[3]

There was also the question of who was going to command Operation Torch. In CCS 94 the British had accepted that this would be an American. After a long talk with Marshall on 30 July, Dill telegraphed Churchill and Brooke urging that Marshall himself was 'clearly the man for the job, and I believe he would accept. Equally clearly, he cannot be spared from here at present, but Eisenhower could well act with his authority.' Because Roosevelt had not yet approached Marshall – which Dill thought 'may be due to the President's fear of losing him' – and the self-effacing general did not wish to canvass for the job, Marshall wanted Churchill and Brooke to initiate discussions. Dill warned that the 'risk of whittling' forces away to the Pacific 'may still exist', but the President was 'entirely sound on this point'.[4]

Roosevelt might well have been sound, but there was a very definite whittling away of resources towards the Pacific going on. Although the Commander-in-Chief made grand strategy, he could not effectively prevent the US Navy pursuing a *de facto* Japan First-Equal policy, and for the rest of 1942 'resources flowed as fast to the Pacific – where the struggle for the Solomon Islands had begun in August – as they did to the Mediterranean, while those to the UK died to a trickle.'[5] Guadalcanal was invaded on 7 August and fierce fighting ensued there until February.

Dill also thought it wise, since Sledgehammer was now moribund, for the Americans 'to delegate the planning and preparations for Sledge-hammer to someone else, obviously a Britisher', so that Eisenhower could concentrate entirely on Torch.[6] (The term 'Britisher' is not one Britons use, making this cable from Dill sound all the more like one initiated by Marshall, unless it was meant facetiously or jokingly, or Dill really had gone as native as some in the War Office thought.) Churchill ordered the lines to Washington to be cleared for a 'most secret, most immediate cipher telegram' to Dill, which stated: 'I am sure that the President's wish is full steam ahead Torch at earliest possible moment. We regard this as decided absolutely with overriding priority.

No one here is thinking of anything else. You should ask to see the President urgently.' Dill replied that the President had 'issued orders for full steam ahead' on Torch, adding that the Americans believed that Torch made Roundup impossible before 1944. This might have been chagrin on Marshall's part, or perhaps Themistoclean foresight in already spotting the way that 1943 would be spent following up Torch in places far from the beaches of north-western France.

Churchill cabled Roosevelt the next day, sending a copy to Brooke, to say that he would be grateful for an early decision about the commanders of Bolero, Sledgehammer, Roundup and Torch. 'It would be agreeable to us if General Marshall were designated for the Supreme Command of Roundup and that in the meantime General Eisenhower should act as his deputy here.' Meanwhile he would appoint General Alexander as the British Task Force commander to work under Eisenhower. 'Both these men would work at Torch and General Eisenhower would also for the time being supervise the Bolero–Sledgehammer business,' wrote Churchill. 'It seems important to act quickly, as committees are too numerous and too slow.'[7] Yet Roosevelt was curiously tardy in making a decision about Marshall and Torch – it was to happen again, in even slower motion, with Overlord – and Churchill got no direct reply to this request even though Roosevelt wanted action in North Africa before the mid-term elections less than four months hence.

Instead, that same day at 12.10 p.m., Roosevelt – who was weekending at Hyde Park – asked Hopkins to put a series of questions to Marshall, who drew up 'a hasty reply' that nevertheless neatly encapsulates the general's strategic thinking at that time.[8] When FDR asked whether there were any moves the United States could make that might favourably affect the situation in the Middle East, Marshall replied, 'No, none that can affect the immediate situation.' He argued that the maximum number of planes was already en route to Cairo and that any more could not be properly serviced by the American personnel there. 'What is your personal opinion about the coming course of events?' Marshall answered that G-2 (US Military Intelligence) estimated that Rommel would be in Cairo in one week, whereas US Army Operations thought two, with one week to refit before he undertook 'the destruction of the remaining British forces'.

With prognostications as doleful as that, it was understandable that Marshall did not want to throw USAAF squadrons into the fray. His view was that he would be able to judge General Auchinleck's position in the Western Desert better forty-eight hours hence, and that if the Auk could check Rommel the long German supply lines from Tunis might place the Afrika Korps in a difficult position. After discussing British plans to block the Suez Canal in the event of defeat – which Dill estimated would take six months to reopen – Marshall suggested that the defeated British would retreat to the upper Nile, Mosul, Basra, Palestine, Aden and Colombo, while the defence of the Iraqi oil fields from Rommel 'would depend upon success of Russian defense in the North'.

To Roosevelt's question about whether America could hold Syria against Rommel, Marshall was frank. With the Mediterranean open to Germany but not the United States, the American Army would have to send nine divisions and about ten air groups, 'an expansion far beyond our capacity'. As for defending Basra and the Black Sea, the Germans would be in a far better position than the Americans, who would have 'long and vulnerable' lines of communication through the Mediterranean. Consequently, 'A major effort in this region would bleed us white.' The conclusion was obvious: the United States could do nothing to prevent Rommel's victory in the Western Desert from denying Middle Eastern oil to the Allies. For America, which took most of her oil from the western hemisphere, this would not be so dire; for Britain it was much more serious.

Marshall's sober assessment of what would happen if Cairo fell was far too pessimistic about Auchinleck's chances of preventing it happening. It nonetheless ought to have enthused Roosevelt all the more for the surprise attack on Rommel's rear in the west and on his hundreds of miles of vulnerable supply lines in the east. Almost every can of petrol poured into panzers close to the Egyptian border had to be taken there by lorry down a very long coastal road through Libya. In the War Office, the Director of Military Operations, John Kennedy, noted that 'Auchinleck is now on the last line of defence for Egypt. And in a war in which the defence has been so unsuccessful this is not a happy situation.'[9] The Second World War had indeed been, at least until the battle that was about to begin at Stalingrad, a conflict where all the laurels had so far gone to those who took the offensive.

It can hardly have come as much of a surprise to Roosevelt when Churchill told him that he was going to Cairo the next day, taking along

Brooke, Smuts and Wavell. He wished to investigate personally why Auchinleck was being so cautious. That same day, Churchill received a message from Stalin inviting him to Moscow 'to consider jointly the urgent questions of war against Hitler', and adding, 'The presence of the Chief of the Imperial General Staff would be extremely desirable.' Churchill and Brooke had never met Stalin, and although they knew that they could expect a freezing reception as a result of cancelling Sledgehammer and postponing Roundup, they accepted immediately. Churchill then asked Eden's advice about whether Beaverbrook should be invited along on the trip to Egypt and Moscow, saying: 'I like to have a pal with me.' Eden advised that since 'Max was an object not only of suspicion but hatred to many, it would not be politic'.[10] Brooke would go, of course, but he could never have been counted as a 'pal' of the Prime Minister.

Jan Christian Smuts, by contrast, was held in very high esteem by Churchill, Brooke and the British public as a whole, and not just because he had managed to bring South Africa into the war against Germany in 1939. The British have long demonstrated a soft spot for brave, defeated former foes, and in 1901 Smuts had been in command of the Boer forces fighting against them in Cape Colony. Smuts was lionized in Britain, in that strange way that also happened to other antagonists such as Napoleon after his surrender, King Cetewayo of the Zulus, Mahatma Gandhi during the 1920s, and even Erwin Rommel during the Desert War. Over lunch at Buck's Club in November 1942 – stout and oysters, steak and kidney pie, two bottles of claret – Churchill told Eden and Lord Cranborne that Smuts was how he imagined Socrates might have been like.

At the War Cabinet of 1 August, Churchill said Auchinleck's report indicating that he would not resume offensive operations before mid-September was 'very depressing' and he was flying out in order to arrange 'a more vigorous handling of matters'. This was a euphemism for Auchinleck returning to his job as commander-in-chief Middle East and someone else taking over the day-to-day command of the Eighth Army. On the morning of 3 August, Churchill and Brooke flew into Cairo West, an airfield on the Alexandria road 25 miles north-west of the Egyptian capital, and stayed at the British Embassy. 'Instead of sitting at home waiting for news from the front,' Churchill later wrote, 'I could send it myself. This was exhilarating.' Jan Smuts arrived in time for lunch, over which he teased the Prime Minister for not giving the British people 'truly spiritual inspiration', such as Gandhi gave the Indians. Churchill replied that he had appointed no fewer than six bishops that year, and 'If that's not spiritual inspiration, what is?'[11] 'But has that done any good?' asked Smuts, whereupon Churchill went on the offensive, saying to the South African Prime Minister: 'You are responsible for all our troubles in India – you had Gandhi for years and did not do away with him.' To which Smuts replied: 'When I put him in prison – three times – all Gandhi did was to make me a pair of bedroom slippers.'[12]

After the war, Marshal of the Air Force Sir Arthur Tedder, who had commanded the RAF in the Middle East, recalled that Churchill, 'fretting that there was to be no offensive action until September', urged Brooke that Auchinleck should turn the Eighth Army over to Lieutenant-General William 'Strafer' Gott, commander of XIII Corps and an outstanding desert fighter. Brooke, who unlike Churchill knew Gott, had the highest opinion of his abilities but judged him 'very tired'. (He based this on a letter Gott had written to his wife, that Brooke had somehow got to hear about.) In the early hours, Churchill offered Brooke himself the Eighth Army command. 'I shall have a job to convince him that I am unsuited for the job,' Brooke recorded at the time, 'having never been trained in the desert.'[13]

In contrast to this laconic, stiff-upper-lipped contemporaneous dismissal of the idea, Brooke admitted years after the war that Churchill's suggestion 'gave rise to the most desperate longings in my heart! I had tasted the thrill of commanding a formation in war . . . For sheer thrill and excitement it stood in a category by itself, and not to be compared to a Staff appointment. Even that of CIGS, when working for a man like Winston, must mean constant frustration, friction, and untold difficulties in achieving the results one was after.' With many of the preparations already in place for what was soon to be the battle of El Alamein, Brooke might well have been in the position of the national – indeed international – hero that fell instead to his protégé Bernard Montgomery.

On the afternoon of 5 August, Brooke visited Eighth Army HQ for tea with Auchinleck. 'I was much impressed by the beauty of the turquoise blue of the Mediterranean along this coast,' he noted. 'The colour is caused by specially white sand along this coast line.' He was less impressed with Gott, whose HQ he had just left and who he thought would not be as energetic as Montgomery in command of the Eighth Army, and equally unimpressed with Auchinleck.

Brooke recorded Thursday 6 August as 'One of the most difficult days of my life, with momentous decisions to take as far as my own future and that of the war was concerned'. While he was getting dressed that morning, 'and practically naked', Churchill suddenly 'burst' into his room 'Very elated' and told him that 'his thoughts were taking shape and that he would soon commit himself to paper!' Brooke 'rather shuddered and wondered what he was up to!'[14] Ten minutes later the Prime Minister 'burst' into Brooke's room again and invited him to breakfast. For an upper-class Ulsterman of conventional mien, one can understand that Brooke found working with Churchill discombobulating at times, but, as Colonel Aubertin Mallaby pointed out, there were no 'off' times for the Prime Minister; he was thinking about the war every waking hour.

Over breakfast, Churchill outlined his plan to split the Middle East Command into two, between a Near East stretching along the coast of North Africa to the Suez Canal and a Middle East comprising Syria, Palestine, Persia and Iraq. He wanted to move Auchinleck to the latter as he had 'lost confidence in him'. He then offered Brooke the Near East Command, with Montgomery as his Eighth Army commander. 'This made my heart race very fast!!' wrote Brooke, who was offered a short time to think it over. He nonetheless declined 'without waiting', giving as his overt reasons his ignorance of desert warfare and the fact that he would 'never have time to grip hold of the show to my satisfaction before the necessity to attack became imperative'. Neither argument was convincing: Montgomery was not a desert general either, but he managed to 'grip hold of the show' quickly enough before El Alamein.

Privately, as he told his diary, Brooke also felt that after working with Churchill for almost nine months he finally believed that he could 'exercise a limited amount of control on some of his activities and that at last he is beginning to take my advice'. By implication, he thought that the Vice-CIGS Archie Nye or someone else might not have been able to restrain the Prime Minister, and he was probably right. Churchill was not pleased with Brooke's refusal, 'but accepted it well'. Only afterwards did Smuts – clearly encouraged by the Prime Minister – take Brooke aside to try to persuade him to take up the offer, telling him 'what a wonderful future' he would have if he defeated Rommel. This was no more than the truth: the achievements of 'Alex' and 'Monty' are known by millions around the world today, that of Brooke only by the cognoscenti of grand strategy.

Brooke was not persuaded by Smuts, not least because, as a gentleman, he couldn't bear the idea that Auchinleck 'might think that I had come out here on purpose to work myself into his shoes!' He thought over the offer throughout the day, but remained convinced that his decision was the correct one, and that he could 'do more by remaining as CIGS'.[15] By putting his commitment to the wider war effort above any personal ambition for fame, or desire for the 'thrill' of independent command, Brooke did his country a very great service. We assume that politicians are driven by personal ambition, but soldiers are too, and although in career terms to swap the job of CIGS for Near East commander-in-chief might have looked like a demotion, in fact it would have afforded, as Smuts intimated, a 'wonderful future'.

At a lunch party of the Army Council at the Dorchester Hotel in November 1943, Smuts claimed it had been his idea to appoint Brooke commander-in-chief Near East, and that Brooke had replied: 'This is a very tempting thing – but my place is by the Prime Minister,' a view Brooke reiterated after sleeping on it. 'That was a great thing to do,' concluded Smuts.[16] One of those present later wondered whether Brooke ever regretted his decision, and concluded that 'Knowing now the victorious campaign that was to follow he would hardly be human if he did not.' The fact that he decided to stay beside a near-unmanageable prime minister, because he felt that no one else could do the job, thereby missing his chance of victorious generalship after a lifetime's training for it, might well explain his exasperation with Churchill on so many occasions thereafter.

Churchill explained Brooke's decision in his memoirs as having been taken because 'he had only been CIGS for eight months, he believed he had my full confidence, and the Staff machine was working very smoothly. Another change at this moment might cause a temporary dislocation at this critical time.'[17] Was Churchill being disingenuous with Brooke, and vice versa? Might Churchill have offered the post because he wanted a more malleable CIGS? The secret reason why Brooke declined was that he feared that might be the case. When the American serialization of *The Hinge of Fate* was published in 1950,

Brooke wrote to Henry Pownall, who was researching the next volume for the former Prime Minister, to say that Churchill had entirely ignored two of the three reasons he had refused the Near East command, so Churchill inserted them in the British edition. As Professor David Reynolds comments, 'It must have been galling to Brooke that Churchill had clearly forgotten "one of the most difficult days of my life".'

On the evening of 6 August, Churchill sent the War Cabinet a telegram whose terms had been agreed by Brooke and Smuts. This proposed an immediate splitting off of Persia and Iraq from the Middle East Command, making them an independent Army command, just as he had proposed to Brooke that morning. This command would be offered to Auchinleck, whom Churchill didn't want to lose altogether. He believed – or professed to – that if Auchinleck had earlier been freed from responsibilities covering the Levant and Caspian Sea he might have been able 'to concentrate his forces in the Western Desert, turned the scale and given us a victory instead of a defeat'. Meanwhile, as Jacob recorded, Brooke agreed with the plan, 'though for a rather different reason. He felt that it was wrong for an area of such vital importance as Persia and Iraq to remain any longer as the Cinderella of either the Middle East or of India.'[18] As so often, when Brooke and Churchill agreed on something, it happened – even if they came to the decision for different reasons.

Brooke and Churchill also agreed that Alexander should succeed Auchinleck in Cairo, Lieutenant-General Thomas Corbett and Brigadier Eric Dorman-Smith were to leave their commands altogether, and Lieutenant-General William 'Strafer' Gott was to lead the Eighth Army, although Brooke had misgivings about this. Yet on his way to take up his new command the very next day, 7 August, flying the Burg el Arab to Heliopolis route, which was considered safe, Gott's slow transport plane was shot down 'in flames' by a lone German fighter. Churchill and Brooke then quickly settled on the man whom Brooke had wanted originally, Lieutenant-General Bernard Law Montgomery.

The War Cabinet meeting in London on 7 August was a good illustration of the way that Churchill and Brooke dominated military policy even in their absence. It had met at 11.15 p.m. to consider the plan to divide the Middle East Command, but before the meeting ended at 2 a.m. on the 8th a telegram arrived saying that Gott had been killed. Archie Nye said that the situation was: 'In [the] hands of [the] PM and CIGS. They have in mind a General Montgomery. Not enough [is] known of the form of commanders to know that any particular man will fit the bill.' The use of the indefinite article before Montgomery's name led Burgis to assume, as he told Churchill's son Randolph years later, that none 'of those present knew him from a crow then'.

At the meeting, Bevin pointed out that it was a 'Strong team. PM, Smuts, CIGS' and for the War Cabinet it was 'Difficult to arrive at a concrete judgment at this distance'. To this Attlee added that he would like to see Alexander running the Eighth Army with Wavell in overall command of the Middle East; however, 'We must either put up counter-proposals or acquiesce.' They acquiesced, telegraphing the Prime Minister to say: 'As you, Smuts and CIGS who are on the spot are all in agreement, we are prepared to authorise action proposed.'[19] Frankly, anything else was unthinkable, and there is no example during the war of a united Churchill and Brooke being overridden on a military issue by the War Cabinet. So Montgomery flew out from Britain, taking up his command on 12 August.

In Washington, meanwhile, Henry Stimson was still deeply pessimistic about any operations in North Africa, and on 10 August he made Marshall promise that he would take a final stand against Operation Torch if 'it seemed clearly headed for disaster'.[20] Marshall had no difficulty in making that promise, which was after all no more than his duty, but it is indicative of the lack of confidence felt by many senior strategists at the time. Stimson's doubts remained, and as late as 17 September he was writing that the undertaking was risky but, 'the Commander-in-Chief having made the decision', it had to be seen through.

Stimson also drew up a sharp note to the President that he did not eventually send, but of which he gave a copy to Marshall. 'The objections to the hazards of Torch had been stated to you in previous conferences with your advisers,' it read, 'and the objection that it was a purely defensive operation instead of an offensive was inserted in the London memorandum on the Chief of Staff's sole insistence and against British opposition.' Marshall and the Staff now 'believe the operation should not be undertaken'. Stimson foresaw a risk of defeat in Africa that would emasculate Roundup until 1944, and thought that Torch wouldn't help Russia either.

In its somewhat formal summation of recent history, and reiteration of what Roosevelt already well knew, the draft read more like the preamble to a resignation, but it merely ended with an 'earnest recommendation' that 'before an irrevocable decision is made upon the Torch operation you should make yourself familiar with the present views of these your military advisers.' Stimson might have been using this unsent letter much as Brooke used his journal – partly to let off steam – and a surprising number of people do write letters they never truly intend to send, for precisely that purpose. Yet Stimson would hardly have written in such terms if Marshall had supported Torch wholeheartedly.

From 12 to 15 August, Churchill and Averell Harriman, Roosevelt's personal envoy to Stalin, conferred with the Russians in Moscow. Because of the danger of a fire in their B-24 Liberator bomber, Brooke, Cadogan, Wavell, Jacob and Tedder had been forced to turn back to Teheran, and only arrived on the 13th. They therefore missed a four-hour meeting with Stalin from 7 to 11 p.m. on Wednesday 12 August, of which, Churchill reported to Roosevelt, the first two hours were 'bleak and sombre'. The Prime Minister explained at length with maps why Sledgehammer – which he and Roosevelt had promised Molotov in writing back in June – had been indefinitely postponed. Stalin argued hard the other way, and as Churchill reported to Washington, 'Everybody was pretty glum. Finally he said that he did not accept our view but we had the right to decide.'

Everyone cheered up once Churchill passed on to what he called 'the ruthless bombing of Germany'. He then brought up Operation Torch, at which Stalin 'became intensely interested'. The conversation ranged over the whole of the rest of the war in the west, with Churchill concluding that once 'Brooke and the others arrive . . . the military authorities on both sides are to sit together and check up both on strategy and technical detail.'[21] The British military authorities arrived safely in a Russian plane at a small aerodrome on the outskirts of Moscow at 7.45 p.m. the next day and were taken straight to State Villa No. 7, where Churchill was staying, for a debriefing. After dinner, the British party and Harriman set off for the Kremlin at 11 p.m. 'It was a dark night,' wrote Jacob, 'and Moscow is completely blacked out. No headlights are allowed on cars, so that we crawled along at a very slow pace. As a result we were half an hour late.'

They were conducted to the 600-square-foot office inhabited by Stalin, whose desk was tucked away on the right-hand side at the far end. Two pictures of Lenin and one of Marx provided the only decoration. Stalin was lounging in a chair sideways on to the table at the head, puffing at a large, curled pipe. After everyone had taken their places, with Brooke next to Churchill and only the interpreter on his other side, the meeting started, badly, with another 'desultory argument about the possibility of a second front and similar matters'.

Jacob wrote that Stalin spoke 'in a very low, gentle voice, with an occasional gesture of the right hand, and never looked the Prime Minister in the face'. The reason he averted his eyes was that 'Stalin was coming out with all kinds of insulting remarks, but one could not really tell whether they were being faithfully put across by Pavlov, because his vocabulary was limited.' Stalin's translator, Vladimir Pavlov, was in fact excellent. At this first meeting with Churchill his English was hesitant, but he would take great care not to distort Stalin's words. Stalin was simply intending to be as rude as possible and 'was suggesting that we were not prepared to operate on the Continent because we were frightened of the Germans'.

According to Jacob's minutes, the entire conversation was carried on by Stalin and Churchill, with only four short interventions by Harriman and one by Tedder. At the post-mortem back at Churchill's villa, Harriman suggested that the explanation 'was probably that Stalin had to adopt an uncompromising attitude at one stage of the negotiations, in order to satisfy his own people'.[22] That too was absurd, but indicative of the way that many Westerners still failed to recognize that Stalin was an all-powerful dictator – indeed, as the title of a recent biography puts it, 'the Red Tsar'.

Four months later, Brooke threw a dinner party in Chelsea where he gave his views of Stalin, saying that the Marshal 'gave him the creeps. He looked pale and even grey with the flesh hanging from the bones of his face. Stalin did not "register" when Winston came into the room – it might have been a footman.' Brooke added that the Russian dictator showed 'no sign of humanity except once when he said to Churchill, before the interpreter could translate an impassioned speech, "I like your sentence although I do not know what it means." '[23]

On 14 August the traffic situation in Moscow could not have been more different. This time in daylight they 'drove through the streets entirely regardless of red and green lights, or of policemen or pedestrian crossings. If there are pedestrians in the way, so much the worse for them.' Brooke's was 'a peculiarly carefree driver. He never actually had a crash, but he ran down one man, who was then extricated from the wheels, and moved away to one side, so that the car could go on. The drivers treat the citizens like so many cattle.'[24] With the experience of the loss of his first wife, Brooke could not have found this pleasant.

The meeting on 15 August went badly. At noon Brooke went to the Soviet Government Hospitality House, 17 Spiridonovka Street, to deliver a statement and discuss the Second Front with Marshal Voroshilov and Marshal Shaposhnikov, the Russian Chief of Staff, who both displayed what Brooke considered an astonishing lack of understanding of how to attack over large stretches of salt water. 'Finally,' recorded Jacob, 'the CIGS told them that the Americans and ourselves had come to very definite conclusions on this subject and were not prepared to alter them.' Voroshilov then refused to discuss the fighting in the Caucasus with Brooke, who in turn replied that he had not been authorized to discuss Torch with him.

It was on this visit that Churchill made the error, while attempting to explain to Stalin the attractions of attacking the Axis from the south before attempting an invasion of France, of drawing a sketch of a crocodile with, he said, a 'soft underbelly'.[25] Once the image was lodged in the Prime Minister's mind, he used the concept of 'attacking the under-belly of the Axis' in a letter to Roosevelt the following month, and subsequently to other audiences on other occasions until it became a well-known phrase associated with him. Given that the future struggles in the south – especially in the Italian peninsula – were to be anything but soft, it was to be a gaffe that would long be held against him. The one disadvantage of having such a vivid, fizzing imagination and verbal dexterity as Churchill was that the law of averages meant that very occasionally it must misfire, and when it did a memorable but ultimately unfortunate phrase was born.

When they were in Teheran on their way back home from Moscow – which also involved travelling to Cairo, El Alamein and Gibraltar – Churchill and Brooke heard of yet another disaster for British Commonwealth forces, to add to Dunkirk, Narvik, Greece, Crete, Singapore and Tobruk. An operation to attack the French Channel port of Dieppe that they had authorized, but then left entirely up to Lord Louis Mountbatten, as director of combined operations, had resulted in catastrophe. At dawn on 19 August, 252 ships, thirty tanks and 6,100 men – two

Canadian infantry brigades totalling over five thousand men and over one thousand Commandos – had taken part in Operation Jubilee. It was intended as a 'reconnaissance in force', but had no clear follow-up plan. Even at this distance of time, it is hard to know what the Dieppe Raid was meant to achieve.

A small German convoy in the Channel alerted the shore defences before the assault could take place, so the element of surprise was lost, yet Mountbatten ordered it to go ahead anyhow. The tanks landed on the shingle beach but could not negotiate the sea wall successfully. German machine-guns accounted for most of the 4,100 Allied casualties, more than two-thirds of the attack force. The Canadians lost 907 killed and 1,874 captured; the Royal Navy suffered 550 casualties; the RAF and RCAF lost ninety-nine planes, the worst single-day total of the war, including during the battle of Britain. The Germans by contrast lost only 314 killed and 37 captured.

Although no German troops were moved from east to west as a result of the débâcle, coastal defences were massively strengthened. 'If I had the same decision to make again,' Mountbatten nonetheless averred, 'I would do as I did before. It gave the Allies the priceless secret of victory.' This is tripe, unless the lesson of not attacking a well-defended town without proper intelligence and a preliminary aerial and naval bombard-ment is a 'priceless secret', rather than the kind of assumption a lance-corporal might have made. Yet even as late as 2003 historians would still take Mountbatten at his word, with one writing: 'The catastrophe provided priceless lessons for a full-scale amphibious invasion.'[26]

(It is surprising how little influence the Canadians enjoyed in the higher direction of the Second World War. They had the world's third largest navy at one point, pushed furthest inland of any of the armies on D-Day, were fabulously generous to British coffers throughout the war, contributing much more than the Americans per capita, and pro-vided the only two armed and trained divisions standing between the south coast and London after Dunkirk. Yet they had virtually no say on the various bodies that ultimately decided how, when and where Canadians would fight.)

The writer Leonard Mosley claimed in 1971 that 'The only people in any way satisfied by the Raid were those advisers of Winston Churchill, like Cherwell and Sir Alan Brooke, who thought it would prove to the Americans once and for all that a Second Front across the Channel was unthinkable for at least another year.'[27] Brooke had served with the Canadians at Vimy Ridge and admired them. The idea that he could have taken any satisfaction in so many of them being killed, wounded and captured is monstrous. 'The casualties were undoubtedly far too heavy,' Brooke commented in his diary; 'to lose 2,700 men out of 5,000 on such an enterprise is too heavy a cost.'[28] Furthermore he did not in fact use the Dieppe Raid as an argument against Roundup, because both he and Marshall knew that in both size and objective a reconnaissance in force was very different from a full-scale invasion, and by the time Dieppe was undertaken Sledgehammer was already off the table for 1942.

On Friday 21 August, Sir John Dill threw a dinner party at his London flat for Eisenhower, Mark Clark and Thomas Handy, the War Office strategists Nye and Kennedy, and the new commander of the First Army in North Africa, Lieutenant-General Kenneth Anderson. Dill told them that Marshall worked from 8 a.m. to 4 p.m. then 'went out on the river with his wife and took a picnic supper or something of that sort', before going back to work. He also said 'what a fine and powerful agent' Marshall was with Roosevelt, and spoke of Brooke's relations with Churchill.[29] Infuriatingly, the next paragraph in Kennedy's diary was later very heavily scored through in ink on the Whitehall foolscap writing paper.

After dinner, the group settled down to discuss Operation Torch. Kennedy said the present plan – to attack Casablanca and Algiers but not Oran – 'would lead to a fiasco' because the numbers involved needed to be trebled. Dill was non-committal, as was Nye, since, Kennedy wrote, 'both have more formal positions to preserve vis-à-vis the Americans than I have.' Anderson sided with Kennedy. Ike was also non-committal, beyond saying that so far no one had said 'anything cheerful' about the plan. The Americans left around 10.30 p.m., after which Kennedy said, 'It was almost incredible that after the Americans had been in the war for a year, their share of this plan should be so small. It is perfectly obvious that their hearts are not in it (anyhow King's) and that the Pacific War is eating up resources that should be here.' He further accused them of not carrying out the agreed 'Germany first then Japan' strategy. Copying the complaint Marshall often made about overall British strategy, Kennedy said that the Torch plan suffered from lack of 'concentration of effort'.

Dill then asked why the War Office had had 'no outline plan' for

Torch ready when Marshall had visited in July, a direct criticism of Kennedy as director of military operations. Kennedy responded that the project had only come up during the visit. 'Before he came we intended only to press for the continuation of the American movement to this country and *then* to decide how to use the forces.' The Torch plan 'had therefore started from the top without a detailed examination'. Now that his War Office Planners had looked at it carefully, Kennedy said, 'we find that the difficulties especially of maintenance and shipping are greater than had been anticipated and that the forces are not nearly big enough.' This brought the conversation round to Churchill, and Dill said that the Prime Minister had ruined Auchinleck by having 'pressed and harnessed' him. 'In fact he has dwarfed him just as he dwarfs and reduces others around him.'[30]

Kennedy's defence of himself to Dill reinforces the suspicion that Torch – then still called Gymnast – was decided upon by Churchill and Roosevelt at Hyde Park, and that they subsequently prevailed upon the Chiefs of Staffs, Brooke because it was the only offensive alternative to Jupiter, Sledgehammer and Roundup, and Marshall because Brooke had blocked Sledgehammer and Roundup. When Kennedy said Torch had 'come from the top' he was more right than maybe even he knew. The reason he did not have a presentable version of it ready for Marshall's visit was perhaps because Brooke was known not fully to approve of it.

On Sunday 23 August 1942 the German Sixth Army launched Operation Blue, the all-out offensive to capture the city of Stalingrad, the industrial (especially armaments) centre on the River Volga, home to six hundred thousand Russians. At 4 p.m. the 16th Panzer Division moved into the outskirts of the city and thereafter a quarter of a million German troops laid day-and-night siege while one thousand German planes bombed the city, which had virtually no anti-aircraft defences, into mountains of rubble and corpses.

Hitherto the Germans, fighting in open country, had managed to force the Soviet Army back further and further, but at Stalingrad house-to-house combat blunted their advantage and played to the strengths of the far more numerous Russians. Instead of tanks and mobile artillery, the weapons that mattered most were grenades, bayonets, sniper rifles, small arms and sometimes even spades as the Russian Sixty-second Army was mobilized to defend the metropolis that bore their leader's name. By 12 September, German troops had got inside the city, and the next day took some key positions, such as the ferry terminal, which changed hands thrice in two hours. (The train station is said to have changed hands no fewer than sixteen times in the course of the battle.)

By 27 September two-thirds of Stalingrad was in the Germans' hands as a result of vicious, merciless fighting they termed *Rattenkriege* (rat warfare). The Russians used the sewers to stage counter-attacks, but by 11 November they controlled only about one-tenth of the west bank of the city. Such waste of strength over a place that was no longer strategically valuable could have only one explanation: prestige. Hitler had publicly promised that Stalingrad would be taken; his eponymous city was equally totemic for Stalin. In mid-November, Russian forces numbering over one million men under Georgi Zhukov smashed through the Roumanian army north and south of Stalingrad and on the 23rd Red Army units met each other at Kalach, thereby trapping the Sixth Army within the city.

The outcome was by no means certain even then, however. The superiority of German combat efficiency over that of the Russians in the early part of the war meant that, on average, 'one German division was a match for three Russian divisions of comparable size and firepower, and that under favourable circumstances of defence, one German division theoretically could – and often actually did – hold off as many as seven comparable Russian divisions'.[31] Nonetheless, attempts to relieve Stalingrad failed, and Hitler refused to countenance a break-out. The stalemate continued through the rest of 1942.

It was Stalingrad that finally, in Stimson's words, 'banished the spectre of a German victory in Russia, which had haunted the Council table of the Allies for a year and a half'. It also greatly reduced the likelihood of a German attack through Spain, cutting off the American forces from their supply lines. Just as Wellington's campaign in the Iberian peninsula had been a small but significant 'ulcer' for Napoleon, but certainly not the Russian 'coronary' that destroyed him, so too the North African and Italian campaigns would be ulcerous for Hitler, but it was the Eastern Front that annihilated the Nazi dream of *Lebensraum* ('living space') for the 'master race'. Four in every five German soldiers killed in the Second World War died on the Eastern Front, an inconvenient fact for any historian who wishes to make too much of the Western Allies' contribution to the victory.

*

Between 24 August, when Churchill received what he called the 'bomb-shell' news that Brooke and Marshall were deadlocked over Torch – with Marshall wanting to attack only Casablanca and possibly Oran, but Brooke wanting Algiers too – and 2 September, when Roosevelt changed his mind and supported the inclusion of Algiers, there was renewed transatlantic struggle between the Staffs. Marshall feared that the American forces would get cut off if they landed too far east; Brooke wanted to try to stop Rommel escaping from Tripoli, and so wanted to land as far eastwards as possible. Moran thought that Churchill 'was never so unhappy as when he was at odds with his military advisers or his American allies', but when he had to make a choice between them, he came down firmly on Brooke's side, not least because he had empha-sized to Stalin the comprehensive nature of Torch.

At 11 a.m. on 24 August, Kennedy and Major-General Francis David-son, the Director of Military Intelligence, were summoned to Churchill's private rooms in the No. 10 Annexe. 'Winston lay in bed in his black dressing gown with dragons, a half-sucked cigar in his mouth which he lit and re-lit during the next hour and a half, without making any appreciable progress with it, a glass of water on the table beside him.' He had just returned from Gibraltar, and told the two men how in Egypt 'with the change of commanders a new wind was blowing, how the Army was all in bits and pieces and that would be put right now' and of 'the terrible wastage the "poor" Army had suffered'.

Reporting on Russia, the Prime Minister said that Stalin had not made out that his situation was bad, 'as he might have been expected to in pressing for a Second Front'. Indeed he had been optimistic enough to remark: 'Dieppe will be explained by Torch.' Churchill added that he had bet Brooke half a crown at odds of evens that the Russians would hold the Caucasus, and boasted: 'I drank as much or more than Stalin and Molotov together – they only sip their liquor you know – and I was in quite good order.'

Kennedy then gave Churchill a blunt assessment on the planning for Torch, specifically the inadequacy of the American contribution, the need for overwhelming force, and a warning about 'the first manifes-tation of divergent strategies'. The bonhomie of the early part of the meeting disappeared immediately as 'Winston's hackles rose at once and his eyes, which were rather watery, began to flash.' Anyone, he said, could make a plan involving overwhelming force, but there could be no delay, especially for further American forces coming from the Pacific. He wanted to bring forward the date rather than delay it, saying that fighting Vichy France 'was a soft job – not like fighting Germans', and that he would even 'be prepared to go ahead without the Americans themselves so long as they had plenty of American flags to wave. What we wanted was a big show in the shop window.'

Kennedy replied that, although three divisions would be ready in three weeks, it would take three or four months to get thirteen together, yet the limiting factor was not troops so much as shipping, naval escorts, aircraft carriers and, as ever, landing craft for the invasion force. He argued that 'the Americans should be in the thing *wholeheartedly* not only at the beginning but subsequently', pointing out that the growing American commitments in the Solomon Islands and elsewhere in the Pacific showed that they were diverting shipping and naval craft away from Germany First. 'Winston was distinctly ruffled' before the meeting ended at 12.45 p.m.[32]

On 25 August, Churchill reported to the War Cabinet about Stalin, whom he described as a 'large man' of 'great sagacity'. His visit had 'Explained some past mysteries' about Stalin's behaviour before the war, and the rebuffing of the British military mission to Moscow in August 1939. Led by Admiral Sir Ranfurly Drax, this had been Britain's last-minute attempt to prevent the Nazi–Soviet Pact taking place. Churchill reported that Stalin had been 'certain Britain didn't intend war . . . This was confirmed by our offers – France of 80 divisions, Britain of 3 divisions. Stalin had been sure Hitler wasn't bluffing. At Munich an effort might have been made, after that nil was our offered strength.' To Churchill, who had denounced Munich at the time and called for a united front with the Soviets against the Nazis thereafter, Stalin's assurances that the weakness of the Chamberlain Government in 1938–9 had left the Russians with no option but to sign the Molotov–Ribbentrop Pact was confirmation of his stance during his Wilderness Years. To those Chamberlainites and *Munichois* still left around the Cabinet table, Stalin's assertions must have been galling. As for Torch, 'Stalin did not exaggerate his plight in order to exploit or extort help from us,' said an impressed Churchill.[33]

That day Churchill and Brooke received a document from the Joint Chiefs of Staff stating that the attack on Algiers would be too risky. 'We are all profoundly disconcerted by the memorandum,' Churchill replied on the 27th. 'It seems to me that the whole pith of the operation will be lost if we do not take Algiers as well as Oran on the first day. In Algiers we have the best chance of getting a friendly reception and even if we got nothing except Algeria a most important strategic success would have been gained.' Not to go east of Oran, he went on, 'is making the enemy a present not only of Tunis but of Algiers'.[34]

'Torch is a great confusion,' wrote Eden's private secretary Oliver Harvey in his diary. 'It is very difficult to make plans on both sides of the Atlantic and expect them to coincide. We are in favour of two prongs – US think they will only have enough for one. We don't like the East prong without the West. Behind and above all this are Winston and Roosevelt goading each other on to fix dates, etc, while all is vague.' Kennedy meanwhile rightly spotted that 'It is a political operation and stands or falls by the correctness of the political appreciations – reactions of the French, Spaniards, etc, etc.'[35]

Some American Planners thought that because the Vichy French were supposed to prefer the United States to Britain, the Stars and Stripes might be welcomed in North Africa whereas the Union Jack would be fired upon. This led to the Americans attempting to persuade the British to play a junior role in the landings, which was resented in some areas of the War Office and Cabinet. Quite why this should be, beyond feelings of national pride, is hard to say. The Americans had diplomatic relations with the Vichy Government, whereas Britain did not, so it made sense for the operation to be presented as an American liberation, and if that required the United States to spearhead it, the Churchill Government should not have baulked at an opportunity to save British lives. If national pride was the reason, as the war progressed there were to be many more such turf wars over symbolism and prestige, which rarely redounded to the credit of those involved.

'We are undertaking something of a quite desperate nature and which depends only in minor degree upon the professional preparations we can make or upon the wisdom of our military decisions,' wrote Eisenhower in his diary that week. 'In a way it is like the return of Napoleon from Elba – if the guess as to psychological reaction is correct, we may gain a great advantage in this war; if the guess is wrong, it would be almost as certain that we would gain nothing and lose a lot.' He feared that there might be 'a very bloody repulse' and that Vichy France and even Spain might enter the war against the Allies.[36] Axis propaganda indeed began to give out that there was a concentration of German forces near the Pyrenees, which there was not; and Marshall's and Eisenhower's worry that the Germans might be invited by Franco to march through Spain and outflank the Allies by closing the Straits of Gibraltar, trapping American forces in the Mediterranean, failed to take into account Hitler's and Franco's considerable mutual mistrust. (After their only meeting, at Hendaye in October 1940, Hitler said that he would rather have three or four teeth pulled out than sit through another conversation with Franco.)[37]

Staying at Chequers on the night of Saturday 29 August, Eisenhower and Mark Clark received a courier from Marshall saying that the President had definitely decided to attack Oran and Casablanca with eighty thousand US troops, but that the British should not arrive until a week afterwards, and the attack on Algiers would be omitted altogether. As Roosevelt was not planning to inform Churchill of this until the following Monday, for Clark 'this admonition to silence came at a difficult moment'. Brooke, Eden, Mountbatten and Ismay were also present, trying to finalize plans for Torch, so while 'Churchill was enthusiastic' and 'Eden expressed optimism,' Clark 'fidgeted and boiled inside, and I imagine Ike did too.' Clark recalled how embarrassing it would have been 'to air the latest word from Washington' and he and Eisenhower left on Sunday the 30th having 'answered no more questions than was necessary'.[38]

It might have been this occasion that Eisenhower recalled in his book *At Ease*, when he wrote of a meeting at Chequers where British and American views were not meshing too well. Brooke said to him: 'Naturally, you cannot be expected to oppose violently something that Washington apparently wants.' Ike recorded: 'Although I am sure he did not mean to imply that I was swayed by fear of a reprimand, I explosively put him right. I told him flatly that only the merits of a proposal, not its place of origin or its sponsorship, mattered to me when the fortunes of nations were at stake.'[39] For all his charm, Eisenhower could be waspish at times, even with Brooke. After the Chequers meeting, Eden wrote in his diary: 'Greatly impressed by Eisenhower and Clark, as I have been before. We are lucky to have them as colleagues.' Clark meanwhile went back to London where he addressed thirty-seven British and American generals, saying: 'Some of you men are less confused than others about Torch. Let's all get equally confused.'[40]

If Brooke assumed that Eisenhower could be swayed by Marshall, Marshall feared that the Torch commander might be swayed by Churchill, warning Admiral Leahy that at Chequers he was 'very much under the guns'. Marshall asked Leahy to use his influence 'to see that the President's message gets off by Monday as the delays are fatal to the completion of the plans and therefore directly affect the date for the operation'.[41] Although many important cables from Roosevelt to Churchill were drafted first by Marshall, they would often be radically redrafted by the President – sometimes in Hopkins' handwriting – before being sent off. Some important messages, such as the one trying to persuade Churchill that British troops should take a junior role in Operation Torch, went through several redrafts over a number of days and emerged greatly different from Marshall's original.[42] This was even truer when Admiral King was let loose upon early drafts, since FDR had a sense of how to turn away wrath in a manner alien to the acerbic, straight-talking head of the US Navy.

When Roosevelt's cable duly arrived on Monday 31 August it caused consternation. 'I feel very strongly that the initial attacks must be made by an exclusively American ground force supported by your naval and transport and air units,' it read. This was because Roosevelt believed that the French would offer less resistance 'to us than they will to the British'. He suggested to Churchill and Brooke that a week after the operation, once French non-resistance was secured, 'your force can come in to the eastward.' The attack should preferably take place before 14 October, thought Roosevelt, but certainly no later than the end of that month. He did not have to remind anyone that the Congressional mid-term elections fell on Tuesday 3 November 1942.

At a War Cabinet meeting that day, Eden said that there was a general impression in the press that the Second Front in Europe had been cancelled for the rest of 1942. Although this was true, Churchill emphasized that it was nonetheless 'Important to play [up] to the Germans, and not let them draw off troops from France.'[43] The last thing Churchill wanted was German troop movements from France either to Russia or to North Africa. If that meant encouraging the British press to believe that a cross-Channel operation was still possible in 1942, it was easily a price worth paying.

Churchill answered Roosevelt's telegram on 1 September, arguing that not to attack Algiers simultaneously with Casablanca and Oran might lead 'to the Germans forestalling us not only in Tunis but in Algeria', and he urged that all three ports be targeted. Roosevelt replied the following day agreeing to this, but demanding that each of the attacking forces be led by American troops, with the United States controlling all relations with the Vichy authorities once they had landed. This was undoubtedly sensible at Oran, where the Royal Navy had sunk much of the Vichy fleet in July 1940.

There was a good deal of doubt over Torch in the British High Command even comparatively late in the planning stage of the operation: on 3 September Dill told Kennedy that he didn't believe in it, and feared it would 'destroy his credibility in the States when it failed', in which event he would have to leave. After a Chiefs of Staff meeting that day, Kennedy told Brooke that the operation 'would have no chance today', but might work in November if Libya was softened up and Stalingrad held out. Kennedy also suspected that Brooke 'is not wholeheartedly behind the plan now that the implications are coming to light more clearly', especially those regarding the Navy and shipping. In reality, Brooke had not been wholeheartedly behind it from the start.

Churchill's and Roosevelt's first serious argument over strategy ended in a compromise whereby they agreed to split the difference, in terms of troops, between Algiers and Casablanca. 'We are getting very close together,' the President wrote on 4 September, offering to reduce the Casablanca force by five thousand men which, as five thousand had already been taken off the Oran operation, released an extra ten thousand for Algiers. 'We should settle this whole thing with finality at once,' he wrote. Churchill agreed the next day, even offering that British troops might wear American uniforms, and alleging that 'They will be proud to do so.' The President signalled the end of the haggling with a telegram simply stating, 'Hurrah! Roosevelt,' to which Churchill replied: 'Okay full blast.'[44]

The next meeting at Chequers with Eisenhower and Brooke was therefore far easier than the last. With Pound and the Minister of War Transport Frederick Leathers present, they decided that Torch must take place on 4 November at the earliest, 15 November at the latest, with Ike's 'best guess' being 8 November.[45] On 12 September Churchill had cause to thank Roosevelt, telling him that the 317 Sherman tanks and 94 self-propelled 105mm guns 'which you kindly gave me on that dark Tobruk day in Washington' had arrived safely in Egypt and 'been received with the greatest enthusiasm . . . As these tanks were taken from the hands of the American Army, perhaps you would show this message to General Marshall.'[46]

Because the system of Allied convoys that were taking large amounts of war matériel to northern Russian ports to help the Soviet war effort was about to be suspended in order to provide shipping for Torch, Churchill argued that further consideration should now be given to his favourite project in the north, Operation Jupiter. In Moscow, Stalin had said that he would contribute three Soviet divisions to seizing northern Norway if Churchill put in two. In a memorandum to the Chiefs of Staff, Churchill reiterated the case for invasion, in order to keep Russia supplied and therefore to prevent 'the whole mass of the German armies' being let 'loose upon us'. He underlined the American aspect first, saying that Roosevelt regarded the maintenance of the convoys as 'an operation of equal magnitude as Torch, although he is ready to skip one or perhaps two for the sake of Torch'. Then he presented his plan to 'clear the Germans out of the north of Norway', which he believed would incur fewer losses than making the Merchant Navy take such lethal risks at least thrice every two months.

Churchill objected to the Canadian First Army commander General Andrew McNaughton's very negative report on the feasibility of Jupiter, complaining that 'the exaggeration of difficulties' seemed to be 'customary' in military reports, and stressing that 'It follows that if Jupiter as well as Torch should get going, there could be no Roundup till 1944. This is already the United States view. But Torch by itself is no substitute for Roundup.'[47] This seems like a more or less blatant attempt to get Brooke to support the Norwegian operation in order to stymie the cross-Channel one for 1943. Churchill brought his plan up at the War Cabinet of 21 September, grumbling that, with Torch under way, the Chiefs of Staff 'took a rather unfavourable view' of providing the necessary shipping for Jupiter too. The phrase belies the genuine strength of feeling the Chiefs of Staff had against attacking Norway, which Brooke hardly ever mentioned in his diary without invective and hyperbole.

As before when repulsed by his own Chiefs of Staff, Churchill turned to Roosevelt. On 21 September he wrote a draft telegram about Jupiter, pointing out that with Stalin, 'simply to tell him now no more [convoys] till 1943 is a great danger.' This was especially serious because Stalin had 'gained the impression' at the Moscow Conference that Roundup was not only 'delayed or impinged upon by Torch but was to be regarded as definitely off for 1943. This will be another tremendous blow for Stalin.' As a result, 'We ought now to make a new programme.'[48]

Churchill predicted that Torch would be successful and 'we might control the whole North African shore by the end of the year, thus saving some of the masses of shipping now rounding the Cape. This is our first great prize.' In that case, he thought,

We might decide to do Jupiter instead of attacking the under-belly of the Axis by Sardinia, Sicily and even possibly Italy . . . To sum up, my persisting anxiety is Russia, and I do not see how we can reconcile it with our consciences or with our interests to have no more [convoys] till 1943, no offer to make joint plans for Jupiter, and no signs of a Spring, Summer or even Autumn offensive in Europe. I should be most grateful for your counsel on all this.[49]

The telegram sent the next day reflected all these arguments and more, but neither Marshall nor King would countenance Jupiter as a result. Churchill had nevertheless allowed the Americans to glimpse the future Mediterranean strategy he intended to adopt if prevented from attempting to liberate northern Norway.

Although Marshall, King and Eisenhower appreciated that undertaking Torch probably meant writing Roundup off for 1943, Roosevelt would not admit as much, at least on paper. Churchill was keen that, despite Torch, large numbers of American troops should continue to come over to Britain under Bolero, not least because 'if things go badly for us' Britain would once again 'have to face the possibility of invasion'. Keeping Roundup an open possibility meant that the United States would continue to reinforce metropolitan Britain, and Churchill asked Roosevelt to send him 'revised programmes of what we may expect in the next twelve months between now and next September under the Bolero–Roundup scheme'. His fear – which was well founded – was that Admiral King was siphoning (or 'whittling') off resources to the Pacific that should have been coming to Britain instead. Meticulous research by Professor Mark Stoler on troop, ship and landing-craft movements during this period suggests that this was indeed the case.[50]

Churchill wrote, in what reads like a begging letter to Roosevelt, that over the next six months 'it will be necessary for you . . . to send at least eight US divisions to the United Kingdom in addition to your air force programme'. These were large numbers, and could be justified only if

Roundup was still a possibility, for as Churchill put it: 'Every argument used for Sledgehammer and/or Roundup counts even more in 1943 and 1944 than it did in 1942 and 1943.' Here, for the first time, Churchill used the word Roundup and the date 1944 together.

The rest of the letter was yet another plea for Jupiter, which only Churchill failed to recognize was a non-starter. He nonetheless continued to promote it right up to the 1943 Quebec Conference, at one point ordering Ismay to 'suspend' the entire War Office Planning Staff for opposing it.[51] 'Winston has been particularly active in suggesting all sorts of schemes,' noted Kennedy on 24 September. 'He always wants to do more than we have resources for and nothing seems to convince him that some things are impossible or that dispersion is a dangerous business and that concentration is a principle of war. Brooke says repeatedly, after seeing him, "I am convinced that man is mad." '[52]

An undated (and eventually unsent) telegram from Roosevelt to Churchill was drafted by Marshall on 25 September, which allowed the US Chief of Staff to explain why Churchill's Jupiter project ran flat against Allied strategy. It was an answer to Churchill's request for 'a new programme', and a devastating one. In the course of a very long exposition of policy, which Marshall must have known would not be sent as drafted, he drew attention to the complete contradiction between Churchill's regular statements about the need for concentration of forces and his tendency to 'advance urgent proposals requiring further dispersion of means'.

Marshall wanted Churchill to be told that Torch must go ahead on time, that the harsh fate of Convoy PQ-18 – thirteen merchantmen sunk out of forty between leaving Iceland on 2 September and reaching Murmansk on the 18th – meant that the northern convoys had to be discontinued, with supplies going through the Persian Gulf and the Alaska–Siberia routes instead, and that the US refused to take part in Jupiter because 'the disadvantages in the plan far outweigh the advantages'. Furthermore, 'The more forces we employ on the perimeter of Continental Europe, obviously the fewer forces will be able to penetrate vital enemy areas.' Marshall even hoped that Roosevelt might say to Churchill: 'I do not believe that Stalin attaches to the Jupiter operation the great importance implied in your message.'

As well as accusing Churchill of misrepresenting the Soviet position,

Marshall hoped that Roosevelt would tell the Prime Minister bluntly that, as Torch had effectively wrecked any hopes of a 1943 Roundup, 'The United States does not plan to send to the United Kingdom during the next ten months landing craft in excess of the number for which there will be operating personnel, and adequate to carry troops for any probable 1943 offensive which might be based on the UK.' Since two paragraphs earlier he had stated that Torch 'definitely precludes' Roundup in 1943, this would have been devastating to Churchill. For all the debates about Roundup versus Torch, by October 1942 only one and a half American divisions had actually reached Britain. This was partly because of the massive amount of food, vehicles and services that went with them. It took 144,000 tons of shipping space to move a US infantry division, and a quarter of a million tons if it was armoured.[53] Though never sent, the draft telegram did set out Marshall's overall strategic thinking unambiguously for Roosevelt:

In the implementation of plans such as Jupiter, Allied military resources would be employed on the perimeter of the enemy citadel [and] . . . the Allied forces would not have sufficient and appropriate means remaining for the initiating of a strong, decisive blow in any selected area. On the other hand, a concentration of our means is most desirable in an area where it will be possible to deal the enemy a decisive blow and come to grips with him.[54]

The tension between Churchill's and Marshall's strategies could hardly be clearer.

Instead of Marshall's draft, Roosevelt sent Churchill a very short telegram merely stating that the next convoy should not sail to Russia.[55] He could see no advantages in a major row with his principal ally only a few weeks before Torch. The suspicion must remain that Marshall wrote the draft more for the President's benefit than for the Prime Minister's.

In 1953, Moran asked Churchill which were 'the two most anxious months of the war'. Without hesitation the Prime Minister answered September and October 1942.[56] On the first day of October, Eden visited Downing Street after dinner, and found Clement Attlee there. 'If Torch fails,' Churchill told the two men, 'then I am done for and must go and hand over to one of you.'[57] With many more Conservative than Labour MPs in parliament as a result of Stanley Baldwin's 1935 election victory, all three knew it would have been Eden rather than the deputy premier.

Later that week Kennedy recorded that Churchill 'was like a cat on hot bricks about the future development of the war'. Lunching with the Prime Minister at Downing Street, he mentioned that he had a tin of snuff to give him, a present from Admiral Richard Stapleton-Cotton. 'I thought of giving up cigars till we were back in Benghazi,' Churchill said on accepting it. 'Then I thought I'd give up snuff. Then I decided to do neither. I didn't see why I should give up anything for any German.' Kennedy later wrote that although Churchill was funny at Cabinet Defence Committee meetings, 'It is rather like the headmaster making jokes to the boys – the laughs come very easily!' This was unfair; Churchill genuinely was funny, and humour was needed to leaven the stress. When he had a sore throat he complained to Brooke that his doctors 'have knocked me off cigars. That is the worst of having a high-class job – you have to go in for high-class cures. I should have said a wet stocking round my neck would cure me in a night.'

After dinner on 6 October, Eden and Oliver Lyttelton had a drink with Churchill in the No. 10 Annexe, where they were joined by Randolph Churchill, who at one point said, 'Father, the trouble is your soldiers won't fight.' Eden was indignant, recording: 'It was a revelation to me that Randolph is so stupid.'[58] Yet it is hard to escape the conclusion that although the British High Command thought their soldiers would fight, there was indeed an underlying fear that the Germans were man for man better soldiers, and this was one of the reasons that the invasion of France was postponed to 1944, until victories had been won over the Wehrmacht in the lesser theatres of North Africa, Sicily and Italy.

On 9 October Kennedy went to see Eisenhower, who had retained his post as commander of the European theatre as well as becoming supreme commander for Torch. 'I found him in a very wrought up state . . . he said he was being continually bombarded with political, operational and administrative problems . . . I was sorry to see that he was feeling the stress so much.' Eisenhower was unhappy with the British Government's instructions to their commanders, which he felt gave them carte blanche to appeal over his head directly to Churchill. Kennedy pointed out that every British commander had always received pretty nearly identical instructions, but it hadn't prevented Lords Haig and Gort from working with foreigners. Eisenhower then told Kennedy that he had 'always considered this operation to be unsound strategically but he had been chosen for a variety of reasons to head it and he would drive it through in a spirit of loyalty and close cooperation'. When this amazing statement was reported to Brooke, the CIGS retorted: 'What a bloody fool the man is!'[59]

Meanwhile, in Washington, Hopkins warned Marshall on 10 October that Roosevelt had received a 'very urgent wire' from Stalin asking that for the next few months deliveries of aircraft to Russia be more than doubled to five hundred per month. The President, through the Soviet Ambassador Maxim Litvinov, sent word that he would look into it at once. That morning Stalin followed it up with 'a very urgent request' for an immediate answer. Although Roosevelt knew that such a figure was completely impossible, he asked if Marshall could send Stalin three hundred extra aircraft over and above what had been agreed in the protocol, beginning immediately and starting with coastal defence fighters. 'The President is anxious to get off a message to Stalin tonight,' he was told.[60]

Marshall replied that same night: 'Any immediate increase beyond the 212 airplanes per month now scheduled for Russia could only be managed by a reduction of planes urgently needed for our units in combat theatres,' primarily Guadalcanal and Torch. US coastal defence units were 'actually operational training units', which had only half their proper complement of planes, and which in any case 'weren't suitable for an active theater'. Moreover, they were an important defence 'against a possible trick carrier raid'. In short, Marshall's answer to Stalin was no. By that stage in the war, he felt secure enough in Roosevelt's regard to be able to take such a firm line, and know that it would be accepted.

Marshall was also subjected to regular demands from Douglas Mac-Arthur in the Pacific, such as one of 17 October about 'the critical situation' in the Solomon Islands and New Guinea, which concluded, 'I urge that the entire resources of the United States be diverted temporarily to meet the critical situation; that shipping be made available from any source; that one Corps be dispatched immediately; that all available heavy bombers be ferried here at once,' and so on.[61] Marshall not surprisingly disliked the high-handed tone of the messages he received from MacArthur, who, rather than being sent 'the entire resources of the United States', had to be content with a heavy bomber group that was flown from Hawaii to Australia.

*

On 14 October Brooke received Montgomery's detailed plan for a great attack to be unleashed on Rommel at El Alamein in nine days' time. He decided that he would not pass it on to Churchill, even though the Prime Minister 'was continually fretting to advance the date' and asking him 'why we were not being informed of the proposed date of attack'. Brooke wanted to protect Alexander and Montgomery from being bothered by Churchill, and subjected to demands that the plans be altered. It was nonetheless a serious and insubordinate decision to have taken. Like Marshall, however, Brooke knew what he could get away with by then.

The searing summer heat meant that there had been little fighting in North Africa since July, and both sides had been able to reinforce themselves, with the Allies being strengthened disproportionately more than the Axis. With two hundred thousand troops and over one thousand tanks, Montgomery had almost double Rommel's forces. The battle front was only 40 miles wide, as the geological phenomenon known as the Qattara Depression closed off Rommel's opportunities for a southern flanking move by fast armour. Montgomery's attack began at 9.40 on the night of 23 October with more than one thousand guns firing the first of more than one million rounds at the German positions, and over the next twelve days a savage battle was fought, costing thirteen thousand casualties on the Commonwealth side and thirty-five thousand on the Axis. After three days, Brooke felt able to give the War Cabinet tentative details of how it was going; five hundred German and one thousand Italian prisoners had been taken, but there had been 'No major clash of armour yet'.[62] Then he filled in the Cabinet on the situation in Russia, New Guinea, the Solomon Islands and Burma.

There followed discussion of the accusation appearing in British left-wing newspapers that Rudolf Hess, the Deputy Führer, had had 'friends' in the British War Cabinet when he flew to Britain in May 1941, which had inflamed Soviet suspicions. Smuts said 'We should ... find out more about H[ess] to find out who were his friends in Cabinet ... Misunderstand[ings are] bad for atmosphere of two allies. They are building up a case and we must meet it before it has gone too far.' Cripps – who had until recently been the ambassador in Moscow – called for a 'simple statement by someone about Hess, clearing up the matter'. Churchill then explained:

Hess arrived, hot from Hitler's entourage, and came to do great service for Germany at great risk. He wanted to be . . . conducted to the King to say that we [that is, the Churchill ministry] had no backing here and to get a Government of the pro-Munich complexion installed. Hess was suffering from melancholia. We tried to make him talk . . . He gave us last chance for peace and the chance of joining the crusade against Russia. But he never said a word about his Cabinet friends who he had come to see. He had once met the Duke of Hamilton.[63]

A minister then suggested that the Government should make the records of Hess' interrogation available to the press, to which Churchill's answer was no. Smuts warned that the 'impression' of the incident might 'seriously' affect Anglo-Russian relations and Cripps added that full disclosure would 'Get rid of an air of mystery'. Churchill, however, believed that the Russians were worried about much more important matters, such as 'their losses', adding that he might consider allowing Cripps to make a digest of the Hess documents for press and parliament, and the Cabinet could then decide whether to hand it to Maisky, the Soviet Ambassador. In the event neither thing happened, and the conspiracy theories about the Hess flight therefore swirled around, inflaming Russian suspicions to the detriment of the British Establishment, until the interrogation reports were finally released somewhat piecemeal half a century later in the 1990s, entirely confirming what Churchill had said.

On Tuesday 3 November, the US Congressional mid-term elections produced the best result for the Republicans since 1928, increasing their representation by ten senators and forty-seven Congressmen. Nonetheless the Democrats still retained a 58–38 majority in the Senate and a 222–212 majority in the House of Representatives. Roosevelt had been in power for nearly a decade, and there was much criticism of the way the war was being fought, but his party still controlled all three branches of the American government. The elections would undoubtedly have gone far better for him had Operation Torch taken place beforehand, but Marshall told four Pentagon historians in 1949, '*off the record*', that the President 'had made no word of complaint when he was told the invasion date although some of his men yelled pretty loud because we could not go in five days earlier'.

*

Montgomery's victory at El Alamein was clear to all by 4 November, when Rommel started his full withdrawal, though hampered by Hitler's policy of refusing to contemplate retreats. Egypt was clear of the Afrika Korps by 10 November. 'Rommel was a fool not to have gone back a month ago,' wrote Kennedy on 2 November, with near-perfect hindsight. 'We should then have been faced with the problem of moving forward and building up again for an attack with a long line of communications exposed to Rommel's raids, etc. Rommel cannot be such a good general as we thought. On the other hand, Montgomery has had colossal luck in arriving at the moment when he did and not sooner.'[64] (Or indeed any later, when his victory would have been ascribed to Auchinleck's dispositions.)

Fourteen years later, Marshall identified this period as the tipping point for the balance of power between the United States and Great Britain:

For a long time they had supremacy and we had a minimum of divisions either organized or overseas. The apex of British supremacy was the victory of the Eighth Army in Africa. Later, their strength dwindled until in the Italian campaign some units wouldn't fight. We had to turn three of our divisions over to the commander there. They had simply lost all their fight. We didn't blame them a bit, because they were completely exhausted and under strength.[65]

On Sunday 8 November, four days after Rommel began his full-scale retreat from El Alamein, he found simultaneous amphibious assaults taking place at eight places several hundred miles behind him, around Casablanca, Oran and Algiers. They were all successful. Eisenhower and his deputy Mark Clark were in overall command, and the American Western Task Force was under the command of Major-General George S. Patton. The Vichy French opposed the landings in all three places in their territory, but with differing intensity. Whereas Algiers had fallen by the first evening, and the fighting in Oran was over by noon on 10 November, the landing at Casablanca was bitterly contested until 11 November. Nonetheless, Torch was a success, and, unlike the Dieppe Raid, lessons genuinely were learnt for combined amphibious assaults in the future.

The counter-attack via Spain did not transpire; the weather was unusually fine; the feared losses to submarines and German bombers did not happen. Stimson 'always believed Torch to be the luckiest operation of the war, although he was prepared to admit that those who had advocated the operation could not be expected to see it in that light'.[66] The President had won his bet.

On the evening of 8 November, as the momentous news about Torch was coming in, Churchill was with Eden, Winant and Bedell Smith. 'The PM was evidently much elevated by the success in Egypt and satisfactory initial stage of Torch and he talked even more frankly than customarily, the conversation lasting the greater part of the night,' Bedell Smith telegraphed Marshall, who immediately passed the message on to Hopkins to show the President. 'He is extremely anxious to have you and probably Admiral King come here at a very early stage for conference to reorient strategy in the light of new Mediterranean situation.' Churchill had given up the Norway idea, thought Bedell Smith (wrongly), but he believed that a properly armed Turkey 'will erupt' into the Balkans against the Germans. (In fact, Turkey only declared war against Germany in late February 1945.) Bedell Smith concluded that Churchill 'seems to be growing colder to the idea of Roundup except as a final stroke against a tottering opponent. As you know, the Pacific to him seems very far away and his constantly reiterated idea is that Russia, Britain and the United States must dispose of Germany and then concentrate on Japan. He hopes to clinch this strategy by your conference here.'[67]

As for France, Winant reported to Roosevelt, and Bedell Smith simultaneously to Marshall, that Churchill 'feels bound in honour to support de Gaulle, with all his faults, as the one man who stuck to the apparently sinking ship and whose name has a following in civilian France'. Churchill feared that the pro-Allied General Henri Giraud, whom the Americans had slipped into North-west Africa during Torch in the hope of establishing him in power there, would turn into a source of difficulty, and he insisted that 'Britain and the United States cannot each have a pet Frenchman.'[68]

In one evening of exuberantly loose talk with two key Americans, therefore, Churchill had effectively warned Roosevelt and Marshall that he wanted to 'reorient' strategy away from Roundup and towards the Mediterranean, do nothing more than contain Japan, and run de Gaulle against their favoured candidate, Giraud. He thereby neatly encapsulated the three next great areas of discord between the Allied grand strategists and told the Americans what he had in mind, much sooner than he needed to have done. Torch had arisen from a negotiated deal whereby he and Roosevelt had effectively split the difference over the numbers of troops needed for each part of the operation, and compromised over the geographical areas to attack.

Yet Churchill deserved his moment of exultation. At the War Cabinet the next day he hailed it as the 'Biggest combined effort since Hitler's attack on the Low Countries, and the largest amphibian operation ever undertaken . . . I beg my colleagues and military authorities to look on this as a springboard. We must look at once at military operations undertaken from there. This is the moment for the offensive.' He added that it would be a 'Tragic mistake to think we can take our time with this war. Hitler is playing now for a stalemate. This is our real danger. Never has there been more need for urgency in the war.' Smuts suggested that the 'real victory front' was to be found 'from the South not from the West', and Churchill agreed, adding: 'President Roosevelt calls this the Second Front. We won't contradict this.' He proclaimed himself 'very anxious' to ring Britain's church bells in celebration the following Sunday; they had not sounded since 1940 because they were to act as tocsins warning of a German invasion.[69]

Churchill also wanted to 'Bomb Italy, bring it forward as fast as possible,' ordering Brooke to 'Have it studied, worked out and report next week.' He declared proudly that the 'British Empire played the leading part of this terrific event,' predicting that it 'meant the obliteration of the German & Italian forces in Libya & Egypt' and announcing that he would 'Mark the victory in respect of Alexander and Montgomery with high reward and promotion', since it was 'One of the greatest victories won by the British Empire in the field. It's a fine story.' He then formally congratulated Brooke and Grigg, the War Secretary, for a 'brilliant showing', remarking that the 'movement of the invasion convoy without loss was a marvellous story with 105 warships, 142 troop and supply ships'. Brooke then filled in the details of the successful operations at Oran, Casablanca and Algiers.[70]

The Cabinet was a long one, a full three hours. 'Winston revelled over our success!' noted Brooke at the time. 'But did not give the Army quite the credit it deserved.' This was unfair: in fact the Prime Minister suggested that the Cabinet should congratulate the CIGS and Grigg 'for the fine performance put up by the Army'. Brooke observed after the war: 'I think this is the only occasion on which he expressed publicly any appreciation or thanks for work I had done during the whole of the period I worked for him.'[71]

As well as being the first significant British Commonwealth land victory of the Second World War, El Alamein was also the last. Henceforth every major engagement was to be fought as part of an alliance. The peal of Britain's church bells ringing to celebrate this great feat of Imperial and Commonwealth arms was also tolling the end of major unilateral military action, at least until the recapture of the Falkland Islands forty years later. Kennedy commented on how 'remarkably thin' the bells in London sounded that Sunday, and it reminded him how many churches had been destroyed.

11

The Mediterranean Garden Path: 'I intended North Africa to be a springboard, not a sofa'
November 1942–January 1943

Prestige has surely been the most fruitful source of military mistakes since the beginning of time.　　Major-General John Kennedy, November 1942[1]

In his speech at the Mansion House dinner in London on Tuesday 10 November 1942, Churchill declared that the victories at El Alamein and Torch represented 'perhaps the end of the beginning' of the war. The phrase was originally thought up by Lieutenant-Colonel (later General Sir) Ridley Pakenham-Walsh when he was advising on Churchill's four-volume life of the Duke of Marlborough in the early 1930s.[2] When Churchill found a vivid phrase it often lodged in his brain to be called forth for later use; 'fight them on the beaches' first appears in relation to seal pups in Rudyard Kipling's *Kim*.

Another famous phrase from that Mansion House speech – 'I have not become the King's First Minister in order to preside over the liquidation of the British Empire' – would only have confirmed Marshall and others in their suspicion that Churchill wanted to recover Malaysia and Singapore in order to preserve the Empire almost as much to defeat Japan. Churchill's researcher William Deakin believed that the Joint Chiefs of Staff 'felt lured into the Mediterranean . . . because they had to admit [not only] that the build-up for a decisive landing in northern France was far from completion, but that Mediterranean operations were essentially designed to further exclusively British interests – to preserve and control from Mediterranean bases the imperial route to Asia and the Far East'.[3] It was undoubtedly what the Americans felt, but to what extent were they right?

King George VI's principal private secretary Sir Alan Lascelles, writing in a diary that he too, as a public servant, ought not to have been keeping, noted of Torch that 'Winston, in his speech, gave the credit for its original conception to Roosevelt; but I believe it belongs more truly to himself.'[4] Rather like some inventions and scientific discoveries, to say nothing of political ideas, it is often hard to pinpoint exactly the authorship of successful military strategies. They cannot be patented, and after the war Brooke strongly denied that it had been Churchill or Roosevelt rather than his own Planning Staff who had thought up the idea. Yet judging by Dill's strictures against Kennedy, Torch does not appear to have been a War Office concept.

Roosevelt himself spoke of the genesis of Torch at a meeting on 10 November with his vice-president Henry Wallace, House Speaker Sam Rayburn and the majority leaders of both Houses, Senator Alben Barkley of Kentucky and Representative John McCormack of Massachusetts. The President mentioned that the Moroccan and Algerian attacks 'had been talked over with Churchill when he was here last June'. Wallace claimed that it had been thought of as early as December 1941, when he had given FDR a Christmas present of a book about the Mediterranean, with certain chapters marked up to imply that an attack in North-west Africa was recommended. Roosevelt replied that it 'began to take more definite form in July of 1942 when Hopkins visited London but it did not really take definite form until . . . August'.[5] In the words of Mussolini's foreign minister Count Ciano, later appropriated by President Kennedy after the Bay of Pigs disaster: 'Victory has a hundred fathers, but defeat is an orphan.'

Of course total victory in North Africa was far from imminent, perhaps partly as a result of the compromises made over Torch. As Ian Jacob told the author, 'We should have insisted on going straight for Tunis and Bizerte . . . It would have saved us six months of battling along the North African coast.'[6] Within hours of Torch, German forces were landing at Vichy-controlled airfields in Tunisia, and on 11 November they landed at Bizerte too, which the British had wanted to earmark as a Torch target only to be overruled by the Joint Chiefs of Staff. Hitler's decision to flood Tunisia with troops led ultimately to calamity after he refused to withdraw them, of course, but at the time they held up Kenneth Anderson's eastward drive towards the Tunis Straits.

The strength of Marshall's voice in the counsels of Allied strategy directly correlated to the numbers of Americans under arms. He kept up unrelenting pressure on Roosevelt over the size of the US Army, claiming in December 1942 that the Bureau of the Budget had privately fixed the strength of the Army in the calendar year 1943 at 6.5 million, whereas the President had approved the Joint Chiefs of Staff recommendation of a planned strength of 7.5 million, and denouncing the way the Bureau of the Budget 'definitely limits the development of the Army'. He rather histrionically told the President that the Bureau's figures 'jeopardize our success in this war and should be revoked immediately'. Roosevelt replied the next day with characteristic sang-froid: 'I wish the Government as a whole would talk in terms either of calendar year or fiscal year but not both! ... Let me put it so clearly that there can be no misunderstanding. For Budget purposes the strength of the Army is fixed for an *average* of 6.5m for the calendar year 1943.' Since the US Army numbered five million in January and would be seven million by December, the President patiently explained, like a mathematics teacher to a plodding pupil, that the average could not exceed 6.5 million. Furthermore, the Director of the Budget understood that he had to budget for equipment for 7.5 million by the end of 1943. 'If the Army and Budget people will only do what I have written,' he concluded, 'they will see there is no argument between them.'[7]

It was a magisterial rebuke, and the figures still have the power to impress. An army of fewer than two hundred thousand when Pearl Harbor was attacked in December 1941 would grow into one of seven million – thirty-five times its size – a mere two years later. In divisional terms, the US Army had 37 trained divisions at the time of Pearl Harbor, 73 by Operation Torch, 120 by the summer of 1943 and 200 by D-Day. By contrast the British Commonwealth had seventy-five divisions by the summer of 1943 and hardly any more the next year.[8] Nor was the American revolution confined to the Army; on 13 November 1942 a US shipbuilding yard built a standard 10,500-ton merchant vessel – a 'Liberty ship' – in exactly four days and fifteen hours. Two days later the ship was fully equipped and ready for service. No other country or alliance could begin to match such efficiency and productive power.

Eisenhower's recognition, immediately after Torch, of the former vice-premier of Vichy, Admiral Jean François Darlan, as political head of French North Africa, raised political hackles in both Britain and America. Nonetheless it had been Darlan who, as titular head of all

French forces, had ordered his troops to stop resisting the Allies, turning Torch, a potentially dangerous operation, into what Kennedy correctly described as 'a walkover'. To recognize someone who had actively collaborated with the Nazis was politically embarrassing for the Americans, however good their *raisons d'état*, and there remained for the next three years deep disagreements between the British and Americans over where true legitimacy lay with regard to the French leadership, state and (especially) Army. Yet as Churchill had intimated with his remark over rival 'pet Frenchmen', and was explicitly to state several times over to his own Foreign Office, he was not about to jeopardize his valuable good relations with Roosevelt over the issue, least of all not in favour of someone as clearly Anglophobic as Charles de Gaulle, however much he admired the general for his actions back in 1940. Churchill, no less than Eisenhower, accepted the iron dictates of expediency in this respect.

'It is a queer situation in North Africa,' noted Kennedy on 12 November. 'The French are allowing the Germans to put forces into Tunisia without firing a shot while they have opposed us in Algeria and Morocco . . . The truth of the matter is that although the French hate the Germans I am afraid they hate us more.' The Germans got over five thousand men into Bizerte and Tunis unopposed immediately after Torch, and their tenacity there had lessons for the Allies over Roundup, underlining the formidable capacity of the Germans to counter-attack invading forces, which were never allowed to keep the initiative for very long.[9]

On Sunday 15 November, a sleep-deprived Anthony Eden was at Chequers with Churchill, Smuts, Brooke and the other Chiefs of Staff for a meeting called by the Prime Minister 'to drive on his ideas of offensive action', as Smuts put it later. 'A bad night,' Eden lamented in his diary. 'I don't know why it is that Chequers never suits me. Cold still heavy, if not heavier and [his physician Dr] Rossdale's cocaine makes me feel giddy.'[10] Just before the conference started at 11 a.m. a telegram arrived from Eisenhower announcing the terms of his agreement with Darlan, against which Eden protested, before the meeting got on to future strategy in general and Italy in particular. The idea that the next stage of Britain's grand strategy of the Second World War, primarily the proposed attack on Italy, might have been discussed with the Foreign Secretary self-confessedly 'giddy' on cocaine might have been a cause for concern had it not been an area where Churchill and Brooke fully agreed. Churchill had also mentioned Italy to Roosevelt in his cable of 22 September proposing action in northern Norway when he had written: 'We might decide to do Jupiter instead of attacking the underbelly of the Axis by Sardinia, Sicily and even possibly Italy.' So the Americans were not unaware of what the British now began to argue was the inevitable next stage of the war.

Captain Basil Liddell Hart believed that it was the tardiness of Eisenhower's advance from Algeria into Tunisia that drew the Germans into moving such large reinforcements southwards across the Mediterranean in order to protect their Tunisian bridgehead, only to be 'trapped with the sea at their back'.[11] Had the entire German and Italian army in Tunisia – approximately a quarter of a million men – not been captured, they might well have stalled the later Allied advances into southern Europe. It might be, therefore, that the very lack of early success immediately after Torch paradoxically increased the successes later, given Hitler's unwillingness to retreat even tactically, a characteristic that Allied strategists were about to note with glee.

At the Chequers meeting, Smuts stated that 'there seemed to him to be no difference of opinion on long-term policy,' even though Churchill had written a memorandum complaining that the Chiefs of Staff were insufficiently aggressive. Churchill and Brooke did disagree on the amount of stores, supplies and general non-military equipment the Army needed, however, with the Prime Minister pressing for a heavy reduction in the administrative 'tail' of the forces in North Africa. 'The Army is like a peacock,' Churchill complained, 'nearly all tail.' Not to be outdone by an ornithological reference, Brooke retorted: 'The peacock would be a very badly balanced bird without its tail.'[12]

On 16 November, Churchill explained the Government's position over Darlan to the War Cabinet. He said that Eisenhower had given a 'convincing' account of the political situation with regard to the French in North Africa, who after all still had four divisions in Morocco, three in Algeria and one in Tunisia. Darlan 'is contemptible figure', said the Prime Minister. 'Whilst the French Navy was fighting, [he] was negotiating. He was now advising the French to fight against Germany.' Churchill equally despised Giraud, who, he said: '1) signed a letter to Pétain saying he would behave, 2) then manoeuvred to get power for himself, 3) now he's accepted a commission from Eisenhower to fight.'

Eden thought Eisenhower's policy towards Darlan would outrage British opinion, and Churchill pointed out: 'He's not our C-in-C – Eisenhower is responsible to the Washington authorities. We can't afford to upset Eisenhower just now.' He concluded: 'Eisenhower is our friend – grand fellow – we don't want to get across him.'

Eden said that nonetheless the US authorities must be told 'fairly soon' that the Darlan position could not be stabilized, and that 'When we get Tunis we ought to get rid of' the admiral. A telegram was drafted to that effect. Brooke then reported on Algiers, Oran and Casablanca and added that the 'French are [now] showing tendency to assist us,' although the capacity of Casablanca harbour was severely restricted. The Eastern Task Force had landed 56,000 men, the Central 51,700 and the Western 37,000, totalling 144,700 men to the west of Rommel.[13]

The success of Torch raised a number of possibilities for the War Office Planners. At last the British Army was on the offensive, a new way of conducting warfare, although Churchill was never satisfied with the amount of ground covered, telling Major-General Sir Noel Holmes, the Director of Movements, 'I intended North Africa to be a springboard, not a sofa.' Among the scenarios that were now subjected to 'urgent examination' were projects for taking Rhodes and the Dodecanese Islands (Operation Mandibles); bombing Italy; invading Sicily; sending air squadrons to help the Russians in the Caucasus; 'getting command of the Aegean again', and using Smyrna and the other Turkish ports further north. 'The Germans are now in a mess, although they are still tough and strong,' concluded Kennedy. 'It is the maddest thing ever done in military history to hold a two-thousand-mile front through the Russian winter.'[14]

Yet what every one of these plans had in common was that they involved what the American Planners were calling – comprehensibly if not felicitously – 'scatterization'.[15] With an estimated total of thirty-six German combat divisions in the west in December 1942, Brooke considered it was still too early for Roundup, and certainly for Sledgehammer, although he could not have known that the number was only set to rise. Writing from the British Joint Staff Mission in Washington, Dill warned that 'There are still a good number of people in authority here, who feel that we have led them down the Mediterranean garden path and although they are enjoying the walk are fearful of what they might find at the end of it.'[16] Foremost among these was George C. Marshall.

For all this urgent new activity at the War Office – 'Winston issues notes urging on action everywhere and very soon,' wrote Kennedy – Brooke still spent his lunch hours prowling around second-hand book-shops and hunting for gadgets for his bird-watching cameras, sometimes not returning until 4 p.m., although like Marshall he often continued working after supper. His contribution to the North African successes was recognized by the award of the Grand Cross of the Order of the Bath. With Churchill and Brooke now tending to agree on the big issue – that the next stage in the war ought to be in the Mediterranean rather than across the Channel – Brooke allowed himself to be irritated only by small issues, such as Churchill's love of rodomontades during meet-ings. At one Defence Committee with 'Winston holding forth', he passed a note to Grigg saying '15 minutes gone and no work done', which he subsequently altered to 20, 25, 30, 35, 40 and then 45, before the real business of the meeting began. 'Winston is really stupid the way he tries his team,' concluded Kennedy after he heard this. 'I wish someone had the courage to tell him about it.' (The only person other than the King who could, Clementine Churchill, had tried it successfully with a very direct note to her husband back in June 1940, but the effect had been temporary.)

At a War Cabinet on 30 November 1942, the crucial issue of the Russian attitude towards future Allied policy was discussed. Churchill read a telegram he had received from Stalin about the Second Front, and commented that the 'Mediterranean must be regarded as an inadequate contribution to the battle. We must not only fool around in the Mediter-ranean. It would be a calamity if it got in Russia's mind we were only going to contribute with Mediterranean next year.' This does not prove that Churchill still felt that Roundup was a possibility for 1943, but more likely implies that he was still hankering after Jupiter, or even a force in the Caucasus to link up with the Red Army. To Stalin's request for aircraft support in the Caucasus, Churchill said Britain should 'make a definite contribution to the fighting. Don't want to dismount our squadrons.'[17] Of course Brooke didn't want to 'dismount' them either, but merely to use them in North Africa.

On Wednesday 2 December, Brooke dined at Kennedy's house, 98 Cheyne Walk in Chelsea, and was 'in great form and kept us interested and amused with graphic stories of his dealings with Winston and his trips to Moscow and America. He is very loyal to Winston and always shows that even when he is telling a story at Winston's expense.' 'Brookie' then described his dealings with Churchill, especially his 'extraordinary obstinacy. Like a child who has set his mind on a forbidden toy. It is no good to explain that he will cut his fingers or burn them. The more you explain the more fixed he becomes in his idea.' When the Prime Minister became 'quite immovable on some impossible project', Brooke said it often meant only that 'he will not give way at that particular moment. Then suddenly after some days he will come round and will say something to show that [it] is all right and that all the personal abuse has been forgotten. Winston is a bully and like all bullies is worse if you don't stand up to him.' Brooke further complained of Churchill's tendency 'suddenly to start using the arguments which have been put to him as his own – even to those who have originally produced them – and as if they were something quite new'. Although many subordinates complain about their bosses in much these terms, it is hard to see how Kennedy could justify describing Brooke's monologue as 'very loyal to Winston'.

Asked by another guest at their dinner *à quatre*, Jean Strathearn, who was 'the biggest man' out of Stalin, Roosevelt and Churchill, Brooke said that since they were so different they could hardly be compared. 'For instance, Roosevelt could never have rallied the country as Winston did after Dunkirk.' This was surprising, and somewhat unfair, considering what Roosevelt had done in the difficult circumstances after Pearl Harbor, when the United States was not in danger of invasion. Brooke went on to say that 'Winston had a very human and lovable side,' and recalled how, when flying back from America early one morning, the Prime Minister had come up the plane to get a first sight of land, wearing his dressing gown and yachting cap and smoking a cigar, 'and they peered down together through the clouds'. Churchill had said: 'Do you know, I feel so thrilled. I can imagine the feelings of those men who first flew the Atlantic.'

What irritated Brooke most, he admitted, was being summoned to No. 10 'when undressed and about to get into his bath at midnight as often happens. Then usually to find that there is nothing definite to answer – only bluster to an indefinite conclusion.' He wound up saying that 'It is very hard to stick to important principles in late sittings. Sometimes you feel you must give way on something. Then in the morning you wake up and think "Now *what* did I do last night – was I too weak?" '[18] With Alan Brooke, the answer was almost certainly no.

In early December 1942 the Allies suffered grave setbacks in Tunisia as the Germans counter-attacked successfully at Tébourba and forced the First Army under Anderson on to the defensive. Eisenhower had poured as many troops as he dared into the eastward move, and another attempt was made later in the month, but rain, mud, lack of supplies but principally exceptionally tough Axis resistance prevented the capture on 22–24 December of the strategically important 'Longstop' Hill which controlled the Madjerda valley in Tunisia. The realization that the race for Tunisia had been lost led to renewed fears in the British War Office that Eisenhower was not 'skilful enough in command'. Nonetheless, Kennedy assumed that 'Rommel would be finished off in a couple of months,' and the Planners needed to agree what to propose to do next.[19]

It was at that point that Churchill made noises favouring Roundup, and produced, in Kennedy's words, 'a directive the gist of which was that we must stop our Mediterranean operations about June next and concentrate here again for an invasion of France'. The Chiefs of Staff were summoned to discuss this with him the same afternoon, where Brooke 'gave it him pretty straight and said he would ruin everything by a premature attack in France'. Brooke claimed that the Germans had forty divisions to oppose Roundup – although some historians today dispute that figure – but just as importantly there was still a serious shortage of landing craft and other shipping necessary to get the forces over in time.

'CIGS is quite determined to go flat out in the Med,' recorded Kennedy.

We can waste the German strength there and tackle him on equal or better terms in outposts like Sardinia, Sicily, tip of Italy, Crete. We cannot develop an offensive on both fronts. The essential condition for France is still a crack in German morale and strength. Italy may be knocked out of the war by a combination of landing attacks and bombing. The Balkans are a weak spot for the Axis. If we can get near enough to bomb the Roumanian oilfields and cut the Aegean and Turkish traffic (chrome, etc) we can go far to hamstring the Germans . . . The Americans must meanwhile pour in here as fast as shipping will allow so that we may seize the opportunity to get back on to the Continent when the time is ripe. There is a real opportunity that the Germans may collapse within a year.[20]

Except for the over-optimistic last sentence, that was a fairly good appreciation of the way the next two years of the Second World War did develop, but at the time it must have seemed to Brooke that knowing Marshall's desire for Roundup he had to work hard to get Churchill back on to the Mediterranean track, hence his giving 'it him pretty straight' at the meeting.

After a conversation with Archibald Clark Kerr, the British Ambassador to Moscow then on home leave, on the evening of 9 December, Brooke noted that it had corroborated 'all my worst fears, namely that we are going to have great difficulties in getting out of Winston's promise to Stalin, namely the establishment of a western front in 1943!' That month, when the Western Allies faced only 6 German divisions in Africa, the Russians were fighting 183 on the Eastern Front.[21] Clark Kerr argued that, without a Second Front to relieve him, Stalin might seek a separate peace agreement with Hitler, which the Führer would be mad to refuse. Brooke utterly discounted the possibility, however, as being domestically impossible for either dictator. In this he was being optimistic: the totalitarian dictators had signed a pact before and could have done so again.

The certainty that there would be another full-scale clash over grand strategy with Marshall was confirmed two days later, when a message arrived from Dill 'giving insight into Marshall's brain'. According to this excellent source, Marshall wanted 'to close down' all offensive operations in the Mediterranean once the Germans were expelled from North Africa, and then concentrate everything on Roundup, combined with a possible move through Turkey.[22] This was to be the next great area of disagreement between Brooke and Marshall. At the time it seemed that, despite Brooke's straight talking, Churchill agreed with the American more than with him. Brooke received a large Smithfield ham from Marshall, 'as a US contribution to your Christmas dinner. I hope it reaches you in good condition. With it go my Christmas greetings with the hope that jointly we may have much to celebrate in the New Year.' Included was Mrs Marshall's recipe for its cooking. Brooke replied on Christmas Day with thanks for both ham and recipe, and added: 'may all your hopes and ours for the New Year be fulfilled.'[23] Sadly, they couldn't both be.

Brooke was right that the titanic struggle between Stalin and Hitler over Stalingrad rendered it very unlikely that they would make peace; it would have meant Hitler giving up his eastern conquests, which was inconceivable. Yet to support the quarter-million-strong German Sixth Army trapped inside Stalingrad from the air would have taken 700 tons of supplies a day, and although Göring vaingloriously promised to provide this, it proved impossible, with only a fraction of that ever arriving. The blizzards and windchill factor reduced General Friedrich Paulus' forces trapped within what they called the *Kessel* (cauldron) to frostbite and starvation, with most of the habitable buildings in the city already destroyed by the German bombing offensive that had rendered it a moonscape wilderness of rubble. Field Marshal Erich von Manstein's efforts to relieve Paulus collapsed by 13 December, having got to within 35 miles of the city.

British strategists were cognizant of the enormous differences of scale between the war in the east and that in North Africa, and thus between the Russian and Western contributions to victory. As Kennedy wrote in May 1943 of the German Army Group A's bridgehead on the Taman peninsula, south of the Sea of Azov, 'It is a curious reflection that the Taman bridgehead is about the same size as the Tunisian final front and the number of Germans in it is about equal. Yet the Taman bridgehead is tiny in comparison with the whole Russian front.'[24] The rout of the Italian divisions on the River Don between 16 and 20 December ensured that Manstein's Eleventh Army would not now have another opportunity to try to relieve Stalingrad. Germany's Operation Winter Storm – an unfortunate title considering what the weather was wreaking on the German defenders in Stalingrad – had failed.

At the War Cabinet of 14 December, Churchill asked if there was any truth to the story of a wholesale massacre of Jews in Poland 'by mass electrical methods'. Eden replied that there was 'Nothing direct, but indications that it may be true', although the method could not be confirmed. Eden said that they did 'Know that Jews are being withdrawn e.g. from Norway and sent to Poland, for some such purposes evidently'.[25] He announced a joint declaration that committed the Allies to punishing Nazi war criminals.

Dill was meanwhile writing to Churchill about Marshall's thoughts, which could not but provide further worries for Brooke. In a 'Personal Important Most Secret Cypher Telegram', he reported that Marshall was 'very encouraged to know that your thoughts and his are running on the same lines, but he has made it clear to me that until he sees the full development of operations in North Africa and has the views of Eisenhower, his opinion as to our future strategy cannot be firm.' Nonetheless Marshall was 'more and more convinced' that they should undertake 'a modified Roundup' before the summer of 1943. This would be possible only if, instead of exploiting Torch in the Mediterranean, 'we start pouring American forces into England'. Marshall thought that that would be 'much more effective' than invading either Sicily (Operation Husky) or Sardinia (Operation Brimstone).

Roundup, Marshall felt, would be less costly in shipping, would be 'more satisfying to the Russians', would engage many more German air forces, and would also 'be the most effective answer to any German attack through Spain'. It sounded as though, far from having no 'firm' idea of the next stage as he claimed, Marshall's plans for 1943 were fairly well advanced. Of course it was precisely the size of the Luftwaffe in the west that made Roundup so impractical for Brooke, who also – rightly as it turned out – discounted the idea of a German advance through Spain, and for whom satisfying the Russians was at best a tertiary consideration, especially now that the German Sixth Army had been comprehensively encircled in Stalingrad with negligible prospects of escape.

'Marshall would, of course, have liked to discuss these questions with you and the Chiefs of Staff,' Dill told Churchill, 'but as American and British ideas now appear to be so close there is, he considers, less need for such personal discussions.' Indeed, Marshall was convinced that another high-level conference would be 'a mistake' because it might 'make the Russians feel that they had been cut out of the decision-making' and 'were being asked to sign on the dotted line', something they would 'resent bitterly' with possible long-term consequences. He also felt that the existing Anglo-American liaison channels – that is, Dill and the Combined Chiefs of Staff – ought to 'clear our Anglo-American strategy in broad outline' before another rendezvous, which Marshall thought ought to be Moscow rather than London. 'I would add', Dill told Churchill, 'that these ideas as well as those on strategy have been given to me most privately "off the record". They have not been discussed with the President.'[26]

Marshall was therefore communicating with Churchill behind Roosevelt's back, but was he being entirely straightforward? Since he knew that Brooke did not support Roundup, either 'modified' or of any other kind, taking place before the autumn of 1943, it was obviously untrue that 'American and British ideas now appear to be so close' that a bilateral meeting was unnecessary. Marshall's stated reason against a meeting, that he did not want the Russians to suspect collusion, was either naive or not the true reason. Of course the Soviets assumed that the Western democracies would forge policy together, indeed they were surprised later in the war when it seemed to be no longer so.

Since Marshall was anything but naive, it is far more likely that he was going behind Roosevelt's back in order to collude with Churchill in agreeing that there should be no meeting between himself and Brooke before the next conference – which would in fact take place in Casablanca rather than Moscow – because he knew that British and American ideas over Roundup were not at all close, and thanks to Dill's briefing he saw a way of splitting Churchill from Brooke. Furthermore, until quite late in the war he feared that Churchill would persuade Roosevelt into further diversionary courses of action. Oliver Harvey recorded on 15 December, 'The PM and the American General Marshall are veering back to a Second Front this year, with increasing bombing against Germany, while leaving the Mediterranean action limited to the bombing of Germany, without invasion, opening the sea lanes to Egypt and bringing Turkey in if possible.' As for a major conference, Stalin 'refuses to leave Russia'; meanwhile 'The PM is hankering after a meeting with the President . . . whilst trying to keep Stalin reassured.'[27]

The next evening saw another major policy discussion between Churchill and Brooke over a 1943 Roundup. With Portal, Pound, Eden, Mountbatten and Ismay also present, Brooke warned emphatically that the rate and scale of the American build-up in Britain were inadequate to allow a successful cross-Channel landing in 1943. He pointed out that the 'magnificent' German rail routes in western Europe would bring many more German troops to north-west France than the Allies could land during the same period. Instead, he preferred to try to hold forty

German divisions in north-west Europe 'by the threat of a cross-Channel operation' while actually pursuing an Italian and 'perhaps' a Balkan strategy (the southern thrust into Austria via Yugoslavia).[28] Churchill fully accepted Brooke's facts and figures, but argued that a cross-Channel landing in 1943 'was still the better strategy, if only adequate forces for a successful re-entry to the Continent could be assembled in this country'. He said he was 'prepared to agree' with Brooke's assessment that the figures proved this could not happen in 1943, and thus agreed with their conclusion, but 'with the proviso' that the figures were 'confirmed in discussion with the Americans'. So there was no hope for an attack unless the Americans 'could vastly improve' on the number of troops, planes and landing craft they provided. They could; Operation Overlord in June 1944 was to involve more than eleven thousand aircraft and over five thousand ships.

'If we were to adopt the Roundup policy we should have to shut down the Mediterranean very soon,' stated Kennedy, who saw Brooke the morning after the meeting. 'Even if we did we could not concentrate a force strong enough in the UK to give us the smallest chance of tackling the Germans on equal terms in France by the autumn of 1943.' The time issue hung heavily on Brooke, because 'In the meantime the Germans would have five or six months' respite.' Brooke concluded, and Kennedy fully agreed, that 'The best and only way to engage the Germans, to wear them down, and to relieve the pressure on the Russians, is actively in the Mediterranean – especially against Italy.' But how was he going to sell this policy to Churchill, Roosevelt and Marshall, who each seemed actively to want Roundup the following year?

The answer was that Churchill was not as wedded to a 1943 Roundup as he seemed. Brooke had gone off to the meeting 'rather apprehensively and armed with quantities of statistics of shipping movement and German facilities for movement', but told Kennedy afterwards that 'to his surprise the PM had accepted the case as put forward'. Kennedy suspected that the Prime Minister had argued against Brooke primarily 'so that he might firstly be sure himself that the calculations had been thoroughly worked out and secondly that he might know the case *au fond* in order to be able to convince the Americans'.[29] If so, it would not have been the first time that Churchill had employed such devil's advocacy. It is perfectly possible that, during the whole period that Brooke felt he needed to convince Churchill, all the Prime Minister was really doing was making the CIGS check and sharpen his arguments, the better for them both to prevail when next they met Marshall.

November and December 1942 saw a series of long, seemingly hard-fought meetings by which Brooke slowly brought Churchill round to the Mediterranean strategy. 'Churchill and Brooke slugged it out, toe to toe, until Brooke won,' records Pound's biographer. 'Churchill was deeply concerned that he had promised Stalin a second front in 1943, and even though Churchill and the Chiefs of Staff had reached agreement, Brooke was continually haunted by the worry that Marshall might win Churchill back to Roundup.'[30] Despite that, Brooke continued to scorn Marshall, telling Kennedy after the Staff Conference of 16 December 'that Marshall, who is still advocating Roundup, was no strategist and that his whole experience had been rather dealing with politicians and in organising the Army. He argued only on what would be put ashore on the beaches and in the ports in France but had no idea what was to be done next and what the plan of campaign should be and what the chances were of carrying it out.'

The trouble with having no set agreement for the next stage of the war, or even an agreement on whether to have a conference in order to hammer out an agreement, was that it impeded all longer-term preparations. In a War Cabinet discussion on building programmes for aerodromes, for example, the Minister of War Production, Oliver Lyttelton, needed to know, as he put it: 'Is Bolero to stand, or be modified?' There were 510 aerodromes then in use in the United Kingdom, with a further 106 under construction and 56 more projected, and important decisions needed to be taken soon. 'Should we make Bolero provision or not?' asked Bevin, who as minister of labour also needed to find an extra 115,000 men if so. Grigg said that Bolero was projected to involve nearly half a million men by April 1943, so the discussion then turned to the competing claims for building resources. At that point, Churchill admitted that the 'Scale of Bolero needs re-examination.'

When Bevin reported that the recruitment of Land Girls – young female volunteers to work in agriculture – was being held up because of lack of housing, Churchill suggested a 20 per cent cut in airfields, and ordered that no new ones be built without further instructions. He stated that the maximum number of US divisions to be stationed in Britain by August 1943 would be fifteen. 'We are preparing for twenty-seven divisions,' he said (the full Roundup figure). 'If they aren't here till August, they won't be any use.'[31] Grigg asked whether, in terms of storage and hospital accommodation, the Planners might 'Assume not more than 65% of Bolero [would be] needed in '43?' and this was then accepted by the War Cabinet. Here was a strong indication that Churchill was not committed to Roundup in 1943, even though he could not intimate that fact to Marshall.

That Christmas Marshall sent Roosevelt and Churchill two identical 4-feet-diameter globes for their offices. The President set his up directly behind his chair, so 'I can swing around and figure distances to my great satisfaction.'[32] On 19 December, Churchill made his Saturday-morning call to Eden, conjecturing about his next meeting with the President. 'The delay is maddening as no plans for the future can be made meanwhile,' he complained. 'I don't know whether to stand on my head or sit on my tail.' Two days later X Corps of the Eighth Army recaptured Benghazi, the first time they had held it since January.

The War Cabinet of 21 December considered the question of making Rome an 'open city' – that is, free from bombing on account of its religious, cultural and historical significance – which Churchill did not mind doing, but only because in his view there was 'Nothing much in Rome worth bombing.' 'Yes, [the] railway,' replied Brooke, and Portal added that vulnerable marshalling yards were just outside the city, which Grigg (wrongly) thought impossible to bomb without hitting the Eternal City itself.[33] Rome itself was spared, in the end, although the marshalling yards were bombed.

On the issue of the internecine loathing among the French leadership in Africa, which he later likened to 'a basket of snakes', Churchill thought de Gaulle was 'missing his market', as his two rivals Giraud and Darlan were both consolidating their positions. As ever, the Prime Minister's underlying worry was of falling out with the Americans, and of their thinking in terms of 'our Darlan and their de Gaulle'.[34] At the next Cabinet meeting, Churchill reported that de Gaulle was 'in a high-stepping mood. Anxious to embroil us with the Americans', which was something Churchill would never allow. Meanwhile Alec Cadogan feared that American policy in North Africa 'is not being directed by State Department, but by Gen. Marshall!'

On Christmas Eve 1942, Churchill, Eden and Attlee lunched together at No. 10 after discussing policy towards Portugal. The Prime Minister 'emphasised how much a key man R[oosevelt] was in the States'. In Britain whatever happened to Churchill himself, 'or even all three of us', there were enough 'resolute men to carry on and see the business through. But in the US?'[35] He didn't think much of Henry Wallace, who had made a bawdy joke – 'and not a very good one' – at their first meeting, which Churchill had taken a few moments even to get.

When the news arrived later that day that Darlan had been assassinated by a French patriot in Algiers, Eden was delighted. 'I have not felt so relieved by any event for years,' he told his diary. The Americans supported Giraud, although de Gaulle still had his own long-term plans. For all that Churchill did not want to row with the Americans over which Frenchman's writ would ultimately run in Africa, on other issues he was willing to stand up to them. In a discussion on a treaty with China over Kowloon, Eden had commented that it was 'going to be tiresome. We shan't get much support from US though they should support us.' Churchill warned him about America's 'Readiness to give away our imperial rights. Don't let us go any further than we have gone.'[36] Squabbling Frenchmen in Africa were one thing as far as the Prime Minister was concerned, the rights of the British Empire quite another.

'I am afraid that Eisenhower as a general is hopeless!' exclaimed Brooke in his diary, provoked by the slowness of the advance in Tunisia that Christmas. 'He submerges himself in politics and neglects his military duties, partly, I am afraid, because he knows little if anything about military matters.' Since Eisenhower had been a soldier all his life, and a notably successful one in all he had undertaken, this can be discounted as a typical example of Brooke's habit of disparagement-therapy. Moreover the political situation in North Africa was almost as important as the military at that point. As minister resident at Allied HQ in Northwest Africa, the Conservative MP Harold Macmillan had an important part to play in advising Eisenhower in his difficult negotiations between Vichy and the Free French. Meanwhile Eden, discussing with Churchill the growing anxiety of the Chiefs of Staff about the American handling of the North African campaign, concluded: 'They are probably right but they are not free from blame themselves for failing to supply our people with good tanks.'[37]

The strategy of North Africa–Italy–France, stated the American historian Rear-Admiral Samuel Eliot Morison in an Oxford lecture series in the 1950s, 'was a perfectly cogent and defensible strategy; but Sir Alan Brooke disclosed it only bit by bit, which naturally gave the Americans the feeling that they had been "had".' He cited Admiral King's prediction that, once committed to the Mediterranean, 'we would be forced to go on and on in that region and never be able to disengage.'[38] Yet if Brooke had proposed North Africa as a stepping stone to mainland Italy and the Balkans, and possibly beyond, right at the start, the Americans would never have undertaken Torch. It is anyway very unlikely that Brooke was really ever seriously considering the Italian peninsula as an alternative when he was arguing against Sledgehammer back in April 1942.

By New Year's Day 1943 there had been what Kennedy called 'a crystallization of the differences between us and the Americans over the strategy for 1943, and the increasing distrust of Eisenhower's conduct of operations in North Africa'. The fact that the British side of the push towards Tunis was under the direct command of his friend and close colleague Kenneth Anderson, but was making slow progress, probably left Kennedy all the keener to blame Eisenhower. Marshall wanted 'to go for Roundup and to cut Mediterranean activity to attacks by air from North Africa and possibly Turkey, while in the Pacific they want to carry on the operations against Japan on a scale that we feel is bound to prejudice the main object of defeating Germany first'.

Meanwhile, by total contrast, Brooke wanted 'to develop operations for the Mediterranean', by which he meant attacking either Sicily, Corsica or Sardinia, and meanwhile put the 'residue' of US forces 'into the United Kingdom pending the time when we can get back into the Continent without the certainty of defeat'. As for the Pacific operations, Brooke wanted 'to cut them to the minimum required to hold Japan until Germany is defeated'. Since the gulf between these conceptions of future policy between Marshall and Brooke was so wide, it was evident that another conference was necessary, indeed overdue. Brooke was however willing to countenance such a thing only once he was confident that Churchill would not waver in his commitment to the Mediterranean strategy. Meanwhile the War Office Planners 'had immense labour working out the shipping and other connected problems' connected with the next stage of operations. This Staff work was to pay off handsomely later that month.

*

The War Cabinet discussion on colonial policy on 7 January 1943 found the Prime Minister in typically pugnacious mood. There had been some pressure for a Government statement on the future of the Empire, somewhat analogous to the Atlantic Charter, yet Churchill severely deprecated any 'Time taken off the war in order to find a formula to gratify the Americans. Why should we apologize? We showed the world a model of colonial development. The only criticism is that we haven't spent enough in the colonies,' he argued. 'Or defended them,' slipped in Eden. 'Not so,' replied Churchill. 'When the war ends, we shall find we have defended them all.'[39] One of the main reasons Churchill wished British forces to recapture Singapore, Malaya and Hong Kong, rather than merely have them returned to him at a peace conference, was so that he could make precisely this argument. It was why he wanted a Far East strategy based on the Bay of Bengal rather than the more direct route to Japan centred in the south-west Pacific.

Lord Cranborne, the Colonial Secretary and a lifelong imperialist, pointed out that the pressure for an official statement had come from within the Government, not from Washington, and that it had been strong ever since the fall of Malaya. 'Pressure, yes,' retorted Churchill, 'from people not in on the war.' At this Eden pointed out that 'Smuts, who is in on the war, favoured it.' (The reason Eden brought up Smuts was that Churchill and Smuts had a relationship akin to a favourite nephew with his favourite uncle, 'with the nephew' in Tedder's estimation 'being Churchill'.)[40] 'Why bring this up now?' asked Churchill. 'We are busy enough with the war. If you do nothing, it will blow over.' Yet the idea that nationalism in the former colonies then occupied by Japan might 'blow over' was sheer wishful thinking on Churchill's part.

The Chiefs of Staff meeting the following day was one of those in which Brooke complained that Admiral Pound was 'asleep 90% of the time and the remaining 10% is none too sure what he is arguing about'.[41] Since the Americans left Washington the very next day for the Casablanca Conference, which was to involve nine days of hard and detailed negotiations deciding the whole future course of the war, it is extraordinary that Pound should have been allowed to stay in his vital role as operational head of the Royal Navy. The task ahead for the Chiefs of Staff – and especially Brooke – was mammoth. Brooke believed that Churchill supported his Mediterranean policy, but he now had to persuade Roosevelt and Marshall of it. If that proved impossible he and

Churchill needed to try to split Roosevelt from Marshall, as they had managed to do over Torch six months earlier.

Locations where Roosevelt, Churchill and Stalin might have met had included Iceland (turned down as too cold by FDR), Khartoum and 'an oasis south of Algiers'. Churchill preferred Marrakesh, where he'd spent a happy month painting in 1936, but the communications were thought not good enough. Stalin then argued that he could not leave Russia at such a vital juncture in the war, and Roosevelt did not want to meet Churchill privately before conferring with Stalin, but Churchill felt that he and the President had to meet beforehand in order to frame a joint answer to the major question that Stalin needed answered: when would there be a Second Front in Europe?

Roosevelt stated that 'for political reasons it would be impossible for him to come to London'. The next presidential elections were two years off, but he did not want to seem to be conforming to British-led strategy. Instead, Churchill accepted Bedell Smith's suggestion of the small port of Fadala, 15 miles north of Casablanca where part of the Western Task Force of Torch had landed under Patton. In the event, the Anfa Hotel and its surrounding villas at Casablanca were chosen. Brooke was very fortunate indeed that Stalin did not attend the Casablanca Conference: had he been present it would have been far more difficult to win the commitment he hoped for, of only 'limited offensive operations' in north-west Europe for 1943, unless Germany suddenly collapsed.[42]

Marshall left for Casablanca at 9 a.m. on Sunday 9 January, flying via San Juan and Trinidad; he was to be away for twenty days in all. With him were Admiral King, Lieutenant-General Arnold, General Somervell, Sir John Dill, Brigadier Dykes and Brigadier-General John R. Deane, secretary to the Joint Chiefs. Meanwhile a second group comprising Roosevelt, Hopkins, Leahy, Captain John McCrea of the White House Map Room, the President's doctor Rear-Admiral Ross McIntyre, Grace Tully and other secretaries and bodyguards left Washington at 10.30 p.m. from a secret railway siding under the Bureau of Engraving and Printing, in the presidential train bound for Miami. This had a drawing room, a dining room that sat twelve, and five state rooms, all protected against machine-gun bullets.[43] In the unlikely event of anyone trying to impersonate Leahy, he was furnished with a letter on White House writing paper signed by the President stating that he was aged sixty-seven, was 5 foot 10 inches tall, weighed 162 pounds and had grey-brown hair and grey eyes.

The weather was too bad for a Clipper passage so it was decided to send Brooke, Pound, Portal, Mountbatten, Slessor, Leathers and Kennedy by two Liberator bombers from RAF Lyneham in Wiltshire. Brooke was placed next to Mountbatten on the overnight flight, which he found very uncomfortable, 'as every time he turned round he overlay me, and I had to use my knees and elbows to establish my rights to my allotted floor space!'[44] This could serve as a metaphor for Sir Alan Brooke's entire war.

12

The Casablanca Conference: 'We go bald-headed for Husky'

January 1943

Each appeared to the other in a romantic light high above the battles of allies or subordinates: their meetings and correspondence were occasions to which both consciously rose: they were royal cousins and felt pride in this relationship, tempered by a sharp and sometimes amused, but never ironical, perception of the other's peculiar qualities.   Sir Isaiah Berlin on Roosevelt and Churchill[1]

When Brooke's party landed at Casablanca at 11 a.m. on Wednesday 13 January 1943, Churchill met them at the aerodrome wearing his air commodore's uniform. 'They arrived dirty, hot and tired,' recalled Joan Bright of the War Cabinet secretariat, who was helping to organize the conference, 'only to see their American colleagues emerge spruce, shaved and fed, from comfortable aircraft into the Moroccan sun.'[2] Churchill was in ebullient mood, crying: 'Now tumble out, you young fellows, and get on parade!'

Elliott 'Bunny' Roosevelt, the President's son, recalled Anfa Camp as 'a pleasant resort hotel, unpretentious but very modern, small and very comfortable', the compound of which was protected by barbed-wire and Patton's troops. The President (codenamed 'Admiral Q') would be assigned Villa No. 2, only 50 yards away from Churchill ('Air Commodore Frankland') in No. 3. The President's villa was far from unpretentious, however. It was decorated with – in Elliott's phrase – 'plenty of drapes, plenty of frills'. It had a bed 'at least three yards wide' and a sunken bathtub in black marble. When the President first saw it, he whistled, and joked: 'Now all we need is the *madame* of the house.'

The morning after Churchill arrived, he asked Sawyers to prepare a bath at 11, after several others had had theirs, and the water was cold. 'You might have thought that the end of the world had come,' recalled Jacob. 'Everyone was sent for in turn, all were fools, and finally the

Prime Minister said he wouldn't stay a moment longer, and would move into the hotel or to Marrakesh.' He only calmed down 'after lunch had had its mellowing influence'. In another of the villas, that occupied by Brigadier Guy Stewart, the Director of Plans, there was discovered 'a large library of decidedly doubtful books. True, they were all in French which made it difficult for some, but many of them were profusely illustrated, in most artistic style, which helped. For some time we despaired of any work being done in that villa, but after a bit the excitement died down.'[3]

The British fielded a large delegation at the Casablanca Conference – codenamed Symbol – comprising Winston Churchill, the three Chiefs of Staff (Sir Alan Brooke, Sir Dudley Pound and Sir Charles Portal), Lieutenant-General Pug Ismay, Sir John Dill, Lord Louis Mountbatten, Harold Macmillan, General Sir Harold Alexander, Air Marshal Sir Arthur Tedder, Admiral Sir Andrew Cunningham, Major-General John Kennedy, John Slessor, the Assistant Chief of the Air Staff in charge of air policy, and Ian Jacob. The Americans meanwhile fielded Franklin Roosevelt, the four members of the Joint Chiefs of Staff (George Marshall, Hap Arnold, Ernest King and William Leahy), King's chief of staff Rear-Admiral Savvy Cooke, General Brehon Somervell of Army Supply Services, Mark Clark, Averell Harriman, Dwight Eisenhower and Albert Wedemeyer. In the lower Planning ranks there were many more Britons than Americans, something Marshall was to come to resent, criticize and ultimately learn from.

The agenda had been fixed with the Americans before they left Washington. 'Actually it contained a list of every topic under the sun,' noted Jacob soon afterwards, 'but the most important thing was to get settled in broad outline our combined strategy for 1943, and then to get down to brass tacks and decide how exactly to carry it out. There were clearly the makings of a pretty vicious little circle here.' Brooke thought it best to put the war against Japan high on the agenda, reasoning that if Admiral King 'was able to get everything about the Pacific "off his chest"', then perhaps he 'would take a less jaundiced view vis-à-vis the rest of the world'. He hoped the same would be true of Marshall over Burma and China, where the Americans wanted a stronger commitment to action than the British, fearful as they always were that the Chinese might be knocked out of the war altogether by the Japanese. It was an interesting psychological move of Brooke's, almost counter-intuitive, but it did not come off.

At 4.30 p.m. that first day, Wednesday 13 January, only a few hours after getting off the plane, the British Chiefs of Staff met Jack Dill to be briefed on the American stance before the first Combined Chiefs of Staff meeting the next day. He told them that the Joint Chiefs of Staff were entirely opposed to further exploiting Torch in the Mediterranean because it would prejudice the chances of Roundup and other, smaller operations against France, including raids against places such as Brest. They further thought it would weaken the bombing of Germany, employ vital naval craft and shipping for inadequate returns, and hinder operations in Burma which they thought important in order to keep China in the war and help them in the south-west Pacific; finally they feared that it would precipitate a German descent into Spain and the closing of the Straits.

Furthermore, Dill warned that the Joint Chiefs of Staff believed that the British did not take the war in the Far East seriously enough, and that a considerable effort would be required to prevent Japan becoming so entrenched there that it would be impossible to defeat her after Germany. He went so far as to say that the Americans had 'a suspicion' that Britain 'would not put our backs into the work once Germany had been defeated'.[4] He went on to speak of clashes between Marshall and King over strategy and allocation of resources: 'The Navy control the landing craft, so that the Army finds it difficult to squeeze out what they want for their own projects.' Dill also told the Chiefs how at Guadalcanal in the Solomon Islands on 7 August, the first American land offensive of the war, 'the US Marines were thrown ashore, and then it was found that there was no follow-up, no maintenance organisation, and no transport'.

The Americans were almost right about the British attitude towards the Pacific, which had been neatly summed up by Kennedy: 'To get Burma (unless the Japs withdraw) we must have Rangoon and while the Japs are in strength we cannot collect the necessary forces for this until the Germans are finished. To get Rangoon we must have naval and air command in the Bay of Bengal – this we cannot get till Germany is out.' But the Americans were undoubtedly wrong to suspect the British commitment to defeating Japan after Germany had surrendered.[5]

Undeterred by Dill's message, Brooke met Churchill at 6 p.m. to try to persuade him of the advantages of attacking Sicily next, rather than

Sardinia. He argued that Sicily would involve crossing a shorter distance from Africa, therefore the landings would be easier to support, and it was also the more direct route into Italy. The US Planners, however, believed that there would be less opposition on the Sardinian landing beaches, and since it was further north its aerodromes would be better placed for the next stage of the bombing offensive against Germany, the unambiguously codenamed Operation Pointblank.

As for Churchill, Ian Jacob noted that 'His view was clear. He wanted to take plenty of time. Full discussion, no impatience – the dripping of water on a stone. In the meanwhile he would be working on the President, and in ten days or a fortnight everything would fall into place.' Churchill was not noted for his patience, but he unquestionably employed it at Casablanca. He was out to get agreement on a programme of operations for 1943 which some Planners thought 'well beyond our powers, but which he felt was the least that could be thought worth of two great powers'. He wanted the expulsion of the Germans from the North African shore to be followed by the capture of Sicily, but he also wanted the reconquest of Burma, and the eventual invasion of northern France, 'on a moderate scale perhaps'. Since something major had to give in order for these huge objectives to be realized, 'Operations in the Pacific should not be such as to prevent the fulfilment of his programme.'[6]

The Joint Chiefs of Staff were therefore right to worry: a large Mediterranean commitment made possible by a Pacific slowdown was precisely what Churchill had in mind, and what they feared. Similarly Brooke was right to worry about an over-hasty return to the Continent in 1943, yet 'The Chiefs of Staff were dismissed . . . and the rest of the evening was given to ice-breaking dinner parties.' Brooke had been ordered by the Prime Minister not to show impatience, though later in the conference this was to prove beyond him. That evening Brooke dined with Marshall, and had a 'long talk with him after dinner'. Sadly we do not know whether they locked horns or skirted the issues, but if they had argued Brooke would probably have mentioned it – doubtless scaldingly – in his diary. Exhausted, Brooke then stayed up late enough to prepare his opening address to the Combined Chiefs of Staff for the following day.

Between 14 and 23 January, the Combined Chiefs were to hold their fifty-fifth through to sixty-ninth meetings in the conference room just beyond the drawing room of the Anfa Hotel; no fewer than fifteen full-scale meetings in ten days. Jacob, who took the minutes for the Chiefs of Staff (his friend Vivian Dykes performed the same role for the Combined Chiefs), found it 'a charming room, the semicircular wall being made chiefly of large windows'. There was a big table seating sixteen in the middle, and two smaller tables seating from four to six on each side.

Brooke set out his view of the global situation at 10.30 the next morning, Thursday 14 January, after which Marshall followed him, showing where they disagreed, which seemed like nearly everywhere. They stopped for lunch and began again at 2.30 p.m., when Brooke asked the Americans for their views on the war in the Pacific. Admiral King answered, and in Brooke's view, 'it became clear at once that his idea was an "all-out" war against Japan instead of holding operations.' King was sixty-four at the time of Casablanca, and Jacob described him as 'active, tall and spare, with an alert and self-confident bearing. He seems to wear a protective covering of horn, which is hard to penetrate. He gives the impression of being exceedingly narrow-minded, and to be always on the look-out for slights or attempts to "put something over" on him. He is secretive, and I should say he treats his staff stiffly and at times tyrannically.'[7] The British had long before correctly identified him as their most serious opponent on the Joint Chiefs of Staff. He proposed an allocation of the total Allied war effort that boiled down to 30 per cent for the Pacific versus 70 per cent for the rest of the world, whereupon Brooke pointed out that 'this was hardly a scientific way of approaching war strategy!' King replied that 'It might be better knocking off the weaker brethren before finally tackling the main enemy,' and maintained that a set percentage of Allied resources should be devoted to preventing Japan getting into an impregnable position. Brooke didn't like the percentages idea, preferring to set specific targets – such as the reconquest of Burma and the capture of Rabaul or the Marshall Islands – for specific dates. Only once the targets were identified could the resources be allocated, he argued, with everything left over concentrated against Germany. He particularly feared that having too many escort vessels and landing craft in the Pacific theatre might 'hamstring all operations elsewhere'.

King's was probably just an initial bargaining position, and other accounts from strategists mention a 25/75 breakdown being acceptable to him, which in retrospect Brooke ought to have seized upon, however 'unscientific' it might have looked to that scion of Britain's finest Staff colleges. Had it been possible to implement statistically, and some kind of formula probably could have been worked out, Britain would have been better off than with the 'whittling' off of resources to the Pacific that did in fact take place in surreptitious defiance of the official agreements made at Casablanca. 'We are in the position of a testator who wishes to leave the bulk of his fortune to his mistress,' said Portal of the situation. 'He must however leave something to his wife, and the problem is how to decide how little he can in decency set apart for her.'[8] It's easy to see why people enjoyed working with the Chief of the Air Staff.

It became clear that the Americans wanted to advance their outpost line beyond the Solomon Islands to broaden their base of action and to get far more depth into their campaign against the Japanese Navy. As Dill had warned might happen, King openly expressed the suspicious (and deeply insulting) view that Britain 'might not be wholehearted in taking [her] share of the Japanese War if it were left till after Germany'. The danger of China being forced out of the war by Japan was also stressed and it was agreed that it would be worth while making an advance into Burma from Assam in India in order to try to open the northern route to China, even though it might prove impossible to keep it open.

After much further discussion it was agreed to instruct the Joint Planners to examine and report on the minimum holding operations required in the Pacific. Hap Arnold remarked of this opening session that it was 'not bad, everything seems to be smoothing out, I hope. British and us have not as yet put all cards on table, perhaps things will get worse then.'[9] The meeting broke up for tea at 5 p.m., after which the Chiefs of Staff had a discussion with the British Joint Planners to instruct them on the line of action to take with their American opposite numbers. After that, Brooke went off bird-watching with Kennedy, and Arnold departed to see the refloated Vichy battleship *Jean Bart*, which had been hit with British 14-inch shells that, he recorded, had left 'holes in bow and stern large enough to take a small bungalow'.

The presidential party – which included FDR's sons Elliott and Franklin Jr, as well as Harry Hopkins and his son Robert – arrived on the afternoon of the 14th in six big planes escorted by fighters. At dinner at the Roosevelts' villa that night, Admiral King, a heavy drinker, 'became nicely lit up towards the end' and got more and more assertive.

With a thick voice and expansive gesticulations, he propounded his views on the best way to organize North Africa politically. This inevitably led to a clash with Churchill, who had failed to spot that King was drunk. 'Most amusing to watch,' was Brooke's laconic comment on the scene.[10] When an air-raid warning sounded at 1.30 a.m., the British and Americans all sat around the table illuminated only by six candles, before finally getting to bed half an hour later. When asked after the war if the Prime Minister ever 'spoke from brandy', Marshall said, 'No, Churchill could hold it,' adding that that was not always true of Admiral King.

The next morning, Friday 15 January, the news arrived that the Japanese were being driven from Guadalcanal; their expulsion from that island would be a significant American victory. At 8.45, 'a lovely sunny morning', Brooke and Kennedy walked down the beach and had a productive hour watching goldfinches, stonechats, warblers of all sorts, white wagtails and several types of waders on the sea-shore, including sanderlings, ring and grey plovers and turnstones. The Chiefs of Staff met later that morning to discuss 'future projects to be put to the Americans', and afterwards they met Churchill again. The Prime Minister advised that all the possibilities should be looked at and not just those for the Pacific, otherwise it might be like 'carrying water from a well in a sieve. There would be nothing left when we got home.'[11] He was positive about the fighting in Tunisia, saying that 'We had to wear the Germans down *somewhere* and it would be better if they came to us. We should have those in the bag eventually.' He was right about that, largely because of Hitler's refusal to entertain even tactical withdrawals. Of the prospect of a German invasion of Spain, Churchill repeated that 'We had to fight the Germans somewhere and he did not mind whether it was France or in Spain or anywhere else.' He approved an attack into Burma from Assam, and said that the Casablanca negotiations could not be rushed and would probably need ten days. Then he went off to lunch with the President. Kennedy meanwhile lunched with Harold Macmillan, who told him that when Churchill had advised him to rejoin the Army and Macmillan replied that he'd be more useful as a civilian, Churchill had said: 'Yes, perhaps you are right. Between the bowler and the baton there is no middle course.'[12]

After lunch the Combined Chiefs of Staff returned to the relative merits of Roundup versus the Mediterranean. Brooke pointed out that the Germans now had forty-four divisions in France against a maximum of twenty-two Allied divisions which could be landed in the early stages, assuming the immediate transfer of landing craft from the Mediterranean, so any invasion would probably be defeated or confined to a beachhead.[13] The Americans naturally did not accept this forecast, and relations were not improved by Brooke criticizing Eisenhower for a lack of co-ordination between the First and Eighth Armies in the failed operations against the Tunisian port of Sfax, which had been seized by the Germans in November 1942 and was held successfully by them until 10 April 1943.

At 5.30 p.m. the Combined Chiefs, along with Eisenhower, Alexander and Tedder, met Roosevelt and Churchill in the first of three plenary sessions – 'at which we did little', recorded Brooke, 'except that the President expressed views favouring operations in the Mediterranean'. Far from 'little', this was the first glimpse that, as with Torch, the Americans were split over strategy, and therefore might be prevailed over by isolating Marshall again. This time it would take much detailed argument, especially by Staffs, rather than the point-blank veto that Brooke had exercised in London back in July.

Wedemeyer, who considered Casablanca an almost personal defeat, claimed in his autobiography that on the second day of the conference Brooke had said that victory in the European Theatre might be possible by the end of 1943, but had then 'let the cat out of the bag a little by saying we could "definitely" count on getting into Europe in 1944'.[14] Yet in January 1943 Tunis had yet to be captured, so the odds on the Allied forces being moved to Britain for a cross-Channel attack before autumn were next to impossible, and the Channel became effectively impassable in late September. The German army in northern France was strong, not worn down by bombing or weakened by withdrawals to the Eastern Front, so Brooke was hardly letting any cats out of any bags with his remark.

Wedemeyer also alleged that Admiral King's fear that strict adherence to Germany First might allow the Japanese to consolidate their gains in the Far East led him actively to support Operation Anakim, a proposed Anglo-American seaborne assault to recapture Burma from the Japanese, to be undertaken in late 1943. According to Wedemeyer, 'King visibly annoyed the British, and they shifted uneasily' at this advocacy of a major campaign there.[15] While it is true that King annoyed the British throughout the Casablanca Conference, and they might well have shifted in their chairs, in fact they had also shifted their position over Anakim, and came out in its support.

Brooke irritated King generally, the admiral complaining that he 'talked so damned fast that it was hard to understand what he was saying'. Jacob thought it more serious than that: 'I think the CIGS's extremely definite views, ultra-swift speech, and at times impatience made [the Joint Chiefs of Staff] keep wondering whether he was not putting something over on them.' In light of what had happened over Sledgehammer, this was understandable.

After dinner that night Hopkins and Harriman visited Brooke, the former 'in rather a bitter mood which I had not yet seen him in', which provoked the CIGS's ambiguous observation: 'There is no doubt that we are too closely related to the Americans to make co-operation between us anything but easy.' Perhaps, as Brooke probably meant to imply, the common language and culture were tending to cause more problems than they solved. Explaining Hopkins' bitterness, his biographer Robert Sherwood wrote that Roosevelt's confidant was 'again disappointed and depressed by the further postponement of Roundup; he was always solidly with Marshall in the conviction that there was no really adequate substitute for the opening of a Second Front in France.'[16] Might there also have been a scintilla of guilt in having effectively undermined the early Second Front when he had visited London with Marshall in July?

Wedemeyer was characteristically forthright on Torch:

After getting the African littoral, then the British maintained that you had to have Sicily because the enemy could occupy that and could jeopardize our lines of communication to the Middle East. Well, I would say . . . I don't care what happens in the Middle East. Rommel could run rampant in Alexandria or Cairo or whatnot. What I wanted to do was to get quickly into the Ruhr and cut off the source of supplies for Rommel. He would die on a limb as I saw it.[17]

*

Meanwhile, over dinner in Roosevelt's villa, Marshall and Eisenhower had to tell their commander-in-chief that he could not visit the front, as he had set his heart upon doing. Elliott recorded the difficult conversation, in which his father reminded his commanders that 'he'd been up front in the last war, as assistant secretary of the Navy', before insisting, 'And I'm going up front in this one too.' In fact Roosevelt had visited the Western Front for only a very few days in the Great War, mostly behind the lines, despite his regularly claiming over the years, quite untruthfully, to have seen a great deal first-hand of the ravages of that conflict.[18] Of course, even had Marshall and Eisenhower known the truth about the President's fantasies, they would hardly have exploded them. Instead the two generals merely exchanged a look, and went on eating. 'Well?' pursued Roosevelt. 'Why the silence?' 'It's impossible, sir,' ventured Eisenhower. 'Out of the question,' agreed Marshall.[19] The President tried to argue the issue, pointing out the lack of real danger and then suggesting that a fighter escort might protect him. The discussion ended only when Marshall concluded: 'Orders are orders, sir. But if you give them, nobody in the US Army from us on down will take responsibility.' Elliott recalled that Marshall 'was very serious, and Father was very disappointed, but forced to agree.' A compromise was reached by which Roosevelt instead reviewed three divisions of Patton's troops north of Rabat.

'There was a curious mixture of holiday and business in these extraordinarily oriental and fascinating surroundings,' recorded Harold Macmillan of Casablanca in yet another against-regulations war diary.[20] With his Balliol College, Oxford, classical education, Macmillan likened the Churchill–Roosevelt meeting to that of the Emperor of the East meeting the Emperor of the West in the later Roman Empire. Meanwhile Ian Jacob took away happy memories of Casablanca, especially of eating large juicy oranges for every meal, which had long been virtually unobtainable in England. He also recalled 'sunshine on a diversity of beautiful colours, and oranges. The red soil, the blue sky, the sea with its perpetual surf, the white villas and farms, and the white mass of Casablanca town, the bougainvilleas, the begonia, and the green of orange grove and palm trees made a picture which one never tired of gazing at from the sunny roof garden.'

Between 10.30 a.m. and 1 p.m. on Saturday 16 January, Brooke tried to extol the positive advantages of his Mediterranean strategy. Of the US Chiefs, he wrote: 'They can't be pushed and hurried, and must be made gradually to assimilate our proposed policy.' Jacob explained after the war that the Joint Chiefs of Staff agreed to the assault on Sicily (Operation Husky) only:

because we were representing to them that it was essential for the clearing of the Mediterranean and saving of shipping, but they could not see the political and strategic importance of knocking Italy out of the war. They thought of the Mediterranean – and the opening of it – as a means of getting a shorter route to Burma and India, and to the Persian Gulf for Lend–Lease supplies to Russia. The idea of exploiting it to threaten Hitler's southern flank appeared to them to be rather ridiculous.[21]

At another meeting from 3.30 p.m. to 5.15 p.m., Brooke thought he could spot 'some progress'. Meanwhile, Churchill planned a reconciliation meeting between de Gaulle and Giraud, another object of the conference. Brooke dined that night with General Patton, whom he thought on first acquaintance 'A real fire-eater and a definite character', a good summation in both aspects.[22]

After the war, returning to a persistent theme, Brooke wrote that his job at Casablanca had been made 'all the more difficult by the fact that amongst Marshall's very high qualities he did not possess those of a strategist. It was almost impossible to get him to grasp the true concepts of a strategic situation.' He yet again repeated the accusation that Marshall 'preferred to hedge and defer decisions until such time as he had to consult his assistants. Unfortunately his assistants were not of the highest calibre, and Cooke was of a very low category.' As for Patton, the post-war Brooke naturally considered him, like so very many other generals of the war, as 'good for operations requiring thrust and push but at a loss in any operation requiring skill and judgment'.[23] So ubiquitous was Brooke's formula in his diaries that it unfortunately undermines the value of his individual estimations of people.

It is hardly surprising, considering these kinds of comments, confidential though they obviously were, that the Americans both detected Brooke's contempt for their strategic abilities and resented it. In a long list of Brooke's faults published in Moran's controversial 1966 journals *Churchill: The Struggle for Survival* – including his matter-of-factness, straitjacketed mind, want of tact, and failure to appreciate Churchill's achievement in 'wooing the fickle Roosevelt' – the doctor–diarist argued that his gravest shortcoming was that 'He did not get on with the Americans. His downright, direct speech, combined with his take-it-or-leave-it manner, did not help him to get his own way.' Moran believed it was left to Dill 'to bring Marshall into line'.[24] There was much truth in this.

'Why did Brooke grate on the Americans?' Moran asked. Whereas Dill managed to make them feel that every word they uttered mattered enormously, by total contrast 'Brooke had an inborn suspicion that there might be an element of insincerity in this kind of approach. He swung instinctively to the opposite pole, throwing down his facts in the path of understanding with a brusque gesture. In his opinion it was all just common sense; he had thought it all out. Not for a moment did it occur to him that there might be another point of view.' Such utter certainty in one's own judgement might have been a necessary part of a grand strategist's mental armament in a world war, but 'He hurled facts at them like hand grenades, it did not matter if they went off and left wounds. Brooke's insensitive handling of his American colleagues had echoes, for it was what was American in Winston that most disturbed him.'

It is doubtful that the Combined Chiefs of Staff sessions of 16 January were a success in terms of mutual comprehension. In a meeting with Roosevelt, King complained that the British 'do not seem to have an overall plan' to win the war, but the Joint Chiefs of Staff would nonetheless go along with Churchill's and Brooke's Mediterranean schemes if the President ordered them to.[25] Kennedy heard from Brooke that 'much progress was made towards seeing each other's point of view' although the main points 'have had to be reiterated a good deal. Marshall gives an impression of great honesty and friendliness and soundness. He takes time to consider everything and although not rigid needs to be thoroughly convinced.' Brooke told Kennedy of a wordplay of Marshall's about American politicians, that 'There were more brass heads among them than brass hats among the soldiers.' Perhaps the truth was that, as Arnold confided to his diary, 'We are getting things done but awfully slow.'[26] His opposite number Charles Portal, meanwhile, explained to Dill, apropos of nothing in particular, how the sap travels in cactus plants.

For all its acerbity, King's complaint to Roosevelt had some validity. 'The "Mediterranean Strategy" was in gestation between September and December 1942,' writes Michael Howard; 'at Casablanca it was born and legitimized. This strategy was not one of manoeuvre, but of attrition.'[27] It had a number of advantages, namely that it could simultaneously release shipping, provide bases for bombing German-controlled Europe, divert troops from the Eastern Front, and knock Italy out of the war. It could not knock Germany out of it, however, which was the principal objection that Marshall and King had, and why the latter was so bitter in his remark to his commander-in-chief.

With large numbers of American troops in North Africa and British forces in the Middle East, the Masters and Commanders had a clear choice. The shipping was not available to transfer them all back to the United Kingdom for a cross-Channel operation, so, as Howard has summed up, 'Unless they were to remain idle for a year while the Russians continued to fight single-handed, some employment had to be found for them in the Mediterranean Theatre.'[28] The idleness option was not politically acceptable to either Roosevelt or Churchill, and while Brooke did not see the region as a soft underbelly he did hope that, were Sicily to fall, Italy would collapse and the Germans would be forced to reinforce Italy, and perhaps also the Balkans and the Aegean, from the Eastern Front. Allied sea and air power could then be concentrated against a long thin peninsula jutting into a sea that contained the Allied strongholds of Gibraltar, Malta, Tunisia, Alexandria and Cyprus.

Any advances made towards getting the Americans to support this strategy on Saturday 16 January, however, were lost the next day. 'We seem to be back at the beginning,' wrote Kennedy. 'Marshall and King expounded all their first ideas of Pacific strategy once again and CIGS repeated our side of the case. At the moment we seem to have reached an impasse, and we are still poles apart.' Brooke agreed, calling it 'A desperate day!' and concluding that agreement was further away than ever. After discussions on Burma and Iceland, it was decided that further conferring was useless until the Joint Planners had made much more headway together.

Yet when Brooke met the British Joint Planners he thought he detected from their remarks that their American counterparts 'did not agree with Germany being the primary enemy and were wishing to defeat Japan first!!!'[29] Well might that have deserved Brooke's treble exclamation mark were it true, but the British Planners had either misunderstood the US Planners' view or had misrepresented it to Brooke, or he had misunderstood it, or had misrepresented it to his diary. What is most likely is that the American Planners, particularly Wedemeyer and Savvy

Cooke, were threatening the Pacific strategy in the same way as Marshall did, 'as a club' to beat the British, but were bluffing. Brooke took his thoughts bird-watching, and saw 'a new white heron, quite distinct from the egret, and a new small owl which we could not place'. It was evident to him that the 'water on stone' technique had not worked so far, and the next morning a more direct approach would be needed. Simultaneously, Marshall was arriving at much the same conclusion.

The key meeting of the Casablanca Conference came at 10.30 a.m. on Monday 18 January, when the contrasting underlying strategic philosophies of the war were debated very openly. Brooke began by saying that Operation Anakim was now definitely on the planning agenda, and 'should be put to the front'. With the assistance of the US Navy in providing landing craft, he said, the amphibious capture of Rangoon was feasible. King said that it should be done in 1943 and the Combined Chiefs of Staff agreed. It was a fine start, and something of a compromise from Brooke, but soon afterwards profound differences emerged.

Brooke stated that the British Chiefs of Staff took exception to the first paragraph of an American memorandum – CCS 153 – which failed to state that Germany must be defeated before Japan. Marshall replied that in his opinion the British Chiefs of Staff 'wished to be certain that we keep the enemy engaged in the Mediterranean and . . . at the same time maintain a sufficient force in the United Kingdom to take advantage of a crack in the German strength either from the withdrawal of their forces in France or because of lowered morale'. He inferred from this that the British Chiefs of Staff would prefer to maintain a dormant force in the UK rather than use it elsewhere, whereas the Joint Chiefs of Staff 'know that they can use these forces offensively in the Pacific theatre'. Marshall's threat was obvious, especially once he added that 'to a large measure the shipping used in the Pacific is already committed and, therefore, could not be made available for a build-up of forces in the United Kingdom.'

Brooke answered plainly by saying that, in that case, 'We have reached a stage in the war where we must review the correctness of our basic strategic concept.' He was personally convinced that the Allies could not defeat Germany and Japan simultaneously and he and his colleagues believed that it would be better to concentrate on Germany first and indeed, 'because of the distances involved', they believed not only that the simultaneous defeat of Japan was impossible, but that 'if we attempt [it], we shall lose the war'. He added that having decided 'that it is necessary to defeat Germany first, the immediate question is whether to do so by an invasion of Northern France or to exploit our successes in North Africa'. The British Chiefs of Staff, he said, 'consider that an all-out Mediterranean effort is best but that it must be "all-out" '. He admitted that American assistance was necessary in any Mediterranean operations, and that failure to 'maintain constant pressure' on Germany would give Hitler 'an opportunity to recover and thus prolong the war'.[30]

Marshall replied to this forthright and all-embracing analysis reciprocally, by saying that the Joint Chiefs of Staff certainly did not propose 'doing nothing in the Mediterranean or in France'. He confirmed that the Germany First policy still stood, but contended that an early end to the war 'cannot be accomplished if we neglect the Pacific theater entirely', adding that he 'advocated an attack on the Continent but that he was opposed to immobilizing a large force in the UK, awaiting an uncertain prospect, when they might be better engaged in offensive operations which are possible'. Brooke replied that the British Chiefs of Staff 'certainly did not want to keep forces tied up in Europe doing nothing'.

To prove this he mentioned 'the desirability of Anakim', which could be undertaken by forces in the local theatre which would not detract from 'the earliest possible defeat of Germany'. He then quoted from a Joint Planning Staff paper which stated that Anakim was acceptable to Britain, 'provided always that its application does not prejudice the earliest possible defeat of Germany'. He was expert at always having to hand the apposite quotations from JPS papers that, since American Planners contributed to them equally, were hard for the American Chiefs to gainsay. At that point Admiral King complained that such wording 'might be read as meaning that *anything* which was done in the Pacific interfered with the earliest possible defeat of Germany' and that the Pacific theatre 'should therefore remain totally inactive'. Portal intervened to say that perhaps the two Staffs had 'misunderstood' each other, and all that the British were saying was that as far as 'getting at Germany in the immediate future' was concerned, 'the Mediterranean offered better prospects than France.' As an airman, he also commended building up a large heavy bomber force, which he pointed out was 'the only form of force that could operate continuously against Germany'.

Attempts to suggest that there are 'misunderstandings' rather than genuine differences of view are often simply a way to proffer an olive branch, so that neither side loses face, but Marshall brushed it aside when he replied that he was 'most anxious not to become committed to interminable operations in the Mediterranean' and he 'wished Northern France to be the scene of the main effort against Germany', something he rightly said 'had always been his conception'. The debate was thus veering back towards those pre-Torch ones held in London the previous year, especially after Portal bluntly admitted that 'it was impossible to say exactly where we should stop in the Mediterranean since we hoped to knock Italy out altogether. This action would give the greatest support to Russia and might open the door to France.'[31]

In response to this, Marshall repeated his view that operations in the Mediterranean and a build-up in Britain 'might well prevent us from undertaking operations in Burma' and therefore he was 'not at all in favour of this'. Moreover, American forces in the south-west Pacific were 'desperately short at present of their immediate requirements'. King of course agreed, stating that 'We had on many occasions been close to disaster in the Pacific.' For him, the real point at issue was to try to 'determine the balance between the effort to be put against Germany and against Japan, but we must have enough in the Pacific to maintain the initiative against the Japanese.' Hence his 30 per cent/70 per cent offer. The nature of these operations, he 'felt very strongly', should be decided not by the Combined Chiefs of Staff but by the Joint Chiefs of Staff alone – that is, a unilateral American rather than a joint Anglo-American decision. This of course flew directly in the face of the whole Combined Chiefs of Staff concept as set up by Marshall at Arcadia.

Marshall then argued that the notion of Germany First had been jeopardized by a lack of resources in the Pacific, for example heavy bombers set up to go to the UK had had to be diverted there, as under CCS 94. 'Fortunately, disaster had been avoided, but if it had occurred, there would have been a huge diversion of US effort to the Pacific theater. The US had nearly been compelled to pull out of Torch.' In order to make the Pacific secure, he said, Anakim and the reconquest of Burma 'would be an enormous contribution to this and would effect ultimately a great economy of forces'. He also mentioned operations to capture Rabaul and Truk.

At this Brooke baulked, seeing a major extension of American commitment far from the Mediterranean. Rabaul was an important two-harbour naval base on the Gazelle Peninsula in New Britain that dominated the entire New Guinea–New Britain–Solomon Islands area, the hub of the region that threatened Australia. Truk – which might be thought of as a Japanese Gibraltar – was a stronghold island in the central Carolines which served as the headquarters of the Japanese Combined Fleet, threatening Allied advances in the central and south Pacific and home port to one of the largest battleships in the world, the *Musahi*. Brooke began by arguing that it would be enough to stop at Rabaul, because to go on to Truk before Germany was defeated 'would take up too much force' and involve 'large shipping losses' which would be 'a continuous drain on our resources'. King demurred, insisting that the same forces could, by stages, liberate the Marshall Islands after Rabaul was captured in May, with Anakim taking place after November 1943.

The two Staffs seemed poles apart, so Portal and Marshall came up with a compromise form of words. Portal said the British would not like to be committed to Anakim, even with forces released after the capture of Rabaul, 'without first reviewing whether some other operation more profitable to the war as a whole might not be desirable'. To take an extreme case, he said, suppose a good opportunity arose, after the capture of Rabaul and owing to a crack in Germany, of attacking France? 'Should we refuse to take advantage of it because we were already committed to Anakim?' Marshall replied that he 'felt that if such a situation arose we should certainly seize the opportunity', agreeing that a further meeting of the Combined Chiefs of Staff might be necessary in the summer to decide.

Pressing the opportunity, Brooke then proposed that 'we should limit our outlook in the Pacific to Rabaul, which should certainly be undertaken, and to preparations for Anakim', with the final decision to launch Anakim being taken later. Any decision on Truk, however, should be 'deferred'. King complained immediately that that would be 'strictly to limit commitments in the Pacific, although the British Chiefs of Staff apparently contemplated an unlimited commitment in the European theater'. Yet Marshall agreed that a decision on Anakim and Truk 'could be left until later'. Brooke had spotted a gap between the positions of Marshall and King and had moved quickly to exploit it.

King argued, with some justification, that 'on logistic grounds alone it would be impossible to bring forces from the Pacific theater to the European theater' and that therefore Anakim was not a genuine alternative to Roundup, as Portal had hypothesized. He was supported in this by the American Planners Somervell and Cooke, but Portal was not to be put off now that an Army–Navy split seemed to be emerging in the American position, and he reiterated that 'it would be unwise to accept a definite commitment for Anakim now, since a favourable situation might arise in Europe'. King countered this by saying that 'favourable opportunities' might arise in the Pacific too. At this point Marshall suggested an amendment to the Joint Planning document, so as to read that seizure of the Gilbert, Marshall and Caroline Islands up to and including Truk would be undertaken 'with the resources in the theater'.

This might have been the point at which, in Wedemeyer's account, Brooke 'was visibly disturbed and impatient with King's position', and Cooke turned to John Deane, who was taking notes for the Americans, and said: 'Nuts!' Sir John Dill, sitting across the table with his British colleagues, overheard this remark and, 'realizing that Anglo-American tensions were becoming acute, skilfully performed his role as peacemaker'.[32] He whispered to Brooke the suggestion that he adjourn the meeting for lunch.

When the two sides resumed their meeting at 3 p.m., the Combined Chiefs of Staff had before them a draft note setting out tentative agreements. A deal had been brokered by Dill during the lunch adjournment, so that the topics under discussion could move on to the much less contentious issues of supply vessels, Polish forces, air raids on Berlin and the naval situation in the western Mediterranean. At one point Brooke revealed that a plan had been drawn up for seizing southern Spain with six divisions if it was deemed necessary to deny it to the Germans. Of the morning arguments with King, Kennedy recorded: 'It is good to blow off steam and probably the process is necessary.'[33] Brooke described the two-and-a-half-hour morning meeting as 'very heated' and thought King was 'still evidently wrapped up in the war of the Pacific at the expense of everything else!'

Years later Brooke showed that much of the credit for the lunchtime breakthrough and subsequent draft note had been down to Dill. 'I was in despair and in the depths of gloom' on leaving the conference room, he recalled, and as he walked upstairs to his hotel room had told Dill: 'It is no use, we shall never get agreement with them!' His friend and mentor replied: 'On the contrary, you have already got agreement to most of the points, and it only remains to settle the rest.' They sat on Brooke's bed after lunch and went through each individual point on the agenda, Brooke occasionally protesting that he 'would not move an inch' on some of them. 'Oh yes, you will,' replied the former CIGS. 'You know that you must come to some agreement with the Americans and that you cannot bring the unsolved problem up to the Prime Minister and the President. You know as well as I do what a mess they would make of it!'[34] This was unfair: the successful Torch had been Churchill's and Roosevelt's operation, carried out against the initial wishes of Brooke and Marshall, but it was undoubtedly this imperative that drove the generals towards compromise.

Brooke authorized Dill to talk to Marshall as a go-between. Portal also helped draw up the compromise formula into the draft note, which both Brooke and Marshall agreed to adopt when the meeting reconvened that afternoon. 'I am certain that the final agreement being reached was due more to Dill than to anyone else,' wrote Brooke after the war, 'acting as the best possible intermediary between Marshall and myself.'[35]

(Not everyone saw Dill as an honest broker: Leonard Mosley claimed that the British Chiefs of Staff lacerated Marshall's plans for a 1943 cross-Channel invasion because Dill had leaked the full US programme for Casablanca to them beforehand. 'They now proceeded to tear it to pieces, not least Marshall's pet plan for the invasion, which was ridiculed out of existence.' Wedemeyer, who spoke into Mosley's tape-recorder at length, told Mosley that Dill 'got extremely close to Marshall, and provided the British Chiefs and the PM with information on Marshall's thinking which Marshall shouldn't have given him' and that Dill had 'sold Marshall short'.[36] Like much of Mosley's history and Wedemeyer's testimony, this was inaccurate and simplistic.)

When the Combined Chiefs of Staff met again at 3 p.m. the compromise paper was accepted with only a few minor alterations. The recapture of Burma through Anakim and a south-west Pacific offensive to Rabaul and then on to the Marshall and Caroline Islands would be conducted with whatever means could be spared without compromising the objective of defeating Germany first. Anakim was promised American assault ships by King, but only once he had protected his Pacific resources from depletion. The British also promised that they would concentrate everything against Japan once Germany surrendered. Truk was left off the agenda. Crucially, Eisenhower would command Husky (the attack on Sicily), and Roundup would be undertaken whenever it was thought likely to succeed.

Meeting Roosevelt and Churchill at 5.30 p.m., Brooke sat next to Churchill, who asked who had chaired the Combined Chiefs of Staff meeting. Brooke said Marshall had, as Casablanca was under US occupation, so the Prime Minister called on him to report. The US Army Chief of Staff instead asked the CIGS to expound upon their report. 'It was a difficult moment,' wrote Brooke that evening, 'we had only just succeeded in getting the American Chiefs of Staff to agree with us.' Churchill did not know how tough it had been to find the compromise, and he could have upset it easily. However, the statement was approved by him and Roosevelt. 'We were then all photographed together!' recorded Brooke, and if some smiles seemed to be through clenched teeth, others were perfectly genuine.

The compromise report based on Dill's draft note was therefore ratified as Allied policy, and it contained the advantages and disadvantages of any such brokered solution. 'Roosevelt', his biographer records, 'was happy with this hard-fought outcome,' as were Churchill and Brooke.[37] Although Marshall also seemed content at the time, many American Planners came to see the Casablanca compromise as a serious setback, because it took large numbers of American troops on to Italian territory in 1943 rather than on to French beaches. So why did Marshall permit a major attack on Sicily, considering he had come to Casablanca opposed to a Mediterranean strategy? Jacob believes that because Roosevelt 'was quite of the Prime Minister's way of thinking, it was not long before everyone accepted Sicily as the thing to do.'[38] Richard Overy argues that Sicily was simply somewhere that the Germans could be convincingly engaged and defeated. In the end, if the British were not going to cross the Channel in 1943, and there weren't enough Americans to do it on their own, what else was there to do, other than drop Germany First and turn over the running of the war to Admiral King? Under those circumstances American troops would probably not get around to a ground assault on Germany until much later in the decade. For Marshall looking at the Sicilian operation in January 1943, most of the alternatives, except the Norwegian adventure which in his view was equally diversionary, had been blocked off by Roosevelt, Churchill and Brooke. Sometimes grand strategy, involving the lives of millions – and the deaths, wounding and capture of tens of thousands – has to be a question of taking the least bad compromise alternative.

There was still much left to agree now that the grand strategy had been set out. At 10 o'clock the next morning, Tuesday 19 January, a revised programme of meetings was disseminated, whereby it became clear that Anakim and the south-west Pacific would be considered in detail only at the very last meeting of the conference, that Saturday. In the meantime there would be meetings on every other conceivable subject, under the headings 'System of Command in French West Africa', 'Turkey and Axis Oil', 'Operation Husky', 'Bomber Offensive from North Africa', 'The U-Boat War', 'Landing Craft', 'The Bomber Offensive from the UK', 'Bolero Build-up' and several others, with Dill, in Jacob's words, 'often acting as go-between and general lubricator'.

Even though consideration of several of these issues obviously over-lapped with Anakim, the decision was taken to relegate the matter most likely to cause another rift with Admiral King to the very end of the conference, and to try to get as much else as possible settled in the meantime. Thus it was agreed that 938,000 US troops would be assembled in the United Kingdom by the end of 1943. Marshall's earlier calculations of attacking across the Channel with four hundred thousand men had had to be more than doubled once the formidable German capacity to resist and counter-attack had been witnessed in Tunisia.[39]

Jacob wrote of this second phase of the conference, conducted in long morning and afternoon sessions:

The remarkable thing about it all was that the gradual education of the Americans to our way of thinking was found to have proceeded even farther than we had thought possible. The beneficial results of holding such a conference for so long a period and on a neutral pitch made themselves clearly manifest. Everyone in these circumstances is freed from the irksome routine of the office, and there is nothing to distract attention from the work in hand ... In Casablanca ... everyone fed, slept and worked in the same building or group of buildings. British and Americans met round the bar, went for walks down to the beach together, and sat around in each other's rooms in the evenings. Mutual respect and understanding ripen in such surroundings, especially when the weather is lovely, the accommodation is good, and food and drink and smokes are unlimited and free.[40]

This implies that in an atmosphere of sweet reasonableness the superior British view naturally emerged triumphant, which is not what happened at all. The American Planners long afterwards counted Casablanca as a defeat at the hands of the British, and universally blamed their own inferior preparation and Staff work. 'The outcome was as we had predicted,' stated Paul Caraway. 'Our people lost their shirts. The only conference we did lose, I might add.' Tom Handy agreed, saying of Wedemeyer and Cooke, 'The British on the planning level just snowed them under ... If a question comes up and you have a paper ready to present on it, you have a big edge on the other guy who hasn't. Consideration can start on the basis of your paper, but the reverse is true if you've got none and if he's got a prepared thing ... These British Planners were just smarter than hell.'[41] Savvy Cooke clearly wasn't quite as savvy as all that.

'We were overwhelmed by the large British staff,' Ed Hull agreed. 'The only staff that General Marshall had was small and the other chiefs of staff were no better fixed. He had Wedemeyer and one assistant ... The British had come down there in droves and every one of them had written a paper about something and that was submitted by the British Chiefs of Staff to the American Joint Chiefs of Staff for agreement ... It taught us a lesson. Never go to a meeting like that without plenty of help, because you need it.'[42] One of the complaints Handy had about the way that Dykes and his team operated was that when they queried agreements that had been minuted, Dykes would say, 'You didn't object,' as 'the British way was to take anything not objected to as accepted.'

The War Office had commandeered a special ship that was moored in Casablanca harbour. HMS *Bulolo* was a 6,000-ton liner converted for use as a floating HQ for combined operations and had taken part in the Algiers landing. Fitted up with an operations room and a complete set of wireless instruments so that contact could be maintained with the landing forces, she was perfect for enabling the Planners to stay in touch with London, requesting any information they needed. 'We could operate exactly as if we were in Great George Street,' recalled Jacob. Harold Macmillan noted: 'In the Bay stood the famous communication ship which can send off as many as thirty wireless messages at the same time, and hosts of cypherers and so on.'[43]

Jacob was surprised by the tiny numbers of Staff officers that the Americans had brought with them, claiming with pardonable exaggeration that when the Joint Chiefs of Staff saw the size of the British operation, 'they went out into the highways and byways of North Africa and scraped together some sort of a Staff ... but ... nearly every paper produced during the Conference had to be produced by our people with little or no help from the Americans.'[44] For all his complaining tone, this was of inestimable advantage to the British in guiding the Staffs to the conclusions they wanted. Matters were made worse by the fact that Cooke and Wedemeyer sat in on all the meetings, meaning that they had little time to do 'solid work' with the British Planners. Furthermore there were personality clashes, since according to Jacob, 'Cooke's personality was so repellent that our people found it hard to get on with him at all.'

Wedemeyer in particular seemed to believe that the British had been somehow cheating with their meticulous preparation for the conference. 'The British brought a much larger team of Planners and advisers,' he grumbled. 'I was the only one that Marshall had, and Hap Arnold didn't have any air men with him at all ... We were overwhelmed by the British. They had so many arguments. This was all pre-arranged obviously, and the strategy was prejudged and they presented a united front concerning continued operations in the Mediterranean area, which I opposed, but only through Marshall.'[45] Wedemeyer was right; of course Brooke had 'pre-arranged' his arguments over future strategy: to have taken any other approach to what was manifestly going to be one of the most important military conferences of the twentieth century would have been profoundly negligent.

Occasionally Wedemeyer took his Anglophobia to truly absurd limits, telling his SOOHP interviewer at his farm in Maryland thirty years later that 'There was a considerable amount of British investment in German industry in the Ruhr during World War II,' and so 'there was a reluctance on the part of the British to bomb certain areas. British commercial interests insisted they be avoided.'[46] It is hard to think of anything more ludicrous, not least because Wedemeyer must have known that the RAF at great cost in lives tried to flatten – and also to flood – the Ruhr. But it does indicate the level of suspicion, indeed of paranoia, that existed in the mind of a senior US Planner whose subsequent writings and appearances on television programmes such as *The World at War* have affected our view of the making of grand strategy.

'There was too much anti-British feeling on our side,' Marshall admitted to his biographer Forrest Pogue after the war; 'more than we should have had. The British were accustomed to Staff business and we were not. When we went to Casablanca the President . . . only wanted about five people. The British had a large Staff; they brought along a ship for them to use. I had few people with me so I was shooting off the hip. Dill had told me the British would be ready.' Marshall privately recalled in 1949 that 'the long ingrained traditional skill of the British in the committee system' had shown up the 'freshman innocence' of the Americans at Casablanca.[47] The Americans would not allow it to happen again.

Wednesday 20 January started early for Brooke and Kennedy, who drove off to go bird-watching just beyond the Casablanca aerodrome. They saw a few larks, including a calandra. 'While we were studying it there was a curious loud gurgling noise in our ears,' recalled Kennedy; 'a camel had strolled up and was looking over our shoulders.'[48]

At 2 p.m. the Combined Chiefs of Staff 'thrashed out' plans for the capture of Sicily. Although it went well, and the Joint Chiefs of Staff conceded almost everything Brooke wanted, he could not help noting in his diary that Americans in general 'are difficult though charming people to work with', and Marshall in particular 'has got practically no strategic vision, his thoughts revolve round the creation of forces and not on their employment. He arrived here without a single real strategic concept, he has initiated nothing in the policy for the future conduct of the war. His part has been that of somewhat clumsy criticism of the plans we put forward.' By contrast, Brooke regarded Ernest King as 'a shrewd and somewhat swollen headed individual' obsessed with the Pacific war, while Arnold concerned himself solely with the air. 'But as a team to have to discuss with they are friendliness itself, and although our discussions have become somewhat heated at times, yet our relations have never been strained.'[49]

Although it had been a full week since the conference began, and the Combined Chiefs of Staff had only just got down to discussing practical plans, Roosevelt was happy with progress, telling his cousin Daisy Suckley that despite what he called 'the Winston hours' – 'sleep to 9 a.m., then morning conferences then a luncheon, then a nap for one hour, then more talk and a dinner at 8 p.m. which lasts to an average of 2 a.m.' – nonetheless 'We are getting on very well with our Staff conferences.'

The next evening, Thursday 21 January, Brooke suddenly found himself in a difficult position since Portal, Pound and much of the Joint Planning Staff now said that they thought Operation Brimstone, the invasion of Sardinia, ought to precede or possibly replace altogether Operation Husky, the invasion of Sicily. The arguments went on for three hours until midnight. 'I have the most vivid recollection of that exhausting evening!' Brooke wrote later. He had only just managed to persuade Portal and Pound over Husky before leaving England; now they sought to reopen the question in what amounted to the most serious mutiny against his authority during the whole course of the war.

'All my arguments with Marshall had been based on the invasion of Sicily and I had obtained his agreement,' recalled Brooke. Yet now the other Chiefs wished to unravel that. As well as the merits of the case itself, Brooke's credibility with the Americans was on the line, something he could not risk losing in the coming no-holds-barred meetings at Anfa Camp. The revolt had started in the Joint Planning Staff, after which Mountbatten – who Brooke believed 'never had any very decided opinions of his own' – had supported them. Mountbatten's support was not significant in itself, thought Jacob, because 'His invariable habit of butting in on detail in the middle of discussion of matters of large principle had destroyed any influence he might have had in the Committee.'[50] Nonetheless, with Portal and Ismay also doubtful about Husky, Brooke had a serious problem even though 'dear old Dudley Pound was, as usual, asleep and with no views either way!' When awake, however, Pound did not support Husky either.

Brooke, who had gone over the strategic and tactical issues in minute detail in London and was as usual certain that he was right, flatly refused to go back to tell Marshall that 'we did not know our own minds', which would 'irrevocably shake their confidence in our judgement'. He thought that soon after the Allies seized the south of Sardinia the Axis would pour into the north, leading to a long and difficult campaign, and losing the chance of taking Corsica easily. 'Being a very obstinate man, further argument only annoyed him,' recorded Jacob, 'and he became more and more rabidly against Sardinia, and in favour of Sicily.' There is a slight indication that Brooke might privately not have considered the Husky plan to have been perfect, writing at the time: 'A good plan pressed through is better than many ideal ones which are continually changing.'[51] He was not about to vouchsafe this view to anyone else at the time.

It was by sheer force of character, therefore, and an implied threat of resignation sooner than go back to Marshall with a change in his proposals, that Sicily was chosen rather than Sardinia. It was a classic case of the influence of personality in strategy-making. Kennedy's diary implies there was also another threat that Brooke used, that 'any variation in the programme at this late stage . . . might result in the Americans doing nothing in the Mediterranean'. As Brooke wrote, 'so few people ever realize the infinite difficulties of maintaining an object or a plan and refusing to be driven off it by other people for a thousand good reasons!' Although he was completely outnumbered by all his British interlocutors, he was on strong ground since Churchill, Roosevelt and Marshall now all supported Sicily over Sardinia. Churchill supported Brooke because whatever was done in 1943 'must look big to Stalin' and he described the capture of Sardinia as 'that piddling operation'.[52] As well as wanting an attack on Sicily, Churchill pressed for one on the Dodecanese soon afterwards, and claimed that he also wanted an assault in northern France in the summer if possible, as well the Anakim attack on Burma in the autumn. Some of these were of course mutually exclusive, but, in Jacob's view, 'He was indulging to the full in his usual pastime of having his cake and eating it – or trying to.'

Friday 22 January saw no fewer than six meetings, as Roosevelt was anxious to leave Casablanca. Arnold noted that at the 10 a.m. Combined Chiefs of Staff meeting, 'much to everyone's surprise', they managed to 'agree on a lot of things'. They concurred on the Husky plans, to the relief of Brooke who had feared the Joint Planners might try a last-minute ambush. Marshall spoke with 'great earnestness and feeling' about 'sticking to the big thing. He said he had come to England before to advocate the invasion of France and had been put off with North Africa.' If Husky fell through, and he had to accept Brimstone instead, then 'he thought they would have to find some one else to sit in his chair'.[53]

For both Brooke and Marshall to threaten resignation over Husky within twelve hours of each other shows how important the capture of Sicily was to both of them, and doubtless explains why the Joint Planning

Staff did not bother to pursue its campaign against the operation. After the meeting Brooke told Kennedy: 'So it is settled that we go bald-headed for Husky.' From one ornithologist to another, the eloquent image was of a bald-headed eagle swooping down on its prey. Before lunch, when they arrived at Roosevelt's villa for photographs, Kennedy noticed the small ramps that had been fixed to allow the President's chair to be wheeled smoothly from room to room. 'What a fine head and face he has,' he wrote. 'It is pitiful to see that great torso and those withered legs.'[54] (Such were Kennedy's powers of observation that he even spotted that Churchill had had a button sewn on to his left shoulder in order to hold his gas mask in place.) Arnold was happy to be summoned to the villa, because in the great wartime conferences he had observed that 'when the photographers appeared, the end was usually in sight.' They were all exhausted, with Brooke noting, 'It has been quite the hardest 10 days I have had from the point of view of difficulty of handling the work.'[55] Kennedy thought that much had been achieved in the time, 'But it is a good thing it is nearly over for we should all be getting on each other's nerves if we stayed here much longer.' He was good at understatement.

The last meeting, at 9.30 p.m. on Saturday 23 January, broke up with effusive words of congratulation, with Marshall speaking of his 'appreciation of the readiness of the British Chiefs of Staff to understand the US point of view and the fine spirit of co-operation which they had shown during the discussions'. Brooke answered in similar vein, saying that 'Mutual appreciation of each other's problems was only possible through personal contacts.' Even Admiral King claimed for the record that 'he fully agreed with Sir Alan Brooke as to the great value of the basic strategic plan which had been worked out at the Conference. In his view this was the biggest step forward to the winning of the war . . . The discussions which had been held had enabled a true meeting of minds to take place between the British and US Chiefs of Staff.'[56]

For all these flowery words, the Americans were convinced that the British had had the best of Symbol. Britain's success there was all the more extraordinary considering the wild mismatch in projected force strengths between her and the United States. When Portal and Arnold compared notes at the conference about the future sizes of their air forces by December 1944, the discrepancy was startling, even if one bears in mind that the figures did not take into account the rest of the

British Commonwealth. Portal, recalled Arnold, 'was shooting for 537 squadrons, or 9,870 airplanes' and a 1.2 million-man RAF. Arnold, on the other hand, was planning on having 52,000 aircraft and a 2.36 million-strong USAAF.[57]

At the concluding press conference on 24 January 1943, Roosevelt stated that General Ulysses S. Grant had been known as 'Unconditional Surrender' Grant, and that the Allies were also demanding unconditional surrender from the Germans and Japanese (but not the Italians) in the present struggle. It is often argued that this insistence led the Germans and Japanese to fight more fanatically than would otherwise have been the case, although it cannot be proven. What is plainly untrue, however, is that the policy merely sprang fully formed from Roosevelt's mind without any consultation with Marshall or Churchill. In fact Churchill had not only given prior approval but had cabled the War Cabinet over the issue four days earlier, and his colleagues had not objected. Although subsequently both Marshall and Churchill claimed they had been surprised by Roosevelt's announcement of the policy at the press conference, they both later accepted that they had in fact been consulted beforehand. On 7 January Marshall stated that Allied morale would be helped by the uncompromising demand, and Stalin's suspicions allayed.[58] 'While the words were obviously of value to Goebbels,' Lord Halifax wrote of this new departure, 'the chaps at the top, knowing that there was only a halter for them at the finish of this business, would have made their people fight to the end anyway.'[59]

The harassed, overworked and fiercely suspicious Wedemeyer was convinced that Casablanca had been a catastrophe which dragged the United States into Brooke's Mediterranean strategy against her will and contrary to her original intentions. 'Our own Chiefs of Staff were not at all in accord with the British,' he recalled fifteen years later. 'But General Marshall's relationship with Roosevelt differed subtly from the relationship that existed between the British Chiefs of Staff and the Prime Minister. They had frequent, almost daily, access to their political leader . . . [which] made for unanimity of purpose and ensured a united front whenever the Britishers marched off to conferences. They always knew in advance what they wanted. They had aims.' Of course, in Wedemeyer's world, these were not the defeat of the Axis powers: 'Usually their aims could be related to Empire or their post-war position in the world of commerce.'[60]

Rather than listening to Marshall and King, claimed Wedemeyer, Roosevelt had been 'surrounded by many drugstore strategists', among whom he numbered Hopkins, FDR's counsel and chief speechwriter Judge Samuel Rosenman, his military aide Major-General Edwin 'Pa' Watson and Averell Harriman. Wedemeyer believed that, when the President demanded Germany's unconditional surrender in January 1943, it was because 'It had been driven home to him by many of his closest cronies like [Justice] Frankfurter, Morgenthau and Judge Rosenman – all Jews who actually felt bitter against the Germans. No question about it, they convinced the President that this time the peace terms must be signed on German soil.'[61] Wedemeyer had clearly imbibed more than just military theory from the time he spent in pre-war Berlin.

On Saturday 23 January the Eighth Army entered Tripoli, the major Axis supply and entry point for North Africa. The retreating Germans had tried to raze the city, harbour and airfield, but these were quickly repaired. After another meeting of the Combined Chiefs of Staff at 10 a.m., there was the third and last plenary session, held in the boiling heat of Roosevelt's villa.

In these somewhat unlikely surroundings, the Chiefs of Staff joined the President and Prime Minister to go through each paragraph of the final report together, occasionally making slight modifications and amendments. 'It has been a wonderfully good conference in the end,' thought Kennedy, as well he might considering how many British desiderata were incorporated into the final text. After lunch that day, Brooke, Dill and Kennedy had motored out to the invasion beach at Fédala, and came away feeling that Torch had been very lucky in terms of weather and lack of opposition; they concluded that more smoke and air cover would be needed in future amphibious operations.

'Finished the Staff conferences,' Roosevelt told Suckley, 'all agreed de Gaulle a headache – said yesterday he was Jeanne d'Arc and today that he is Georges Clemenceau!'[62] Although de Gaulle's handshake with Giraud had been duly photographed, neither would agree on subordination to the other. The French North African Army therefore stayed separate from the Free French forces, an absurd arrangement militarily.

At Casablanca Churchill believed – or professed to believe – that Roundup could still be mounted in 1943. That placed him on the same side as Roosevelt and Marshall, and put Brooke in the perennially dangerous position of being out of step. Fortunately, however, Churchill's strong desire to see the Allies marching into Rome and overthrowing Mussolini meant that Roundup considerations were kept, in the American phrase, 'on the back-burner' during 1943. By contrast, Roosevelt did not support Marshall to the extent that he needed him to in order to defeat the Mediterranean strategy. At Casablanca, no less than in London the previous July, Marshall was the odd man out of the quartet of power, and his strategic views suffered as a result. The fact that he went to Casablanca opposed to any further action in the Mediterranean, but left it having threatened to resign if Sicily were not attacked, shows how far his views had been brought around by circumstances, primarily by lack of presidential support.

'Brooke's personality and drive have accomplished great things,' concluded Kennedy, who was inclined to a degree of hero-worship of his boss. 'Both sides are really convinced now that we are on the right lines and we can now drive ahead with the war on a co-ordinated plan.' The historians of the Chiefs of Staff Committee, Field Marshal Lord Bramall and General Sir William Jackson, believed that the whole Allied grand strategy of the war – the WW1 Arcadia plan as updated by CCS 94 – might have been overturned by the Americans at Casablanca if Brooke had not been by Churchill's side there.[63] The stakes had therefore simply been too high for Brooke to indulge his own first desire – indeed any general's – which was to lead an army into battle.

Brooke had got almost all he wanted from Casablanca, but at a very high price. For him it was never glad confident morning again. Too many Americans in very senior positions were certain that he had driven too hard a bargain over too many policy areas – and from too weak a position – to permit another conference like that. Portal explained this feeling shortly after the war, saying of Symbol that:

One of the greatest snags was that the American Chiefs of Staff were always looking for hidden motives, whenever we put up a plan. They were the victims of the common American impression that the British are frightfully cunning and will do you down at every turn. They were aware of the popular American conception that the British are much more clever than the Americans in diplomatic negotiation, and they always seemed to us frightened of being trapped . . . They were convinced that we wanted to use American troops to further our own political ambitions, and we never really allayed this fear.[64]

13
The Hard Underbelly of Europe:
'Total War requires total mobilization'
January–June 1943

Logistics Churchill disdained.   Sir John Colville, May 1982[1]

'It's difficult to argue against success,' wrote General Hull in his unpublished autobiography, 'for the North African operation was a complete success. However, I still believe it was strategically a mistake.' In order to capitalize on Torch's success, he argued, the Allies were 'practically forced' into the long struggle 'over extremely difficult terrain' up the Italian peninsula. He went on to argue that there were twice the number of German divisions in France and the Low Countries in 1944 that there had been in the summer of 1943, which was not quite accurate but not far off. The phrase 'mission-creep', like 'group-think', is a modern one, but they both have applications for the Second World War. Under mission-creep, operations continue to be undertaken for reasons far removed from the original intention, because the conduct of the conflict has imperceptibly outgrown the reasons it was undertaken. That is what was to happen in Italy.

Ian Jacob believed that being expelled from North Africa 'would be shattering for the Italians. Their vitals would be exposed to attack.'[2] The surrender of Italy would present Hitler with a tough choice: either let her go or else reinforce her by taking troops from elsewhere, such as Russia and the Balkans. There was an aspect to the Führer that was only just becoming apparent to the Allied High Commands: it seemed clear from the orders that he gave both to Paulus in Stalingrad and to Rommel at El Alamein (and again in Tunisia) that he could not countenance even strategically justified withdrawals. This psychological disorder on his part – the result, perhaps, of going from corporal to commander-in-chief without the intervening stage of divisional command or Staff college –

was to be a strategic blind-spot that was to be greatly exploited by the Allies over the coming months and years.

Interviewed by NBC television in 1958, Brooke explained how the British strategy for the Mediterranean was designed to draw out the Germans:

The soft underbelly of Europe was the whole of southern Europe including a portion of southern France, the whole of Italy and the whole of Greece, all of which Germany was defending, and all of which is difficult to defend. It's like a series of fingers spread out into the sea. In order to defend it you've got to disperse your forces through it . . . We crossed over into Italy by defeating the Italian forces and wiping them off the map, forcing German detachments to take over the jobs that the Italians had been doing and to detain forces in Italy. That was the idea.[3]

On the day the Casablanca Conference ended, Churchill and Roosevelt drove 120 miles to Marrakesh, having a picnic lunch on the way, and stayed the night at the Villa Taylor, which had been built in 1927 by the American industrialist Moses Taylor and which had subsequently become the residence of the US vice-consul. It had a 'Marvellous view of snow-capped Atlas' mountains, as FDR told Daisy Suckley.

Churchill insisted on the President being carried up on to the roof of the Villa Taylor, 'his paralysed legs dangling like the limbs of a ventriloquist's dummy, limp and flaccid' in the words of an onlooker, and together they watched the purple mountains changing colour in the setting sun. It was from that roof that Churchill painted his only picture of the war, despite taking his canvases and paint box on several trips. He gave it, a view of the minaret of Katoubia Mosque in Marrakesh with the Atlas range beyond, to Roosevelt. Inspector Thompson, who admittedly was a bodyguard rather than an art critic, remarked that 'The whole scene was a riot of the colour from which he draws his inspiration.'[4] After the President had been carried back down the tower, Churchill walked with Charles Moran among the orange trees in the garden. 'I love these Americans,' he said. 'They have behaved so generously.' It is unclear whether he was referring to their vice-consul providing the villa, or their agreeing to his Mediterranean strategy.

'I saw him in bed and I think it was one of the most marvellous sights I've ever seen in my life,' said Brooke of Churchill at the Villa Taylor, before going on to describe the lights on either side of the 'ornate

Moorish' bed and how he was 'wearing one of those two marvellous dressing gowns, which suited the ceiling, full of red dragons'.[5] Brooke himself was staying at the Mamounia Hotel, where Jacob joined him for coffee – there was no tea on offer – in the garden. 'The Hotel is said to be the finest in North Africa,' enthused Jacob, 'and I think it could lay great claim to the distinction. The public rooms are large and airy, and the bedrooms on the second floor are a dream, as they open onto balconies from which it almost seems as if one could touch the Atlas snows.' The next day the CIGS went bird-watching. Churchill was still wearing his bedragoned dressing gown when he saw the President off from the aerodrome on the morning of Monday 25 January, and he himself left with Brooke shortly afterwards for Cairo and then Turkey.

It was while he was in Cairo that Brooke learnt from Portal that one of the Liberators returning home from Gibraltar with Planning Staff on board had crashed. The outer starboard engine had caught fire an hour away from the Welsh coast, and then it and its surrounding framework had plunged into the sea. When the pilot tried to land at Haverfordwest aerodrome, the inner starboard engine had cut out and then the bomber somersaulted and smashed into fragments. Vivian Dykes and Guy Stewart were asleep in the hold when they died instantaneously, and two others were seriously injured. Astonishingly, thirteen people on board escaped with minor injuries.

On Brooke's instructions, the news was kept from Churchill 'in view of tomorrow's journey'. When it was broken to him once he had landed in Turkey, his thoughts naturally turned to what would happen were he himself to die. 'It would be a pity to have to go out in the middle of such an interesting drama,' Jacob heard the Prime Minister ruminate. 'But it wouldn't be a bad moment to leave. It is a pretty straight run now, and even the Cabinet could manage it.'[6] The war saw a large number of prominent people die in aircraft crashes or shootings-down, including 'Strafer' Gott, General Sikorski, the Duke of Kent, Orde Wingate, Admiral Yamamoto, Sir Trafford Leigh-Mallory, Marshal Italo Balbo, Major-General Vyvyan Pope, Sir Bertram Ramsay, Subhas Chandra Bose and of course Major Glenn Miller. Roosevelt and Churchill were brave men to undertake such arduous journeys at their ages. In all, Churchill made no fewer than nineteen journeys outside the United Kingdom between August 1941 and March 1945, none of them free from hazard.

Speaking to his assistant Bill Hassett, who had asked about the risks involved in flying to and from Casablanca, Roosevelt said the decisions could not have been arrived at any other way. 'We were not getting anywhere in our plans for operations,' he said. 'The British joint chiefs would agree among themselves, but they could not reach an accord with our joint Staffs.' Furthermore Roosevelt had to consider 'the personal equation; the *prima donna* temperament'. If he sent Stimson to London, then the Secretary of the Navy, Frank Knox, 'would have thought he should go. It was the same between the Army and Navy in London. Churchill and I were the only ones who could get together and settle things.'[7]

At 7.45 a.m. on Sunday 31 January 1943, Field Marshal Friedrich Paulus (promoted by Hitler the day before in an attempt to dissuade him from surrendering) destroyed his wireless equipment – it must have been something of a relief to know there would be no more 'Stand fast' messages from the Führer – and was captured in Stalingrad after 162 days of the most gruelling campaign in the history of warfare.[8] Hitler's refusal to allow any attempted break-out from the 15-by-9-mile *Kessel* area had helped seal the Sixth Army's doom. What happened next was almost biblical in its apocalyptical ferocity: of the 91,000 survivors of the Sixth Army and allied detachments who surrendered and became prisoners-of-war in Russia – more than 170,000 had already died in the fighting – only nine thousand ever returned home, many of them as late as the mid-1950s.

The immediate effect of the surrender was to make Roundup less likely in the short term as a necessity for relieving Russia, but more likely in the medium term as a means of exploiting the loss of German morale attendant upon the slow realization by the more perceptive of them that the odds had now tipped significantly against their winning the war.

On 3 February Churchill wrote to Roosevelt to say that Stalin was entitled to more precise information about the Casablanca conclusions, since 'no one can keep secrets better'. He suggested that they should tell Stalin that they hoped to 'destroy or expel' a quarter of a million Germans and Italians in eastern Tunisia by April; that in July they would attack Italy with three or four hundred thousand men with the object of promoting an Italian collapse 'and establishing contact with Yugoslavia', and that 'we are aiming at August for a heavy operation across the Channel' of between seventeen and twenty divisions, supported by almost the entire metropolitan air force of Great Britain. Churchill concluded by proposing that they also 'say that in accepting the conclusions of our Combined Chiefs of Staff, the President and the Prime Minister have enjoined upon them the need for the utmost speed and for reinforcing the attacks to the extreme limit that is humanly and physically possible'. Churchill added: 'I have talked it all over with the CIGS who is in agreement.'[9]

Yet Brooke was simply not in agreement with the concept of an August 1943 date for Roundup, one month after Operation Husky, not least because some of the same landing craft were thought to be needed for both operations. Unless Churchill took refuge in the wording of the sentence – that these operations were being 'aimed at' – it was disingenuous. Roosevelt was equally sanguine, however, replying: 'I wholly approve of your view.' He did suggest changes, such as replacing the word 'Italy' with 'Sicily' and saying that Roundup 'must of course be dependent upon the condition of German defensive possibilities', but otherwise he was just as over-optimistic as Churchill. His military aide, Pa Watson, commented to Marshall the next day that over Roundup the President 'promises *much more* than can be done, even though the word used is "aiming" '. It seems that Churchill, Roosevelt and Brooke, and possibly Marshall, were taking refuge in lexical exactitudes in order not to fall out at that stage either with each other or with Stalin over the realities governing Roundup. Promising Stalin an August 1943 Roundup – or seeming to – suited each of their agendas at the time. They had done the same with Molotov the previous June, but in the long run it was very bad policy to mislead Joseph Stalin.

The first full Cabinet after Churchill's return from his four-week journey on 7 February dealt with the threat of a fast by Mahatma Gandhi. Taking Brooke's place at the meeting was John Kennedy, who recorded: 'Winston thundered at the Cabinet and said that the line proposed by them and the Viceroy was much too weak – Gandhi should not be released on the account of a mere threat of fasting. We should be rid of a bad man and an enemy of the Empire if he died.' Grigg then said that Gandhi was getting glucose in his orange juice, and another minister suggested that he was having 'oil rubbed into him which was nutritious', allowing Churchill to claim that this 'is apparently not a fast, merely a change of diet'.[10] In the event Gandhi was not released and completed his three-week fast in good health, though if he had died in gaol on a fast-cum-hunger-strike their decision would have haunted the Government for years.

The mood then lightened, and Churchill told the Cabinet a little of his African and Turkish visits. He and Brooke had failed to persuade Turkey to enter the war, which was the primary purpose of their visit. He said that British troops got a bad exchange rate for sterling in Algeria, but, after all, they 'had nothing to spend their money on except oranges and women'. He teased Herbert Morrison by saying that the recent criticism of the former Vichy Interior Minister Marcel Peyrouton, then a Giraudist governor of Algeria, was not wholly justified: 'After all, it may be a misfortune to be Home Secretary, but it is not a crime.' He said he did not know whether Eisenhower was a good general or not, but he supposed that good generalship consisted largely of 'sleight of hand'. He then reminisced about how he had had the President carried up the tower and on to the roof of the Villa Taylor in Marrakesh to see the sunset over the Atlas mountains: 'One of the most lovely sights I know in Africa.' All of this charm and good-natured badinage allowed him to vouchsafe the barest minimum to his colleagues in terms of the actual strategy agreed at Casablanca.

Between 16 and 21 February, Churchill suffered a bad bout of pneumonia, a serious problem for someone in their sixty-ninth year. It was during one such illness that Marshall told Churchill that he nicknamed pneumonia 'the old man's friend'. Churchill said: 'Pray explain?' to which Marshall replied: 'Oh, because it carries them off so quickly.' Moran recalled how jokes like that established Marshall 'high in the PM's favour'.[11]

Discussing the ban on visitors in 'Invasion Areas' along the south coast that had been in force since 1940, the War Cabinet later in February slipped neatly from maintaining the ban because of fear of a German invasion of Britain to maintaining it because of the hope of an Allied invasion of France. Herbert Morrison argued that there was 'No justification for maintaining this ban unless very strong military reasons for it. [We] Don't want to destroy the "invasion mentality" – but [the] public aren't fools.' At this point Grigg enlightened the Home Secretary: 'Our main point is not that: it is preparations for offensive action. We shan't want people unnecessarily to see what we are doing.' Brooke then pointed out that 'We shall later have to deny entry to [the] coastline: and over a wide area to provide cover.'[12] These, again, are not the actions of officials who – as many American Planners would have had it – 'never' wished to re-enter the Continent. Neither was the RAF's systematic bombing of the European railway system which started on 6 March 1944.

The Allied offensive that took place in the gap in the Tunisian mountain chain called the Kasserine Pass opened on 14 February. It ended in an American defeat at the hands of Rommel five days later, although the Allies subsequently rallied to a fine defence. For the cost of 989 German casualties (and 535 captured Italians), the US II Corps suffered 6,600 killed, wounded and missing. It proved that there was still much to learn about equipment, generalship, tactics and above all the capacity of the Wehrmacht to counter-attack. All these had serious implications for a summer 1943 Roundup, to which all of the Masters and Commanders, even including Brooke, continued to pay at least a form of lip-service.

In early March several newspapers ran stories saying that Marshall was shortly to be appointed to command the Allied forces in Europe. Katherine Marshall later recalled that she read these comments and editorials and 'was determined not to be caught napping. George had said nothing and I asked him no questions.' Nonetheless she bought a second-hand trailer in readiness for any move. Marshall was meanwhile testifying before Congress trying to get a Manpower Bill passed that would increase the US Army to 8.2 million men. 'Emergency manpower shortage cannot be met overnight,' he told the special committee. 'The 1944 army must be trained in 1943 . . . Total war requires total mobilization of resources. This is not being done.'

Brooke and Marshall were meanwhile arguing over the American desire to ship arms to Giraud's forces in Africa, which Brooke saw as a waste of shipping resources considering that the French 'can play no part in the strategy of 1943'.[13] The situation was not eased by the antics of de Gaulle, and on 3 March Churchill even threatened to have the Free French leader arrested. According to the verbatim minutes taken at the War Cabinet meeting by Norman Brook that day, Eden reported that de Gaulle had asked to visit his troops in North Africa. At a sensitive time between the Giraudist and Free French forces in the French colonial empire, it was feared that his always combustible presence might wreck the delicate balance of power there. When de Gaulle was told by the Foreign Office that the moment was 'unsuitable', he enquired whether he was 'to regard himself as a prisoner'. Churchill suggested a form of words by which he should be told that it was 'Not considered in [the] interests of the United Nations at this stage . . . that he should leave the country.' As usual, he worried about Roosevelt's reaction, urging Eden to 'Think of the Americans, who believe us responsible for all de Gaulle's acts.'[14]

Brendan Bracken then suggested that they should prevent de Gaulle from broadcasting from London, but Eden pointed out that the general couldn't in any case 'without pre-censorship by me'. Churchill said that the press should be asked not to discuss the issue, as the 'Free French will try [to make] propaganda about [de Gaulle being a] "prisoner".' Attlee believed there would be trouble anyway, since 'his reputation is higher than ever'. Churchill was adamant. 'Put it quite bluntly,' he said to Eden. 'And arrest him if he tries to leave e.g. by a French destroyer. Security measures should be laid on to prevent that.'[15] When discussions resumed on 15 March, Churchill said that de Gaulle was 'actuated by personal motives', and believed he had the 'title-deeds of France in his pocket'. The Prime Minister concluded that Giraud was 'a much better man: and de Gaulle was probably a bitter enemy of Great Britain.'[16]

On the same day that Churchill threatened to arrest de Gaulle, 3 March, 111 adults and 62 children were killed in Bethnal Green tube station in the East End of London, the worst non-military single-incident death toll of Britain's entire war. When on 6 April the report commissioned by the Government was brought to him, Churchill was determined that it should be suppressed until after the end of hostilities. It revealed that public panic during an air raid had been responsible, rather than a direct hit by the Luftwaffe, which had been the official explanation. Churchill declared himself:

Against giving such limelight to this incident . . . It would give disproportionate importance, and be meat and drink to the enemy and an invitation to repeat. We will say the Report was received and considered: no need to publish it: and all its lessons are being vigorously applied. Why publish? The Government's position is unassailable. Moreover, we said earlier there had been 'no panic': this makes clear there was panic and it was partly the cause: and this we are withholding.

The Home Secretary, Herbert Morrison, disagreed, contending that the large number of deaths 'shook the public', and that refusing to publish the report would be taken as an admission that there was something to hide. He added: 'We held off the discontented locals by the promise to publish the results.' Attlee, who sat for the nearby Limehouse constituency, was concerned to 'Make it clear that panic was not due to Jews and/or Fascists,' which were two of the rumours swirling about at the time.[17] In the event the report was not published until 1946.

Between 8 and 12 March 1943 the Eighth Army repulsed heavy German counter-attacks in Tunisia, foreshadowing the end of the German presence in North Africa. Reporting to the War Cabinet about the successes in the Western Desert, Brooke stated that the Churchill tank was doing well, and that the American forces in the middle sector were 'almost back to where they were before the German attack'. As for the Eighth Army, there were 'indications of German columns turning round and going back'.[18] It was not all sweetness, however; on 22 March after the War Cabinet heard from Nye about the land advance on the west Burmese port of Akyab in the Arakan – in which nine Japanese battalions were fighting fifteen Allied ones – Churchill pronounced it 'Very unsatisfactory. Though we outnumber [the Japanese] they outmanoeuvre us. No unit ascendancy on our part.' Nye had to 'Admit we are not as good as Japs at jungle warfare', even though the Allies had superior weapons and, at 1,500 versus 2,500, fewer battle casualties.[19] In the event Akyab stayed in Japanese hands until January 1945.

By 29 March the battle in Tunisia seemed to have turned against the Allies once again. After Brooke had given his report to the War Cabinet, Churchill said: 'Not happy about fighting in the north, we uniformly had the worst of it. Looks as if Germany is beating us unit for unit, despite the fact that we have greater artillery than the enemy.'[20] The terrible truth was – and it was not just true of the Tunisian campaign – that unit for unit the Wehrmacht regularly did indeed beat the British and American armies. The statistics are incontrovertible; in his intensely detailed studies of several hundred individual military engagements, the historian Trevor Dupuy has concluded that:

In 1943–44 the German combat effectiveness superiority over the Americans and British was in the order of 20–30%. On a man-for-man basis, the German ground soldiers consistently inflicted casualties at about a 50% higher rate than they incurred from the opposing British and American troops under all circumstances. This was true when they were attacking and when they were defending, when they had local numerical superiority and when, as was usually the case, they were outnumbered, when they had local air superiority and when they did not, when they won and when they lost.[21]

It was an astonishing achievement, and one that Brooke and Churchill reluctantly recognized privately but of course could never publicly acknowledge. The battles at Kasserine Pass, Anzio, Monte Cassino, Caen, Arnhem and the Ardennes forced the deeply uncomfortable fact upon both the British and American High Commands that the Germans, even in defeat, were formidable fighters against whom significant numerical superiority on the ground and in the air was needed.

The same day that the War Cabinet assessed Eisenhower's slow progress in North Africa, Marshall met Harry Butcher who was on his way back to Algiers and told him to tell 'General Eisenhower' – it was never 'Ike' with Marshall – to ignore the criticisms that were being made by the press and politicians, and also *sotto voce* in the armed forces. Marshall said he was 'not to waste time on any effort to defend any of his past actions' and 'not to waste his brain power' over it. 'The General said Ike's rise or fall depended on the outcome of the Tunisian battle,' recorded Butcher. 'If Rommel & Co. are tossed into the sea, all quibbling, political or otherwise, will be lost in the shouting of a major victory.' It was doubtless meant to be highly supportive, but if Marshall really did use the phrase 'or fall', it would have concentrated Eisenhower's mind wonderfully.

On 30 March Churchill was asked in the Commons how many fronts His Majesty's forces were engaged upon, to which he replied 'three fronts – in North Africa, in Burma and in the South-west Pacific . . . His Majesty's ships have to operate continuously on all the oceans of the globe. The areas in which our air forces are engaged may be defined as follows: Western Europe, the Atlantic, the Mediterranean, India and Burma, the Pacific.' Another member asked whether, therefore, the use of 'that very misleading phrase the Second Front ought to be discontinued'. Churchill was not about to fall into any trap over this, since any public criticism of the phrase might be misinterpreted by the Soviets, so he said: 'No, sir; I do not want to discourage the use of it, because our good friends, fighting so hard, know very well what they mean by it.'

An aspect of foreign policy that was to bedevil both Britain and America – naivety about the true post-war intentions of the USSR – emerged at the War Cabinet of 13 April when Anthony Eden reported on his recent visit to the United States. The Foreign Secretary said that Roosevelt had asked him whether he 'thought that Russia would want to "Communise" Europe after the war', and he replied that he 'did not think so, and that he thought one of the best ways of avoiding this was that we should do what we could to keep on good terms with Russia'.[22] As the verbatim reports make clear, the Churchill ministry was just as naive as the Roosevelt Administration about Stalin's true purpose, which was to bring as much of Europe under the Soviet heel as he could. Kennedy recorded that one delegate from a British military mission to Moscow in December 1941 had declared that Stalin was 'like a clergyman, another [that he was] like a respectable old farmer, another that he was like a great cat. All agreed that he was quiet and shrewd and absolutely ruthless.'[23] It seems that far too many people in the higher directorate of the war concentrated too much on the worthless first parts of this analysis rather than the accurate last four words.

In the course of a War Cabinet discussion about the command structure of Husky, Churchill reported that the Joint Chiefs of Staff wanted an American-led enterprise. 'We have suggested it would be administratively convenient to have a British commander and joint staff all under Eisenhower. But we may have to argue on the basis that command goes with major forces . . . And no US Navy goes inside the Mediterranean. On this, Husky couldn't be more than equal share. I don't feel we should give way.'[24] The Cabinet agreed. The result was that Montgomery and Patton were given equal billing in the invasion of Sicily under Eisenhower, which was to lead to much competition and animosity between the two generals, and just as many headaches for their superiors.

By late April 1943, Brooke and Churchill were considering what should happen after Sicily fell, fearing that Marshall might refuse to attack mainland Italy, but instead demand that everything should be concentrated on a cross-Channel attack. 'I was conscious of serious divergences beneath the surface,' Churchill wrote later, 'which, if not adjusted, would lead to grave difficulties and feeble action during the rest of the year.' So he wrote a telegram to Roosevelt offering to be with him by 11 May, along with Brooke, in order to discuss the exploitation of Husky, the future of Anakim, the shipping situation and 'a number of other burning questions'.

The month originally envisaged for Roundup – May 1943 – saw forty-two German divisions stationed in the West. Although the Roundup plans envisaged forty-eight divisions landing eventually, that would not now be enough, even supposing 'there had been sufficient assault shipping to lift them, or merchant shipping to supply them, or aircraft to cover them, or signals intelligence to guide them, which in each case there was not'.[25]

After tea on Saturday 17 April, Churchill called Brooke about a wire he had received from Marshall suggesting that the attack on Sicily should take place even before Tunisia was finally cleared of Axis forces. 'Quite mad and quite impossible,' was Brooke's reaction to the idea of simply undertaking Husky seemingly regardless of the fact that over two hundred thousand enemy troops were still holding out astride the Allied supply lines, 'but PM delighted with this idea which showed according to him "a high strategic conception". I had half hour row with him on the telephone.'[26] North Africa was only finally cleared of Germans and Italians in May, just two months before Husky was launched. The whole discussion merely confirmed Brooke in his low appreciation of Marshall's grasp of strategy, and of Churchill's. The advent of the 'Trident' Conference in Washington therefore filled Brooke with foreboding. 'I do NOT look forward to these meetings,' he wrote, 'in fact I hate the thought of them.' It had been less than four months since Casablanca, and Brooke was not constitutionally attuned to debate, preferring to give orders rather than to discuss them. He did not trust the Americans' good faith, fearing that the meetings would involve many hours of argument and hard work defending Germany First, after which 'they will pretend to understand, will sign many agreements and . . . will continue as at present to devote the bulk of their strength to try to defeat Japan!! . . . It is an exhausting process and I am very *very* tired, and shudder at the useless struggles that lie ahead.' After the war, Brooke blamed the diversion of American effort into the Pacific for the fact that the war went on as long as it did. The accusation of prolonging the war was a surprisingly common one among strategists, even though it was about the most serious one imaginable.

Trident was necessary, however, as the prevailing sense between the British and American Staffs at this period was one of deep mutual suspicion. Just as Marshall and his Planners in the OPD of the US War Department suspected that the British never wanted to invade France, so too Brooke and the War Office Planners in Whitehall suspected that Marshall was resiling from the Casablanca final report, and sending resources to the Pacific that should have been allocated to Bolero and the Mediterranean. King, meanwhile, was still suspicious that the British would drop out of the war against Japan once the United Kingdom was finally made safe by the defeat of Germany. Conversely, Wedemeyer, Stilwell and others were suspicious that if Britain pursued the war against Japan vigorously it was merely to re-establish her Empire and block off American interests in the Far East.

Such was the atmosphere when at 6.30 p.m. on Tuesday 11 May the British party reached Washington, where it was met by Roosevelt, Marshall, King and Dill. They all went straight into a cocktail party in a large hotel, before Brooke moved into Dill's house. The British again fielded a large team: Sir Alan Lascelles thought it dangerously so, complaining that Churchill 'took with him, in the *Queen Mary*, the three Chiefs of Staff, Cherwell, all the Planners, the three Commanders-in-Chief in the East (Wavell, Peirse and Somerville), Leathers, Ismay and Jacob, the secretaries to the War Cabinet, and Beaverbrook. "*Was für Plunder*", as any young Blücher in an Atlantic U-boat might exclaim.'[27] Although the great liner sailed far faster than any German submarine, zig-zagging for further protection, it was a reasonable criticism. To confuse the enemy, notices in Dutch had been posted around the ship, intending to deceive spies into believing that it was Queen Wilhelmina of the Netherlands rather than Churchill who was the mystery VIP passenger.

On the day the Trident Conference opened, Wednesday 12 May, General Jürgen von Arnim surrendered in Tunisia, and over 230,000 Axis troops passed into captivity, though not Rommel himself, who had been recalled to Germany shortly beforehand 'on health grounds'. It was a great victory after thirty-two months of fighting backwards and forwards along the North African littoral; it had proved that major opposed amphibious assaults could work, and that the Western Allies could co-operate successfully on the field of battle. Yet could they still co-operate successfully in the conference chamber? Between 13 and

25 May there were fifteen meetings of the Combined Chiefs of Staff – the eighty-second through to the ninety-sixth – as well as six plenary sessions of military officials with Churchill and Roosevelt at the White House.

'The PM spoke at length on the advantages that would accrue to the Allied cause by a collapse or a surrender of Italy through its effect on the invaded countries of the Near East and Turkey,' wrote Leahy of the opening meeting in the White House on 12 May. 'In regard to a cross-Channel invasion in the near future it is apparently his opinion that adequate preparations cannot be made for such an effort in the spring of 1944, but that an invasion of Europe must be made at some time in the future.' Churchill spoke of the psychological effect of 'a definite break in the Axis conspiracy', and of the withdrawal of Italian troops from the Near East, and of the chances of bringing Turkey on to the Allied side.

Churchill's rhetoric about how he 'passionately wanted to see Italy out of the war and Rome in our possession' shows the degree to which grand strategy can be directed by considerations other than strict military logic.[28] Sir Michael Howard has written that 'For Mr Churchill himself, and perhaps for the commanders of the victorious British armies in Africa, the impulse to carry the battle into Italy was emotional as well as strategic.' To force Italy out of the war and overthrow the strutting, bombastic Mussolini, the butt of so many of the Prime Minister's best jibes – 'This whipped jackal Mussolini is frisking up by the side of the German tiger' – was a principal British war aim, just as a modern-day triumph through the streets of Rome could not but appeal to the historian and romantic in Churchill.

Leahy's diary reveals his own profound suspicions of the British in his complaint that Churchill had 'made no mention of any British desire to control the Mediterranean regardless of how the war may end, which many persons believe to be a cardinal British national policy of long standing'. In fact control of the Mediterranean had not really been a British policy objective since Nelson's day. Leahy nonetheless seemed impressed that Churchill had described the group assembled in the Oval Office as 'the most powerful group of war authorities that could be assembled in any part of the world' (except, of course, in the Kremlin on any day of the week).

In reply, Roosevelt advocated a cross-Channel invasion at the earliest practicable date. He expressed opposition to any Italian adventures beyond the seizure of Sicily and Sardinia, and stated that the air transport line from India to China must be opened without delay, and that China must be kept in the war at all costs. He directed his Staff to look into the military possibilities of invading Bulgaria and Roumania via Turkey, however, promising to 'investigate the political possibilities of such a move' himself. Churchill added the idea of an attack on Sumatra, which he described as 'lightly garrisoned'.[29]

Churchill opened his remarks by saying that the last time he and Brooke had been in the President's Office was when Tobruk had fallen. 'It was not a very happy beginning,' Hopkins told Moran.

The Americans had not forgotten the occasion. They had gone to the White House to clinch the plan for the invasion of France, when news had been brought to them of the disaster. Then in some manner – they were even now not quite clear how – they found themselves agreeing to the diversion of ships and troops to North Africa that were meant for the invasion of France. They could not help admiring the PM's gift of dialectic, but they had made up their minds that it was not going to happen again.[30]

The memory of the Tobruk news affected Brooke too. 'I could see us standing there and the effect it had on us,' he later wrote. 'I felt rather as if in a dream, to be there planning two stages ahead, with the first stage finished and accomplished.'[31]

Just like Churchill, Brooke started off his meetings with Marshall badly. He ought to have reassured the Americans by making positive noises about some aspects of a future Roundup in 1944, but instead he left them deeply suspicious by painting a vista of victory in the Mediterranean. He also went badly wrong by mentioning various Greek and Turkish islands in the Aegean as places that might be captured next. This merely increased American suspicion that the British were looking to their supposed post-war eastern Mediterranean and Middle Eastern interests and ambitions, rather than concentrating on early re-entry into France.

At one Combined Chiefs of Staff meeting Marshall likened Brooke's strategy to 'a suction pump' that would – if the Allies were to land in mainland Italy – suck enough troops from Roundup to leave it a mere Sledgehammer-sized operation. In order to prevent this, Marshall forced Brooke to agree at Trident that if the mainland of Italy were invaded after Sicily, three British and four American divisions would be withdrawn from the Mediterranean for Roundup by 1 November 1943. Without such an agreement, Marshall intimated, there could be no invasion of mainland Italy. Brooke duly promised. He was to regret it ever after.

When Moran, who seems to have conducted extraordinarily sensitive conversations on strategy even though he was only Churchill's doctor, asked Hopkins what Roosevelt made of Churchill's belief that Italy's surrender might be the beginning of the end for Germany, the answer came: 'Not much. This fighting in Sicily does not make much sense to him. He wants twenty divisions, which will be set free once Sicily has been won, to be used in building up the force that is to invade France in 1944.' Churchill nonetheless convinced himself that, if he could only get Marshall out to Eisenhower's headquarters in Algiers, 'it would all be plain sailing' and the general would agree with him about the need to invade mainland Italy. Moran personally believed that Churchill's optimism was 'interfering with the cold functioning of his judgement', and in this he might well have been right.[32] Even though Moran's diaries need to be treated with a degree of wariness, since he wrote them up after the war from contemporaneous notes, this sentiment rings true. Marshall was far too objective a judge to be swung around by a trip to Algiers.

The next day, Thursday 13 May, Brooke gave the Combined Chiefs of Staff a presentation on global strategy that indicated for Leahy 'that the British will decline to engage in 1943 in any military undertaking outside the Mediterranean Area'. Since Roosevelt had directed Leahy to press for an Anglo-American invasion of Europe 'at the earliest possible date', disagreements flared up immediately.[33] The meeting left Brooke 'thoroughly depressed', which a tour of the brand new Pentagon building with Marshall and a quiet dinner alone with Dill did little to alleviate. Roosevelt, vigorously supported by Marshall, said he 'feared that this [the British approach] meant a lengthy pecking away at the fringes of Europe'. Presidential elections beckoned in November 1944, and he wanted a definite commitment to a cross-Channel operation taking place before then.

Brooke opened the Combined Chiefs of Staff meeting at 10.30 a.m. on Friday 14 May by telling the Americans frankly that the British

Chiefs of Staff did not agree with their views on global strategy. Stilwell followed, saying that he disagreed with most of what the British thought too. They then all lunched together, and went to the White House to discuss Burma with Churchill and Roosevelt, who both made opening remarks. Then Wavell spoke, followed by Somervell who contradicted him, followed by Stilwell who disagreed with them both. 'I remember feeling the absolute hopelessness,' wrote Brooke after the war. 'The Americans were trying to make us undertake an advance from Assam into Burma without adequate resources.' Of 'Vinegar Joe' Stilwell, Brooke was predictably dismissive ('a small man with no conception of strategy'), noting that he had a 'deep rooted hatred of anybody or anything British!'[34] Stilwell's biographer, Barbara Tuchman, described Brooke as 'a small, dark, unamiable man who disliked Americans and vice versa', which was wrong in almost every instance. He did not dislike Americans in general, just Stilwell (and a few others, such as Cooke) in particular.

Meanwhile Stilwell recorded in his own diary that he had 'locked horns with Brooke to King's delight', on specifics rather than over general policy.[35] The British attitude towards Operation Anakim was characterized by Stilwell as 'can't – can't – can't'. Since Roosevelt, Churchill and Brooke were lukewarm towards Anakim, no amount of eloquence from Stilwell could have brought the Allied High Command behind it. By the end of the day, relations between the two negotiating teams had clearly become fraught, so it was very welcome when, at the weekend, Marshall and McNarney invited the six British Chiefs of Staff and Commanders-in-Chief as their guests to Colonial Williamsburg in Virginia. Today such 'getting to know you' weekends in hotels among work colleagues – sometimes nicknamed 'awaydays' – are a common business practice, but in the 1940s they were not. Yet that is essentially what this was intended to be, rather than the more traditional 'Friday to Monday' of the British country-house weekend. The idea was that the British and Americans would relax and interact socially, thereby lowering the temperature and helping to dispel the mutual suspicion that was tending to poison their counsels. If they got to know one another as individuals, so the reasoning went, they would be able to trust one another as comrades in the great combined purpose. Ismay said afterwards that the weekend was 'beautifully arranged, had given them all a chance to get to know one another, and there had not once been a mention of the war. Otherwise, during a fortnight of meetings, discussions had been frank, at times bitter.'[36]

It was organized by Marshall's personal aide, Frank McCarthy, who had a meticulous eye for detail. McCarthy had been born in 1912 near by in Richmond, Virginia, and like Marshall he attended the Virginia Military Institute, after which he became press agent to the theatrical producer George Abbot in New York. He joined the US Army Reserves on hearing an inspirational radio broadcast by John Wheeler-Bennett on the night of France's surrender in June 1940, which the young assistant director of the British Press Service in New York had begun with the words: 'Tonight my country stands alone – *alone* – before the embattled might of totalitarian Europe.' By 1941 McCarthy was Marshall's military secretary and in 1943–5 he was secretary of the War Department General Staff. After the war he co-produced magnificent war movies such as *MacArthur*, *Decisions before Dawn* and the 1970 Oscar winner, *Patton*. (Marshall, naive in some social matters, constantly but in vain introduced the handsome young McCarthy to attractive single women such as Joan Bright, not realizing that he was homosexual.)

The Williamsburg Inn where the two parties stayed had been built in 1936 by John D. Rockefeller Jr, who ordained that it should combine 'comfort, convenience and charm', but that it must not compete in splendour with the Governor's Palace or the Capitol in Colonial Williamsburg, which he had been restoring at vast expense. McCarthy allocated Marshall Room 212, one of the smallest in the Inn, because 'General Marshall especially desired that no particular attention be paid to him.' Brooke was three doors down in 215. A new table was built for the thirteen (clearly unsuperstitious) guests who sat around it on the Saturday night, and the pitch of the swimming-pool lights was altered so as not to throw a glare in diners' eyes. William Johnson, Rockefeller's butler, was specially brought down from New York to supervise the food and service. The terrapin for Saturday's dinner had to simmer in its own juices for two days in the Union Club in New York, before Johnson carried it down to Richmond on the upper berth of a Pullman.

Marshall's plane arrived at Langley airfield at 3.52 p.m. and the British party five minutes later. From Langley they drove up Route 17 to Yorktown, hardly the most tactful of destinations for British guests. Marshall later recalled that, after stopping the car at the Yorktown

Victory Monument, 'I announced that they were engaged in a peaceful, and I hoped pleasurable, visit. I did not want them to pass by the historic spot that marked the virtual close of our American Revolutionary War – Yorktown – the site of Cornwallis' surrender to George Washington. There was much laughter, but I was somewhat humbled to find one or two who had no recollection of that event.'

In his car, General Marshall's conversation turned to duck-hunting and he told Pound and Wavell that he had found time to go only twice in the previous two years. Pound swanked about a hunt in India where his party had shot 1,656 ducks. Meanwhile, in the other car with McNarney, 'General Brooke remarked about the birds and endeavoured to identify as many as he could.' They spotted a hawk on the way into Williamsburg, which afforded the falconry expert Charles Portal a good line of conversation. On the way to tea at the Raleigh Tavern, Marshall was told that the large numbers of people lining the route had turned up in the hope of seeing Winston Churchill.[37]

Before dinner at the Williamsburg Inn on Saturday night, some British officers went for a swim in the pool – Portal dived in and momentarily lost his over-large swimming trunks – while others played croquet on the south terrace.[38] By the time they reconvened in the hotel lobby for mint juleps at 7.45 p.m., in front of roaring fires, it was clear that the weather meant they would have to eat inside. Marshall sat at the head of the table, with Pound on his right, Brooke on his left and McNarney at the far end. For Britons who had survived two-and-a-half years of rationing, dinner was sumptuous. A crabmeat cocktail was followed by Terrapin à la Maryland, drunk with Harvey's amontillado. For the rest of the dinner they drank 1929 Heidsieck Dry Monopole champagne. There was fried chicken in cream gravy, fresh asparagus, a Virginia ham, Canadian cheddar, and strawberry ice cream. The conversation at dinner was 'lively and interesting', and studiously avoided anything to do with the war. Marshall told of the time in April 1942 when his plane had been grounded at Bermuda on the way to the Modicum Conference and he read the lesson in church the next morning. Afterwards a woman 'threw her arms around his neck and kissed him roundly'.

After dinner Marshall invited the guests to retire to the drawing room for coffee and brandy, and 'urged all of them to forget about the future for the time being and let their thoughts dwell on that interesting period in the past when Williamsburg was the capital of England's most important colony'. At 11 p.m. they visited the palace of the royal governors, admiring the authenticity of its furnishings, the beauty of the flower arrangements and the charm of its gardens. Marshall played a tune on an antique spinet. Admiral Pound got lost in the maze and had to call to his colleagues to rescue him, a couple of whom promptly got lost too.[39] When they visited the Governor's council chamber, Marshall remarked: 'Gentlemen, why don't we just sit down here and continue the meeting where we left off a few hours ago?' But of course that wasn't the point of the exercise at all.

'I kept on feeling I had been transported back to the old days, and expected the Governor to appear at any moment,' Brooke told his diary. He was also enchanted by the flower arrangements in the mansion, done 'on historical principles', and by the way clothes were laid out, the chessmen on a board, gloves on a table and books pulled out for reference, all 'as if the house was inhabited'. There was undoubtedly a romantic, almost whimsical side to the flinty Ulsterman. They left at midnight, each in turn shaking hands with Fleming, the palace's usher. Marshall later told Admiral Stark that his guests 'appeared to enjoy it thoroughly, examining everything minutely and at great length. The Sea Lord and the Air Marshal even went swimming, in water too frigid to tempt anyone else.'[40]

On Sunday morning, after ham and eggs on the terrace, Marshall and Brooke became engrossed watching a robin redbreast feed her young in a pine tree. Marshall asked the innkeeper whether there were any orioles in that part of the world, and was told there were. They made a tentative plan to view some later. Marshall's modesty also led to his asking for a less conspicuous pew than the Governor's in Bruton Parish Church, where they all attended the Sunday-morning service. Admiral Pound read the lesson, St Matthew 6 verses 19 to 34, from a Bible donated by King Edward VII on the tercentenary of the founding of the Episcopalian Church in America. 'Take therefore no thought for the morrow,' he read; 'for the morrow shall take thought for the things of itself.' It was a perfect summation of the weekend.

Unfortunately, before Marshall and Brooke could go looking for orioles – 'Behold the fowls of the air' had featured in Pound's reading – Marshall and Portal were called off at 10.50 a.m. to see the President and Churchill at Shangri-La (now Camp David). They flew to Washington and then on to Hagerstown, Maryland, where they motored up into the mountains. One wonders what Brooke felt about the others conferring without him. Since the discussions were about Eisenhower's reception of Marshall's plans for Sicily, they did concern him, but Churchill probably did not want to interpose the CIGS into what was primarily an American matter. In 1955, Marshall recalled to a would-be biographer that he was 'most impressed' by Brooke and the other Britons that weekend.

Some swam, some occupied themselves in photography (Lord Wavell), but Alan Brooke, with his field glasses, devoted his time to a study of the Virginia birds in that locality. His persistence and his pleasure in his task were very appealing. We had been having a very hard time in Washington reaching agreements, but the weekend in Williamsburg with no business discussions cleared the air entirely.[41]

Back at the Inn there were mint juleps before Sunday luncheon – lobster salad, cold Virginia ham, cold roast turkey and crabflake soufflé – and then they returned to the airfield for the 160-mile flight back to Washington and a Chiefs of Staff meeting at 5.30 p.m.

In a novel, the Chiefs' sojourn together would have brought a miracle breakthrough in negotiations, with spontaneous Anglo-American trust and unity breaking out the very next morning. The reality of global war was different. 'Another very disappointing day,' recorded Brooke, after a long strategy meeting on Monday 17 May, 'which led us nowhere.' He put this down to national characteristics, and the supposed fact that 'the American mind likes proceeding from the general to the particular, whilst in the problem we have to solve we cannot evolve any form of general doctrine until we have carefully examined the particular details of each problem.' Brooke had to admit, however, that the major problem 'really arises out of King's desire to find every loophole he possibly can to divert troops to the Pacific!'[42]

Trident is regarded as one of the most ill-tempered and rancorous of all the wartime summits, not least because after their perceived 'defeat' at Brooke's hands at Casablanca the Americans were determined not to lose out again. Leahy had to admit to the Chinese representative in Washington, D. R. Soong, that information regarding aircraft shipments to China could not be given him 'because of the present unsettled state of the Staff conversations'.[43] Only very slowly did the bare outlines of an agreement for a spring 1944 cross-Channel attack, to take place after an invasion of mainland Italy, emerge from the hard-fought negotiations. 'The Americans are now taking up the attitude that we led them down the garden path taking them to attack North Africa!' noted Brooke on Tuesday 18 May. 'That at Casablanca we again misled them by inducing them to attack Sicily!! And now they are not going to be led astray again.' Only half jokingly he added, 'before long they will be urging that we should defeat Japan first!' Re-reading his diaries after the war, Brooke admitted that it was 'evident' that he 'went through a phase of deep depression' during Trident. He blamed this on the Americans who 'still failed to grasp how we were preparing for a re-entry into France through our actions in the Mediterranean'.

Yet again he listed the advantages of the Mediterranean strategy as having opened up that inland sea, 'and in doing so had regained the equivalent of about a million tons of shipping', taken a quarter of a million Axis prisoners in Tunisia – roughly the same number as were killed or captured at Stalingrad – and inflicted heavy losses at sea and in the air, opened the way for an attack on Sicily and Italy, forcing the enemy to defend southern Europe – 'a region of bad intercommunication' – and allowing the bombing of Germany from the south. Few of these, except the first and last, would have impressed Marshall, however. Prisoners would have been captured and damage inflicted, he believed, in a cross-Channel attack, where the communications in southern Europe would have been irrelevant. Similarly, if the Allies had won a foothold in Normandy, the bombing of Germany could have been carried out from airfields just as close as Foggia in Italy. The great advantage of Foggia, however, was that the Roumanian oil fields of Ploesti could be attacked, which was not possible from bases in Britain and Normandy. Most of the Wehrmacht's oil came from Roumania.

American resentment manifested itself at Trident when, walking to a Combined Chiefs of Staff meeting with Brooke and Dill, Marshall said: 'I find it very hard even now not to look on your North African strategy with a jaundiced eye.' Brooke asked what he would have preferred. Marshall answered: 'Cross-Channel operations for the liberation of France and advance on Germany, we should finish the war quicker.' 'Yes, probably,' came Brooke's rejoinder, 'but not the way we hope to finish it!'[44] Sharp-tongued ripostes like that were something of a speciality with Brooke, honed by his regular verbal jousting with Churchill.

'It was quite evident that Marshall was quite incapable of grasping the objects of our strategy,' Brooke noted for about the dozenth time as he re-read his diaries, 'nor the magnitude of operations connected with cross Channel strategy . . . Any attempts to unduly push our strategy on Marshall had a distinct tendency to drive him into King's Pacific Camp.' This was wildly unfair: Marshall understood perfectly what Brooke was telling him over and over again; it was just that he disagreed with it. Furthermore, except for very briefly in July 1942 – which, as we have seen, was almost certainly a bluff anyway – Marshall never reneged on Germany First.

Part of the explanation for Anglo-American tension might have lain in something as prosaic as the seating arrangements at Combined Chiefs of Staff meetings. At the back of the room sat large numbers of Planning officers and other experts, up to sixty of them, whose mere presence increased the antagonistic, even gladiatorial, nature of the encounters. The atmosphere has been compared to 'a centre court tennis match at Wimbledon with tiers of supporting staff seated behind their principals'.[45]

Wednesday 19 May saw the toughest day of a tough conference. Both sides began the 10.30 a.m. Combined Chiefs meeting by criticizing the other's papers on future strategy. Marshall then suggested the conference room be cleared for an off-the-record discussion between the two Staffs alone. Rather like a boxing match where the seconds leave the ring, the room emptied; tennis was replaced by pugilism. The six Chiefs remained, plus Leahy, Dill and a secretary whose pen was firmly holstered. 'We then had heart to heart talk and as a result of it at last found a bridge across which we could meet!' wrote Brooke. 'Not altogether a satisfactory one, but far better than a break up of the conference!'

Brooke's reminiscences about these off-the-record meetings, given to NBC in 1958, are worth reproduction *in extenso*:

At a long table on one side sat the American Chiefs of Staff with their secretary, on the other side we sat with our secretary and in the back of the room were a rather large number of staff officers – possibly thirty on either side. It rather weighted the atmosphere of the conference when you had sixty onlookers . . . A lot of the items on the agenda were settled quite easily but occasionally amongst the more difficult ones we argued up to a certain point where it became clear that we were going to have difficulties at arriving at agreement. Well, then, depending on who happened to be in the chair, either Marshall or I would suggest

THE HARD UNDERBELLY OF EUROPE

that we should have an off-the-record conference, and clear the room. By clearing the room of this rather heavy atmosphere, and by being able to take our hair down then and really thrash out the matter, we knew that we had to arrive at an agreement. It was no good going up and meeting the President and the Prime Minister and telling them we had failed to agree . . . Well, by approaches then of a rather more intimate [nature] – and that is a time when General Marshall always shone, he was always perfectly ready to discuss – he put his cards on the table and we put our cards on the table and we shuffled them round until we got some pattern out of it. And I felt always in those discussions there how fortunate we were to have a man of Marshall's temperament, integrity and outlook generally, to arrive at these decisions, and how difficult it could have been had we had someone else in the chair . . . We both spoke the same language, but rather more than that . . . There's rather more than a language between English-speaking people. I think it's an English way of thinking that we have, and I found that Marshall had the same way of thinking that I had.[46]

Of course this was far from Brooke's earlier phrenological view about how British and American minds worked in completely different ways.

Sure enough, by 6 p.m. on 19 May, before the American Chiefs entertained their British counterparts to dinner at the Statler Hotel, they had agreed that an initial assault force of nine divisions, growing quickly to twenty-nine, would be prepared for entry into France by 1 May 1944. Meanwhile, Eisenhower would exploit victory in Sicily in ways that 'are best calculated to eliminate Italy from the war and to contain the maximum number of German forces'.[47] Although this did not quite amount to a written commitment to invade mainland Italy after Sicily, and the wording left Eisenhower as the ultimate arbiter, it seemed obvious that the only way to force an Italian surrender was to land troops. Even so, Brooke considered that the agreement on Italy had been 'a triumph as Americans wanted to close down all operations in Med after capture of Sicily'.

It was at Trident that the British made the first of three binding decisions to launch a cross-Channel invasion in May 1944, and the price the Americans seem to have paid – off-the-record meetings really were just that – was an Italian campaign. Marshall finally had a firm date for Roundup – from now on called Operation Overlord – but it looked as if he and King might be reluctantly dragged behind Brooke's chariot wheels, step by step all the way up to the Eternal City. Brooke was also bound, in the black-and-white of the final report, to allow the seven divisions to be removed from the Mediterranean on 1 November 1943.

Their first off-the-record meeting without advisers had been successful, and the technique was later to be used on several occasions when Marshall and Brooke, in Brooke's words, 'arrived at loggerheads. It always helped to clear the air.' Brooke believed that, without the vast staffs present, Marshall felt less uneasy about 'shifting from some policy he had been briefed on by his Staff lest they should think he was lacking in determination'. This too was unfair: Marshall was not the kind of man to be unduly influenced by a desire to retain the esteem of his Staff rather than doing what he thought right, and it ignores the fact that Brooke shifted significantly too, in a way his own Planners might deprecate, especially in agreeing a definite date for Overlord and relinquishing the seven divisions from Italy. Roosevelt was keen on getting a date for the cross-Channel operation by the end of the conference, so Brooke knew he could hardly hold out over furnishing one. Just as the 1942 mid-terms had influenced the Torch decision, so the looming 1944 presidential elections affected the Trident ones.

Brooke accepted that the Allies should discuss how far Italy needed to be penetrated, which proved a contentious issue for the rest of the war, the British generally wanting to push much further north up the peninsula than did the Americans (except Mark Clark).[48] Whereas the Americans only ever saw the Italian campaign as a way of drawing off German divisions from France and Russia, and for setting up air bases in the Foggia Plains, the British believed it had a further inherent value as a gateway into various other places in Europe, including France, Austria and the Balkans. Yet they knew that when on 1 November 1943 Marshall came to demand his seven divisions for Overlord, the push northwards up Italy would necessarily be undermined.

The off-the-record discussion method was used again over Burma at the Combined Chiefs of Staff meeting on Thursday 20 May, after there was a 'complete impasse' over the American desire to attack that country from the north-eastern Indian province of Assam, 'contrary to all administrative possibilities', so the British thought. The deal struck on that occasion was that ten thousand tons of supplies a month would be sent to China from India, and that an air offensive would be conducted against Burma from Assam, but there would be no land offensive in the short term. This agreement was reached just in time for the Combined

Chiefs of Staff members to be serenaded by a Marine band at the White House, prior to the British Chiefs giving their American counterparts dinner. Further agreements were made the next day on American plans against Japan in the central Pacific, this time all on the record.

In his memoirs, Leahy listed the agreements made at Trident as: a twenty-nine-division cross-Channel invasion from England by 1 May 1944; an intensified combined bomber offensive against Germany; the attack on Sicily; the destruction of the Ploesti oil fields; the occupation of the Azores; ten thousand tons of supplies a month to be flown from India to China; the seizure of New Guinea and the Marshall, Caroline, Solomon and Aleutian Islands, and an air campaign against Burma.[49] There was no written commitment to invade Italy in the final report of Trident, however, nor for any action elsewhere in the Mediterranean once Sicily had fallen.

When Brooke spoke of Allied troops' inexperience and lack of man-power for Overlord, compared with possibilities in the Mediterranean, Marshall replied that the invasion of Italy would create a 'vacuum' for 1944. He bluntly stated that if the British 'were committed to the Mediterranean, except for air power alone, it meant a prolonged struggle and one that was not acceptable to the United States.'[50] The deal was clear if not written down: either both Italy and Overlord were to be undertaken, or neither was. Leahy might have been biased, but he was right in thinking that a pendulum had swung since Casablanca, when the British had gained all their desiderata to the Americans' intense chagrin. Now, only four months later, 'Roosevelt, who seemed to domi-nate the conference, finally obtained British approval in principle of his plans, including the 1944 invasion. He also succeeded in getting approval of a plan that was effective in keeping China in the war against Japan.'[51] Roosevelt's change of mind on Overlord at Trident led directly to the change of Allied policy, and the setting of the May 1944 date for the operation.

From 4.30 p.m. to 7 p.m. on Monday 24 May the Combined Chiefs of Staff were at the White House presenting their joint final report of agreements reached during the conference. Yet one aspect, an American proposal to attack Sardinia, Churchill now refused to accept, and instead spent an hour trying to extend the whole Mediterranean theatre into Yugoslavia and Greece, something the Americans would not entertain. He asked that the final decision be postponed until the next day. This hugely increased the Americans' suspicion of British double-dealing, knowing as they did how closely and often Churchill and the British Chiefs of Staff conferred. 'The PM's attitude is in exact agreement with the permanent British policy of controlling the Mediterranean Sea regardless of what may be the result of the war,' a profoundly distrustful Leahy told his diary. 'It has been consistently opposed by the American Chiefs of Staff because of the probability that American troops will be used in the Mediterranean Area at the expense of direct action against Germany.'[52]

Brooke was livid with Churchill, describing his behaviour as 'tragic'. Having agreed the final report, and Roosevelt even having congratulated the Combined Chiefs on it, 'Now at the eleventh hour he wished to repudiate half of it.' Worse, some of his proposed alterations were on points that Brooke had conceded to Marshall in the course of their hard-fought and detailed negotiations in order to secure better ones. High-level pacts on global grand strategy were multi-layered and multi-faceted. 'He had no idea of the difficulties we have been through,' complained Brooke, 'and just crashed in "where angels fear to tread".'

Brooke feared that Marshall and the Joint Chiefs would assume that he had 'gone behind their backs in an attempt to obtain those points through Winston . . . and it was not possible to explain to them how independent Winston was in his actions.' After the war Brooke recalled Churchill saying that he 'always feared that we should "frame up" (he actually accused me in those terms one day) with the American Chiefs of Staff against him! He knew the Americans could carry the President with them, and he feared being opposed by a combined Anglo-American block of Chiefs of Staff plus President.'[53] Far from 'framing up' with Marshall, Brooke said he was usually 'at loggerheads with him over Pacific and cross Channel strategy', but 'Under such circumstances it may be imagined how complicated matters became!' While Churchill feared Brooke and Marshall 'framing up' against him, Brooke was anxious lest Marshall think that he (Brooke) was *not* supporting Roosevelt and Marshall against Churchill, when in fact he was. Matters had indeed become complicated, and were about to get more so.

The next day Churchill accepted the Combined Chiefs of Staff report, which must to a degree have allayed Marshall's suspicions. However, as a consolation prize from the President, Churchill was told that Marshall could join him and Brooke on a journey to Algiers, which the Prime

Minister hoped would enthuse the Army Chief of Staff about the advantages of an Italian campaign. Marshall, who was due to take three days' rest after the rigours of Trident, was instead given only six hours' notice – between 2 and 8 a.m. – before Churchill's plane took off. He 'ruefully' remarked to Stimson 'that he seemed to be merely a piece of baggage useful as a trading point'.[54] He felt understandable chagrin at the President not consulting him, but as Henry Morgenthau recorded in his diary, Roosevelt 'was tired. He has had ten days of arguing with Churchill, and the man is exhausted.'[55]

The President's evident weariness worried Churchill. 'Have you noticed that the President is a very tired man?' he asked Moran. 'His mind seems closed; he seems to have lost his wonderful elasticity.' That might have been a euphemism for his ability to be persuaded by Churchill; certainly Hopkins no longer worried about the President being left alone with him. Later that day, Churchill articulated his concerns again. 'The President is not willing to put pressure on Marshall,' he said. 'He is not in favour of landing in Italy. It is most discouraging. I only crossed the Atlantic for this purpose. I cannot let the matter rest where it is.'

In his memoirs, Churchill claimed that he wanted to take Marshall to Algiers in order to show that he had not 'exerted an undue influence' in favour of his Mediterranean strategy on the various commanders that he and Brooke were meeting in North Africa. Stimson was furious at this hijacking of Marshall, however, and accused Churchill of taking him along 'in order to work on him to yield on some of the points that Marshall has held out on'. It was a mistake to do this because 'To think of picking out the strongest man there is in America, and Marshall is surely that today . . . and then to deprive him in a gamble of a much needed opportunity to recoup his strength by about three days' rest and send him off on a difficult and rather dangerous trip across the Atlantic Ocean where he is not needed except for Churchill's purposes is I think going pretty far.'[56] Marshall thought so too, but of course he obeyed his commander-in-chief.

Sicily was not even slated for invasion until 10 July, and Marshall's extreme reluctance to discuss any post-Husky strategy with Churchill – let alone to get into detailed conversations regarding specific objectives such as Rhodes, Sardinia, Italy and Corsica – until he had first sounded out Eisenhower led to a comic situation on the plane. As Marshall told

Pogue of Churchill, 'Well, all during the earlier part of the trip he was so busy with his own state papers, which he'd gotten far behind in, that the hazard of such a conversation didn't arise. But as we were approaching Gibraltar, Mr Churchill ran out of work and came back and sat down with me, and then I knew I was in for it.' In order to steer the subject off future strategy, Marshall 'hurriedly thought up something to talk about'. Lord Halifax had just lent him a biography of the former Governor-General of India, Warren Hastings, so he chose that as a topic, which Churchill talked about for twenty minutes before 'he suddenly ran out of soap'. After that Marshall asked what had really happened over Rudolf Hess' flight to Britain in 1941. 'I know that Brooke had never heard this and he was fascinated. I again was overly fascinated and he got to the end of that – that was about fifteen minutes.' Just as Churchill had finished, Marshall asked about the Abdication Crisis, whereupon Churchill said that the King should have just gone ahead and married Mrs Simpson, which took up another twenty minutes. 'It was a marvellous lecture, just marvellous. Then the steward, thank God, announced supper – and it was all over.'[57]

(It was a trick Marshall was to use on other occasions. At dinner the night before one conference's plenary session, when he wanted to avoid talking strategy, Marshall asked Churchill about the Victorian historian Lord Macaulay. The Prime Minister leapt up and strode around the room reciting passage after passage, until 'the lateness of the hour resulted in General Marshall respectfully saying goodnight and departing', leaving Churchill 'well satisfied about Macaulay but entirely frustrated about working on General Marshall'.)

On the earlier part of the journey to Algiers, from Washington to Botwood in Newfoundland, where they refuelled before crossing to Gibraltar, Marshall redrafted a communiqué to Stalin the wording of which Roosevelt and Churchill had previously disagreed over. Once Marshall had finished, Churchill telegraphed the President: 'I agree with every word of it, and strongly hope that it can be sent to Stalin by the Chief of the United States Staff, concurred in by the CIGS, and that it has our joint approval.' Churchill was thus hoping that the Western Allies were united on a message which stated that the aims of Allied strategy were 'to give priority to the control of the submarine menace' and 'next in priority, to employ every practicable means to support Russia'.

The rest of the priorities Marshall set out were to try to get Turkey into the war, to keep up pressure on Japan by attacking the Aleutian Islands, to maintain China, and to prepare the French forces in Africa for eventual fighting in Europe. Roosevelt's sole alteration was diplomatically to combine the first two, so that the goal of supporting the Russian war effort equalled that of defeating the U-boat threat. Stalin was also told that a full-scale invasion of the Continent would now definitely be launched 'at the peak of the great air offensive' in the spring of 1944.

'I was immensely impressed with this document, which exactly expressed what the President and I wanted,' Churchill later wrote of Marshall's draftsmanship, 'and did so with a clarity and comprehension not only of the military but also of the political issues involved. Hitherto I had thought of Marshall as a rugged soldier and a magnificent organizer and builder of armies – the American Carnot. But now I saw that he was a statesman with a penetrating and commanding view of the whole scene.'[58] (Lazare Carnot was the French Minister of War whom Napoleon had dubbed 'The organizer of victory'.) Of course when Churchill was writing *The Hinge of Fate*, Marshall was the US Secretary of State who was saving Europe with his eponymous reconstruction plan, so the foresight to spot a great statesman would have redounded well on Churchill. Nonetheless it was a fine encomium, and in the plane from Gibraltar in May 1943 Churchill had written to his wife, 'I got the President to let General Marshall come with me in order that the work I am now about to do at Algiers should run evenly, and that there should be no suggestion that I exerted a one-sided influence. I think very highly of Marshall . . . There is no doubt he has a massive brain and a very high and honourable character.'[59] Another occasion saw Churchill describe Marshall to Moran as 'The greatest Roman of them all'.[60] This is worth bearing in mind when considering Elliott Roosevelt's accusations that Churchill disliked and 'can't abide' the US Army Chief of Staff.[61]

On 28 May 1943 Churchill, Marshall, Brooke and Ismay landed in Algiers, on the same day that the conclusions of Trident were announced to the press in a single sentence: 'The recent conference of the Combined Chiefs in Washington has ended in complete agreement on future operations in all theatres of the war,' which had the advantage of brevity despite its mendacity.

There were three formal meetings at Eisenhower's villa in Algiers to map out post-Husky strategy, between late May and 3 June, but plenty of other, extraneous meetings too. In his 1967 book *At Ease*, subtitled *Stories I Tell to Friends*, Eisenhower related how at this Algiers mini-conference 'It developed that General Brooke . . . had never really liked the Overlord idea . . . He came to me privately and argued that all Allied troops should stay in the Mediterranean, chipping away at the periphery of the Axis empire. But we should avoid any commitment of major ground forces.' To Ike's question about how to rid central and western Europe of the Soviets in the absence of an Anglo-American invasion in the west, Brooke is reported to have 'thought that the Soviets would not try to maintain such an extended empire and would retire back into the limits of Russia once the war had been won'.[62] Of this, Eisenhower wrote: 'I was reasonably confident that both President Roosevelt and General Marshall were determined to take no chances on such an outcome and I must say I agreed.'

Yet little about this reminiscence rings true: firstly, Brooke was not absolutely opposed to Overlord ever taking place: he always knew the Continent had to be invaded from the west eventually for Germany to be defeated. Secondly, there is no record of this important conversation in Brooke's private diary, though he had no reason to exclude it. Thirdly, there is no record in Butcher's diary, which covered Eisenhower's every move at that time. Fourthly, Brooke was not so naive when it came to Soviet ambitions, as his subsequent behaviour proved. It might be that Eisenhower, in this rather tall 'story I tell to friends', was 'remembering with advantages' as sometimes old soldiers are capable of doing, especially once their supposed interlocutors are safely dead.

At the first strategy session at Eisenhower's villa his naval, army and air force commanders, Admiral Sir Andrew Cunningham, General Sir Harold Alexander and Air Marshal Sir Arthur Tedder – all Britons – insisted that mainland Italy should be invaded so that the Germans would be obliged to reinforce Italy and the Balkans after Rome had surrendered. Forces in the Mediterranean, they said, could not be kept idle between the end of Husky, which they hoped would be in August, and the start of Overlord nine months later.[63] Not to keep the initiative was to ask for the Axis to counter-attack, and they needed little invitation. Churchill then offered Marshall eight Commonwealth divisions in any push to capture Rome. Since in 1943 the British Army never fielded more than twenty divisions, this was a significant amount. (The Germans meanwhile had 156 divisions fighting full time on the Eastern Front alone.)[64]

Eisenhower accepted that, if Sicily fell soon, an invasion of Italy could indeed be advantageous, although Marshall resolutely refused to commit himself until they knew how it had gone. At the second strategy session on 31 May, Marshall asked Eisenhower's chief of staff, Bedell Smith, how many extra troops would be needed to attack Italy. Bedell Smith estimated thirty-three thousand American and thirty-three thousand Commonwealth troops needed to come from beyond North Africa and Sicily. Marshall said this might upset the agreements made at Trident for Overlord, but Churchill said the shipping could be made available from Britain.

Churchill was optimistic that he had persuaded Marshall, telling Moran that he 'doesn't, for the moment, want to make up his mind what we ought to do once Sicily is taken. But he is ready to accept my plan. He is not opposed to the invasion of Italy now.' When Moran asked what had changed Marshall's mind, Churchill seemed taken aback. 'The merits of the case', he expostulated, 'are surely beyond question.' Two months after that, however, Moran got a very different version from Marshall himself, who said he hadn't thought then that the moment had come for a decision:

It would be better, I said to the Prime Minister, to decide what to do when Sicily was well under way. I wanted to know whether Germany was going to put up a stiff resistance in southern Italy or whether she would decide to retire to the Po as Winston suggested. I wanted more facts. I wanted to ask Winston a dozen questions, but he gave me no chance. He kept telling me what was going to happen. All wishing and guessing. When I did get a question in, Winston brushed it aside . . . I said to the Prime Minister that I would be content if Sardinia were taken before the invasion of France. He replied that the difference between taking southern Italy and Sardinia was the difference between a glorious campaign and mere convenience.[65]

The way that the great protagonists were clearly already buffing up their anecdotes while the war was going on – this one only a couple of months after the event – indicates the degree to which they knew that their deeds would interest historians for centuries to come.

Moran described how 'Marshall's long upper lip stretched in amusement' as he told the story, and commented that Churchill had obviously 'talked at the American' rather than to him. 'I have never heard anyone talk like this before,' Marshall said. 'I'd never met anyone like Winston. He is a very wonderful man, but he won't look at things like a man who has been all his life a soldier. I must have facts.'[66] Yet facts were impossible to glean about a future operation such as the invasion of a mainland five weeks before the operation to take its adjacent island had even begun. Informed, calculated 'wishing and guessing' had to play an important part in the creation of future grand strategy, and in this particular case the quest for non-existent facts materially damaged the next stage of the campaign.

The third and last strategy session at Algiers, on Thursday 3 June, agreed the bombardment of the railway marshalling yards on the outskirts of Rome. Marshall recalled to General John R. Deane after the war that he had 'favored very much' bombing Rome itself, believing 'the blood of the present' completely outweighed 'the desire to preserve the historical treasures of antiquity'.[67] Nonetheless, the minutes show that Churchill 'expressed satisfaction at the great measure of agreement which he had found in these meetings', and indeed the entry in Butcher's diary covering the meeting was headed 'Love Fest', with more of a sense of relief than of sarcasm. Marshall made generous remarks about the British support for Eisenhower, and spoke of the 'greatest discomfort' that the Germans must be feeling about the Anglo-Americans working 'so well as a team'. He did not, however, commit himself to any Mediterranean operation after Husky. Churchill's memoirs therefore contained an outrageous misrepresentation of Marshall's position when he summed up the meeting with the words: 'I felt that great advances had been made in our discussions and that everybody wanted to go for Italy' (unless by Italy Churchill meant just Sicily, which is unlikely).

Eisenhower meanwhile merely set up two Planning groups to investigate operations against Sardinia and southern Italy. 'The curious situation obtained, therefore,' records Michael Howard, 'that when the Allied armies landed in Sicily, nobody had yet decided where they were to go next.' Marshall went back to Washington via Accra, Ascension Island, Recife and Belem in Brazil. By 7 June he had covered 14,000 miles in eleven days, without yielding any significant strategic ground whatever. Brooke left Algiers, where he had spotted some crossbills and little bustards, for a weekend in the country with his wife.

Reporting to the War Cabinet in London on 5 June, Churchill commended the Trident final report, saying that his:

journey had been justified because of the Anglo-American difference in point of view. The US masses' attention was turned mainly on Japan and tended to think it more important to keep China in the war than Russia. At the outset there were sharp differences between the Chiefs of Staffs. Theirs suggested that concentration on Italy and the Mediterranean would interfere with Bolero and would even prolong the war. But personal contacts and personal friendships broke this down and agreement was reached. This document agrees on Italy being the target, but we've undertaken to move some troops back at intervals for Bolero. Over Anakim there were differences also. We came to the conclusion there was no reason to re-open the Burma Road until mid '45. But all the same we must fight, wherever we can engage the enemy ... The US public hadn't realized until I said it to Congress that the greater part of US forces are deployed in Pacific. The US Executive treats Congress as an enemy, and was surprised at the sort of speech I made, though I said it was common form for the House of Commons. US opinion is quite cool about North Africa. Strategic issues were settled in broad outline. Whereas a year ago we had to say 'Hitler first and Tojo after', there is now enough force to take a rather different view – a matter of emphasis now, not of choice ... Eisenhower was inclined to go for Sardinia before Italy: I strongly advocated the second, then the first will fall in. I therefore went to see Eisenhower and took General Marshall with me. Marshall rewrote the paper for communication to Russia ... this document is evidence of his great mental grasp. His visit to Africa with me has done him good – widened his appreciation of the African campaign.[68]

Churchill had misled either the Cabinet or himself: Marshall was careful that neither the Trident final report nor any other document committed the United States to invading mainland Italy. Of Giraud and de Gaulle and the French National Committee, Churchill told the Cabinet that he 'gave them all lunch yesterday, and a speech in my best anglicised French'.[69]

Returning to the War Office, Brooke summed up the Trident and Algiers talks as holding operations, telling Kennedy that the Americans had 'slipped right back to their old conception of the invasion of Europe, and were most unwilling to be drawn into large and unknown commitments in the Mediterranean'. At Algiers, Marshall had even told Brooke that he still felt Torch had been a mistake. Brooke had no doubt that

Trident had been 'badly needed', but another conference would be required soon. It was a prospect that filled him with dread.

Churchill introduced General Alexander to the War Cabinet on 7 June, after Brooke had summed up the general situation on all the fronts of the war. The Prime Minister began by expressing his 'admiration for a great military achievement', before Alexander, far more modestly, reported that 'The situation is in a bit of a tangle. The first thing was to stabilize the front, tidy it up and separate the US and ourselves.' Thus far he might have been talking about a complicated piece of knitting, but it got more martial when he spoke of how the Germans were attacking to regain the initiative, and he ended his long exposition by saying that his troops were 'in terrific heart – never had such a good army as we have today. The Germans are not as good as they were.'[70] This weakening of German combat effectiveness was the news that the British High Command had waited nearly four years to hear. Before it came, they rightly thought a return to the Continent suicidal. Once it was certified on Italian soil, however, it would lead to the beaches of Normandy.

14

The Overlordship of Overlord: 'A balance of disguised bribes and veiled threats' June–August 1943

High level war planning is an awesome responsibility, yet the whole business is carried on in an impersonal way . . . Plans were made on the basis of numbers, not individuals. General Charles Donnelly's unpublished autobiography, 1979[1]

At 5 p.m. on Tuesday 15 June 1943, Winston Churchill did something that would colour his relationship with Brooke for the next twenty years. Just before the War Cabinet that day, he told the CIGS that he wanted him to 'take the Supreme Command of Operations from this country across the Channel when the time was suitable. He said many nice things about having full confidence in me, etc.' This was the first of three occasions that Churchill expressly offered Brooke command of Operation Overlord. Sworn to secrecy, Brooke did not even tell his wife, although when Churchill next met Benita he mentioned it to her too. Brooke later wrote that the offer 'gave me one of my greatest thrills during the war. I felt that it would be the perfect climax to all my struggles to guide the strategy of the war into channels which would ultimately make a reentry into France possible, to find myself ultimately in command of the Allied forces destined for this liberation!'[2]

Yet from the very beginning, Roundup and then Overlord – as distinct from Sledgehammer – was always going to have significantly more American than British Commonwealth troops, twenty-seven divisions to twenty-one at its first conception. Furthermore, the whole political impetus for early implementation of the operation had come from Washington, indeed from Marshall personally, while the incubus had always been seen to be Brooke. Although senior Americans such as Leahy were calling him 'Brookie' by 1943, they still thought him 'a somewhat forbidding personality', something Churchill can hardly have failed to spot.[3] Whereas Eisenhower, Montgomery, Bradley, Clark, Patton and

Alexander had all commanded troops successfully in North Africa, Brooke had not taken the field since the two ill-fated BEF campaigns in 1940. He was undoubtedly qualified for and equal to the task of Overlord, but it was hardly a recommendation.

The Americans had not been officially consulted by Churchill about Brooke's appointment, but the post of supreme commander of Torch and Husky had gone to Eisenhower after Marshall refused them. In retrospect, therefore, it seems surprising that Churchill should have built up Brooke's hopes without first clearing the appointment with Roosevelt and Marshall, and almost incredible that Brooke did not spot that the job was certainly not by then in Churchill's sole gift.

Perhaps Churchill offered it in the hope that Brooke would become more enthusiastic about Overlord as a result. The Briton Lieutenant-General Sir Frederick Morgan had been appointed chief of staff to the Supreme Allied Commander (COSSAC), at Casablanca. Although there was still no supreme allied commander appointed, he had got on with drawing up the plans for Overlord at Norfolk House in St James's Square in London. His first set were ready by July 1943 and approved the following month. With remarkably few alterations, these were the plans that were put into operation on D-Day the following year.

The American view of Brooke was represented by Elliott Roosevelt, who wrote in *Rendezvous with Destiny*:

The overall commander of the operation would not be Churchill's candidate, General Sir Alan Brooke, who did not want a second front anyway. Marshall was the man who could be counted on to hold his own against Winston and strike at Germany where the Nazis were strongest, not temporize, as the British would have preferred, and delay landings in France until the Reich had been brought to its knees.[4]

This was a completely inaccurate representation of Brooke's view: he believed an eventual Overlord was necessary, and had accepted May 1944 as its launch date at the Trident Conference. Yet it does help explain why he would not secure its command.

Operation Citadel, the Germans' fifty-division assault on the Kursk salient in Russia, was launched on Monday 5 July 1943 and became the largest single battle in history, with more than two million men engaged on both sides. The German forces facing Kursk comprised nine hundred thousand troops, two thousand seven hundred tanks and assault guns, ten thousand artillery pieces and two thousand aircraft. Marshal Zhukov decided to reply with Operation Kutuzov, aptly named after the hero of the 1812 campaign since it involved allowing the Germans to attack first before unleashing a massive counter-stroke a week later, on 12 July. The mammoth battle continued over an area roughly the same size as the United Kingdom, before it was finally won by the Russians on 17 August. During a key moment of the Kursk struggle, on 17 July, Hitler withdrew II SS Panzer Corps from an important part of the line and sent it to Italy, striking confirmation of Brooke's hopes for the Mediterranean strategy.

On the same day that Hitler launched Citadel, Churchill telegraphed Roosevelt to protest against his proposed draft proclamation to the Italian people on their liberation, stating that it did not mention the British contribution enough, and that, although he had acted as 'your Lieutenant throughout' Torch, as far as Husky and post-Husky were concerned 'we are equal partners' in terms of numbers of troops, ships and aircraft, and 'I fully accepted your dictum that "There should be no senior partner".' The amendments Churchill suggested – 'in all the frankness of our friendship' – were minor, such as adding Alexander's name to the proclamation as Eisenhower's deputy and asking that the last three words in the phrase 'the vast air armadas of the United Nations' be replaced with 'the United States and Great Britain', explaining 'After all it is the United States and Great Britain who are virtually doing the whole thing.' Roosevelt concurred in these various alterations, and forwarded a copy to Marshall with the message 'I think the Prime's point is well taken.'[5]

The next day Churchill invited John Kennedy and Major-General Francis Davidson, Director of Military Intelligence at the War Office, to discuss troop strengths, although he was not up from his afternoon nap when they arrived at the No. 10 Annexe at 6 p.m. After talking about Kursk, which took precedence, he asked about Husky, and was told that the British and Americans would each land a quarter of a million men on Sicily in four days' time. 'The enemy forces should be outclassed quickly once we get a footing on the island,' said Kennedy, which is indeed what happened.

Churchill didn't like the way Syracuse was printed 'Siracusa' on the map that Kennedy produced, and as a great stickler for English pronunciation

of foreign place-names he said 'he didn't like it and we should change it
or the BBC would be getting hold of it next'.[6] Later on in the campaign
he objected to the foreign pronunciation of any place-names, as Cadogan
noted, 'calling it Pantellārea, and ridicules Pantelleíra; asks if we are
going to talk about Afreéca and Parée! CIGS plays up nobly by calling
Porto Empedocle Porto Empedoakle. PM didn't bat an eyelid.'[7]

In a Cabinet discussion on war criminals on 7 July, Churchill reported
that 'FDR [was] inclined to let our troops shoot them out of hand! I
suggested the United Nations [should] draw up list of fifty or so who
would be declared as outlaws by the thirty-three nations. (Those not on
the list might be induced to rat!) If any of these were found by advancing
troops, the nearest officer of Brigade rank should call a military court
to establish identity and then execute without higher military authority.'
The Lord Chancellor Lord Simon pointed out that Roosevelt had signed
an Allied public declaration which said that the terms of an armistice
would include provisions for surrender. 'My scheme would be a refine-
ment on that,' argued Churchill. This did not persuade Simon, who
mentioned that there was a UN Commission for Investigations of War
Crimes. Attlee thought it might be convenient to hand over the worst
Nazis to the 'most injured nations like Norway and Poland' for their
own form of (rough) justice.[8]

After a dinner for the King at Downing Street that evening, Churchill
took Brooke into the large garden of No. 10 at 1.30 a.m. and, standing
'in the dark', again told him that he wanted him to be the supreme
commander for Overlord, this time even naming January or February
1944 for an announcement. 'He could not have been nicer and said that
I was the only man he had sufficient confidence in to take over the job.'
In fact Churchill could not have been more heartless. After the war,
Brooke recalled that he had been 'too excited to go to sleep when I
returned home, and kept on turning the thought over and over in my
mind. Was fate going to allow me to command the force destined to
play the final part in the strategy I had been struggling for?' This time,
however, Brooke 'realized well all the factors that might yet influence
the final decision and did not let my optimism carry me off my feet'.[9]

On Saturday 10 July 1943 the Allies landed eight divisions across a
100-mile front in southern Sicily, using several new kinds of landing

craft which were successful despite rough seas. By nightfall the beaches were secure and the campaign against the 230,000 German and Italian defenders could begin in earnest. Some advocates of an early Overlord point out that the Allied sea invasion force for Husky was larger than that for D-Day, and the Allies took fewer casualties than at Salerno that September. Once it became clear by the second half of July that the Allies would win in Sicily, the next issue was whether the enemy could be prevented from fleeing the island over the 2-mile Strait of Messina.

Churchill welcomed the 'Great Husky success' at the War Cabinet on 12 July. 'Generally speaking,' he said, 'we are over the first and most deadly phase. A letter from Eisenhower says after first 48 hours most critical time will come ... Think myself very satisfied indeed. Syracuse captured before daybreak yesterday. Nothing like it I've ever read of.'[10] Brooke was particularly delighted with the progress as he had strongly pushed for Husky at Casablanca against Portal, Pound, the British Planners, Mountbatten, Ismay and many American Planners, though not Marshall himself. Later in the Sicilian campaign, especially once the Allies got to Catania, they met tough German resistance, but the opening salvoes of it merited Churchill's and Brooke's satisfaction.

The next day Churchill's thoughts moved towards the boot of Italy. 'Why should we crawl up the leg like a harvest bug from the ankle upwards?' he asked the Chiefs of Staff. 'Let us rather strike at the knee ... Tell the Planners to throw their hat over the fence; they need not be afraid there will be plenty of dead weight to clog it.'[11] This idea, of leapfrogging up the leg of Italy, was one that Marshall also supported, albeit without the gloriously clashing Churchillian metaphors. On 16 July he suggested to the Combined Chiefs of Staff that Eisenhower should take Naples with an amphibious operation, which should then turn into a march on Rome. An attack on Salerno, just south of Naples, was immediately put into planning mode, codenamed Operation Avalanche.

For Marshall, Avalanche was intended solely to remove Italy from the war and make the Germans draw away forces from their Western and Eastern Fronts towards the Southern, to defend against a serious attack that was not in fact going to come. For Churchill and Brooke, by contrast, such an attack most definitely was contemplated. 'I will in no circumstances allow the powerful British and British-controlled armies in the Mediterranean to stand idle,' the Prime Minister told Smuts. 'I shall go to all lengths to procure the agreement of our Allies. If not, we have ample forces to act by ourselves.'[12] This was sheer bravado; by that stage of the war an all-Commonwealth invasion of Italy without American participation was unthinkable.

On his trip to London in the second half of July, Henry Stimson found plenty of support for his view that a cross-Channel operation was feasible sooner rather than later. (Marshall told Pentagon historians in 1949 that the US War Secretary had quite literally included the early adoption of the cross-Channel operation in his nightly prayers.) Stimson spoke to Lieutenant-General Morgan and his American deputy, Major-General Ray W. Barker, both of whom feared that the Mediterranean campaign might well delay Overlord, which Stimson very confusingly called 'Roundhammer', yet another hybrid name despite the official change of nomenclature. He also met American Planners who said that Overlord was achievable as it stood, and that the RAF were confident of being able to drive off any German counter-attack against the Normandy bridgeheads.

Stimson accompanied the Churchill family on a weekend trip to Dover in the Prime Minister's special train on Saturday 17 July to visit the Overlord preparations. As he reported to Roosevelt on his return, Churchill had brought in 'with evident delight' a telegram from Dill telling him that Marshall had proposed that a study be made of Avalanche. Churchill took this, according to Stimson, 'as an endorsement by Marshall of his whole Italian policy and was greatly delighted'. Stimson was not about to permit this interpretation to be put on Marshall's action, however, and as he told Roosevelt, 'I pointed out to him that it probably meant that Marshall had proposed this as a short-cut intended to hasten the completion of the Italian adventure so that there would be no danger of clashing with the preparations for Round-hammer.'[13] The use of the word 'adventure' was an indication of Stimson's feelings about any more diversions from Overlord.

Speaking to Marshall by scrambler telephone, Stimson established to his own satisfaction that Avalanche was only being investigated in order 'to obviate the danger of a long slow process "up the leg" which might eliminate Roundhammer altogether'. When he reported to Churchill that Marshall had supported his interpretation of the Avalanche message, and that therefore the Salerno landings were not intended to spearhead a full-scale invasion of Italy, the Prime Minister immediately 'broke out into a new attack' on Overlord, referring to a setback the British had received at Catania in Sicily during the past few days. Churchill even 'praised the superlative fighting ability of the Germans. He said that if he had fifty thousand men ashore on the French Channel coast . . . the Germans could rush up sufficient forces to drive them into the sea.' The Prime Minister then resurrected an image he had used before with Stimson and Marshall, the haunting one 'of having the Channel full of corpses of defeated Allies'.[14]

This started an outright row, and 'for a few minutes we had it hammer and tongs', as Stimson proudly reported to Roosevelt. Churchill denied Stimson's accusations that he opposed Overlord, asserting that Britain would 'go through with it loyally', which was hardly a ringing endorsement. He added that he was not insisting on going further north than Rome, 'unless we should by good luck obtain a complete Italian capitulation, throwing open the whole of Italy'. He also – at least according to Stimson's contemporaneous report – said 'that he was not in favour of entering the Balkans with troops but merely wished to supply them with munitions and supplies'.[15] Was Churchill deliberately misleading the US Secretary of War about his true intentions? Or did he merely wish to wait and see what the next stage might bring? To discuss the Balkans before Palermo had even fallen was more phantasmagorical than foresighted, and the whole conversation was later to have deeply baleful long-term effects, especially for Alan Brooke personally.

Stimson flew on to see Eisenhower in Algiers. He found the general in favour of a limited attack on mainland Italy, one which captured the air bases in the Foggia Plains that were needed in order to prosecute the bombing campaign against south-eastern Germany and Roumania, which could not easily be bombed from Sicily owing to the distance and lack of airfields. He concluded to the President that the American and British conceptions of the Italian operation were wildly at variance, that the British were hoping to 'supplant' the cross-Channel operation altogether and 'neutralize' any invasion of France, and that the two invasions could not be conducted simultaneously. Subsequent events were to prove him wrong, but there can be no doubt that when his report was delivered by Harry Hopkins to Roosevelt at Shangri-La, it caused great consternation.

Churchill stated his view to the Chiefs of Staff Committee on 19 July: 'I have no doubt myself that the right strategy for 1944 is maximum post-Husky, certainly to the Po, with the option to attack westward in the South of France or north-eastward towards Vienna, and meanwhile to procure the expulsion of the enemy from the Balkans and Greece.' This was very different indeed from what he told Stimson, if Stimson's report to Roosevelt was accurate. However, Stimson would have recognized the next sentiment: 'I do not believe that twenty-seven Anglo-American divisions are sufficient for Overlord in view of the extraordinary fighting ability of the German Army and the much larger forces they could so readily bear against our troops even if the landings were successfully accomplished.'[16] Although twenty-seven was not the full complement of what was intended for Overlord, that was the number the Staffs thought would in practice be available by the target date.

Brooke spoke to Kennedy soon after the 'harvest bug' memorandum, believing there to be four or five German divisions in Italy. The Avalanche landings at Salerno just south of Naples would have limited air cover because of their distance from Sicily, and at Rome it would be non-existent, beyond what could be provided by aircraft carriers. The Germans were putting up strong resistance in Sicily – Catania didn't fall until 7 August – which did not bode well for Italy either. 'We might easily be outnumbered and outmatched in a landing as far north as Naples,' feared Kennedy, and furthermore the Navy might suffer heavy losses. To lose large numbers of landing craft would also mean operations in the Mediterranean grinding to a standstill for months. Kennedy therefore preferred to land around the largely ungarrisoned toe of Italy or down by the heel and instep at Taranto.

'Another great difficulty is that if we go for the mainland of Italy we shall have to break it to the Americans that resources will be swallowed up which they want directed to the problematical invasion of France next spring, and to operations against Burma,' Kennedy warned Brooke. 'Of course the correct strategy is to continue to hammer Italy till she drops out of the war. It is childish to give up this object for anything else at the moment. But the Americans will not take this easily and another conference is essential to flog it all out.' Brooke told Kennedy that at a Chiefs of Staff conference the previous week, 'Winston held forth on this and said the Americans considered we had led them up the garden path in the Med – and a beautiful path it has turned out to be. They have picked peaches here, nectarines there, and we have done it all for them, etc, etc.'[17] It was a joke that Churchill was to make quite often, but inadvisedly not always to Britons.

'We are prepared to jump a bridgehead on the mainland at the earliest opportunity,' General Alexander wrote to Churchill and Brooke on 22 July. Yet Marshall had still not been persuaded. Alexander's biographer believes that the absence of any agreed post-Husky strategy meant that the wholesale capitulation of huge numbers of Germans in Tunisia was not repeated at Messina. Forty thousand Germans could be prevented from escaping across the Strait of Messina only by occupying the toe of Italy behind them, but 'agreement to overlap the end of one campaign by the beginning of the next was not reached in time'.[18] Because there was no written commitment at Trident to invade the Italian mainland, and Marshall had been reserved and cryptic about it at Algiers, the Germans managed to cross the Strait and escape. On 14 August Eisenhower admitted to Butcher that he had made a 'mistake' over 'our super-cautious approach to Italy', where 'we should have made simultaneous landings on both sides of the Messina Strait, thus cutting off all Sicily and obtaining wholesale surrender.'[19]

Such an operation would not have been too difficult, either. In late July there were landing craft available and Allied air superiority was complete; the toe was barely defended except by some coastal guns. Had a single Allied corps advanced across the narrow peninsula and seized the ferry terminals north of Reggio, the whole of the Sicily garrison might have been captured.[20] Yet any such landing on the Italian mainland would have involved a psychologically new departure of the war, one that needed the full agreement of the Combined Chiefs of Staff, which in effect meant Marshall.

As it was, the American and British strategists did agree on joint planning for Avalanche, much further up the leg of Italy, which Eisenhower put into active consideration before Palermo fell to Patton on 23 July. That day, Churchill rang Kennedy for casualty figures for the Sicilian campaign. Kennedy got them from the Adjutant-General's statistical branch and read them over to the Prime Minister – 1,000 killed, 1,700 missing, 4,000 wounded – saying it made a total of 5,600. 'No, it doesn't,' said the Prime Minister immediately. 'It is one thousand out.' In fact it was 1,100 out, but Kennedy was impressed that Churchill's mental arithmetic was fast enough to add the figures up while he was reading them out. (It later turned out that only seven hundred men were in fact missing.)

Mussolini fell on 24 July; isolated, bitter and depressed since El Alamein, he was voted out of office by his own creature, the Fascist Grand Council, by nineteen votes to seven. When he reported the result to King Victor Emmanuel, he was arrested, and Marshal Pietro Badoglio was asked to form a government. This released Churchill from the obligation made to Stimson not to move further north than Rome, as suddenly there was indeed the prospect of being able to 'obtain a complete Italian capitulation, throwing open the whole of Italy'. After tortuous negotiations with the Italians, in the armistice finally signed at Cassibile on 3 September both Churchill and Roosevelt insisted that the Badoglio Government declare war on Germany, which it did six weeks later, to Hitler's great fury.

On 30 July Brooke received from Kennedy an appreciation of the size of the Allied forces required if from thirteen to eighteen German divisions chose to make a stand on the 100-mile-wide Ravenna–Pisa line. Kennedy estimated that sixteen Allied divisions would be needed for Italy, including one in Sicily, assuming that the Germans needed to keep four or five in the Po Valley guarding their lines of communication through a hostile Italian population. Brooke added six divisions to this total, on the basis that 'We had a similar number on the Tunisian front' which was roughly the same length, but here the Germans were fighting closer to home and therefore could more easily reinforce Italy. These calculations meant that he would have to discuss with Marshall the rescinding of their agreement to return seven divisions from the Mediterranean to the UK on 1 November. 'He is certainly against taking any chance and no doubt he is right,' thought Kennedy.

In early August Kennedy was worried that 'it will be extremely difficult to get the Americans to agree to an advance to the north of Italy. Marshall has always held that it would be a bottomless pit into which to fling resources and is still obsessed by the idea of invading France.' Brooke never accepted that Marshall understood the British concept for the Mediterranean. Even in 1958, when NBC made a radio programme about Marshall, Brooke told the interviewer: 'I look upon him more as a great organizer than a great strategist.' (He pronounced it 'strateegist', with a long 'e'.) He did at least add: 'He was amongst the biggest gentlemen, and using that term in its very best sense, that I've ever met. [He had] a sense of extraordinary integrity. One could trust him with anything, simply ... He treated his inferiors almost the same as his peers.'[21]

The need to agree the next stage of the Italian campaign meant that another major conference was badly needed, and in July the long-serving Canadian Prime Minister William Lyon Mackenzie King generously offered to host it in Quebec the following month, to be codenamed Quadrant.

The night before he left for Canada on the *Queen Mary* on Thursday 5 August, Churchill spoke after dinner to Eden, who said that he thought the conference would 'infuriate Uncle Joe'. Churchill accepted that it would but thought that 'If we could persuade the Americans to help us form a line in the valley of the Po ... and thus open a real Second Front, Joe might become more amenable again.'[22] In fact, as we have seen, Stalin took Anglo-American co-operation for granted. Churchill's answer showed what he wanted out of Quadrant. It was very substantially what Brooke wanted too. On their five-day Atlantic journey, the two men reviewed every aspect of their plans, 'knowing that their reception by their American colleagues, although physically warm, would not necessarily be strategically welcoming'.[23] Churchill also brought the Chindit leader Orde Wingate along to persuade the Americans that the British were serious in Burma.

On the train to the port of Faslane, 25 miles west of Glasgow where they were to board the liner, 'a furious' Brooke told Joan Bright, who had organized the accommodation, 'with sharp clarity *never* again to allocate him a sleeping compartment right above the grinding train wheels'. He did not give her time to reply that she had naturally accorded him Compartment A as the right one for someone of his seniority. Minutes later Portal complained that she had put Wing-Commander Guy Gibson VC, the hero of the Dambusters Raid, into a second-class compartment, but, as she recalled, 'the effects of this attack did not last so long as those produced by General Sir Alan Brooke.'[24]

During the voyage, Churchill invited the War Office Planners to prepare a strategy for the invasion of northern Sumatra, codenamed Operation Culverin. This was to be a key aspect of his strategy centred on the Bay of Bengal, based on retaking Rangoon, northern Sumatra and Singapore, which was completely at variance with the Far Eastern strategy that Brooke and the Chiefs of Staff wanted to pursue, 2,000 miles to the east in the Pacific. 'Great show,' recalled one of the Planners, Captain C. E. Lambe RN, of Churchill's request, 'he was drinking white wine, dressed in black dressing gown with golden dragons. Gave us a fine feed – everything just right – then patted us on the shoulders and said be good boys and write me a nice plan.' They delivered it when they got to Quebec, along with a covering letter to Ismay declaring that the scheme was 'no good', signed by them all. Somehow Ismay failed to detach the letter from the plan and both went on to the Prime Minister. 'When Winston saw our names,' recalled Captain Lambe, 'he roared: "The Joint Planners are suspended." '[25] They asked Ismay what this involved, and learnt that they ought to carry on exactly as before, but just to regard themselves as suspended. (As we have seen, this suspension story has also been told in relation to Jupiter. Whichever operation it was, the episode clearly didn't affect the future career of Admiral of the Fleet Sir Charles Lambe, First Sea Lord and Chief of the Naval Staff in the 1950s.)

On the journey over, Brooke persuaded Churchill not to present their Po strategy in terms of ultimately going either to Vienna or into the Balkans, which would only result in Marshall turning against the concept completely. The British argument for the advance to the Po was thus made on the sole ground of drawing off German divisions from Normandy. It was a sensible recommendation, and served Churchill well.

The *Queen Mary* docked at Halifax, Nova Scotia, on the afternoon of Monday 9 August, and the British party took a special train to Quebec, arriving at 6 p.m. the next day. They occupied the fifteenth floor of the Château Frontenac hotel, a magnificent edifice which still completely dominates the Quebec skyline, and then every odd-numbered floor down to the lobby; the Americans took the sixteenth floor, and then every even-numbered floor downwards. The Combined Chiefs met in the Salon Rose on the second floor, a pink room with a fine view over the St Lawrence River, and press conferences were convened in the ground-floor coffee shop. Oranges were flown in from New York for the delegates' juice, and the Canadian Government paid for everything. 'We were spoiled,' recalls Joan Bright, 'even to a running buffet being set up in the foyer of the dining room for all those who could not forgo their tea and coffee breaks.'[26]

With a few days to go before the conference opened, Churchill headed off to Hyde Park for three days between 12 and 14 August, via the Niagara Falls. Since all three of the British Chiefs of Staff and Mountbatten were keen fishermen, they went to the lakes north of Quebec. At one point Mountbatten (he claimed) saved Dudley Pound from falling into a ravine. It was on that trip that the First Sea Lord's failing powers became unavoidably apparent. 'On the way back we had great difficulty in getting him back to the car,' noted Brooke. 'He seemed completely exhausted.' It was considerably worse than that.

'Both sides approached Quadrant in an exasperated mood,' Field Marshal Lord Bramall and General Sir William Jackson write in their authoritative history of the Chiefs of Staff Committee, 'verging upon outright mistrust of each other.'[27] It was the third major Anglo-American strategy conference in only seven months, so exasperation might have played a part. That was no excuse for the unnamed Planner, however, who absent-mindedly left a file of top-secret strategy documents in the Salon Rose after a meeting. All the hotel employees had already been security-screened by the Canadian intelligence services, but nonetheless the blameless hotel bell-hop, Frank Brettle, who had found the file and immediately handed it to a Mountie, noticed that he was being followed for months afterwards.

One of the few fictional accounts of the Combined Chiefs of Staff to appear in literature can be found in the American author J. G. Cozzens' 1949 novel *Guard of Honor*. In it, a USAAF general Joseph Nichols watches Roosevelt and Churchill confer with their Staffs at Quadrant, and notices how:

The object could not be simply to concert a wisest and best course. The object was to strike a bargain which was the congeries of a thousand small bargains wherein both high contracting parties had been trying, if possible, to get something for nothing; and if that were not possible, to give a little in order to gain a lot ... Agreement was ordinarily resisted by mutual misrepresentation, and obtained by a balance of disguised bribes and veiled threats.[28]

This might have seemed a cynical view so soon after the war, but it was surprisingly accurate. Indeed, one of the complaints that Brooke made about Churchill's sudden volte face on the penultimate day of Trident had been that by questioning his small deals the Prime Minister was jeopardizing the larger concessions he had wrung from Marshall.

Stimson, the US Secretary of War, was scheduled to be on holiday during the Quebec Conference, which shows how completely he had been eclipsed by Marshall in strategic decision-making. But he was due to meet Roosevelt at 1 p.m. on Tuesday 10 August, before the conference opened. After breakfast that day, he dictated a memorandum to Roosevelt which might have doubled as a resignation letter had discussions gone awry. It repeated the conclusions of the memorandum drawn up after the disastrous Dover train journey with Churchill, and insisted that Marshall rather than Brooke be given command of Overlord. 'We cannot now rationally hope to be able to cross the Channel and come to grips with our German enemy under a British commander,' the War Secretary wrote. 'The Prime Minister and his Chief of the Imperial Staff [sic] are frankly at variance with such a proposal. The shadows of Passchendaele and Dunkerque still hang too heavily over the imagination of his government. Though they have rendered lip service to the operation, their hearts are not in it.' By total contrast, 'General Marshall already has a towering eminence of reputation as a tried soldier and as a broad-minded and skilful administrator.'[29]

Once the letter was typed up he signed it, and then showed it to Marshall, 'in case he had any objections to it'. Marshall merely said that 'he did not want it to appear' that Stimson had consulted him about it, and the War Secretary replied that that was why he had signed the paper before showing it to him and he would not show it to anybody else. So armed, Stimson went to the White House and had 'one of the most satisfactory conferences I have ever had with the President'. The reason was obvious: Roosevelt also preferred Marshall to Brooke.

The President invited Stimson to stay for a meeting with the Joint Chiefs at which he 'went the whole hog' on the subject of Overlord, insisting that 'We should have more soldiers in Britain dedicated for that purpose than the British.' They then discussed the best stance to take at Quebec, and Roosevelt made it clear that he wished to go 'no further into Italy than Rome and then for the purpose of establishing [air] bases'. He confirmed that he wanted an American commander for Overlord, which must have pleased Marshall, who knew what Stimson had been discussing with FDR only moments earlier.

The Joint Chiefs emerged from the meeting 'astonished and delighted at his definiteness', upon which Stimson naturally congratulated himself. Brooke seems to have had no comprehension of this lobbying for a position he hoped was his, even though Frederick Morgan, who ran COSSAC, recalled that 'throughout the summer there was continual hardening of unofficial opinion that the Supreme Commander would be George C. Marshall', opinion of which Brooke could not have been unaware.[30]

The first Combined Chiefs of Staff meeting at Quadrant took place on Saturday 14 August, before which Dill had given Brooke prior warning that Marshall and both the other American Chiefs of Staff had 'a feeling that the British are not standing firm enough to considered decision of Trident, and are tending too readily to depart from these decisions and to set aside the operations already agreed upon'. In particular Dill wanted Brooke to appreciate that, although the Joint Chiefs of Staff wished to force Italy out of the war, they would not countenance Pacific or Overlord operations suffering because of it. The conference thus started off with a Joint Chiefs of Staff position paper demanding a reiteration of the Trident promise to pull out the seven divisions by 1 November 1943, and stating that 'The US Chiefs of Staff believe that the acceptance of this decision must be without conditions and without mental reservations.'[31] For Brooke, the next day was even worse. Indeed, apart from the death of Janey, it was most probably the worst of his life.

Just before lunch on Sunday 15 August, Churchill sent for Brooke to come to the Citadel, where he was staying for the conference. Not far from the Château Frontenac, the Citadel was a military stronghold from the eighteenth century, 200 feet above the St Lawrence River with magnificent views. To the south-west are the Plains of Abraham, where Wolfe defeated the Marquis de Montcalm in twenty minutes in a surprise attack that won Canada for the British Empire in September 1759, both of them dying in the process. Neither the view nor the glorious history was to provide much comfort to Brooke, however.

Churchill had just returned from a meeting with Roosevelt at which Hopkins had pressed hard for the appointment of Marshall as supreme commander for Overlord. 'As far as I can gather Winston gave in,' Brooke told his diary, 'in spite of having previously promised me the job!!' The Prime Minister asked Brooke how he felt about it and was told, with some understatement, that he 'could not feel otherwise than disappointed'. They then discussed other appointments, including Eisenhower's becoming US Army chief of staff in Marshall's place, and Brooke said that he thought Mountbatten 'lacked balance' for the job of supreme commander of the newly formed South-East Asia Command.

Brooke then realized that Mountbatten's appointment had actually been a *quid pro quo* for his not getting command of Overlord, which only made the situation worse, as the two posts were in no way analogous.

After the war Brooke wrote of this shattering blow to his hopes, which had been so assiduously stoked by Churchill for so long. 'I remember it as if it was yesterday as we walked up and down on the terrace outside the drawing room of the Citadel,' he recalled. 'Looking down on to that wonderful view of the St Lawrence River, and the fateful scene of Wolfe's battle for the heights of Quebec. As Winston spoke all that scenery was swamped by a dark cloud of despair.' He remembered how two months before El Alamein he had given up the chance of taking over Auchinleck's command because he needed to guide Churchill, but with the broad global strategy now agreed he 'felt no longer necessarily tied to Winston, and free to assume this Supreme Command which he had already promised me on three separate occasions'.[32]

Nor had Brooke ever been philosophically opposed to the Roundup–Overlord operation, as Americans like Stimson, King, Handy, Wedemeyer, Hull and Elliott Roosevelt alleged; he had just been opposed to crossing over too early. Furthermore, he had resolutely not done what a number of other generals might have in his position, which was to change his tune and extravagantly champion Overlord in the hope of securing its command for himself. He had put his professional judgement first, and had paid a very high price for it. If he had commanded on D-Day, it is likely that the name of Brooke would be as famous as that of any Allied general of the Second World War, indeed as famous as Marlborough's or Wellington's.

What made it worse for Brooke was Churchill's reaction. 'Not for one moment did he realize what this meant to me,' he wrote afterwards. 'He offered no sympathy, no regrets at having had to change his mind, and dealt with the matter as if it were one of minor importance!' Churchill, who always wanted to lead great armies himself, simply must have felt Brooke's disappointment, not least because it was verbally expressed to him on the Citadel terrace that afternoon. Stiff-upper-lipped, upper-class, Ulster-born senior soldiers of the early 1940s were perhaps the very last people in the world ever to allow their emotions to show; they were about as far from the touchy-feely as it is culturally, socially and generationally possible to be. Yet Churchill cannot have failed to notice Brooke's profound dismay just because he did not remonstrate with him, or remind him that he had thrice offered him the post.

Of course we only have Brooke's word for it that the Prime Minister offered no word of consolation, but it is clear that the CIGS deeply resented it. For him the incontrovertible fact that many more Americans were due to take part in Overlord, and should therefore be commanded by one, 'did not soften the blow, which took me several months to recover from'. In fact Brooke never truly recovered from it at all, certainly not by the time of the publication of *The Turn of the Tide* in 1957, which contained severe criticisms of Churchill that effectively severed their relations. For the legendarily tough-minded 'Colonel Shrapnel', who suffered fools not at all, it might be thought that he only received what he had spent years dishing out, but as Moran put it: 'In that moment there was revealed to Brooke the crushing indifference of these monolithic figures to the lower forms of life.'[33]

It was not only the deed itself that Brooke minded so much, but the fact that 'the only reference to my feelings' in the fifth volume of Churchill's war memoirs, *Closing the Ring*, published in 1952, was the single phrase that he 'bore the great disappointment with soldierly dignity'.[34] In an earlier draft Churchill had written: 'I had to break the news to Brookie. He was bitterly pained, but bore it all as a soldier should. Not one word escaped him.'[35] Churchill explained in his memoirs that politically Roosevelt could not have allowed a foreign commander for an army that was expected to exceed two million men, a majority of them American. As Leahy put it in his autobiography, 'I would have had no personal objection to Brooke, but if he or any other Englishman had been named to the post, there would have been a storm of criticism in our country.' In the earlier draft of his memoirs, Churchill also gave a more Machiavellian explanation for the decision:

I had the fear that if a bloody and disastrous repulse were encountered, far bigger than the first day's battle in the Somme in 1916, there might be an outcry in the United States. It would be said that another result would have attended the appointment of an American general. I therefore made my mind up on the voyage over to propose to the President that he should choose the general to whom so much would be confided. If he refused, our position would be invulnerable.[36]

These two passages were edited out just before publication, which was very damaging to Brooke's subsequent relations with Churchill.

Stimson, called from vacation to Quebec on 22 August, was told by

Roosevelt that 'Churchill had voluntarily come to him and offered to accept Marshall for the Overlord operation.' In another conversation, Churchill 'said he had done so in spite of the fact that he had previously promised the position to Brooke and that this would embarrass him somewhat, but he showed no evidence of retreating from his suggestion.' Stimson recalled, 'I was of course greatly cheered up.'[37]

Would Brooke have made a good supreme commander? Sir James Grigg was doubtful, telling an interviewer after the war that his utter disdain for popularity or public relations – let alone public opinion – would have counted against him. 'A successful commander in the field must be able to command the imagination of his troops and impress his personality on them,' argued Grigg. 'It was doubtful if Brooke had the patience or understanding to do this; rarely did he inspire affection because he was too insular and rarely proffered friendship.'[38] The lengths to which some commanders went to create charismatic personae for themselves – Montgomery, Mountbatten, MacArthur and Patton foremost among them – were always dismissed by Brooke as mere 'stunts'. Meanwhile, several rumours went around the War Office about why Brooke had not been chosen, the most fanciful of which was recorded by Kennedy, who wondered 'if it can be true that' Brooke had 'offended Mrs Roosevelt by some remark about niggers. Less surprising things have happened before.'[39]

Four decades after the decision, Jock Colville recalled that 'Roosevelt was determined the commander should be American, even though neither Marshall nor Eisenhower had Brooke's experience or strategic brilliance.' A devil's advocate – and there was no shortage of them in the Pentagon – might have pointed out that Brooke's last two forays on the Continent had both ended in humiliating evacuations. The restriction of the choice of supreme commander of Overlord to Americans was a clear signal that the Atlantic balance of power had shifted, and for all his enthusiasm and bulldog spirit, Churchill was simply not in a strong enough political position vis-à-vis Roosevelt to award the post to Brooke, even though he had thrice promised it him. When it came to the ultimate decision-making moment, however, as is clear from several sources, including Ismay's interview with Pogue – 'Churchill on his own initiative told FDR that the commander should be an American' – there was no contest.[40] Brooke's subsequent behaviour at the time of the publication of his memoirs was self-defeating – even, to some, reprehensible – but it was entirely understandable.

PART III
Estrangement

15

From the St Lawrence to the Pyramids: 'All this "Overlord" folly must be thrown "Overboard"'

August–November 1943

I suppose that when working with allies, compromises, with all their evils,
become inevitable.   General Sir Alan Brooke, August 1943[1]

Only moments after being dealt his terrible disappointment at lunchtime on Sunday 15 August 1943, Brooke had to attend a Combined Chiefs of Staff meeting at 2.30 p.m., which lasted for '3 very unpleasant hours'. Marshall argued that, if Overlord did not have 'overriding priority' over Italy and everywhere else, then 'in his opinion the operation was doomed and our whole strategic concept would have to be recast'. This was of course simply the same old threat to dump Germany First in favour of the Pacific. He further insisted on the seven divisions being removed from Italy by the agreed date. Brooke reiterated that the Mediterranean and cross-Channel operations were not in competition but rather intimately connected, the former being intended to draw off German forces from the latter.

The fact that he had just heard that an American – therefore probably Marshall – would take the post he coveted, and had doubtless made private plans for, cannot have helped Brooke's temper, and he predictably noted in his diary of Marshall: 'It is quite impossible to argue with him as he does not even begin to understand a strategic problem!' Both Marshall – the boy crying 'Wolf!' over Japan yet again – and Brooke, mistaking plain disagreement for strategic ignorance, were starting to sound like stuck gramophone records by the late summer of 1943. That evening Brooke dined alone with his thoughts, after which he spoke to Dill, who had earlier found Marshall 'most unmanageable and irreconcilable, even threatening to resign if we pressed our point'. That 'point' was explained in one sentence by Brooke: 'By giving full priority to the cross Channel preparations you might well cripple the Italian theatre and thus render it unable to contain the German forces necessary to render the cross Channel operation possible.'[2]

Marshall definitely himself dated what he later called his 'big break' with Brooke to his own insistence that the seven divisions come out of Italy in order to 'solidify on a practical basis for the landings in Normandy'. With Overlord intended to number from twenty-seven to twenty-nine divisions, these therefore made up one-quarter of the entire force, and were thus indispensable. He nonetheless told Pogue years later that he had 'a great sympathy for the British in their situation', because 'there was the fact that I hadn't commanded troops. Brooke had commanded II Corps in France when he was sent over after the first withdrawal to establish a line to defend the Brest peninsula. He had done all these things and, while I had been chief of operations in an army in the First War, I had done nothing like that. So they felt I didn't understand the problems.'[3] Marshall was sensitive enough to spot this, yet tough enough not to let it affect him. (In fact Brooke had commanded II Corps in the Dunkirk campaign, not the Brest peninsula, when he had commanded the whole of the Second BEF.) Asked by NBC in 1958 what would have happened if Marshall had been in charge of Overlord, Brooke took a few seconds before answering, diplomatically: 'That's a very difficult question to answer. There are so many ifs . . . You get led on from one if to another if, and I don't think one gets very far with them.'

On Monday 16 August, Brooke and Marshall returned to the Trident system of off-the-record meetings. The secretaries and Planners left the Salon Rose, and for three hours after 2.30 p.m. the Combined Chiefs undertook 'the difficult task of finding a bridge'. These discussions were 'pretty frank', with Brooke opening by saying that 'the root of the matter was that we were not trusting each other'. He went on to accuse the Americans of doubting the British commitment 'to put our full hearts into the cross Channel operation next spring', while for their part the British were not certain that the Americans 'would not in future insist on our carrying out previous agreements irrespective of changed strategic conditions'. This was a veiled reference to the seven divisions due to be withdrawn from the Mediterranean theatre only eleven weeks hence.

'In the end I think our arguments did have some effect on Marshall,' noted Brooke, though not on Admiral King. Brooke was feeling the strain, stating that since that was his sixth meeting with the American

Chiefs, 'I do not feel that I can possibly stand any more!'[4] At one stage during the Quebec Conference, Marshall and Dill even discussed the possibility of the Combined Chiefs of Staff taking a vote every time a division needed to be moved anywhere.[5] It was never tried, but shows how far down the atmosphere of mistrust had descended.

Captain Lambe pointed out the best riposte to the American accusation that the British were never serious about Overlord, telling Pogue in February 1947:

Vast amounts of construction work had to be done – hard roads, railways to beaches, exits, fuel and storage tanks, railway sidings. The amount of construction in southern England was terrific. It is interesting to note that millions of pounds were spent from early 1943 onwards, when there was only COSSAC Staff; millions spent on a plan which had not been approved.

Lambe believed that if the Supreme Allied Commander had chosen to attack anywhere other than Normandy, the logistics by then dictated that 'he couldn't have'.[6]

Roosevelt arrived in Quebec on the evening of Tuesday 17 August, the same day that the US 3rd Division under Patton took Messina, ending the Sicilian campaign. Basil Liddell Hart believed that the mainland of Italy was invaded primarily because Sicily was cleared in mid-August, by which time it was too late for anything in the Channel, and Sicily was so close to Italy that it really chose itself. 'It was the logic of events resulting from loss of time more than logic of argument,' he argued, 'which swung the Allied strategy.'[7] There were other reasons too. When Lieutenant-General Morgan presented COSSAC's plans for Overlord at Quebec, Brooke observed that they required the Luftwaffe to be drastically weakened in France, the number of German divisions in France and Holland to be reduced, and communications between Germany's two fronts to be severely disrupted. Simultaneously, Arnold supported Brooke's plan for invading Italy by saying that the capture of the huge Foggia airfields near Naples would allow the USAAF and RAF to bomb all of southern Germany, 60 per cent of German fighter-production factories and all the major east–west road and rail connections through Germany, while drawing off Luftwaffe units from the west. Morgan and Arnold thus effectively made Brooke's case for him.

Although Brooke constantly admonished Marshall in his diary for failing to appreciate the connection between the Mediterranean and cross-Channel strategies, as early as the first day of the conference Marshall emphasized the connection between the European and Pacific theatres. For him the reconquest of Burma was always a far higher priority than it was for Brooke, despite Burma being a British colony and the gateway to India. This was because Marshall believed that China was on the verge of being forced out of the struggle altogether, with potentially disastrous long-term implications for the war against Japan. Only by opening up and keeping operational the northern Burmese–Chinese connection known as the Burma Road, he thought, and by flying in supplies to Chiang Kai-shek's Nationalist Army, could China be sustained. Brooke, by contrast, believed China had no choice but to stay in the war almost come what may, considering what Japan had inflicted on her since 1931, that the Chinese could absorb almost any amount of punishment, and certainly any supplies that were sent. He was not impressed with Chiang Kai-shek's contribution to the war, and thought that major efforts in Burma should wait until Germany was defeated. His agreement to undertake Anakim had been reluctant, hedged with reservations, a *quid pro quo* for other concessions, and conditional on the United States providing most of the shipping and landing craft.

On 17 August, Brooke also stated that MacArthur's plan to thrust up the northern tip of the Vogelkop – the peninsula of western New Guinea – ought to be curtailed in order to release men and matériel for Overlord. Since, together with Admiral Nimitz's attack from Hawaii through the Gilbert and Marshall Islands to Palau, MacArthur's was the major offensive towards recapturing the Philippines, Marshall and especially King profoundly differed from Brooke over this too.

Meanwhile, Churchill was still pressing hard for an attack on the northern tip of Sumatra, codenamed Operation Culverin. Rather condescendingly Brooke wrote that 'Winston . . . had discovered with a pair of dividers that we could bomb Singapore' from Sumatra, 'and he had set his heart on going there.' Brooke believed Sumatra to be an unsuitable place for any long-term projects against the Malay States, and told the Prime Minister at a meeting at the Citadel at noon on 19 August that 'when he put his left foot down he should know where the right foot was going to'. In the cold black and white of print, that does not look too rude, but we cannot know the tone of voice and the body language that accompanied it. The result was that Churchill lost his temper completely and shook his fist in Brooke's face, saying: 'I do not want any of your long term projects, they cripple initiative!'[8]

To have a fist shaken in one's face is even more of an aggressive act than breaking pencils in half during meetings, but Brooke kept calm and 'agreed that they did hamper initiative', but nonetheless told him that 'I could not look upon knowing where our next step was going as constituting a long term project!' Coming from that meeting, Brooke went straight into one of the Combined Chiefs in which, because it became so 'heated', he and Marshall ordered their Staffs to leave yet again for an off-the-record discussion. 'After further heated arguments in our closed session we ultimately arrived at an agreement,' Brooke recalled at the end of what he called 'Another poisonous day!'[9]

There then occurred one of the classic moments of the history of the Combined Chiefs of Staff, which also underlines how intractable, hard fought and confrontational these meetings had become. After the agreement was reached with Marshall, Mountbatten asked to be allowed to demonstrate an invention codenamed Habbakuk, a self-propelled floating airfield made entirely of Pykrete, a mixture of ice and wood pulp named after its British inventor Geoffrey Pike. These artificial islands, it was hoped, could be used by fighter squadrons to give close support to the invasion troops in Normandy. To demonstrate Habbakuk's superior attributes over a normal iceberg, Mountbatten had one block of ice and one of Pykrete rolled in to the Salon Rose, whereupon he theatrically produced his gun and announced that he was going to fire at each of the blocks in turn, in order to demonstrate their different defensive qualities. 'As he now pulled a revolver out of his pocket,' recalled Brooke, 'we all rose and discreetly moved behind him.' Firing at the ice merely produced 'a hail of ice splinters', as expected, but when Mountbatten shot at the Pykrete the bullet ricocheted off, and 'buzzed round our legs like an angry bee!' When the shots were heard outside the room, one of the Staff officers who had left at the start of the off-the-record meeting exclaimed: 'Good heavens, they've started shooting now!!'[10]

Like every oft told anecdote, there are a number of slightly different punch-lines and attributions of it, but the waggish remark wouldn't have been retold – and indeed wouldn't have been funny – had the Combined Chiefs of Staff not by then had a reputation for acrimony. Although the joke sounds too good to be true, it was confirmed to

Charles Donnelly by Colonel Andrew McFarland, secretary of the Planning committee of the Joint Chiefs of Staff, one of those turfed out of the conference room. What is still inexplicable even today is what Mountbatten could possibly have been thinking of in demonstrating the ricochet-inducing qualities of Pykrete in front of the entirely unprotected Combined Chiefs of Staff.

Marshall's and Brooke's off-the-record meeting and other similarly frank discussions meant that by Thursday 19 August much had been agreed at Quadrant: Overlord was to be 'the primary US–British ground and air effort against the Axis in Europe' with 1 May 1944 reiterated as the definite launch date; an attack on southern Italy would be undertaken by the forces agreed at Trident 'except insofar as these may be varied by decision of the Combined Chiefs of Staff' (Marshall would have preferred no strings attached, but Brooke got that wording added); there would also be a landing on the south coast of France (codenamed Anvil) to coincide with Overlord, and the defeat of Japan was planned within twelve months of victory in Europe, which was assumed would come in the autumn of 1944.[11]

The Manhattan Engineer District project was also beginning to bear fruit. This was the codename given to the creation of the atomic bomb that had been developed jointly at Los Alamos in New Mexico by British and American scientists, under the terms of the agreement Churchill and Roosevelt had come to at Hyde Park in June 1942. At Quadrant it was agreed not to deploy the weapon without the consent of both powers, and that Marshall would chair a committee to control the project.

(The prospect of the Bomb actually working filled Churchill with joy. On 23 July 1945 he told Brooke – 'pushing his chin out and scowling' – that the imbalance of power was now redressed with the Russians, who did not have it, and that he could finally dictate to Stalin: 'we could say if you insist on doing this or that, well we can just blot out Moscow, then Stalingrad, then Kiev, then Kuibyshev, Karkhov [sic], Stalingrad, Sebastopol, etc, etc. And now where are the Russians!!!'[12] One wonders what the poor inhabitants of Stalingrad did to deserve a double dose of vaporization, despite the flattening of their city by the German Sixth Army. As it was, Churchill lost the general election three days later and neither Britain nor the United States so much as considered employing nuclear blackmail against their still domestically fêted ally, which lost over twenty-seven million dead in its Great Patriotic War.)

At Quadrant, long-range strategy for the Pacific was put back to a future conference, and the US was left to attack the Philippines in its own way, with Rabaul to be 'neutralized rather than captured'. A South-East Asia Command (SEAC) was to be created, separate from the India Command, although it was not long before Pentagon wags nicknamed it 'Save England's Asian Colonies'. All these agreements, some going against the grain of the British strategic thinking, disprove Wedemeyer's suspicion that 'Quebec was a repetition of Casablanca, of course.'[13]

On the penultimate day of the conference, Brooke once again exploded against Churchill in his diary, writing on Monday 23 August that it was bad enough having to face the American demand for the transfer of the seven divisions in ten weeks' time, 'But when you add to it all the background of a peevish temperamental prima donna of a Prime Minister, suspicious to the very limits of imagination, always fearing a military combination of effort against political dominance, the whole matter becomes quite unbearable! He has been more unreasonable and trying than ever this time.' Some historians believe that Brooke was on the verge of a breakdown at this period; his own diagnosis, recorded on 24 August, is revealing:

The conference is finished and I am feeling inevitable flatness and depression which swamps me after a spell of continuous work, and of battling against difficulties, differences of opinion, stubbornness, stupidity, pettiness, and pigheadedness. When suddenly the whole struggle stops abruptly and all the participants of the conference disperse in all directions, a feeling of emptiness, depression, loneliness and dissatisfaction over results attacks one and swamps one! After Casablanca, wandering alone in the garden of the Mamounia Hotel in Marrakesh, if it had not been for the birds and the company they provided, I could almost have sobbed with the loneliness. Tonight the same feelings overwhelm me, and there are no birds!

At the very least he was in desperate need of rest.[14] The very next day he wrote that he had been feeling 'liverish' and would have liked 'to remove' the criticism of Churchill, but that would have meant rewriting several pages of his diary, and he didn't have the time.

At noon on Tuesday 24 August, Quadrant ended with a press conference. During it, as Churchill launched into a speech, Roosevelt leant over to Leahy and whispered: 'He always orates, doesn't he, Bill?' The next day Brooke and Portal went fishing for two days in the Lac des Neiges, but after driving 60 miles 'through the bush', and going 2 miles up the lake by motor launch, just as they had started fishing, to Brooke's 'horror', Churchill turned up with his Canadian host, Colonel Clarke, and Brooke was obliged to leave his beat. 'I could have shot them both I felt so angry,' he wrote. This does not mean, as some have speculated, that Brooke was genuinely feeling murderous towards the Prime Minister, but was merely an expression of furious frustration at having his brief respite from him temporarily ruined. 'What made it worse was that Winston couldn't fish and lost a good trout through having his line slack,' he later told Kennedy. Brooke was clearly not long disturbed, because in two days he caught 93 trout and Portal 104, averaging 1½ pounds, with the biggest at 3¾ pounds.

When Churchill was on his boat, unwittingly disturbing his service Chiefs' well-deserved relaxation, Royal Marine Lance-Corporal Emerson, who was fishing with Inspector Thompson, fell into the lake. 'Don't expect your prime minister to come and fish you out!' Churchill called, but he did lend him one of his suits to wear.[15] Admiral Pound did not feel well enough to fish. Ismay had remarked to Joan Astley about Pound's 'lethargy, and the lack of his former crisp grasp of essentials' while the conference was going on. 'Little did I realize on saying goodbye to old Dudley Pound', Brooke noted in his diary, 'that I should never see him again!'[16]

Back in London on 30 August, Brooke and Kennedy went through Eisenhower's plans for Operation Baytown, landings on the toe of Italy on 3 September, also Operation Avalanche, the landings in the Gulf of Salerno just south of Naples on 9 September, and the simultaneous Operation Slapstick, the attack on Taranto in the heel of Italy. They thought them 'sketchy', and the fact that the British Chiefs of Staff couldn't issue instructions direct to Eisenhower was 'a confounded nuisance'. The Joint Chiefs of Staff tended not to demand as much information from their commanders in the field as the British did, which Kennedy put down to the fact that 'The Washington machine is not so highly organized as ours.' Overall they considered that these operations were 'in the nature of a gamble – which might be justified'.[17]

Brooke has been criticized for not interfering enough with his generals' battle plans. 'Errors in the planning of Tunis, Sicily, Salerno, Cassino and Anzio might have been averted by the Chief's timely advice,' argued the historian Nigel Nicolson.[18] Since none of those was solely a British operation, any advice to commanders such as Eisenhower and Clark would have had to be directed through the Combined Chiefs of Staff; Brooke was no longer able to give 'timely advice' that had the force of authority, and anyhow he tended to think it best to trust to the judgement of the man on the spot. He did, however, write regular personal letters to commanders like Wavell, Auchinleck and Montgomery that had all the force and authority of the CIGS, who each of those men accepted was a master strategist.

Discussing the upshot of Quebec – 'Good in so far as they left most of the big things to be settled in the light of developments' – Brooke complained to Kennedy about the Americans' lack of trust, especially over the return of the seven divisions in the autumn. He said that it had emerged that this distrust stemmed from 'something the PM had said to Stimson during his visit here'.[19] This was doubtless the Dover train conversation in which Churchill had criticized 'Roundhammer' and spoke of thousands of Allied corpses floating in the Channel. He had also mentioned supplying the Balkans from Italy. Small wonder the Americans had got alarmed whenever Brooke tried to renegotiate the seven-divisions deal.

Kennedy believed that once the Americans were committed on the Italian mainland, 'the problem would solve itself,' since 'If we needed the divisions, they would stay there.' He recalled Foch being asked in 1914 how many British soldiers he needed for his campaign, and the marshal answering: 'One, and I shall make sure he is killed.' Brooke agreed that once American troops were in action it might prove near impossible to withdraw too many men from a continuing battle. On the issue of the Pacific, he complained that Churchill had been 'extremely trying' over the operations against the Japanese from India and had 'brushed aside' Chiefs of Staff appreciations and demanded attacks that were 'quite beyond the capacity of the forces available. His free discussions with the American Chiefs of Staff, Louis Mountbatten and Wingate had embarrassed the Chiefs considerably.'[20]

For all that the British deplored American distrust of them, they too distrusted the Americans. Inveighing against Eisenhower's Italian plans, Kennedy wrote: 'The real fault goes back to the American unwillingness to devote great strength to the Mediterranean effort. We do not feel that the Americans fully grasp the realities of operations against the Germans and how seriously they have to be taken.' This, only days after Sicily had been wrested from the Axis grasp, seven months after the Kasserine Pass débâcle, and less than forty-eight hours before the Baytown landings, was deeply unjust.

When Mark Clark's Fifth Army landed at Salerno on Thursday 9 September 1943, the Germans had fourteen divisions in Italy. By the end of October they had no fewer than twenty-five, with reinforcements having been drawn equally from Russia and north-west Europe. Professor Sir Michael Howard, who won the Military Cross with the Coldstream Guards at Salerno, considers that number 'more than enough to pacify the country and compel the Allies to fight hard for every yard of their advance on Rome'.[21] Moreover, because the invasion of Sicily opened up all sorts of other possibilities for the Allies in Europe's 'underbelly', the Germans had to cover the entire area. By the end of 1943 there were a further twenty German divisions stationed in Yugoslavia, Greece and the Aegean islands, for example.[22]

With forty-five divisions thus standing guard on Germany's southern flank, protecting south-east France, Austria and the Balkans, the Wehrmacht could not help but be weakened on the two Eastern and Western Fronts that were indeed to mean life or death to the Reich. The reason that the Germans did not simply pull back to an easily defensible line north of Rome was that their commander Field Marshal Albert Kesselring – a former chief of staff of the Luftwaffe – well understood the use that could be made of the airfield complex around Foggia in south-east Italy. He was also under orders to contest every inch of Italy. After the war, Major-General Walter Warlimont of Hitler's military staff wrote that the Führer's Mediterranean strategy 'threw a far greater strain upon the German war potential than the military situation justified and no long-term compensating economies were made in other theatres.'[23] Brooke could hardly have hoped for a better encomium. 'One elementary principle governed everything but there was no great lofty idea behind it and no thought of concentrating upon essentials,' continued Warlimont; 'instead the supreme command had only one object, to defend the occupied areas everywhere on their outermost perimeter.' This was precisely how Brooke had hoped Hitler would try to cope with the defeat in Sicily: by replicating the Tunisian and Stalingrad errors in Italy.

It fitted into the Nazi philosophy that willpower was all in warfare as in politics, but it made for terrible strategy.

Mark Clark's Fifth Army, comprising one American and one British corps, landed at Salerno, a port on an inlet of the Tyrrhenian Sea. With air cover provided from British aircraft carriers and Sicily, the troops got ashore in Operation Avalanche, but heavy German counter-attacks spearheaded by three panzer divisions from 12 to 14 September came close to flinging them back into the sea. Amphibious operations were risky ventures, and had it not been for the battleships HMS *Warspite* and HMS *Valiant* arriving from Malta during the six-day battle and training their 15-inch guns on the German positions, events might have gone differently. The Germans also used heavy bombs guided by radio for the first time, and two Allied cruisers were severely damaged on 11 September. *Warspite* herself was knocked out of action on 16 September, and had to be towed back to Malta. By then, however, contact had been made with the Eighth Army, which had marched up the 150 miles from Reggio. The near-disaster of Avalanche militates against the idea that Sledgehammer or Roundup could have worked in 1943.

The next row between the British and Americans was started by Churchill putting forward various schemes for Italy designed to take advantage of her surrender, but which Roosevelt and Marshall feared would divert resources from Overlord. Brooke, meanwhile, thought that the campaign to take Rome needed a major concentration of effort. It started a month-long argument between Churchill and the others over the recapture of Rhodes, in which Churchill recognized that he was in a minority of one.

In the first fortnight of September, after Quadrant, Churchill went to Washington, Boston and Hyde Park. Brooke was spared this trip and returned to London. Eden thought Roosevelt's 'determination not to agree to a London meeting for any purpose, which he says is for electoral reasons, is almost insulting considering the number of times we have been to Washington ... We are giving the impression, which they are only too ready by nature to endorse, that militarily all the achievements are theirs and Winston, by prolonging his stay in Washington, strengthens that impression.'[24] By early October Churchill had to accept that his scheme to attack in the Dodecanese – Rhodes, Leros, Samos and Cos – had failed to win over anyone else.

Once back in Britain, Churchill wired the President asking for Marshall to come out to the Mediterranean for a conference in Tunis to settle the matter, hoping 'in his heart' to be able to swing the meeting by his force of personality, as he had convinced himself he had done at Algiers. Roosevelt replied: 'Frankly I am not in sympathy with this procedure under the circumstances. It seems to me the issue under discussion can best be adjusted by us through our Combined Chiefs of Staff set up in better perspective than by the method you propose.'[25] Brooke thought this a 'very cold reply', but Churchill wired back again asking Roosevelt to reconsider, leaving Brooke to comment: 'The whole thing is sheer madness, and he is placing himself quite unnecessarily in a very false position! The Americans are already desperately suspicious of him, and this will make matters far worse.' Marshall flatly refused to allow any troops to be spared for the Dodecanese operations, and this largely unnecessary row, brought on by Churchill, was perhaps the first clear indication that the Americans were beginning to recognize what had been true since Brooke was denied the post of supreme commander for Overlord, that they had become the senior partner in the relationship.[26]

The Prime Minister had not helped his own case by including in his original telegram to Roosevelt the sentence 'Even if the landing craft and assault ships on the scale of a division were withheld from the buildup of Overlord for a few weeks without altering the zero date it would be worthwhile.'[27] Over Rhodes, Marshall recalled telling Churchill, using a profanity that was all the more powerful because it never usually passed his lips: 'God forbid if I should try to dictate, but not one American soldier is going to die on that goddamned island!'[28]

'I doubt if I did anything better in the war', stated Marshall in 1956, 'than to keep him on the main point. I was furious when he tried to push us further into the Mediterranean.' He remembered many 'hectic scenes' with Churchill over the Dodecanese, recalling that the Prime Minister 'Could be strong and loud. Churchill, however, once he accepted a point, would not hold it against him. Would put his arms around him.' Marshall also appreciated the British Chiefs of Staff's position over that issue, which he described as 'extraordinarily difficult. Mr Churchill was very intense when he got a certain idea and he did business with them every day, where sometimes I didn't see the President for a month.'[29] As well as being a strength, of course, this could also be a weakness, because, as Marshall admitted, he was still fearful – though less so now – that the President might be 'inveigled' or 'palavered' by Churchill into 'side shots'.

In early drafts of his war memoirs, Churchill was bitter about the Americans baulking his Dodecanese enterprise, arguing that, 'but for pedantic denials in the minor sphere', the Allies could have controlled the Aegean. From there he had hoped to strike north, and pressurize Turkey into entering the war. Chapter 12 of *Closing the Ring* is entitled 'Island Prizes Lost', and he blamed the 'prejudice' of the President's advisers, though not Marshall by name. He also had some criticisms of Eisenhower, and Roosevelt was described as 'ungenerous'. All these were removed before publication, however, probably because Ike was supreme commander of NATO's Allied Command Europe at the time. Perhaps Churchill used the early proofs of his book as a safety-valve, rather as Brooke used his diary and Roosevelt his draft telegrams, because throughout the extensive redrafting process sharp comments about individuals were constantly toned down, often on the advice of General Pownall, Norman Brook, William Deakin and other members of his 'Syndicate' of drafters and fact-checkers. What Churchill also omitted was that Brooke had been just as resolutely opposed as the Americans to what the CIGS described in his diary as 'Rhodes madness'.[30]

Even though the Eighth Army took Taranto on 10 September and there were four divisions ashore at Salerno by 13 September, with the 7th Armoured Division arriving that day, Alexander, who as commander-in-chief of Allied forces in Italy was Clark's superior officer, was sending 'somewhat gloomy' telegrams stating that the Salerno bridgehead saw 'the Germans concentrating faster than we can'. The War Office were ready with the blame. 'Eisenhower never produced a proper appreciation,' wrote Kennedy; 'one doubts if he even made one. It is very hard to understand why he cannot deploy more of his colossal forces. It all seems to indicate rather a muddle at Algiers.' The speed with which the Planners got these excuses arranged would have impressed a politician.

'The only good feature so far is that it will bring home to the Americans a truer sense of the realities of these landing operations against Germans,' Kennedy continued, 'which should have a wholesome influence on their future plans and make our task of rubbing it into them less thankless. One must admire their drive and fearlessness in planning, but without due caution as well, disasters are bound to happen.' Partly because of the two battleships from Malta, disaster did not in fact happen, but it is easy to guess the extent of the mutual Anglo-American recrimination that would have exploded if it had. But a few days later Kennedy was even prompted to observe: 'If we had a walkover at Salerno we should almost certainly have had a bigger disaster later on. This is all very salutary.'[31]

Press rumours that Marshall was about to be promoted emerged in early September, with opposition to such a plan emerging from various quarters, all of them flattering to him. Some papers said that he was too precious to be wasted on a theatre commandership, others that Roosevelt was trying to remove the one strong man capable of preventing the reorganization of the Army Service Forces, for various supposedly nefarious reasons of his own. Three ranking senators on the Senate military affairs committee had to be personally placated by Stimson over this. On 16 September, Pershing himself wrote to the President from his hospital bed opposing Marshall's transfer, describing his 'deep conviction' that it would be 'a fundamental and very grave error in our military policy' to lose someone so talented from the General Staff, because in a global war the most accomplished officer should be the Chief of Staff.[32]

If Marshall had asked for the job, he could have been supreme commander of Overlord, which after all won Eisenhower the presidency in 1952. Since Marshall served as US secretary of state after the war, it is not inconceivable that he might have run for the White House had he commanded the Normandy invasion. He is famous today, of course, but primarily because of his post-war European reconstruction plan. Just like Brooke, had he become supreme commander in the autumn or winter of 1943, Marshall would now be seen as one of the greatest generals of history, and the books, statues, aircraft carriers and street names presently dedicated to Eisenhower would today bear a different name.

Did Marshall glimpse any of this back in 1943? Whether he did or not, he put what he perceived to be his duty before hopes of lasting global glory, just as Brooke had in the Western Desert in August 1942. Eisenhower, a Marshall appointee and acolyte, had no hope of competing for the job against his mentor. He told his naval aide and confidant, Captain Harry Butcher, that 'while his record as Allied Commander in North Africa would cause his name to be considered, he doubted if anyone except Generals Marshall and Brooke could be assigned.' He also claimed that he liked his 'semi-independence' in Algiers, and 'didn't relish' moving to a headquarters in London or Washington where he would be subjected to much more political interference.[33]

The War Cabinet of 20 September had to consider what would happen to the five battleships, eight cruisers, eight destroyers and twenty-one submarines of the Italian Navy that had surrendered to date. If there was any dispute with the Americans over which navy got which ships, Churchill thought, they should be divided equally. 'What right does Italy have to pretend to be a modern power that has a future?' he mused. Of the Italian ships that had already taken refuge in Fascist Spain, Churchill said there must be 'no nonsense about that'.[34] Very often, as on this occasion, Brooke gave long and detailed analyses of the global position on all fronts with no comment at all from Churchill. For the most part they worked closely with one another over the three-and-a-half years they were together; anything else would have been insupportable over that period of time.

On 28 September, Marshall reported in almost lyrical prose to Roosevelt all the benefits that had already accrued from the Italian campaign:

The fall of Foggia has come exactly at the time when it is needed to complement our bomber offensive now hammering Germany from bases in the UK. As winter sets in over northern Europe, our heavy bombers operating from the dozen or more (13) air bases in the Foggia Area will strike again and again at the heart of German production not only in Germany proper but in Austria, Hungary and Rumania. For our bombers operating from England, this aerial 'Second Front' will be a great assistance.[35]

It was typical of Marshall's exactitude, even in an almost elegiac letter such as this, that he should still need to give the exact number of air bases that justified his phrase 'dozen or more'. He went on to say that the Germans now had to provide 1,200 further miles of air defences, and 'In a matter of days we will be in a position to strike into the soft side of Germany.' This was the closest he got to accepting Churchill's stance that Europe had a 'soft' side.

That same day, on the other side of the Atlantic, Churchill was 'shouting a lot' at a Defence Committee meeting on the Far Eastern situation. 'Brookie obviously anxious lest too much were diverted from Mediterranean,' noted Eden.[36] After it, Eden, Attlee and Mountbatten stayed up until 1.30 a.m. discussing with Churchill whether Admiral Sir Andrew Cunningham should replace Pound as first sea lord. On 8 September, Pound had visited Churchill in his big bed–sitting room in the White House to say that he had suffered a stroke and his right side was largely paralysed. 'I thought it would pass off, but it gets worse every day and I am no longer fit for duty.' When he signed his resignation letter back at the Admiralty, his secretary had to guide his hand. In his memoirs, Churchill claimed that, after Admiral Sir Bruce Fraser had refused the promotion, 'the obvious choice' had been Cunningham. In fact, however, Churchill put up a series of objections to the man who had stood up to him over Greece and Crete in 1941 and he only accepted Cunningham – who was indeed the obvious choice – with some reluctance. 'Winston was loud in cries that he would not appoint any First Sea Lord who didn't accept his Far Eastern plans,' wrote Eden.[37]

At the end of the long session, Churchill joked that 'he would end by killing us all by these late hours', to which Eden laconically added in his diary, 'which may well be true'. Three days earlier, Eden had worked hard on Saturday morning, then lunched with the Churchills and motored down to Cranborne. In his own words he was 'Very tired after dinner and suffered from a complete "black out" which happily nobody noticed.' The following year, when he slept badly, he 'had to use heavy ration of dope to produce any effect'.[38] A week later he was taking a pink pill to help him sleep. The physical and psychological pressures on the senior wartime decision-makers need to be borne in mind when considering their occasional bouts of furious ill-temper with one another.

The King's Dragoon Guards took Naples on 1 October 1943. Dill wrote from Washington that day to congratulate Brooke on the Italian successes. 'I don't believe it was ever possible to make the Americans more Mediterranean-minded than they are today,' he claimed, saying the Joint Chiefs of Staff 'have given way to our views a thousand times more than we have given way to them'. It was true, but it was all about to change. When Alexander wrote to Brooke on 17 October asking for landing craft to put ashore a force further north on the west coast of Italy, behind the German lines, Brooke replied that he 'feared that any opinion from our side would be suspected by the Americans who seemed to have an ineradicable impression that our hearts were not in Overlord and that we took every opportunity of directing to the Mediterranean resources which they considered should be concentrated in Great Britain'.[39] He therefore suggested that it would be better to go via Eisenhower and the Combined Chiefs of Staff than direct to him.

That same day Kennedy made a diary note which explodes Brooke's later claims that he had no interest in operations in the Balkans, an issue that was to become the next major bone of contention between the Masters and Commanders. 'There is still a very distinct cleavage of opinion between us and the Americans as to the correct strategy in Europe,' wrote Brooke's colleague, closest aide, Director of Military Operations and fellow bird-watcher.

CIGS feels very strongly that we should exploit the openings in the Med and extend the range of our offensive operations to the Aegean and Balkans. The Germans are sitting on a volcano in the Balkans . . . The PM has come round to this point of view too, and has just said that he would like to tackle the Americans again upon it. But I must say I see no chance of converting them – especially in view of Marshall's impending appointment for Overlord.

Just as in the Pentagon, the assumption until December 1943 at the War Office was that the post would go to Marshall.

Far from his holding Churchill back from the Aegean and Balkans, therefore, Kennedy believed – and was in a prime position to know – that Brooke was in fact building Churchill up to believe that the Mediterranean strategy might have much further to go northwards and eastwards. Unlike Churchill, however, Brooke was always intensely conscious of the importance of not disclosing future stages of his strategic ambitions to the (understandably) suspicious Americans.

Where Churchill, Roosevelt and Brooke were wrong, and Marshall right, was in rating highly the chances of bringing Turkey into the war against Germany. The British calculated that this was possible, even likely, and Roosevelt wanted to plan for it too, but Marshall never thought it likely, and did not want to design strategy around such an eventuality. In March 1943, Churchill said to Brooke of the Turks: 'We must start by treating them purry-purry puss-puss, then later we shall harden!'[40] Yet the Allies had no opportunity of doing either, as Marshall was correct: the Turks felt no inclination to declare war on a power as geographically close and still capable of lashing out as viciously as Nazi Germany. If purry-purry puss-puss did not work, neither did threats; in

November Eden had asked Churchill what he should tell Turkey to try to coerce her into the Allied camp. Churchill replied: 'Tell Turkey Christmas is coming.'[41]

There was only one occasion during the war when Brooke's commitment even to an eventual Operation Overlord in 1944 seems momentarily to have wavered. Having received a note from Churchill at the morning Chiefs of Staff meeting on Tuesday 19 October, expressing the wish 'to swing round the strategy back to the Mediterranean at the expense of the cross Channel operation', Brooke noted: 'I am in many ways entirely with him, but God knows where that may lead us to as regards clashes with Americans . . . I shudder at the thought of another meeting with the American Chiefs of Staff, and wonder whether I can face up to the strain of it.' That evening at 10.30 there was another important meeting to discuss Overlord. As well as the Chiefs of Staff, Churchill had brought in Smuts, Attlee, Cadogan, Lyttelton and Leathers, a sure sign that he wanted political muscle behind him for a major démarche.

The Prime Minister criticized the May 1944 target date for Overlord; Cadogan suggested 'stirring up action in the Balkans' instead; Brooke complained of fighting a war based on 'lawyers' contracts' (a reference to returning the seven divisions); Smuts spoke of 'a clear run in to victory' in the Mediterranean, at which Portal warned that the Americans would instead divert larger forces to the Pacific. The Prime Minister said he would be willing to risk that, and summed up in Utopian terms: 'if we were in a position to decide the future strategy of the war' the Americans should agree to 'reinforce the Italian theatre to the full', 'enter the Balkans', 'hold our position in the Aegean Islands', intensify air attacks on Germany, and build up US forces in Britain for an operation which 'might not occur until after the spring of 1944'. Churchill therefore called for another full-scale Combined Chiefs of Staff conference in early November to try to sell this strategy to Roosevelt and Marshall.

Brooke put his name to this wish-list, knowing that that was all it could be, since Britain was no longer 'in a position to decide the future strategy' almost unilaterally. Roosevelt and Marshall were certainly not about to sign up for the first three desiderata, suspecting that it was merely yet another way of postponing the last. So the meeting served no useful purpose other than blowing off prime ministerial steam. It was not mentioned at all in Churchill's war memoirs, probably because he did not want readers to appreciate how doubtful he still was about Overlord.[42] Yet he was, and so – at least on this occasion – was Brooke. It was during the month of October 1943 that Brooke's commitment to a spring 1944 Overlord, which had always been genuine, contrary to US suspicions, was seriously questioned for the first and last time. It had probably been the correct decision not to appoint him as its supreme commander after all.

On 21 October 1943 – appropriately enough, Trafalgar Day – Sir Dudley Pound died in a London hospital. He had worked himself into an early grave. Commander Thompson recalled how deeply 'distressed' Churchill was by the death of one of his few personal friends in the higher command of the war.[43] With Smuts often abroad and Beaverbrook in the Government but not the War Cabinet, Churchill had plenty of colleagues to work with, but few close friends. Brooke and Portal acted as pall-bearers at Pound's funeral at Westminster Abbey, another was Pound's successor, Cunningham, a man Jock Colville believed to be 'impervious to Churchill's spell'.[44] A true naval hero (his biographer subtitled his book 'The greatest admiral since Nelson'), Sir Andrew Browne Cunningham has been described as 'a smallish man with spark-ling blue eyes, not particularly robust physically, but with a will of iron and no respecter of persons'.[45] Or prime ministers.

'A.B.C.' had entered the Navy in 1897, serving ashore in the Boer War before specializing in destroyers. He was commander-in-chief Medi-terranean when he sank three 10,000-ton cruisers in the close-range night action off Cape Matapan in March 1941, without loss to the Royal Navy, and when, two months later, he accepted sinkings, against the General Staff's advice, in order to bring twenty-two thousand troops off Crete rather than leave them to be captured. 'You have said, General, that it will take three years to build a new Fleet,' he had told Wavell on that occasion. 'I will tell you that it will take three hundred years to build a new tradition.'[46] He also kept Malta supplied by sea, despite heavy Luftwaffe bombing. Cunningham had been Allied naval com-mander for both Operations Torch and Husky. Eisenhower described him as vigorous, hardy and straightforward: 'A real sea-dog'. The Navy meant everything to him. 'He appears to have had no inner life and no intellectual interests,' writes one historian. 'His idea of a good time was to lie on his back and throw ping-pong balls into a chandelier.'[47] (Which doesn't sound like such bad fun, in fact.)

It took a man like Cunningham, who had crippled the Italian fleet at Taranto harbour with Swordfish torpedo bombers in 1940, putting three Italian battleships out of action, to stand up to Churchill's constant demands for daring action. Yet if anyone deserved Churchill's cruel jibe about the Chiefs of Staff Committee demonstrating 'the sum total of their fears', it was he. 'In unexpected contrast to his powerful leadership at sea,' admits his biographer,

A.B.C. himself was curiously passive as First Sea Lord. In the Mediterranean he had been master of men and events. He made things happen. In Whitehall, he let things happen to him. By his own testimony there were Chiefs of Staff meetings at which he made no contribution, or remained neutral on a particular subject because in his opinion it did not concern the Navy.[48]

*

After Stalin had agreed to meet the other two members of what was to be called the 'Big Three' at Teheran in late November 1943, Churchill telegraphed Roosevelt on 23 October begging for another Anglo-American meeting soon, saying that they ought not to meet the Russians without first agreeing about future Anglo-American operations. He pointed out that by their Teheran meeting it would be ninety days since Quadrant, and in that time Italy had surrendered and been invaded 'successfully', and the Germans were gathering twenty-five divisions there. 'All these are new facts,' he concluded. He went on to state quite bluntly that 'Our present plans for 1944 seem open to very grave defects.' Putting fifteen American and twelve British divisions into Overlord while there were six American and twelve British and Commonwealth divisions in Italy meant that 'Hitler, lying in the centre of the best communications in the world, can concentrate at least forty or fifty divisions against either of these forces while holding the other . . . without necessarily weakening his Russian front.' (This was an exaggeration; Hitler had indeed weakened his Russian front in order to defend central Italy.) Instead of strategic need dictating the disposition of forces between the Mediterranean and the Channel, Churchill alleged, it was down to 'arbitrary compromises', by which he meant the seven-divisions deal. 'The date of Overlord itself', he pointed out, 'was gained by splitting the difference between the American and British view.' This telegram merely confirmed the suspicions of Americans such as Marshall and Hopkins that the Prime Minister had gone cold on the whole Overlord project once again.

Churchill went on to state that the date was looming when decisions over Overlord's landing craft needed to be made, which if they were taken from Alexander 'will cripple Mediterranean operations without the said craft influencing events elsewhere for many months'. Ditto the 50th and 51st Divisions then in Sicily, which were close to the Italian battle but would be out of action for seven months if they were transferred to Overlord instead. Speaking for Brooke and himself, Churchill wrote: 'We stand by what was agreed at Quadrant but we do not feel that such agreements should be interpreted rigidly and without review in the swiftly changing situations of war.'[49] There was only one way to interpret it: Brooke – many of whose favourite phrases appeared in the telegram – wanted to renege on the promise to withdraw the seven divisions, whatever effect that would have on the May 1944 date for Overlord.

Quoting the German general Wilhelm von Thoma, whom secret microphones had picked up telling fellow POWs in Hertfordshire, 'Our only hope is that they come where we can use the Army on them,' Churchill said he didn't doubt that Overlord would be successful in getting ashore and deploying, but he was 'deeply concerned' with what would happen between the thirtieth and sixtieth days thereafter. 'I have the greatest confidence in General Marshall and . . . if he is in charge of Overlord we British will aid him with every scrap of life and strength we have,' Churchill claimed, before imploring Roosevelt, 'My dear friend, this is much the greatest thing we have ever attempted, and I am not satisfied that we have yet taken the measures necessary to give it the best chance of success. I feel very much in the dark at present, and unable to think or act in the forward manner which is needed. For these reasons I desire an early conference.' He suggested 15 November 'at latest'. Roosevelt, who was suffering from influenza, replied on 25 October, suggesting a meeting, with small staffs, at the Pyramids.

Simultaneously Churchill telegraphed Marshall: 'Naturally I feel in my marrow the withdrawal of our 50th and 51st Divisions, our best, from the very edge of the battle of Rome in the interests of distant Overlord. We are carrying out our contract, but I pray God it does not cost us dear. I do hope to hear of your appointment soon. You know I will back you through thick and thin and make your path here smooth.'[50] Meanwhile Brooke was furious that operations in Italy seemed to be 'coming to a standstill'. He did not credit Kesselring's fine resistance, but blamed Marshall's 'insistence to abandon the Mediterranean operations for the very problematical cross Channel operations'.[51]

Brooke feared a major German counter-offensive on the Eastern Front unless the Russians were successful – which they were being, breaking into the Nogaisk Steppes on 27 October – and complained that the Allied build-up in Italy was both slower than the German and far slower than he had expected. 'We shall have an almighty row with the Americans who have put us in this position,' he predicted, and then wrote sarcastically: 'We are now beginning to see the full beauty of the Marshall strategy!! It is quite heartbreaking when we see what we might have done this year if our strategy had not been distorted by the Americans.'[52] Rarely was his diary safety-valve used to greater effect.

In fact the Americans were not insisting on 'abandoning' the Mediterranean. All that they had insisted on at Quadrant was sticking to the Trident agreement that seven divisions should be withdrawn from the Mediterranean for Overlord by 1 November. Now, within a week of that date, the British were frantically attempting to renege. Rather than 'distorting' the British strategy, the Americans had gone along with it to a remarkable extent, and the fruits were evident: the defeat of Italy and diversion of many German divisions southwards. Brooke was thus not only indulging in hyperbole, but attempting to turn what was originally intended to be a subsidiary front into a major one.[53] For all that the Third Reich could be harried from the south, it could only be killed stone dead from the west, by taking the Ruhr, and the east, by taking Berlin. With the time for that approaching, the seven divisions were needed more in Britain than in Italy.

Yet senior British figures could not appreciate that at the time. If Alexander's pessimistic prognostications about the Germans massing to crush him in Italy proved accurate, thought Alec Cadogan, 'all this "Overlord" folly must be thrown "Overboard".' After Alexander's doleful report, Cadogan recorded that the whole War Cabinet were 'definite' that Britain could not be tied to the 1 May 1944 timetable for Overlord, and 'Winston will fight for "nourishing the battle" in Italy and, if necessary, resign on it.'[54] Of course Churchill's resignation was unthinkable, and was merely a negotiating tool or rhetorical device. Michael

Howard has written of Churchill and the Italian campaign, 'over whose destinies he brooded with such possessive passion', that this was a case where emotion took over in strategy-making. Churchill spoke of the Italian campaign, where there were twelve British and Commonwealth divisions under their own commanders Alexander and Montgomery, in a more personal and emotional way than he ever did of the far larger Allied campaigns in France and Germany the following year. Howard rightly puts this down to 'sheer chauvinism'.[55]

On 26 October, the Joint Chiefs of Staff began selecting attendees for the coming Cairo Conference (codenamed Sextant) and were not going to make the same mistake that they had at Casablanca. Seventy-five officers and thirty warrant officers and enlisted men were chosen for the military delegation, with twenty more officers coming in from other theatres, not including further signals, medical and supply personnel. Furthermore Eisenhower, Stilwell and General Claire L. Chennault, the USAAF commander in China, would also be attending along with their Planning Staffs. Nor did that include the political and diplomatic group personally attending the President. The Americans would arrive at the Pyramids inundated with Staff, ready to deal with anything the British might spring on them.

'I feel that Eisenhower and Alexander must have what they need to win the battle in Italy,' Churchill wrote to Roosevelt about the importance of taking Rome, 'no matter what effect is produced on subsequent operations.' On 27 October Roosevelt drafted a reply to Churchill that he never sent, but a copy of which he passed to Marshall, stating that preparations for Overlord – which he was 'anxious' should take place on 1 May 1944 – 'seem to have reached a stage from which progress difficult unless and until [the] Commander is appointed. As you know [I] cannot make Marshall available immediately.' FDR suggested that Churchill 'may care to consider early appointment of British Deputy Supreme Commander', before going on to suggest Brooke, Dill or Portal.[56]

Brooke would not have entertained the idea after having been turned down for the top job, and it anyway seems slightly absurd that Roosevelt should have encouraged the appointment of a deputy before he himself got round to choosing the supreme commander, which was perhaps why the message was never sent. After the war, Lieutenant-General Morgan told Pogue that the President had revealed to him in November 1943 that he needed an Army chief of staff 'who could handle MacArthur'. Eisenhower had served on MacArthur's staff for nearly nine years in the 1930s, and might therefore have adopted a deferential attitude, whereas the President knew that Marshall could handle the difficult, occasionally irascible but undeniably brilliant MacArthur, whom Marshall had never liked. That might have been the real reason why Eisenhower became supreme commander instead of taking Marshall's place as US Army chief of staff. Another possibility is that health reasons intervened; in June 1954 Marshall told Moran that his heartbeat had been 'all over the place' at Quadrant.[57]

'During the past week,' Kennedy confided to his journal on 30 October, 'matters have moved steadily to a head with regard to the divergence of view between us and the Americans on future strategy.' Alexander had sent Eisenhower his appreciation of how his advance on Rome had slowed up for lack of landing craft to attack round the German flanks, and asking for more troops. (Very few generals ever write to their superiors requesting fewer.) The problem was, in Kennedy's words, that:

if we give Alexander what he wants, and if we allot further resources for operations in the Aegean and Balkans as we should do to take full advantage of the situation, Overlord must perforce be postponed. The American Chiefs take the view that this is a breach of contract and almost dishonourable. The impasse arises from a fundamental difference in the British and American points of view as to what is possible in a combined operation.[58]

Kennedy believed that history proved that combined operations often failed through 'slowness of "buildup" on the part of the attacker'. Well-planned and well-executed plans generally succeeded, he believed, citing Gallipoli as an example of a plan that was badly executed. (In fact Gallipoli had been executed as well as it could have been; it was the plan to send thousands of men again and again up precipitous slopes over a rocky peninsula far from Constantinople in the teeth of well-entrenched and dogged Turkish resistance over eight months that had been the real problem.)

'We have never yet carried out a successful combined operation in the face of strong opposition *on the beaches*,' wrote Kennedy. 'Sicily and Salerno were *not* strongly opposed. In France we may expect to meet such opposition.' The Americans had a faulty appreciation of the problems involved, he believed, and instanced Eisenhower's steep learning curve once he had arrived in North Africa, and the way that he had declared the Sicilian landings 'impossible' if the Germans had more than two divisions on the island. American reliance on overwhelming air power in breaking up enemy concentrations was, in Kennedy's view, 'illusory', since 'Air power has never yet in this war interfered seriously with the Germans' power to move their armies.' Despite the Allies' overwhelming domination of the air over Sicily, two German divisions had managed to escape, but Kennedy was wrong to deprecate the importance of air power, which was indeed seriously to interdict the Germans' capacity to counter-attack at Normandy. 'In the end,' he predicted, 'I suppose we shall probably go into France with little opposition and then the historians will say that we missed glorious opportunities a year earlier, etc, etc. It will be the Easterners and the Westerners as in the last war but with locations reversed!'[59] (Kennedy was referring to the controversy within the British High Command in 1915 over Gallipoli versus the Western Front.)

The prospect of yet another Combined Chiefs of Staff conference filled Brooke with weariness and dread, as usual. 'The stink of the last one is not yet out of my nostrils!' he wrote on 1 November. 'I now unfortunately know the limitations of Marshall's brain and the impossibility of ever making him realize any strategical situation or its requirements. In strategy I doubt if he can ever, ever see the end of his nose.'[60] Brooke went on to blame himself for his failure to make the Joint Chiefs of Staff 'see daylight', which was really a disparagement of them, posing unconvincingly as self-criticism. Had the Americans gone along with the Dodecanese projects, 'We should have been in a position to force the Dardanelles by the capture of Crete and Rhodes, we should have the whole Balkans ablaze by now, and the war might have been finished in 1943!!' Instead of that golden prospect, however, Brooke believed that to satisfy 'American shortsightedness' the seven divisions had to be sent off 'for a nebulous 2nd Front' which has 'emasculated our offensive strategy!! It is heartbreaking.' Did Brooke really mean all this, since he had not supported the capture of Crete or Rhodes – let alone the Dardanelles – at the time, and certainly not in the Chiefs of Staff Committee? He was also writing on 1 November, the date the divisions were due to be withdrawn, yet five of the seven were still fighting in Italy.

Brooke's outburst – written after the strain of a long Chiefs of Staff meeting and a Cabinet meeting and before yet another night meeting, of which he was 'sick unto death' – was, as Michael Howard points out, 'neither accurate nor fair'.[61] In fact, the aims for the Mediterranean theatre agreed in the three 1943 conferences had all been achieved by the end of the year, including clearing the sea for Allied shipping and tipping the anti-U-boat campaign in the Allies' favour. Around forty-five German divisions were now operating in Italy and the Balkan peninsula; Allied planes from Foggia were bombing central and south-eastern European targets nightly, and Yugoslavian partisans were on their way towards receiving huge amounts of supplies to tie down almost as large a number of German divisions as those operating in Italy. That was about as much as the Mediterranean campaign could ever realistically have achieved.

The Americans – who were indeed probably wrong to consider invading France in 1942 or 1943 – were right to insist on doing so by the late spring of 1944. Furthermore the British, who were right to call for the invasion of Italy up to Naples in 1943, were similarly wrong to continue the campaign up to Rome, let alone beyond. The Allies had already achieved their aim of luring large German reinforcements into Italy; they did not then need to play to the Germans' defensive strengths by trying to break through the Gustav Line and subsequently the Gothic Line much further north. There had to be a Salerno to take Naples, but once the Foggia airstrips were secured there did not need to be an Anzio landing to help take Rome, nor the many costly assaults on Monte Cassino. If a moment needs to be pinpointed when the British started to get strategy wrong, and the Americans started to get it right, it was in mid-October 1943 when Churchill successfully persuaded Brooke to try to postpone Overlord. Brooke seems to have overcome his month of autumn madness, however, and never returned to the supposed attractions of attacking Rhodes and Crete or 'forcing' the Dardanelles. Indeed he later came to deny they ever existed.

In late October 1943 Dill wrote to Marshall and King suggesting that the British and Americans should each have an officer attached to the other's Planning Staff. Predictably, King opposed this, so Marshall wrote him a memorandum saying, 'It seems to me that Dill's proposal is sound unless we assume an attitude of suspicion in relation to this matter. We have to work with these people and the closer the better, with fewer misunderstandings I am certain.' He instanced the success of having General Morgan at COSSAC. Morgan had been working in Washington since Quadrant, and Marshall believed he had gone so native that 'it may be embarrassing in his relation to the British Chiefs of Staff in London.'[62]

Marshall then pointed out that, when it came to 'the other side' – meaning Britain, not Germany – 'We are fighting battles all the time, notably in regard to the Balkans and other places, and the more frankness there is in the business at the lower level the better off I believe we are; particularly because it seems to me in a majority of cases the younger elements on the British side favour our conceptions rather than those of the Prime Minister.' King once again replied negatively, arguing that such an arrangement 'would permit us no privacy in the consideration of problems which are purely those of the United States'. Marshall forwarded this reply to Dill, and the project was stillborn.

Just as some Americans retained their suspicion of Britons, so some Britons preserved their equally unjustified sense of superiority over the Americans. After dinner at 10 Downing Street with the King and the Chiefs of Staff on 3 November, Sir Alan Lascelles reported that the Chiefs':

great problem at the moment is to teach the Americans that you cannot run a war by making rigid 'lawyers' agreements' to carry out preconceived strategic operations at a given date (i.e. 'Overlord'), but that you must plan your campaign elastically and be prepared to adapt it to the tactical exigencies of the moment. They don't seem to grasp that a paper-undertaking made in the autumn to invade Europe (or any other continent) in the following spring may have to be modified in accordance with what the enemy does or does not do in the intervening winter.[63]

Since these are all views and phrases that crop up in Brooke's diaries, memoranda and conversations of the time, it is not difficult to work out Lascelles' *placement* at dinner that right. Needless to say, had the Americans twice promised something in writing that Brooke still very much wanted, and had they threatened to renege on it, Brooke would probably not have been so aristocratically disdaining of 'lawyers' contracts'. Marshall did not want the seven divisions out of pedantic literalism but because he felt they would do more good in France than in Italy. And by then he was right.

A proposal from Admiral Leahy that Marshall should take over both the Mediterranean and the Western Fronts caused consternation in London on 8 November. Dill warned Brooke who warned Churchill, and, according to Brooke's report to Kennedy, the Prime Minister threatened that he would not agree to it 'while he remained in office'. Kennedy, who had only that week been made assistant CIGS, with control over the Operations, Intelligence, Plans and Maps departments, saw it as an attempt to stymie the Mediterranean campaign in favour of Overlord. As it turned out, the following day Marshall contacted Churchill and Brooke via Dill to say that Leahy's proposal had come from Roosevelt, but it 'had never received any proper consideration by the US Chiefs of Staff. I do not think you need to take it too seriously.'[64] It was a relief to the British to hear that Marshall was not interested, but not that Leahy had been encouraged by the President.

With only days to go before the Cairo Conference began, the British crystallized their ideas about what they wanted out of it. The main desiderata would be to continue the offensive in Italy, to increase the flow of supplies to the partisans in the Balkans, to try to induce the Balkan powers to break away from Germany, to induce Turkey to enter the war, and to accept a postponement of Overlord. Of these five British hopes, only the first two were adopted in the short to medium term, far below Brooke's normal hit-rate. For by the time of the Cairo Conference, the balance of power had decisively shifted to the Americans, and it was never to shift back. In terms of economic might, industrial and military production, troops under arms, sheer numbers of ships and aircraft, and almost every single other measurable criterion of power, the United States had comprehensively overtaken Great Britain by November 1943, and was also overtaking the British Commonwealth as a whole.

16
Eureka! at Teheran:
'I wish he had socked him'
November–December 1943

*You can use a brilliant but lazy man as a strategist, a brilliant but energetic man
as a Chief of Staff, but God help you with a dumb but energetic man!*
General Henry 'Hap' Arnold[1]

'We must not regard Overlord on a fixed date as the pivot of our whole
strategy on which all else turns,' wrote Brooke in a memorandum for
Churchill before the Cairo Conference. 'We should stretch the German
forces to the utmost by threatening as many of their vital interests and
areas as possible and, holding them thus, we should attack wherever we
can do so in superior force.'[2] Brooke suggested advancing beyond Rome
to the Pisa–Rimini Line, intensifying aid to the Balkan partisans,
pressing Turkey to open the Dardanelles and doing everything possible
to 'promote a state of chaos and destruction in the satellite Balkan
countries'. Yet, as was now very apparent, areas east of the Adriatic
'were regarded by American strategists with something akin to the
superstitious dread with which medieval mariners once contemplated
the unknown monster-infested reaches of the Western Ocean'.[3] Marshall
was foremost among these, and with Overlord slated for less than six
months away, only one strategy could prevail.

On Monday 15 November 1943 Roosevelt held a meeting at Shangri-La
attended by Harry Hopkins and the Joint Chiefs of Staff, at which he
indicated the very firm stance he intended to adopt towards Churchill
and Brooke at the coming Cairo (Sextant) and especially the Teheran
(Eureka) conferences, the second of which Stalin was also going to
attend. The Chinese generalissimo Chiang Kai-shek would also be
coming to Cairo, and Far Eastern strategy was another area at which
the Americans expected clashes with the British.

Although 'the British wanted to build up France into a first-class

Power, which would be on the British side,' the President said that 'It was his opinion that France would certainly not again become a first-class Power for at least twenty-five years.' To a Joint Chiefs of Staff memorandum opposing involvement in the Balkans strategy, all FDR said was 'Amen', adding that the United States should 'definitely' take a stand against it at the earliest opportunity. The minutes also record that Roosevelt 'said it was his idea that General Marshall should be the commander-in-chief against Germany and command all the British, French, Italian and US troops involved in this effort'.[4] It is evident from his mention of the Italians – who had no part to play in Overlord – that he meant Marshall to accept an inclusive European command that encompassed the Mediterranean theatre. Given that Marshall could be guaranteed to push for Overlord rather than Italy were he to assume the overarching supreme command Roosevelt wanted, and given that the previous month Churchill had threatened to resign if the Italian battle were not 'nourished', trouble clearly loomed.

Over the future of King Victor Emmanuel III of Italy, Roosevelt told his advisers: 'The British are definitely monarchists and want to keep kings on their thrones. They are monarchist-minded,' whereas 'We would like to get the King out.' Although Leahy and Admiral King could not see what difference it made, and thought that the situation would 'solve itself', the President insisted on a national plebiscite on the issue. Whether this was more for ideological republicanism, to punish Victor Emmanuel for waiting so long before acting against Mussolini, or to irritate the British, must be open to doubt. Finally he said that he would like his meeting with Chiang Kai-shek 'to be separate from and precede any meeting with the British'. Churchill's telegram resiling from the May date for Overlord meant that the Anglo-American relationship was in for a testing time the following week.

'The coming conference will be a difficult one,' judged Kennedy, accurately.

The Americans seem to think we have acted in an almost underhand way over the Mediterranean and have been guilty of unilateral action to implement our belief that the Mediterranean should have priority over Overlord, in spite of signed agreements in the contrary sense. This is curious because we have felt exactly the same about them. CIGS feels that the war may have been lengthened by the American failure to realize the value of exploiting the whole

Mediterranean situation and of supporting Turkey strongly enough to bring her into the war.

Kennedy concluded that 'The time has now come for plain speaking on both sides.'⁵ In fact, as we have seen, there had already been a good deal of plain speaking up until then. Cairo was to hear much more.

Just as Roosevelt was gradually becoming less enamoured of Churchill, so Churchill was finding an anti-American streak in himself that had not been apparent since the naval disputes of the mid-1920s. A new tone of asperity had entered the 'Former Naval Person' correspondence on both sides, and on the evening of 18 November, when the Chiefs of Staff met around the Prime Minister's bed in Malta after his nap, Churchill 'gave long tirade on evils of Americans and of our losses in the Aegean and Dalmatian coast'. Of course by losses he meant missed opportunities, and Brooke was worried that the line the Prime Minister was intending to pursue at the conference would be 'all right if you won't play with us in the Mediterranean we won't play with you in the English Channel.' After the war Brooke put what he called Churchill's 'new feelings of spitefulness' down to the fact that 'the strength of the American forces were now building up fast and exceeding ours. He hated having to give up the position of the dominant partner which we had held at the start. As a result he became inclined at times to put up strategic proposals which he knew were unsound purely to spite the Americans. He was in fact aiming at "cutting off his nose to spite his face".'⁶

Brooke believed that the primary attraction for Churchill of an Austrian or Balkan front lay in the fact that it would be 'a purely British theatre when the laurels would be all ours'. Objectively, such nationalistic atavism – if Brooke was right in attributing such feelings to Churchill – was of course completely out of place in a war that had to be fought as a combined effort, but both Roosevelt and Churchill had their re-election to consider: Roosevelt in November 1944 and Churchill as soon as the war in Europe was won, which was then expected to be at about the same time. Just as it was important for Roosevelt electorally not to be seen as holding on to Churchill's shirt-tails, so Churchill had proclaimed only a year before that he had 'not become the King's first minister' in order to see the empire he loved liquidated. Churchill could see the way events were headed, and did not like it. Overlord would be a majority American operation, so prestige dictated that the British Empire should be able to claim the lead role in destroying Nazism in the Mediterranean.

The minutes of a strategy meeting held at 3 p.m. in the admiral's cabin of Roosevelt's ship on Friday 19 November make it clear quite how deeply conscious the Americans were of their emergent superiority by the time of Sextant. Right at the start of the discussion about having a supreme Allied commander for the west, covering Overlord, the Mediterranean and the Middle East, the question was raised whether Churchill would accept such a proposition.[7] Marshall pointed out that 'at times the British Cabinet has overridden Mr Churchill's decisions', but it was the President who put his finger on the main point at issue, when he asked what total forces Britain and the US would have at home and abroad by 1 January 1944. The figures produced were as follows: the British Army, Navy and RAF would total 3,822,000 men plus 1,070,000 Dominion forces, which for some reason was rounded down to four and a half million. The United States, meanwhile, would have 3,779,600 men serving overseas but enough in uniform in America to bring the total to 10,529,400, which for some equally mysterious reason was rounded up to eleven million. Therefore, the President observed, 'We are definitely ahead of the British as regards the number of men we have overseas at the present time and we will soon have as many men in England for Overlord as the total British forces now in that place.' Marshall added that 'he felt that we were already ahead of the British in England,' where there were only five operational divisions. Arnold added that overseas, 'with regard to air, we have passed the British rapidly. By 1 January 1944 we will have over twelve thousand operational planes, while the British will only have about eight thousand.' The implications were obvious: Roosevelt and Marshall could speak with a far stronger voice at Cairo than at any other conference hitherto.[8]

Roosevelt said he 'felt that the overall Mediterranean command proposed by the British might have resulted from an idea in the back of their heads to create a situation in which they could push our troops into Turkey and the Balkans'. King pointed out that this would be impossible because the Combined Chiefs of Staff would still have the final say in grand strategy. The President replied that he would veto any plan of Alexander's to use American troops and landing craft against the Dodecanese if the Prime Minister brought up the subject again. The meeting then got around to the question of Churchill's proposed Balkan initiative. Here Marshall was adamant, telling the President:

We must see the question of this Balkan matter settled. We do not believe that the Balkans are necessary. To undertake operations in this region would result in prolonging the war and also lengthening the war in the Pacific. We have now over a million tons of supplies in England for Overlord. It would be going into reverse to undertake the Balkans and prolong the war materially ... The British might like to ditch Overlord at this time in order to undertake operations in a country with practically no communications. If they insist on any such proposal, we could say that ... we will pull out and go into the Pacific with all our forces.[9]

This time, unlike in the summer of 1942, Roosevelt did not demur. Churchill's complaints about Overlord had gone too far. The Pacific threat, which in most of the Anglo-American strategic discussions up to this point had been implicit rather than explicit, was now becoming real. This time the Americans were not going to be bluffing.

Roosevelt evinced yet more hostility towards Britain in the COSSAC proposals for the division of post-war Germany into zones, codenamed Rankin. He believed that 'The British wanted the north-western part of Germany and would like to see the US take France and Germany south of the Moselle River. He said he did not like that arrangement.' Other than mentioning its Roman Catholicism, the President did not explain what he had against 'southern Germany, Baden, Württemburg, everything south of the Rhine', but he clearly preferred America to control the Protestant north-west of the Reich. The reason was doubtless because that was generally where the manufacturing industries were located. But once again the President was wrong to ascribe sinister intent to British policy-makers. Marshall explained that the geographical breakdown sprang from the simple logistical fact that the British were going to be on the left (that is, northern) flank of Overlord, which would take them into northern France and hence northern Germany, whereas the United States on the right (that is, southern) flank would necessarily – unless there was an administratively nightmarish cross-over of forces – wind up to the south. King added that the military plans for Overlord were too far developed to permit any change in deployment. Roosevelt then astonishingly suggested that American forces might instead be sent around Scotland and land in northern Germany, adding that 'He felt that we should get out of France and Italy as soon as possible, letting the British and the French handle their own problem together. There would definitely be a race for Berlin. We may have to put the US divisions into Berlin as soon as possible.'[10]

At this stage of the war, therefore, it was not Russia that Roosevelt was hoping to beat to Berlin, but Great Britain, and he was suggesting completely altering the entire Overlord planning in order to effect it. As the OPD Planner Lieutenant-General Charles Donnelly recorded: 'To have carried out Roosevelt's [first] idea, there would have to have been a massive crossing of the lines by Army groups which would have played no end of havoc with transportation and logistic dispositions. Roosevelt finally gave in on this later when the logic became clear to him.'[11]

It was at this meeting too that Roosevelt said one of the strangest things of the entire war, when he 'envisaged a railroad invasion of Germany with little or no fighting'. Marshall disagreed, saying that the land advance would have to be done by motor trucks as there was unlikely to be much rolling stock, whereupon Hopkins 'suggested that we must be ready to put an airborne division into Berlin two hours after the collapse of Germany'. Leahy felt that with civil war likely in France it would be best if the British were left with the problem rather than the United States, since 'The Germans are easier to handle than would be the French under the chaotic conditions that could be expected in France.'

Roosevelt concluded the session with the observation that 'the British would undercut us in every move we make in the southern occupational area [of Germany] proposed for the United States. He said that it was quite evident that British political considerations were at the back of the proposals in this paper.' As the Sextant Conference opened in Cairo, therefore, Roosevelt evidently felt deep suspicion of and even hostility towards Britain, although often on absurdly illogical grounds.

'We felt almost from the outset of the meetings at Cairo that our American associates, so many of whom we looked upon as close friends, had worked themselves into a state bordering on self-deception,' wrote Air Marshal Sir William Sholto Douglas, the Commander-in-Chief of the RAF in the Middle East. 'They seemed to be determined not to listen to our ideas . . . there was more of a negative, automatic rejection of the views of the British by the Americans than any positive approach to a consideration of new ideas.' British commanders felt 'something approaching wonderment over the blank refusal by the Americans to listen to what we had to say'.[12]

Mutual suspicion was not helped by the fact that by mid-November 1943 the Allied push up Italy had frustratingly stalled in bad weather at the Gustav Line, which ran through the town of Cassino, 87 miles south of Rome close to the Rapido river. Hitler had imposed a draconian no-withdrawal policy on Kesselring, even though the same strategy had cost Germany a quarter of a million men both in Tunisia and at Stalingrad. Cassino's sixth-century Benedictine abbey – once it was severely damaged by Allied aerial bombardment on 15 February 1944 – provided in its rubble a highly effective strongpoint to hold up the Allied advance. The battle of Monte Cassino was one of the most hard-fought engagements of the Second World War, after the flooding of the Rapido meant that tanks and motorized equipment could not be employed.

The Mena House Hotel in Giza, in the suburbs of Cairo, was built in the early 1890s, almost in the shadow of the Great Pyramid of Cheops. Although the Combined Chiefs of Staff held twelve meetings there – their 127th through to 138th between 22 November and 7 December – its members were distributed in villas along the road to Cairo. There were also five plenary sessions held in Roosevelt's Villa Kirk in the grounds of the hotel. The British Chiefs' villa looked imposing from the outside, but Hap Arnold decided he preferred his rooms because although the British had three baths each, they had no hot water. 'It was said to belong to a princess,' recalled Cunningham of his villa, 'but if so she had rather skimped the plumbing.'[13]

The British delegation nonetheless moved in en masse, their eight hundred items of baggage weighing 35 tons. Everything was provided by the Royal Marines, including a barber, orderlies (whose average age was eighteen and a half, since higher gunnery ratings could not be spared), and batmen for Women's Royal Naval Service officers, 'whose duties included washing their "smalls"'.[14] Part of the job of the security detail included burning all jottings and blotting paper left over at the end of each day's deliberations.

The Americans arrived at Sextant with even more Staff officers than they had had at Quadrant. As well as Marshall, Leahy, King and Arnold and five aides, additional officers included Lieutenant-General Somervell, Major-General Handy, Rear-Admiral Cooke and eleven other senior officers, plus three from the Joint Strategic Survey Committee; three from the secretariat; six Joint Staff Planners; the 'Senior', 'Red',

'Purple' and 'Blue' teams of the Joint War Plans Committee, numbering twelve; three from the Joint Logistics Committee; nine Logistics Planners; four from the Planners' secretariat (including Donnelly); two from Intelligence; eight Theater Representatives; two from War Shipping Administration (including Lew Douglas); two from the Assistant Secretary of War's Office and two from the Civil Affairs Division, totalling seventy-nine officers in all.[15] When the British asked difficult questions this time, there would be someone on hand with the answers.

Between Monday 22 and Friday 26 November the two sets of Chiefs of Staff met separately in the mornings, and then the Combined Chiefs of Staff met formally together in the afternoons. On the Tuesday and Friday the discussions got rather acrimonious. Although the British wanted an agreement on Overlord and the Mediterranean before they all met the Russians at Teheran, the Americans needed a decision on south-east Asia immediately, but wanted to discuss Overlord and the Mediterranean only at Teheran, where they knew they would be supported by Stalin, who was as desperate for Overlord as he was opposed to a Western presence in the Balkans. Furthermore, Roosevelt and Marshall rated Chiang Kai-shek highly and saw China as a post-war great power, whereas Churchill thought him a peripheral figure and Brooke considered that he 'Evidently [had] . . . no grasp of war in its larger aspects but [was] determined to get the best of the bargains' and 'never did much against the Japs during the war'.[16] The scene was thus set for another titanic clash.

The plenary session with Chiang Kai-shek on the morning of Tuesday 23 November was not successful, despite his wife Madame Chiang appearing at the meeting wearing a clinging black satin dress with 'a slit which extended to her hip bone and exposed one of the most shapely of legs', as a result of which Brooke thought he 'heard a suppressed neigh' coming from a group of the younger Staff members.[17] Over the question of air operations from India and China into Burma, Arnold wrote in his diary, 'Before we finished, it became quite an open talk, with everyone throwing his cards on the table, face up.'[18] Another problem centred around ships: were there enough for an expedition to Rangoon? Or to land troops near Aykab? Would aircraft be available to attack Rangoon? The British were more interested in keeping shipping in the Mediterranean than in the Indian Ocean, and little progress was made. 'Brooke treated the Burmese campaigns with an indifference so cold that it might have been culpable if they had not been so strategically irrelevant,' wrote a reviewer of his biography many years later.[19]

The next Combined Chiefs of Staff meeting became, in Brooke's classic understatement, 'somewhat heated' over his proposal to divert landing craft from a project supported by Roosevelt, Marshall and especially King – an attack on the Andaman Islands in the Bay of Bengal codenamed Operation Buccaneer – to the Aegean. Donnelly recalled that 'things became so hot that Admiral King and General Brooke traded insults'. Stilwell's diary entry for that meeting read: 'Brooke got nasty and King got good and sore. King almost climbed over the table at Brooke. God he was mad. I wish he had socked him.'[20] Sholto Douglas found it an 'unhappy – though perversely stimulating – experience' to hear Brooke and Marshall have 'the father and mother of a row'.[21] The upshot was that the landing craft went to neither the Andamans nor the Aegean but stayed in Italy for use in the Anzio landings now scheduled for January 1944.

It is still one of the great mysteries of the Second World War that while the United States could spend $350 billion on the conflict, and was capable of building a Liberty ship in one hundred hours, there never seemed to be enough landing craft to go around all the major theatres. Marshall blamed officials in the Bureau of Yards and Construction for deliberately engineering the shortages, but without producing any evidence for the accusation.[22] 'After Midway,' concludes Roosevelt's biographer,

it should have been possible to direct more men and landing craft to the European theatre. By the summer of 1943, the US Navy had twice as many battleships and three times as many aircraft carriers as Japan. In May 1944, the United States had thirty-one thousand landing craft, but only 2,500 assigned to D-Day. Two-thirds of the landing craft on that occasion were provided by the British.[23]

Immediately after the Combined Chiefs of Staff meeting ended, a trilateral one began attended by four Chinese generals, at which Brooke further infuriated the Americans by cross-examining them mercilessly about their plans. Stilwell thought it a 'Terrible performance. They couldn't ask a question. Brooke was insulting. I helped them out.' When the Chinese were asked about the Yunnan force protecting the Burma–India–China communications link in south-west China, Stilwell replied for them. 'Brooke fired questions,' he recalled, 'and I batted them back.'[24]

In this he was helped by Chennault, but since Brooke thought Stilwell 'was nothing more than a hopeless crank' and Chennault 'a very gallant airman with a limited brain', not much was achieved. 'The meeting came to a standstill!!' wrote Brooke afterwards, and as he left it he told Marshall: 'That was a ghastly waste of time!'[25]

On no evidence whatever, Stilwell's biographer Barbara Tuchman put Brooke's tough cross-examination down to his being 'the kind of Englishman who considered a foreigner to be snubbed and if non-white to be stepped on'.[26] Regardless of how Brooke would have taken to being called an Englishman, it was surely within his rights as Chief of the Imperial General Staff to ask the Chinese searching questions about the defence of Yunnan, the key province where China's training facilities and supply depots were located and where the airlift over the Himalayan 'Hump' terminated. Meanwhile Stilwell 'continually made snide remarks concerning the British', recalled a doubtless delighted Wedemeyer, 'criticizing their inability or unwillingness to fight'.[27] For the most distinguished member of the Fighting Brookes, this too must have been infuriating.

Marshall called for a major attack in Burma in early 1944, using British, Indian and Chinese troops. He said that Chiang Kai-shek's support for Operation Tarzan – capturing Upper Burma in order to protect Hump supplies and open the Burma Road – 'constituted a milestone in the prosecution of the war'.[28] As with Sledgehammer, however, Brooke tended to look askance at Marshall proposing major attacks that were not to be carried out by significant bodies of Americans.

'The British used one word repeatedly which irritated our side,' wrote Donnelly; 'when they wanted some aircraft or other matériel they submitted their "demands"; we would have said "requests" or "needs". The British did not use the term in any harsh sense but did not understand the difference in emphasis we attached to the term.'[29] Although the British and Americans spoke a common language, Ed Hull explained in his unpublished autobiography that 'We didn't always mean the same thing with what we said, however, and this sometimes led to misunderstandings. The British use the expression "to table" a matter when they want to bring it up for immediate discussion, while we use it normally to put the question aside or defer action on it.' At one Combined Chiefs of Staff meeting a rancorous discussion ensued after Portal said he wanted to table the issue of intensifying the bombing campaign against Germany, and Arnold thought he wished to postpone it.

Furthermore, the British included a broader group in their need-to-know list than the Americans. 'When they were trying to swing us over to their point of view in some particular matter,' recalled Donnelly, 'it was not uncommon for them to bring the matter up in a number of different committees and discussion groups. Our representative often did not know anything about the matter, not having been briefed or included on the need-to-know list, with the result that on several occasions someone expressed a view that was opposite to our official position.' The British would then naturally fasten upon whichever answer was more amenable to them. 'Even General Arnold', Donnelly recalled, 'got talked into a commitment to furnish three hundred C-47 transports for a Far Eastern operation before he talked it over with his logistics experts. It took him a couple of months to get off this hook.'[30]

At the Sextant and Eureka conferences, wrote Arnold, 'We found it very difficult to reconcile the conflicting racial and national aspirations: Chinese, Russian, British, American.' Through it all, however, he noted how 'General Marshall was increasing in stature, in comparison with his fellows, as the days went by. He had more mature judgment, could see further into the future.' When they went to view the Sphinx together at Giza, Arnold told Roosevelt that, although Marshall would be the best supreme commander for the European theatre, he would 'very much dislike to see him go', not least because he was the best adviser the President had.[31]

At lunch with Eden on Thursday 25 November, Churchill complained that poor progress was being made at the conference. Cadogan believed this was because Roosevelt was 'a charming country gentleman' who had no methods of organization, and 'the Prime Minister had to be a courtier, seizing opportunities when they arose.'[32] He might have been right about Churchill's seizing opportunities, but Roosevelt's façade of charming country gentleman should not have fooled the permanent under-secretary of the Foreign Office. Little progress was being made at Cairo because the President did not want it to be until they met Stalin at Teheran.[33]

The Americans vastly overestimated Chiang Kai-shek's capabilities and importance at Cairo, while showing much suspicion of British intentions in the Balkans but surprisingly little of Russian ambitions there or elsewhere. Roosevelt had wanted to invite Molotov to Cairo, but the Russians wouldn't meet the Chinese generalissimo for fear that it might compromise the uneasy truce Russia had maintained with Japan since 1941. Churchill also had to put up with presidential joshing over Russia such as: 'Winston, you have four hundred years of acquisitive instinct in your blood and you just don't understand how a country might not want to acquire land somewhere if they can get it. A new period has opened in the world's history and you will have to adjust yourself to it.'[34] Considering that Churchill had signed the first article of the Atlantic Charter, which stated that Great Britain desired no territorial changes that did not accord with the freely expressed wishes of the people concerned, it couldn't have been easy remaining the ever attentive 'lieutenant' during such lectures as that.

Hopkins, telling Moran about a meeting on 25 November, reported that Churchill had hardly stopped talking, mostly about 'his bloody Italian war'. The President's confidant was dry and 'full of sneers and jibes', remarking: 'Winston said he was a hundred percent for Overlord. But it was very important to capture Rome, and then we ought to take Rhodes.' He made it clear to Moran – perhaps with the intention of having it passed on – that if Churchill was going to adopt this stance at Teheran, 'The Americans will support the Russians.'[35] The threat could not have been more explicit. 'All Hopkins' views on strategy come, of course, from Marshall,' concluded Churchill's doctor, who added a medical analogy, 'but in changing hands they seem to go sour, as a microbe gains in virulence when it passes from one host to another.'

Moran, an acute observer who (it must be emphasized again) wrote up his contemporaneous notes much later, noticed at Cairo 'a certain hardening of purpose in the American camp. They left Quebec in great heart, assured that everything was settled for good. And here is the British Prime Minister at his old game again. There is an ominous sharpness in their speech when they say that they are not going to allow things to be messed about in this way indefinitely.'[36] When considering the testimony of any doctor, of course, it is worth bearing in mind that he sees his subjects most often when they are least well.

Churchill's doctor even wondered whether the invasion would ever come off, believing that his patient had changed his mind since Quebec, especially after the near-defeat at Salerno, and had 'grown more and more certain that an invasion of France as planned must fail'. Churchill also supposedly told Moran at this time that Roosevelt 'is the most skilful strategist of them all . . . better than Marshall'. Roosevelt certainly got his way more often than any of the other Masters and Commanders, despite self-confessedly knowing the least about grand strategy. As for Marshall, Churchill's effect on Roosevelt meant that 'He can never be sure what will happen when Winston and the President get together. With the President wobbling, he and Admiral King fear that the Prime Minister may, after all, get his own way.' As Hopkins aggressively reiterated to Moran: 'Sure, we are preparing for a battle at Teheran. You will find us lining up with the Russians.'[37]

Hopkins – ever FDR's assiduous courtier – undoubtedly seems to have viewed Churchill as an opponent rather than a friend at this time, telling the American editor and publisher Ralph Ingersoll that the reason Marshall had to become supreme commander in the West was because he was the only man qualified to 'stand up to Churchill', who dominated his own generals. As already noted, it was Hopkins' view that 'He is the only general in the world whom Churchill is afraid of,' for, 'when Churchill gets oratorical, Marshall just listens and then brings the conversation back to earth with just the right facts and figures to destroy the PM's case.'[38] Marshall was not the 'only' general like that; he might also have been describing Sir Alan Brooke.

Thanksgiving Day, Thursday 25 November, was dedicated to trying to improve relations between the American and British Chiefs, because, as Arnold wrote, they 'had had hard sledding over many fundamentals and weren't getting along very fast'. The meeting of the Plans Committee that evening was postponed and instead they concentrated on socializing. However, Brooke thought the church service before their Thanksgiving Dinner was 'a sad fiasco and abominably badly run . . . I was in a cold sweat of agony throughout.'[39] Whatever the state of Anglo-American relations, those between the British Army and RAF can hardly have been perfected by Air Marshal Tedder opining that 'We should decorate Rommel for teaching the British Army how to fight.'

The next day, Friday 26 November, saw Brooke and Marshall have what the former called 'the father and mother of a row!' at a 2.30 p.m. off-the-record Combined Chiefs of Staff meeting. Given that they knew they had to meet both their own political masters almost immediately, and then Stalin in forty-eight hours' time, they nonetheless 'made more progress', recorded Brooke. 'In the end we secured most of the points we were after.' A pattern had thus definitely emerged, of initial disagreements and intense brinksmanship, with a big row clearing the air in the final hours of the conference and a compromise deal covering most of the theatres agreed only moments before the generals were due to present their final report, knowing that if they could not agree one then Roosevelt and Churchill would be free to settle matters between themselves, which both sets of soldiers automatically assumed would be far worse.

'We discussed the war in all its aspects and phases, and came to full agreement on all matters of moment without any difficulty,' wrote Cunningham of the Cairo Conference in his 1951 memoirs, *A Sailor's Odyssey*. 'There was no dissimilarity of views at this conference.'[40] This was completely untrue, and Cunningham must have known it to be, writing only eight years later and having been present at all the key meetings. In the Sextant compromise, the Joint Chiefs of Staff accepted the British view that the Mediterranean command would not be subordinated to the Supreme Allied Commander, who would be responsible for Overlord alone, thus stymieing Marshall's ability to wind down the Italian campaign.[41] They also accepted the British proposals for the Mediterranean 'as a basis for discussion with the Soviets' and further agreed that the date of 1 May 1944 was not completely sacrosanct for Overlord. The Americans nonetheless refused to postpone, let alone cancel, Operation Buccaneer, the attack on the Andaman Islands.

After the war, Brooke summed up Cairo by complaining that the three fronts of Russia, Italy and the Channel 'might have been separate wars' for Marshall, who he claimed refused to accept their intimate interrelation or the way in which strategy 'had become a delicate matter of balancing'. He wondered what might have happened had MacArthur – 'the greatest general of the last war' – been Army chief of staff instead. (MacArthur more than returned the compliment in 1962, describing Brooke as 'undoubtedly the greatest soldier that England has produced since Wellington'.) 'I must . . . confess that Winston was no great help in the handling of Marshall,' continued Brooke, 'in fact the reverse. Marshall had a holy fear of Winston's Balkans and Dardanelles ventures, and was always guarding against these dangers even when they did not exist.'[42] Yet for all his post-war protestations over this, the dangers did indeed exist, and Brooke was a leading progenitor of them.

According to Elliott Roosevelt's tendentious 1946 memoir of his father, *As He Saw It*, the President grew exasperated with Churchill during the Cairo Conference. 'Believe it or not, Elliott,' FDR is supposed to have said there, 'the British are raising questions and doubts again about that western front.' Elliott claims he replied, 'But I thought that was all settled at Quebec!' 'So did we all,' replied his father. 'It is, too. It's settled. But Winston keeps on making his doubts clear to everybody.' Of course one must read with scepticism claims of perfect verbatim recall of three-year-old conversations, but Elliott then reports his father telling him that over the issue of Churchill's desire for a Balkan offensive: 'I think Winston is beginning not to like George Marshall very much. He finds that no matter what tactics he uses, whether it's wheedling or logic or anger, Marshall still likes best the strategy of hitting Hitler an uppercut right on the point of the jaw.'

In fact there is no evidence that Churchill held Marshall in anything other than high regard, and he was perfectly capable of feigning anger to get results, and was completely able to compartmentalize his life. 'The morning at Mena when the Prime Minister was really on the warpath,' recalled his Royal Marine chief orderly Major Buckley, 'he growled at Eden; he told off the Chiefs of Staff; he moaned at General Ismay and, firing a real broadside at John Martin, disappeared into the grounds.' Later Martin discovered him 'apparently in high good humour', arguing with his orderly, Lance-Corporal Wright, over the respective merits of first-class versus village-green cricket, with 'Wright arguing back'.[43]

On Saturday 27 November the entire conference decamped to Teheran, flying over the Suez Canal, Bethlehem, Jerusalem, the Dead Sea and the Syrian desert. Stalin had arrived the day before. Iran was under Allied – mainly Russian – control, and security was intense. It was the first time that the Big Three had met together; Churchill was nearly sixty-nine, Stalin sixty-four and Roosevelt sixty-one. (They were to die in reverse order of age, Roosevelt in 1945, Stalin in 1953 and Churchill in 1965.)

Three plenary sessions were held on Sunday 28, Monday 29 and Tuesday 30 November at the Russian Legation and there was one military conference on the middle day. For all their vast secretariat, Marshall and Arnold missed the first plenary session due to a scheduling error; they went on an automobile tour of the mountains north of Teheran instead. From the first moment of the conference, Brooke convinced himself that Stalin had 'a military brain of the very highest calibre.

Never once in any of his statements did he make any strategic error, nor did he ever fail to appreciate all the implications of a situation with a quick and unerring eye.' By contrast Brooke characteristically thought Marshal Voroshilov had 'nothing in the shape of strategic vision'.[44] Nonetheless, everything Stalin said that day would have delighted Marshall and was utterly opposed to Brooke's strategic views.

The Russian dictator stated unequivocally that Overlord should be the overriding priority for 1944, that the Italian campaign was a mere diversion (and an unimpressive one at that); that Turkey would not enter the war, so Britain's Aegean plans were stillborn, and that southern France needed to be invaded before Overlord.[45] Arnold's notes of Stalin's remarks on how to win the war – 'Hit her where the distance to Berlin is the shortest. Don't waste time, men or equipment on secondary fronts' – was Marshall's policy distilled to its essentials. It is impossible not to notice that when Marshall said exactly these kind of things, and far less didactically, Brooke derided him as a worthless strategist, yet when Stalin said them he was accorded Brooke's ultimate accolade.

Stalin, who had not left the USSR since 1918, was of course the cynosure of all eyes at Teheran. Other than Harriman and Hopkins, the Americans were almost all meeting him for the first time. He was also disrespectful about allies in an overt way that sent a frisson of excitement through the conference. 'When he talked about the British, the Prime Minister, and the CIGS,' recalled Arnold, 'he was half humorous, half scathing.'[46] He dressed in a light-brown uniform with red stripes on his trousers and gold epaulettes on his shoulders which featured a large gold star and the insignia of a Red Army marshal. He wore the Gold Star medal, was about 5 foot 4 inches and Arnold thought him 'a fine-looking soldier'.

There was a cult of personality about Stalin among Western strategists almost as powerful as the one that his propagandists had ordained for him back in the USSR. 'Stalin is more of a hero than the King or even Winston,' marvelled Kennedy, who was shocked to notice that many of the Scots Guards stationed at the Tower of London had a picture of the marshal over their beds. At the end of the plenary meetings at Teheran, Major Buckley recalled 'a neck-and-neck race round the table between Cunningham and Hollis to secure the latest of Stalin's famous doodles, with the Royals as ever winning by a short head'.[47] (Buckley and Hollis were Royal Marines.)

Roosevelt later explained to his long-standing labor secretary, Frances Perkins, how he had melted the ice with Stalin by making common cause with him against Churchill, at least in a social sense. 'I had come there to accommodate Stalin. I felt pretty discouraged because I thought I was making no personal headway,' he said. 'I couldn't stay in Teheran forever. I had to cut through this icy surface so that later I could talk by telephone or letter in a personal way.' Roosevelt said he had an inkling that the Russians did not feel happy that he and Churchill were 'conferring together in a language which we understood and they didn't'.

Therefore, at the next session, Roosevelt joked in a whisper to Stalin, 'Winston is cranky this morning, he got up on the wrong side of bed.' After Stalin smiled at that, Roosevelt decided he 'was on the right track. As soon as I sat down at the conference table I began to tease Churchill about his Britishness, about John Bull, about his cigars, about his habits. Finally Stalin broke out into a deep, hearty guffaw, and for the first time in three days I saw light. I kept it up until Stalin was laughing with me, and it was then that I called him "Uncle Joe" . . . that day he laughed and came over and shook my hand.'[48] The light Stalin saw was a chink through the Anglo-American façade, and that was what made him happy quite as much as Roosevelt's jokes. Far from not 'feeling right' about the Anglo-Americans conferring in English, the Russians assumed the capitalist nations would provide a united front; when Stalin saw that this was not necessarily true he was naturally delighted.

'Churchill, lighting up his cigar,' recalls his interpreter, Hugh Lunghi,

at first seemed not unduly embarrassed by the fairly heated arguments between the Americans and British over strategic priorities now being played out in front of Stalin. As the debate developed, the Prime Minister increasingly appeared on the defensive, still arguing strongly for his vision of the military options. At the start, regardless of Roosevelt's 'jokes' at Churchill's expense, Stalin seemed puzzled at the open display of disunity between the Americans and the British. Then . . . he allowed his normally inscrutable face a rare smile. Stalin spoke – as always – softly, briefly, to the point, completely in command of facts and statistics, hardly ever looking at a note, asking pertinent, awkward questions. At times we could hardly make out his words, with their marked Georgian accent.[49]

Lunghi was also 'struck by the yellow whites to his greenish-brown, cat-like eyes'.

Even though dangerously outnumbered, Churchill could not be shifted from some stances. After Roosevelt told Stalin that Overlord was targeted for 1 May 1944, Churchill said that the operation would indeed take place some time in 1944, and that the Mediterranean operations were 'always regarded' as 'stepping stones', but added: 'I wish to place on record that I could not, in any circumstance, agree to sacrifice the activities of the armies in the Mediterranean in order merely to keep the exact date of May 1 for Operation Overlord.'

Writing in 1951 in *Closing the Ring*, Churchill said that although the cross-Channel assault 'was the greatest event and duty in the world', a million Allied troops were fighting in Italy and he did not want to see them sabotaged. 'Here the American, clear-cut, logical, large-scale, mass production style of thought was formidable,' he wrote. 'In life people have first to be taught "Concentrate on essentials".' As a result of these battles with the Americans, he recalled, 'Twenty or a dozen vehicle landing-craft had to be fought for as if the major issue turned upon them.'[50] Marshall's reply would be that it was he who was concentrating on the essential operation of Overlord, and Churchill who was not, and at that stage of the war he would have been right.

On Tuesday 30 November – Churchill's sixty-ninth birthday – Voroshilov mounted a full-scale attack on Brooke over his supposed lack of commitment to Overlord, especially with regard to shipping, landing craft and air cover. It was the first time that the Soviets had met the Combined Chiefs of Staff for Staff discussions on all the war fronts, and Leahy recalled that although Marshall was 'inclined to go along' with Voroshilov about Overlord, 'Sir Alan Brooke insisted stubbornly that all available Mediterranean forces should be used in the Italian and Eastern Mediterranean campaigns, including the pet project of . . . the capture of the island of Rhodes.'[51] Since Marshall had already stated that no American would be taking part in that 'goddamned' operation, the battle-lines were drawn. It was soon clear that Hopkins had not been bluffing when he threatened that the Americans would join the Russians in putting maximum pressure on the British.

When Voroshilov asked Brooke point-blank if he attached the same importance to Overlord as General Marshall, Brooke replied that he did, but added that he knew how strong the German defences were in France and that 'under certain circumstances Overlord could fail.' To this Voroshilov 'admitted the difficulties of a trans-Channel operation', but said that the Russians 'had encountered comparable difficulties in the crossing of wide rivers and had overcome them because they had the will to do it'.[52]

At that point Marshall decently came to Brooke's defence, pointing out the obvious difference between crossing a river and undertaking an amphibious invasion of a country. Failing to cross a river might be a reverse, he said, whereas 'the failure of a landing operation is a catastrophe'. Marshall went on to say that, although his Great War experience had all been about roads, rivers and railroads, in the last two years it had been based on oceans and seas. 'Prior to the present war I never heard of any landing craft except a rubber boat,' he joked. 'Now I think about little else.'[53] He added that the US had plenty of men – 1.6 million stationed in Europe and 1.8 million in the Pacific – but his primary problem was moving them.

Brooke told Voroshilov that of the twenty-seven German divisions in Italy, eleven could be destroyed or captured by an amphibious attack just south of Rome behind the Gustav Line, and the need for landing craft for that would push back Overlord by one month to 1 June. He tried to explain amphibious operations to the Russian, but with limited success. (Hugh Lunghi, who was translating for Brooke, recalled Voroshilov being dismissive about landing craft during the Channel crossing, saying 'We usually managed to find local resources like trees, timber to make rafts. Red Army men can use their initiative.')[54] The Combined Chiefs of Staff then agreed to sixty-eight landing craft staying in the Mediterranean until 15 January 1944, and a general push being made up to the Pisa–Rimini Line. In return, the British Chiefs of Staff agreed that an attack in the south of France – Operation Anvil – should coincide with Overlord, disregarding Stalin's wish that it precede Overlord as strategically impractical.

'My British counterparts will tell you that we didn't have the landing craft,' sneered Albert Wedemeyer in 1973. 'They made a great evacuation from Dunkirk without the landing craft, particularly craft that would permit it. We could have if we had the spirit and the will, and if the Navy had not begun to take surreptitiously some of our landing craft out to the Far East if they were convinced they were going across earlier.'[55] What he said about the US Navy was true, but so senior a Planner ought to have recognized the very different types of craft needed when assaulting a beach and when evacuating from one.

The Americans were deeply suspicious of Churchill's eloquence and persuasiveness at Teheran, while also admiring it. In 1946 General John R. Deane, who had been secretary of the Joint Chiefs of Staff before being sent to head the US military mission to Russia in 1943, published a book about US–Russian wartime co-operation in which he recalled of the Eureka conference: 'Churchill used every trick in his oratorical bag, assisted by illustrative and emphasizing gestures, to put over his point. At times he was smooth and suave, pleasant and humorous and then he would clamp down on his cigar, growl, and complain.'[56] It seems hard on Churchill that he should be expected to have identical moods throughout these exhausting, stressful and vital times, but equally he was the master of the theatrical, which is partly what makes him still so fascinating.

Hugh Lunghi recalls of Cairo that 'Brooke was like the headmaster of our group, beside whom Portal and Cunningham seemed like sixth-formers on holiday. He was the least easy, he was strict, serious about everything, didn't relax. But he was a joy to interpret for, as he was a linguist and so he stopped every few sentences and was completely logical.'

At the end of the afternoon plenary session on the second day, Stalin put the key question: 'Who will command Overlord?'[57] He appreciated that it would come off only if someone senior, talented and utterly committed was in charge. Roosevelt – from whom a decision was by now long overdue – answered that the commander had yet to be appointed, while Churchill said that although a Briton had been respon-sible for the planning, he was willing to see an American take the job, which must have been galling for Brooke to hear once again. 'Stalin made it plain', Harriman later recalled, 'that until the supreme commander was appointed he could not take seriously the promise of a cross-Channel invasion. For him the appointment was a specific assurance that the invasion would take place.'

Roosevelt and Churchill both had Marshall in mind at this stage, and Stalin also believed that Marshall would get history's ultimate call-up. 'They considered him the one soldier pre-eminently qualified to com-mand what all agreed was likely to prove the most difficult operation in the history of warfare,' averred Harriman. Roosevelt also believed that, with his headquarters in London, 'Marshall, alone, with his granite integrity, was equipped to resist any eleventh-hour manoeuvres by Chur-chill and General Brooke to delay or divert the cross-Channel invasion.'[58]

Yet at Cairo both King and Arnold had protested that Marshall simply could not be spared, and if Eisenhower was to take over as Army chief of staff while Marshall was in London, the wilfulness of his former boss MacArthur might cause serious problems. Harriman's views on the issue, as stated in 1975, are instructive: 'I know that General Marshall wanted more than anything else to command this historic military action, and I have no doubt that Roosevelt would have appointed him if he had given the slightest indication of his personal desires. But he left the decision entirely to the President. It was the most selfless thing any man could do.'[59]

With an American commanding Overlord, a Briton – either Alexander or General Sir Henry 'Jumbo' Maitland Wilson, the Commander-in-Chief in the Middle East – could command in the Mediterranean. No one seems to have considered promoting a Canadian to an important international command, despite the number of their troops, the quality of their commanders, the size of their Navy and the generosity of their exchequer. A more assertive prime minister than Mackenzie King could probably have secured a better deal for Canada in particular and the British Dominions in general.

Was Marshall somehow shirking the awesome responsibility of commanding Overlord? There is no evidence to suggest that he would have done anything other than thrive on it. He had faith in Eisenhower's abilities, of course, otherwise he would certainly have taken it on himself. Had he done so, then Eisenhower, Bradley, Clark, Patton and MacArthur could not have complained. He probably felt, like Brooke the previous year, that he had a duty to keep his master in check, and that no one else was likely to do it so well. Because he vouchsafed his private thoughts to no one on this matter – not even to his wife or Forrest Pogue – we only have Marshall's own high character as circumstantial evidence on which to base a judgement.

The most likely explanation for his great act of self-abnegation is the straightforward one: that he put his country's best interests before his own ambition, despite the fact that, as he had told Stimson, 'a soldier's first wish is to serve in the field'. Brooke was right: when asked after the war whether Marshall had stayed because he was 'the one man they had who could get on with the British', the field marshal answered, 'Well, that may have been the case but I think it would have been a great pity to move General Marshall ... as he had a complete grasp by that time of the functions of the Staffs, and to take him away for a command would have removed one of the lynchpins of the higher direction of the war.'[60] He might just as easily have been describing his own decision to stay on as CIGS, and probably was.

Churchill's birthday banquet at the British Legation in Teheran on Tuesday 30 November 1943 afforded Stalin an opportunity to chaff Brooke. Arnold thought he counted a total of one hundred toasts drunk at the dinner, each after a short speech. Major Buckley, who swiped one of the candles from Churchill's birthday cake as a memento, recalled Marshal Semën Timoshenko 'finished under the table'. At one point Roosevelt proposed a toast to Brooke, making a reference to their fathers having known one another. Stalin then chipped in, saying that as a result of this conference, and 'of having come to such unanimous agreement', he hoped that General Brooke 'would no longer look upon Russians with such suspicion' because if he really got to know them, as the CIGS himself paraphrased it, 'I should find that they were quite good chaps!!' For some reason Brooke blamed mischief-making by Harriman for what he called in his diary this 'most unexpected and uncalled for attack'. To modern ears it sounds as if Stalin was teasing Brooke rather mildly, not making a serious accusation, but of course we cannot know the tone in which he uttered his remark.

Roosevelt's interpreter, Charles 'Chip' Bohlen, took it seriously too, recalling in his autobiography *Witness to History* how Stalin's mock-toast caused 'some consternation among the British because Sir Alan was known to be an Irishman with a quick temper, and it was feared that he might destroy the friendly atmosphere with an angry reply to Stalin's gratuitous insult'.[61] Instead Brooke waited for a propitious moment to answer – 'It was rather nervous work, considering what the audience was!' – and made a graceful speech implying that he had only been feigning his anti-Bolshevism, just as Stalin himself had created 'masses of dummy tanks and aeroplanes on the fronts [where] he was not going to attack'. This went down well and the buttoned-up Briton was surprised to be virtually hugged by Stalin after dinner, 'almost with our arms round each other's necks!' (It is quite untrue, as Bohlen claimed, that Brooke had said in his speech that he had feigned his anti-Soviet feelings in the same way that Stalin had feigned his anti-Nazi feelings at the start of the war, and that 'the dictator [took] the jibe in good humour'. Bohlen might have been good at interpreting Russian into English but he couldn't interpret English into American very well, because to have mentioned the 1939 Molotov–Ribbentrop Pact to Stalin in semi-public would have been an unthinkable diplomatic faux pas.)

'Overlord and Anvil are the supreme operations for 1944,' stated the Combined Chiefs of Staff final report of the Eureka Conference. 'They must be carried out during May 1944. Nothing must be carried out in any other part of the world which hazards the success of these two operations.'[62] This seemed to go back upon the agreement over landing craft needed for the Anzio operation, codenamed Shingle, and the Andamans attack, codenamed Buccaneer. Stalin promised to declare war on Japan after Germany surrendered, and to launch an offensive during Overlord to discourage the Wehrmacht from moving troops westwards during its initial stages. The minutes of the discussions regarding Poland read: 'The Prime Minister demonstrated with the help of three matches his idea of moving Poland westward, which pleased Marshal Stalin.' Put simply – and with three matches there could hardly have been any other way – Churchill wanted the old Curzon Line to be the future Russo-Polish border, so Poland would be compensated with east German territory for the loss of land to the USSR. The outlines of a proposed new world organization, to be called the United Nations, were discussed, and agreements were reached on Iranian post-war independence.

Brooke was content with the final report, and soon afterwards managed to get the Overlord date pushed back a month so that it would not hobble the campaign to take Rome, and the south of France attack turned into something more elastic which he believed at the time 'can be adjusted without affecting Italy too seriously'.[63] By the time of D-Day in June 1944, the Allies had twenty-seven (fuller-strength) divisions in Italy. Military historians such as Basil Liddell Hart and Richard Holmes have legitimately questioned, therefore, quite who was pinning down whom in that wasp-waisted peninsula.[64] Once the far higher Allied populations and sizes of armies are taken into account, however, Churchill's and Brooke's Italian strategy was worth while, up to a point. The problem was that they took that point further north than the Gustav Line, with fewer returns than the great effort and loss of life strategically justified, especially after the fall of Rome on 4 June 1944.

After a day's rest and relaxation at the King David Hotel in Jerusalem and visiting the holy sites, the Staffs returned to the Mena House Hotel in Cairo by lunchtime on Thursday 2 December 1943. For the next three days the Combined Chiefs of Staff debated hard whether to use the available landing craft for attacking the Andaman Islands, or for the Anzio operation scheduled for the following month. Marshall feared that cancelling Buccaneer might encourage Chiang Kai-shek to stop fighting in Upper Burma, which would in turn threaten the air routes to China and possibly allow the Japanese to transfer forces to the Pacific. Finally, on the evening of Sunday 5 December, Roosevelt supported Brooke over Marshall and agreed to postpone the Andaman assault.

A newspaper in Tucson, Arizona, with reputed connections to Pershing, claimed that 'the British were trying to kick [Marshall] upstairs' in order to get a more pliable chief of staff, which Marshall dismissed as 'absurd'. Conspiracy theorists also claimed that Roosevelt wanted to replace Marshall with General Somervell and ensure the use of Army contracts to secure re-election, which was equally ludicrous. Yet the decision over who was to be supreme commander for Overlord – with the British refusing to have an American supreme commander over the whole of the west because of the effect on Italy – could not be long delayed after Stalin's enquiry.

There is no reason to doubt Marshall's own account to Pogue about how he visited the President at Cairo where Roosevelt, 'after a great deal of beating about the bush', asked him what he wanted to do. Marshall repeated that he would 'cheerfully' do whatever the President said. He could advise the Commander-in-Chief on all Army appointments, which would usually be accepted unhesitatingly, but not his own. Roosevelt finally said, 'Well, I didn't feel that I could sleep at ease with you out of Washington,' so Eisenhower was appointed instead. Marshall himself recalled that Churchill, sensing Roosevelt's hesitation in confirming his appointment, offered the job to Brooke yet again, and was 'very much embarrassed' when it went to Eisenhower.[65] There is no confirmation for this from any other source, and Brooke would certainly have fumed to his diary if it had happened. No other Americans but Marshall and Eisenhower were in the running, because Marshall championed Eisenhower. After the war Marshall said that if he had taken the Overlord command he probably wouldn't have been able to put up with Montgomery's 'overwhelming egotism' in the way that Eisenhower did.

Discussing the situation with Stimson afterwards, Roosevelt said that 'he had got the impression that Marshall was not only impartial between the two but perhaps preferred to remain as Chief of Staff.'[66] Of course this was inherently contradictory: Marshall could not be both impartial between the two and also prefer to remain, so was Roosevelt salving his conscience in persuading himself that Marshall preferred the post that he, Roosevelt, wanted him to stay in? Stimson believed he knew from Marshall's phrase about 'any soldier preferring a field command' that Marshall actually wanted Overlord.

The realities were spelt out to Stimson by Roosevelt after Marshall had specifically refused to ask for the Overlord post: 'The President said that he had decided on a mathematical basis that if Marshall took Overlord it would mean that Eisenhower would become Chief of Staff.' Yet Eisenhower was unfamiliar with the war in the Pacific and, in Stimson's view, he 'would be far less able than Marshall to handle the Congress', both of which were vital aspects of the Chief of Staff's duties. Roosevelt told Stimson that he 'would feel far more comfortable if he kept Marshall at his elbow in Washington and turned over Overlord to Eisenhower'.

Afterwards, as Stimson attested, 'never by any sign did [Marshall] show that he was not wholly satisfied with the President's decision.' That in itself shows tremendous strength of character if indeed he wasn't satisfied. Stimson himself was deeply disappointed, and always regretted not going to Cairo where he thought he could have emphasized that Marshall had definitely wanted to command Overlord and was the right man for the job, because he could 'push through the operation in spite of the obstacles and delays' which he feared Churchill and Brooke were putting up against the operation.[67] Handy, who ended the war as a four-star general and worked closely with both men as assistant chief of staff in the OPD up to 1944 and then as deputy chief of staff thereafter, thought that Eisenhower was chosen over Marshall because 'There were situations back home and a lot more people had confidence in [Marshall] than they had in the President ... They were very strongly anti-New Deal.'[68]

As they drove together to see the Egyptian monuments during this Second Cairo Conference, Roosevelt remarked 'almost casually' to Churchill that 'as Marshall was not to have the Mediterranean and Overlord in his hands, he would prefer to keep him in Washington. He could not spare him except to have this supreme direction of the final phase of the war.' In refusing to allow Marshall to take both western commands, therefore, Churchill was being blamed by Roosevelt for blocking Marshall's chances of taking on Overlord alone, yet the final draft of Churchill's war memoirs makes almost no mention of this. It was unmerited, in any case, because Churchill had been enthusiastic about Marshall commanding Overlord, just not both theatres simultaneously.

Elliott Roosevelt also blamed Churchill for effectively preventing Marshall from taking command of the cross-Channel attack. He stated that his father had said: 'It seems pretty clear that Winston will refuse absolutely to let Marshall take over . . . It's not that he's argued too often with the PM on military matters, it's just that he's won too often.' Elliott's own view was that Marshall's great qualities as a commander were the same ones 'which had made him an enemy in Winston Churchill'.[69] However ridiculous that might sound, it was Elliott's genuine view. The publication of these judgements in his bestseller *As He Saw It* in October 1946 understandably pained Churchill.

17

Anzio, Anvil and Culverin: 'The inevitable stumbles on a most difficult course'
December 1943–May 1944

There was no division, as in the previous war, between politicians and soldiers, between the 'Frocks' and the 'Brass Hats' – odious terms which darkened counsel.
Winston Churchill, *Their Finest Hour*[1]

Having read the minutes of the Eureka Conference, Major-General John Kennedy concluded that the Russians and Americans had got their way on Overlord at Teheran, an operation that Stalin wanted even if it failed, and one that if the Americans had 'had their way they would have launched more than once in the past when they would certainly have failed'.[2] One can see from this response why the Americans felt the British were lukewarm about the operation.

Eisenhower's appointment as supreme commander for Overlord meant that changes were needed elsewhere. Churchill and Brooke conferred on 18 December 1943 and asked Roosevelt to approve Jumbo Maitland Wilson to take Eisenhower's place as supreme commander Mediterranean, with an American commanding in Algiers, Alexander in Italy, Tedder as Eisenhower's deputy and Mark Clark commanding Anvil. 'We understand that this was what you and General Marshall had in mind,' Churchill wrote to Roosevelt, copying his message to Brooke: 'If so, we concur.' Maitland Wilson, despite being sixty-two, 'has all the qualifications and the necessary vigour. This is also the opinion of the CIGS. When I mentioned this idea to you at Cairo you seemed to like it.'[3] Meanwhile Walter 'Beetle' Bedell Smith would stay on as Eisenhower's chief of staff and move to London with him. Roosevelt and Marshall approved all the major appointments except Clark, who stayed in Italy. The Anvil attack was to be commanded by Lieutenant-General Jacob L. Devers of the US Sixth Army.

Although Tedder was to be Eisenhower's deputy, an acknowledgement of the importance of air power in the operation, Britons also took the next three most important roles, much to American ire. Montgomery was given the subordinate command on land, Admiral Sir Bertram Ramsay at sea, and Air Chief Marshal Sir Trafford Leigh-Mallory in the air. Ike had initially wanted Alexander as his land commander, but the post went to Montgomery. 'The support of one man, the CIGS Sir Alan Brooke, had carried Montgomery first to the army command in which he gained fame in the desert, and then to the principal British role in Overlord,' records the historian of the operation. 'Without Brooke, it is unlikely that Montgomery would ever have gained the chance to display his qualities in the highest commands.'[4]

At a military conference at his Wolfsschanze (Wolf's Lair) headquarters deep inside East Prussia on the evening of 20 December, Adolf Hitler, after considering the situation in Russia and Italy, stated that he had studied all the intelligence files, and 'There's no doubt the attack in the West will come in the spring; it is beyond all doubt.' He worried that too many troops were stationed in Norway – at 430,000, he was right – and considered whether diversionary attacks might be made in the Bay of Biscay and the Balkans. He added, 'We have one advantage; [the Allies] will come with units that are not combat-experienced.' In answer to General Walter Buhle's complaint that they were taking too many forces from the west to reinforce the east, Hitler said: 'Who are you saying that to? . . . It is hard for me as well. Every day I see the situation in the East. It is horrible . . . But I've always had these concerns in the West.'[5] These comments would have delighted Brooke, whose role as a master of strategy was acknowledged on New Year's Day 1944 when Churchill personally insisted that he be immediately raised to the rank of field marshal, the highest in the British Army.[6]

Operation Shingle was a daring plan of Alexander's to land troops on the beaches of Anzio, in the rear of the Gustav Line and only 20 miles south of Rome, thus it was hoped forcing the enemy to abandon first one and then the other. On closer investigation in December 1943, the fear that the beachhead could not link up with the Allied armies further south meant that Eisenhower tried to shelve the operation, but it was revived once he had left the Mediterranean command. On 6 January

15. At Allied HQ in North Africa on 8 June 1943. From left to right around Churchill are Anthony Eden, Brooke, Tedder, Andrew Cunningham, Alexander, Marshall, Eisenhower and Montgomery

16. A Combined Chiefs of Staff meeting at the First Quebec Conference, codenamed Sextant, in August 1943. From the left: Brooke, Brigadier-General John R. Deane, Arnold, Marshall, Admiral William Leahy and Pound

17. The Château Frontenac hotel towers over the Second Quebec Conference, codenamed Octagon, in September 1944. Sitting at the nearby Citadel are, from left to right: Clementine Churchill, the Earl of Athlone, President Roosevelt, HRH Princess Alice, Countess of Athlone, Winston Churchill, Eleanor Roosevelt and the Prime Minister of Canada, William Mackenzie King

18. Churchill and Roosevelt greet each other at the Second Quebec Conference, 1944

19. Field Marshal Sir John Dill, Andrew Cunningham, Brooke, Portal and Ismay at Quebec, 1944

20. Planning campaigns in smoke-filled rooms: the British Joint Planning Staff at the Château Frontenac in September 1944, Quebec Conference. Lieutenant-Colonel Aubertin Mallaby is sixth from the left

21. Marshall's stepson, Lieutenant Allen Tupper Brown, who was kill by a German sniper near Rome in May 1943

22. Brooke (*centre*) and his friend, flatmate and aide-de-camp, Barney Charlesworth, duri exercises in October 1941. Charlesworth was later killed in an air crash in February 1945

23. Churchill with his friend and surrogate uncle Field Marshal Jan Smuts on the lawn of the British Embassy in Cairo in August 1942; behind them are Tedder and Brooke

24. Churchill's chief of staff, 'Pug' Ismay, thanks a lady called Hildegarde for singing a song about England, 1942

25. The Anglophobic General Albert C. Wedemeyer with General Marshall: he bugged his office desk to record British officers' indiscretions, with Marshall's knowledge

26. General Sir Archibald Wavell at the Viceroy's House in New Delhi, with the American General Joseph W. Stilwell, another Anglophobe

27. The British Chiefs of Staff on the eve of Victory in Europe: Cunningham, Brooke (with jabbing finger in motion) and Portal in April 1945

28. The US Joint Chiefs of Staff: Leahy, Arnold, King and Marshall in a celebratory mood at the end of the War

29. Lawrence Burgis of the War Cabinet secretariat, who took verbatim notes of the War Cabinet throughout the conflict, and his friend Brigadier Leslie 'Joe' Hollis in the Map Room of the Cabinet War Rooms

30. Major-General John Kennedy, Director of Military Operations and later Assistant Chief of the Imperial General Staff, who had much in common with his boss Alan Brooke, including keeping a forthright daily diary

31. Major-General Thomas Handy, Marshall's assistant chief of staff, whose recollections shed light on the inner working of the Operations Division of the US Army after 1942

1944 the Prime Minister tried to persuade Brooke to fly out to visit him in Marrakesh, where he was recovering from pneumonia, saying, 'We must get this Shingle business settled, especially in view of the repercussions of the new proposals about Anvil which will certainly make the US Chiefs of Staff Committee stare.'[7]

Because the Germans had fiercely defended the Gustav Line that winter, Anvil started to resemble not an associated but a rival operation to the Anzio attack, to both Churchill's and Brooke's chagrin as they had never thought its strategic value matched the investment it would require. Although Brooke did not fly out, Bedell Smith, Alexander and Maitland Wilson all conferred with Churchill in early January, and Shingle was resuscitated, in conjunction with an attempt to smash through the Gustav Line to the Liri Valley, which led to Rome. (On his return from Marrakesh, Churchill insisted that a Customs official came to Downing Street in order to assess the duty on everything he had brought home; Lawrence Burgis saw the cheque duly made out to HM Customs and Excise.)

Marshall later acknowledged that the struggles over the size, composition and timing of Operation Anvil had constituted 'a bitter and unremitting fight with the British right up to the launching'.[8] The mutual suspicion was evident at the time, and even in 1949, when Marshall was asked by Pentagon historians whether the British had attempted to use Anvil in order to secure additional resources for the Mediterranean theatre, 'although they never seriously considered actually invading Southern France', he replied that 'this was the case' and 'that's what the British always were doing.'

As Eisenhower's Planners in London increased the number of divisions needed in the initial Overlord assault from three to five, so pressure mounted for extra landing craft and naval assault vessels to come from the Mediterranean. Montgomery and Bedell Smith, who both worked under Eisenhower, agreed in early January that Anvil would be greatly reduced in size as a result. Eisenhower, who like Marshall saw Anvil as an important concomitant to Overlord which would hopefully draw away German troops from northern France, complained vociferously to Washington on 17 January, saying that at Teheran the Combined Chiefs of Staff 'definitely assured the Russians that Anvil would take place'. Since French, British and American troops 'cannot profitably be used in decisive fashion in Italy', Anvil must go ahead, although he accepted that it had to be postponed until early June, to coincide with the new date for Overlord.

Both Churchill and Brooke believed that Allied troops could be used more profitably in Italy than on the French Riviera; the scene was thus set for another titanic clash between Marshall and Brooke, and not one in which Marshall would this time accept compromise, not least because January 1944 was the first month of the war when more American than British Commonwealth troops were engaged fighting Germans in the European theatre.

Yet not all Americans agreed with Marshall and Eisenhower. 'The weakening of the campaign in Italy in order to invade Southern France, instead of pushing on into the Balkans, was one of the outstanding political mistakes of the war,' wrote Mark Clark in his 1951 autobiography, *Calculated Risk*. His Fifth Army had been trying to break through the Gustav Line for several months, with mixed results.

I am firmly convinced that the French forces alone, with seven divisions available, could have captured Marseilles, protected Eisenhower's southern flank, and advanced up the Rhone Valley to join hands with the main Overlord forces. The American VI Corps, with its three divisions, could then have remained in Italy . . . and we could have advanced into the Balkans.[9]

The very mention of an Allied offensive in the Balkans, which Churchill saw as the natural next step after the Germans were expelled from northern Italy, was anathema to Marshall. Michael Howard believes that minds in the OPD were completely closed over the Balkans, 'with its overtones of European subtlety and intrigue'.[10] They also suspected British neo-imperialist designs there, rather as they did in the Far East, however absurd that might have been for the area north-east of the Adriatic Sea in the mid-1940s.

Where did Roosevelt stand? In October and November 1943, the US Planners feared that Overlord might be lost altogether because the President seemed to be interested in Churchill's ideas about the Balkans. 'We were always scared to death of Mr Roosevelt on the Balkans,' Marshall told Pogue frankly in 1956. 'Apparently he was with us, but we couldn't bet on it at all.'[11] There was always the possibility that the President might do over the Balkans in late 1943 what he had done over North Africa in the summer of 1942. It is clear from a telegram Churchill sent Roosevelt in late June 1944 – 'Please remember how you spoke to me at Teheran about Istria' – that the two men had been at the very least 'shooting the breeze' together about a Balkan campaign. As for Brooke, after the war he wrote of the Americans, 'At times I think that they imagined I supported Winston's Balkan ambitions, which was far from being the case. Anyhow the Balkan ghost in the cupboard made my road none the easier in leading the Americans by the hand through Italy!'[12] In fact Brooke had on occasion supported a Balkan campaign, whatever his later protestations.

The Anzio landings of the Allied VI Corps on Saturday 22 January 1944 – initially comprising one British and one American division – might have succeeded had its American commander Major-General John Lucas got inland fast enough to capture the Alban Hills just south of Rome. He had come ashore with minimal opposition because the Germans had sent two reserve divisions from the Rome area to reinforce the Gustav Line, but he decided to get reserves, equipment and supplies ashore first, which proved a costly mistake. Kesselring despatched troops from central Italy to protect Rome, and then further reinforcements from France, Germany and Yugoslavia hemmed VI Corps into a beachhead of only 8 miles, which was defended gallantly for the next four months as Clark fought northwards to relieve it.

'If we succeed in dealing with this business down there,' Hitler told Warlimont, 'there will be no further landings anywhere.'[13] The Führer sent Eberhard von Mackensen's Fourteenth Army, with its crack panzer, panzer-grenadier and paratroop units, to try to destroy the Allied beach-head, leaving the Tenth Army to hold the Gustav Line. The battle-grounds of Anzio and Monte Cassino were constantly reinforced by Hitler in early spring 1944, thereby denuding himself of divisions that he would need to deal with Overlord three months later. Marshall could not understand why Hitler did not merely withdraw his forces to the impregnable Alps, but it was evident from Ultra decrypts that he wanted to defend every inch of Italy instead.

This was Brooke's plan for Italy, and disproves Basil Liddell Hart's theory that it was the Germans who successfully diverted the Allies in Italy rather than the other way around. Throughout 1944, from nineteen to twenty-three German divisions – one-seventh of the entire Wehrmacht – were stationed in Italy, unable to operate in Normandy. In 1943, a full one-third of all Luftwaffe losses were sustained in the Mediterranean theatre, and in all the Italian campaign was to cost the Germans 536,000 casualties against 312,000 Allied.[14] It was far harder to supply the Allies, of course, but the campaign was well worth undertaking in its earliest stages. It certainly tied down far more Germans than Anvil ever could have. The problem was that once committed emotionally – and in Churchill's case chauvinistically – the British carried on fighting for objectives far removed from the central one that had taken them there in the first place.

According to Beaverbrook, who was lord privy seal at the time and had good access to his friend the Prime Minister, Anzio was 'definitely an attempt to re-open the Mediterranean theatre in the hope that such progress might be made there that the Americans could be persuaded to delay D-Day until it would be little more than a mopping-up operation'.[15] He claimed that at Marrakesh Churchill had been talking in terms of 'driving the Germans headlong over the Alps and capturing Vienna'. It is most unlikely that Churchill referred to Overlord as a mere mopping-up operation, however, a phrase which smacks of Beaverbrook's *ex post facto* rationalizations in favour of an early Second Front, of which he had been a chief advocate. For all that, Churchill did write a minute on 25 January saying that it was 'very unwise to make plans on the basis of Hitler being defeated in 1944. The possibility of his gaining a victory in France cannot be excluded.'

It was not long before the failure of the break-out at Anzio became apparent, along with the failure of the Allied forces in the south to link up with the beachhead. On sending Roosevelt birthday greetings on 27 January, Marshall said: 'I anticipate some very hard knocks, but I think these will not be fatal to our hopes, rather the inevitable stumbles on a most difficult course.'[16] The next day Eden, after he had attended a Staff Conference, noted that 'Our offensive seems to have lost its momentum.' When Churchill suspected that he was going to get into a row with the Chiefs of Staff, he used to invite Eden along to give moral support. Even when the Foreign Secretary was recuperating from a cold, sore throat or insomnia at Binderton, he always turned up. Since Churchill had been ill at Marrakesh for as long as a fortnight over the New Year, and Eden was prime-minister-in-waiting, it was a sensible precaution.

On Monday 31 January 1944 Churchill told the War Cabinet of:

Serious disaffection about the Anzio landings. First phase has not yielded brilliant results . . . German offensive started. Great disappointment so far . . . Remarkable limitations of air, unable to prevent enemy from flinging his troops from one Front to another . . . A great opportunity has been lost, but may be regained . . . We have got a lot to learn in the way of seizing opportunities before we can beat these people.[17]

*

Meanwhile in Washington, the Joint Chiefs of Staff expanded the plans for the naval and air support of the Overlord invasion (Operation Neptune). In order to obtain the necessary landing craft for the new plan, and also to help Anvil, the Americans wanted to halt all Italian offensives northwards once Rome had fallen. The British instead wanted Anvil ditched and to press on up the Italian peninsula. Roosevelt supported Marshall against the combined, determined but ultimately doomed attempts of Brooke and Churchill to get the plan changed. The British were as one in believing that terrain, distance and the Germans' defensive tactics meant that Anvil could be safely disengaged from Overlord, with Italy providing fine opportunities for the Allies instead. Churchill and Brooke were never convinced, for the rest of their lives, that Anvil had been worth while.

It did not help that Churchill, even in the early spring of 1944, would sometimes speak, 'as if he were addressing a multitude', of his fears regarding Overlord. 'When I think of the beaches of Normandy choked with the flower of American and British youth,' Eisenhower recalled him saying, 'and when, in my mind's eye, I see the tides running red with their blood, I have my doubts.'[18] Although reminiscences at sixteen years' distance are necessarily suspect, in 1960 Ike reminded Ismay of these remarks without being corrected, and Churchill's faithful Pug was always assiduous in defence of his former boss's reputation, even against a president of the United States.

On 6 February 1944 – two days after Allied troops reached Monte Cassino – Eisenhower asked Marshall from London for his views on retaining Anvil in the face of Churchill's and Brooke's united opposition. Of Overlord he wrote: 'I honestly believe that a five division assault is the minimum that gives us a really favourable chance for success. I have earnestly hoped that this could be achieved by 31 May without sacrificing a strong Anvil.' Eisenhower thought that 'Some compensation would arise from the fact that as long as the enemy fights in Italy as

461

earnestly and bitterly as he is now doing, the action there will in some degree compensate for the absence of an Anvil.'[19] He was thus coming round to the possibility of cancelling Anvil, whereas Marshall certainly was not.

On the first of twenty days of strong German attacks on the Anzio beachhead, Marshall wrote 'For Eisenhower's eyes only' from Washington: 'Count up all the divisions that will be in the Mediterranean, including two newly arrived US divisions, consider the requirements in Italy in view of the mountain masses north of Rome, and then consider what influence on your problem a sizeable number of divisions heavily engaged or advancing rapidly in southern France will have on Overlord.' The fact that there were also the mountains of the Massif Central north of Provence was not mentioned. Instead Marshall concluded: 'I will use my influence here to agree with your desires. I merely wish to be certain that *localitis* is not developing and that the pressures on you have not warped your judgment.' *Localitis* was cod-Latin for 'going native', and since Marshall's 'influence' in Washington was of course enormous, he was effectively advising Eisenhower to stick to his pro-Anvil, anti-Italy position and promising that, if he did, all would be well against Churchill and Brooke.

Eisenhower could not leave the *localitis* accusation hanging, and replied the next day to say that, although the British were opposed to Anvil, he had to compromise occasionally as part of a coalition. Nonetheless, 'So far as I am aware, no one here has tried to urge me to present any particular view, nor do I believe that I am particularly affected by *localitis*.' That Marshall was indeed worried about pressure being put on Eisenhower by Brooke, and more particularly by Churchill, was spectacularly demonstrated the following month at Malta.

On the same day that Marshall wrote to Eisenhower, Sir John Dill told Brooke that he had been 'in and out of Marshall's room lately trying to get him to see your point of view regarding Anvil–Overlord and trying to get his point of view'. He reported that the Joint Chiefs of Staff had delegated their power to Eisenhower on this issue and were 'engaged in a great battle regarding Pacific strategy', which boiled down to 'King in particular v. the Rest'. Dill believed that Marshall was 'somewhat afraid that some of their higher commanders had failed in Italy', doubtless meaning Lucas, who was replaced shortly afterwards, but possibly also Clark, whose progress was painfully slow. Over the post-war occupation zones for Germany, Dill told Brooke that it was, 'of course, the President who won't play. The better I get to know that man the more superficial and selfish I think him. That is for your eye alone as of course it is my job to make the most and best of him.' As for Admiral King, Dill believed 'his war with the US Army is as bitter as his war with us'.[20]

On Thursday 10 February, Brooke lunched at the Fleet Street offices of the *Daily Telegraph* with its proprietor Lord Camrose, as well as the National Labour MP and BBC Governor Harold Nicolson and Lord Ashfield, chairman of the London Passenger Transport Board. Teased about the Anzio reversals by Camrose as he entered – 'Well, what about the bridgehead?' – an irritated Brooke poured himself 'a sulky glass of sherry' and said, 'It's difficult to judge such matters at this distance.' Nicolson recorded that after they had taken some claret in the dining room, 'things brighten up, and a slow flush spreads over the handsome face of the CIGS.' Brooke said that he had first noticed that 'Winston was on the verge of a great illness' at Cairo, when he seemed more interested in swatting flies than in listening to him, and 'then they had great difficulty in preventing him leaving for Italy and were almost relieved when he developed fever.'[21]

Brooke added that, when he visited Italy that December, 'The terrain defies description. It's like the North-West Frontier; a single destroyed culvert can hold up an army for a day.' He then went on to talk about the Germans, saying they were fighting magnificently: 'Marvellous it is, perfectly marvellous.' Hitler's strategy was all wrong, however, in trying to establish a front in Italy so far south while simultaneously holding Nikopol on the lower Dnieper, for 'While one is on the wave of victory one can successfully violate all the established rules of war. But when one starts to decline, one cannot violate them without disaster.'

The fact that Hitler was reinforcing Italy from southern France and the Balkans indicated to Brooke that he was running low on reserves in Germany itself, and was probably hoping for the 'wet period' in Russia from mid-April to early June to move troops from Russia to the west using the 'good transport facilities for a shuttle in the east–west direction'. The Germans did not seem to be running out of oil, and 'the morale of their troops is still admirable and only a slight change can be noticed in the quality of prisoners captured.' From these remarks it is understandable if Brooke was still deeply apprehensive about Overlord.

In the days before lunches at newspapers offices were assumed to be on the record, the Chief of the British Army could make remarks about the 'perfectly marvellous' Wehrmacht that would have gravely embarrassed him were they ever to appear in print.

On 14 February Churchill reassured the War Cabinet about the Anzio beachhead, where there were 130,000 men and 20,000 vehicles, as well as local superiority in artillery and tanks. 'No reason to suppose the situation dangerous,' he said. 'Must keep good nerves this year.' Brooke then informed them further about the heavy fighting at and bombing of the Benedictine abbey of Monte Cassino. 'Nothing to regret that the scale of the fighting has magnified,' added Churchill.[22]

On 19 February, as London was subjected to the heaviest air raids since May 1941, Churchill warned the Chiefs of Staff that 'In the event of Overlord not being successful or Hitler accumulating forces there quite beyond our ability to tackle, it would perhaps be necessary to adopt the flanking movements both in Norway and from Turkey and the Aegean in the winter of 1944–45.' Of course it was important to look at every scenario, but the resuscitation of both his favourite northern and southern schemes in the event of defeat in France shows how doubtful Churchill still was about the coming invasion.

Yet the Overlord build-up was continuing apace, with Marshall informing Roosevelt that by 1 June 1944 there would be forty-one American divisions operational in the US, twenty-one in the United Kingdom, eight in the Mediterranean, and nineteen in the Pacific, and 'there will be a total of 1,514,700 US soldiers in the UK, 2,804 four-engined bombers, 711 medium bombers and 4,346 fighter bombers or fighters.'[23] It sounded formidable, but Eisenhower was meanwhile informing the British Chiefs of Staff that although there were just enough resources for Overlord and a two-divisional Anvil, nothing would be left over for further Italian operations. He also crucially held out the possibility of the abandonment of Anvil if it was for any reason reduced below the level of two divisions. The Italian campaigns, Eisenhower told Brooke, 'have been leading me personally to the conclusion that Anvil will probably not be possible'.[24] Leaping upon this admission, Brooke lost no time in informing the Joint Staff Mission that in the opinion of the Chiefs of Staff the prospect of launching Anvil was now 'exceedingly remote', and recommending its cancellation.

Marshall hand-drafted a letter to Eisenhower two days later, saying that as far as the British High Command was concerned, 'we have no clear cut statement of basis of your agreement or disagreement with them and the situation is therefore seriously complicated. Please seek an immediate conference and reach agreement or carefully stated disagreement, and the Joint Chiefs of Staff will support your decision, subject of course to the approval of the President.'[25] Eisenhower met the British Chiefs of Staff the next day, when they agreed to maintain the status quo until a review on 20 March, and Ike accepted that Anvil would be contingent on the situation in Italy.

'If vehicles could have won the war we'd have won it long ago,' Churchill told the War Cabinet on 21 January, adding that they still could 'Lose this war by running short of unimportant people.'[26] (He was joking that people were unimportant compared to vehicles, not in themselves.) The next day Maitland Wilson wrote to the British Chiefs of Staff recommending that Anvil be cancelled, because until the southern front at Monte Cassino had joined up with the Anzio beachhead, 'the withdrawal of forces cannot be risked from the battle front in Italy.' Maitland Wilson anticipated that the need for extra divisions for Italy, along with a high rate of expected casualties that spring, would drain the Mediterranean theatre of operational reserves in such a way as to render Anvil unviable.[27]

Maitland Wilson's suggestion was fully accepted by Brooke and Churchill, but not by Marshall and Roosevelt, and so the scene was set for a full-scale standoff, with the Americans having the Eureka conclusions from Teheran to support them. In Marshall's view it was obvious that a landing in the south of France was likely to help Overlord more than a continuation of the Italian slugging match; he also considered that the Germans might make a sudden withdrawal northwards in Italy, which would wreck Allied hopes for German dispersion. With Toulon and Marseilles in Allied hands – far better ports than a couple of Mulberry harbours off Arromanches in the English Channel – large quantities of supplies from North Africa could be landed in France rapidly.

On 25 February, a day of difficult discussions covering seven-and-a-half hours, Churchill and Brooke fell to talking first about their children and then about 'The President's unpleasant attitude lately'. The exhaustion of an ill Roosevelt might have been part of the explanation, for late February also saw a visibly tired President leave Washington for a few days' rest, prompting paragraphs to appear in the press touting Marshall for the top job should he resign. When the respected columnist David Lawrence even suggested that this would be a preferable outcome, 'friends' of Marshall told the *New York Herald Tribune* that the general's own view was: 'I'll be in my grave before I go into politics.'[28]

In March and April 1944, discussions in the War Cabinet, Staff Conferences, Chiefs of Staff Committee and Defence Committee centred on how to break through the Gustav Line and link up with the Anzio beachhead in the hope of capturing Rome, preferably before Overlord and instead of Anvil. As Michael Howard states, 'The possibilities beyond that – a breakthrough into the Po Valley, a landing in the Gulf of Genoa, a landing in Istria, a massive switch in forces to the South of France – still lay in the realm of speculation.'[29]

Simply because Brooke and Churchill chatted about their children and agreed over Anvil and the President does not mean they were getting on with each other any better. On 6 March the Prime Minister submitted to the Chiefs of Staff what Brooke called a 'desperate' – meaning hopeless, rather than fraught – memorandum concerning future operations in the war against Japan, and resuscitating the Culverin plan for attacking the northern tip of Sumatra. Brooke was angry that Churchill wanted to bring Eden, Attlee, Lyttelton and Leathers – whom he considered a chorus of yes-men – along to support him over this. (Although Brooke could not have known it at the time, Attlee would be prime minister for the last part of the war.) 'It will be a gloomy evening,' predicted Brooke, 'and one during which it will be hard to keep one's temper.' Sure enough, during the two-and-a-half-hour meeting he 'went at it hard', arguing against all four of them, whose points, he recorded, 'were so puerile that it made me ashamed to think they were Cabinet Ministers!' He did not get much support from his fellow Chiefs either, as Portal was 'as usual not too anxious to argue against PM, and dear old Cunningham so wild with rage that he hardly dared let himself speak!!'[30] Once again the Prime Minister chose not to overrule the Chiefs over Culverin.

Returning to Washington from a fact-finding trip to Algiers in mid-March, Major-General John 'Ed' Hull reported that, although the situation in Italy had effectively stalemated, Anvil should nonetheless go ahead. He feared that without it the Germans could divert forces from Italy, southern France and the Balkans against Overlord, and recommended that all available forces not required for Italian operations be allocated to Anvil. Marshall agreed and told Handy that they must not 'permit our effort to be boxed up in Italy where the geographical situation and the character of the terrain would permit the Germans to play us a scurvy trick to the great disadvantage of our principal effort in the war: Overlord'.[31] Reinforcing a threatened front from a stationary one is hardly 'a scurvy trick', but Marshall was right that attempting to smash through the Gustav Line was a waste of effort by the Allies considering that they had already pinned down as many German divisions as they could, and there was no realistic strategic advantage to being in the Po Valley once Overlord had begun. His alternative – Anvil – was hardly the answer, however, as it turned out to be at best a minor sideshow launched too late to make much difference to Overlord.

Soon afterwards Marshall sent Eisenhower a radio message, reminding him that the 20 March deadline for decision on Anvil was only four days hence, and that 'There is nothing to indicate a sufficient break in the German resistance to permit a further advance on Rome during March.' He feared that German divisions freed up by withdrawal to the Riga Line (also known as the Sigulda Line) might crush Overlord. Even were Alexander's twenty-one divisions in Italy increased to twenty-eight, against the Germans' twenty-four, all but five of which were in the south, he could be held up in the Apennine mountain range, Italy's backbone. Therefore more than ten German divisions might be freed up for France, perhaps up to fifteen, 'to your great disadvantage'. Marshall ended by saying that he left the Anvil decision entirely up to Eisenhower, but he had made his own views very clear.

On 15 March the Allies launched further heavy attacks on Monte Cassino, virtually levelling the abbey – a treasure-house of ancient Christendom – which was nonetheless still fiercely defended and which did not fall for another two months. In the east, the Red Army crossed the River Dniester on 19 March, making it less likely that Hitler might be able to withdraw significant forces for the coming Western Front, although the Riga Line was not crossed until 18 October. The Wehrmacht's capability for inflicting damage was still awe-inspiring, however, even – perhaps especially – in retreat.

It is estimated that in the calendar year 1944 the German Army 'on a man-for-man basis inflicted more than 300% more casualties than they incurred from the opposing Russians'.[32] Brooke's remarks at the *Daily Telegraph* luncheon were therefore fully supported by the statistics. Yet the mechanical balance of the war had changed just as much as the strategic. During 1944, when Germany and the USSR each produced forty thousand aircraft, and Britain twenty-eight thousand, the USA made no fewer than ninety-six thousand, more than the Germans and Soviets combined. Throughout the war, the United Kingdom produced 123,819 aircraft but the United States more than double that, at 284,318, even though she was not a belligerent until December 1941.[33] During the month of April 1944 alone, the Allies dropped 81,400 tons of bombs on Germany and Occupied Europe. As well as a strategic *coup de main*, Overlord would require a huge logistical effort: on 27 April the Chiefs of Staff heard from the Quartermaster-General that 1,600 tons per division would have to be landed every single day for the first forty-two days of Overlord.

The American emphasis on the war in the west was also finally becoming pronounced. By the end of 1943, the United States had 1.41 million men and 8,237 aircraft in Europe against 0.91m men and 4,254 aircraft in the Pacific, despite having two major allies in Europe and no strong one in the Pacific.[34] In the first nine months of 1944, a further 1.8 million men were shipped overseas, of whom over three-quarters went to Europe, and by 1 October 1944 the United States had forty divisions in Europe and the Mediterranean, with four more en route, whereas it had only twenty-seven fighting the Japanese. With 2.75 million troops in Europe and the Mediterranean versus 1.31 million in the Far East, the emphasis of the American war effort was plainly for Germany First. These figures represent 67.7 per cent for Europe against 32.3 per cent for the Pacific, confirming that Brooke should have seized upon King's 75/25 offer. As well as producing armaments for herself, the United States also produced 27 per cent of all munitions used by Commonwealth forces in 1943 and 1944. Overall, Lend–Lease aid to the UK reached a total value of $27 billion, plus an added $6 billion of purchases made in the US before the Act was passed. It was another factor giving ever increasing weight to Roosevelt's and Marshall's views in the councils of the Western Allies over those of Churchill and Brooke.

The lowest point in relations between Churchill and the British Chiefs of Staff during the Second World War was reached in late March 1944. The issue was a vital one: long-term strategy in the war against Japan. Whereas Brooke and his colleagues – but especially the CIGS himself – wanted to approach Japan from the south-west Pacific and Australia in close conjunction with the Americans, Churchill wanted a much more British-led attack from the Indian Ocean, recapturing former British colonies as it moved eastwards. Operation Buccaneer against the Andaman Islands in the Bay of Bengal, which the Chiefs of Staff wanted postponed as its landing craft were needed in the Mediterranean, meant much to the Prime Minister, as did the Culverin operation against northern Sumatra. He felt that in getting Buccaneer effectively cancelled for the rest of 1944 the Chiefs of Staff had gone behind his back while he had been convalescing in Marrakesh, and he therefore wrote a highly abrasive five-page memorandum that he copied to each of them.

'I very much regret that the Chiefs of Staff should have proceeded so far in this matter and reached such settled conclusions upon it without in any way endeavouring to ascertain the views of the civil power under which they are serving,' he began. 'They certainly have the duty of informing me as Minister of Defence, and making sure I understand the importance they attach to the issue.'[35] To lecture on their constitutional duties Chiefs who, in Portal's case, had been four years in the job, and in Brooke's three, was as otiose as it was, frankly, pompous. The reference to 'under which they are serving' was nonetheless a reminder of who held the ultimate sanction over their appointments.

'Considering the intimacy and friendship with which we have worked for a long time in so many difficult situations,' Churchill wrote, changing tack on to the personal, 'I never imagined that the Chiefs of Staff would get into a great matter like this of long-term strategy into which so many political and other non-military considerations enter without trying to carry me along with them, so that we could have formed our opinions together.' Here, all in one sentence, was an appeal to comradeship, an accusation that they were trespassing into areas over which they had no authority, a lament that Churchill was effectively being cut out of the decision-making process, a hint that he could have been persuaded anyhow, and lastly a warning that they needed to speak with one voice to the Dominions and the Americans.

Next came the direct accusation that 'The serious nature of the present position has been brought home to me by the reluctance of the Chiefs of Staff to meet with their American counterparts for fear of revealing to the United States their differences from me and my Cabinet colleagues.' Churchill added that the Defence Committee 'are convinced, and I am sure that the War Cabinet would agree if the matter were brought before them, that it is in the interests of Britain to pursue what may be called the "Bay of Bengal Strategy" at any rate for the next twelve months'.[36] His threat to turn the issue into a political one – 'frock coats' versus 'brass hats' – was explicit, and he was right to assume that his Cabinet colleagues would have supported him, spelling defeat for the service Chiefs.

Pressing home his point, but taking it from the general to the particular, Churchill then gave five direct orders, again cloaked in constitutional terms. 'I therefore feel it my duty', he wrote, as 'Prime Minister and Minister of Defence', to give the following rulings:

(a) Unless unforeseen events occur, the Bay of Bengal will remain, until the summer of 1945, the centre of gravity for the British and Imperial war against Japan

(b) All preparations will be made for amphibious action across the Bay of Bengal against the Malay Peninsula and the various island outposts by which it is defended, the ultimate objective being the re-conquest of Singapore

(c) A powerful British Fleet will be built up based in Ceylon

(d) The plans of the South-East Asia Command for amphibious action across the Bay of Bengal shall be examined, corrected and improved with the desire of engaging the enemy as closely and as soon as possible

(e) The reconnaissance mission to Australia shall be sent as soon as I have approved the personnel.

Churchill then stated that he was willing to discuss these 'rulings' only in order that 'we may be clear in our minds as to the line we are going to take in discussions with our American friends'. His final sentence was similarly unbending, yet contained a typically Churchillian call: 'Meanwhile, with this difference on long term plans settled, we may bend ourselves to the tremendous and urgent tasks which are now so near, and in which we shall have need of all our comradeship and mutual confidence.'[37]

Brooke certainly did not consider these 'rulings' as 'settled'. He knew he had to tread very carefully if he was not to be forced to resign, ultimately over a landing-craft issue that the public – and history – would struggle to understand. Churchill would undoubtedly have prepared the ground politically, indeed as Portal said of this incident four years later: 'If it had come to the point of him wanting to sack me he wouldn't have said "I dismiss you." He would have said "I must tell Parliament about this." [38] Faced with a choice between finding another prime minister or three more chiefs of staff, the Commons would undoubtedly have kept the former.

In an early draft of his *Closing the Ring*, whose title emphasized the correctness of the Mediterranean strategy, with its implication of Nazi Germany being encircled with an ever tightening ligature, Churchill ended the relevant chapter by saying that his 'rulings were accepted and the subject dropped', but as that was quite untrue alternative endings were proposed by those devilling on his manuscript. Churchill refused them, but changed the draft at the last moment. The paragraph accusing the Chiefs of Staff of reaching 'settled conclusions' without trying 'to carry the views of the civil power' was (rightly) considered libellous, so it was excised at the very last moment before publication, leaving an empty space on page 579 of the first edition. [39]

'What the hell of a time you must be having,' Dill wrote sympathetically to Brooke from Washington. 'It is a thousand pities that Winston should be so confident that his knowledge of the military art is profound when it is so lacking in strategical and logistical understanding and judgment.' He reported a great struggle going on over Pacific policy between the American Army and Navy, and commented: 'I hope the US Chiefs of Staff will take the abandonment of Anvil quietly.' He estimated that Marshall's 'greatest fear' was that the Germans would 'within the next month or so, give up Italy to all intents and purposes'. [40]

Dill was doing his best to persuade Marshall that the Germans could take very few divisions out of Italy to oppose Overlord without the Italian front collapsing altogether. He thought that the Joint Chiefs of Staff believed that by the end of May 1944 Germany would be on the verge of collapse, 'and that we should therefore hit her everywhere'. The Americans' view that Hitler wished to withdraw from Italy failed to take into account the Führer's own psychology and the philosophy of Nazism, while their belief that Germany would collapse before Overlord was even launched underrated the extraordinary capacity of the Reich to fight on against what must by then have seemed like an overwhelming, avenging Fate.

On 21 March the British Chiefs of Staff unanimously agreed on 'how best to deal with Winston's last impossible document'. Brooke thought it 'full of false statements, false deductions and defective strategy' and concluded: 'We cannot accept it as it stands, and it would be better if we all three resigned rather than accept his solution.' Resignation in wartime was a very serious matter indeed; as Churchill often used to tell Beaverbrook early in 1942, 'People don't resign in war; you either die or are sacked!'[41]

'It was no courageous thing to resign,' Ismay told Pogue. 'When men were dying they had to have better reasons than pique to cut it.'[42] Did Brooke and his colleagues – who knew it was dangerous to try to bluff Winston Churchill – have a better reason than mere pique? For Churchill to lose another CIGS might be interpreted as a personality clash that any prime minister must win, yet to lose all three Chiefs of Staff simultaneously would have indicated that there had been a profound disagreement over grand strategy, with Churchill pitted against the top three, highly esteemed experts in their field. The effect on national morale three months before Overlord would have been devastating, as Churchill knew. It was probably because the Chiefs agreed to hang together that none was to hang separately.

On 27 March Churchill complained to Portal in the War Cabinet that, for all his nightly bombing of German cities, the RAF was 'not able to knock out Monte Cassino', which had been bombed for nearly a fortnight, yet German units were still holding out there in the rubble. During Brooke's contribution, the Prime Minister announced that he had written to Alexander privately the previous week to say he found it 'puzzling why no attacks on the flanks. Why was Cassino the only point of attack? Please explain why no flank movements can be made. We've broken the teeth of six divisions.' Alexander had answered that Monte Cassino 'blocked and dominated' the main valley leading to Rome, which anyone who has been there will immediately recognize. The conversation then turned to the question of what would happen if Church House, where the House of Commons had sat since the Blitz, were hit by a bomb, prompting Churchill to joke: 'The world would go on. I seldom go there myself.'[43]

Brooke asked Archibald Nye to draft a reply to Churchill's Bay of Bengal memorandum, pointing out five 'fallacies' and concluding that the Chiefs of Staff should 'discuss the subject with the PM and to suggest to him that his action is precipitate, is taken without full knowledge of all the factors and is, in any case, quite unnecessary at this stage'.[44] This formed the basis of the Chiefs' considered response. It was an important moment for Brooke, which he fully recognized could possibly cost him his job were it mishandled. Ismay was allowed to see the reply before it was sent, so Churchill was doubtless forewarned of what it contained, a sensible step for all concerned.

After a seemingly conciliatory opening paragraph – 'We feel sure that there is still some misunderstanding as to our views and proposals, and we welcome the opportunity of a further discussion with you on the whole subject' – the 'Private and Top Secret' reply of 28 March categorically rejected each of the accusations Churchill had made. Its wording has the unmistakable imprint of Brooke upon it, not least in its readiness to trade accusation for accusation. 'We cannot accept the charge that you make,' it stated. 'We did our best to explain our views on long-term strategy for the war against Japan to you before Sextant, but your other preoccupations, both before and after the Conference, precluded this. We were therefore at pains to ensure that the conclusions of the Combined Chiefs of Staff were couched in the most non-committal terms.'[45]

All that had been agreed with the Americans, the Chiefs pointed out, was that joint strategy in the Pacific was to be 'approved in principle as a basis for further investigation and preparation, subject to final approval', wording that was about as nebulous as it was possible to have. They denied that they had yet reached any 'settled conclusions' about the Bay of Bengal strategy because there were still three factors at play, namely Australia's and India's capacities as bases and the shipping situation. They then went on to argue that the south-west Pacific was superior to the Bay of Bengal approach because it 'should, in our view, lead to a substantial shortening of the war against Japan'; it would 'enable us to use the forces and resources of the Empire in a more closely related and concentrated effort than would the Bay of Bengal strategy; and it should not delay the recapture by our own forces of our own territories in Malaya and the Far East.'

This was about as bald a statement as the Chiefs of Staff could deliver, and Churchill was under no illusions that the severest consequences would result in overruling their collective decision, which Brooke, Portal and Cunningham each signed at the foot of the document complete with their titles of CIGS, CAS and CNS. To the document was attached the conclusions of the Staff Conference of 8 March, which stated that there were 'insufficient data upon which to base a decision as to whether the centre of gravity of the main British effort against Japan should or should not be shifted from the Indian Ocean to the South-west Pacific'.[46]

Brooke and his colleagues were calling Churchill's bluff, never a safe option with a statesman so headstrong and unpredictable.

Two days later Brooke wrote to Dill in a blatant attempt to get Marshall, and possibly also Roosevelt, to support him in his struggle against Churchill. 'I have just about reached the end of my tether and can see no way of clearing up the frightful tangle that our Pacific strategy has got into,' he wrote. 'In fact we feel that the Indian Ocean policy will result in our walking round with the basket picking up the apples whilst the Americans climb up into the tree and shake the apples off by cutting [Japanese] lines of communication. The PM, on the other hand, remains as determined as ever to do Culverin, and has got very little else as a plan beyond the capture of Culverin!'

Since the US Chiefs had not approved the Sumatran attack, 'We might be able to fight this situation out as we have others before now,' Brooke told Dill, 'if it was not for all the other complications that I am coming to.' The first was Mountbatten, 'who is determined to do something to justify his Supremo existence', just as much as Churchill was keen to justify 'creating Dickie and his command'. The second was John Curtin's government in Australia, which under MacArthur's influence did not want a British or Commonwealth force operating from Australia as a self-contained whole, even under American overall command.

Washington presented Brooke with 'further difficulties', because he could not detect 'any great urge' from the Joint Chiefs of Staff for co-operation with Britain in the Pacific. 'In fact we have grave doubts as to whether King is not opposed to such a strategy.'[47] Admiral King was indeed opposed, as they were about to discover very soon. Despite their pressure to recapture northern Burma, the American Chiefs disapproved of Culverin, and Brooke declared himself 'quite clear in my own mind that strategically it is right for us to use all our forces in close cooperation from Australia across the Pacific in the general direction of Formosa' (modern-day Taiwan), but admitted that 'Unless we get 100% support and drive from the American Chiefs of Staff I rather doubt where we may finish up!' He was thus effectively asking Dill to get Marshall to intervene with Churchill over Culverin.

It was at around this time that, as Joan Bright later related:

Brooke in the company of other ministers was far more rude to the PM than he had any right to be – and Churchill was shocked. He broke up the meeting and said to Ismay: 'I have decided to get rid of Brooke. He hates me. You can see the hate in his eyes.' Ismay said: 'I think that he behaved very badly at the meeting but he is under terrific strain. He is bone honest and whatever else his views may be, he doesn't hate you.' Ismay left then to see Brooke and said: 'The PM is frightfully upset and says you hate him.' Whereupon Brooke said 'I don't hate: I adore him tremendously; I do love him, but the day that I say that I agree with him when I don't, is the day he must get rid of me because I am no use to him any more.' Asked if these words could be repeated to the PM he said 'Yes'. Ismay went back and told Churchill what had been said and his eyes filled with tears. 'Dear Brookie.' That was the last row they ever had.

Barring the last sentence – there were plenty more rows over the next sixteen months – this is an accurate summing up of an incident that is also recorded in Ismay's memoirs in much the same emotional and personal terms. 'When I thump the table and push my face towards him,' Churchill said of Brooke, 'what does he do? Thumps the table harder and glares back. I know those Brookes – stiff-necked Ulstermen and there is no one worse to deal with than that!'[48]

Whether Churchill really did intend to sack Brooke, or more likely to rap him over the knuckles via Ismay, is impossible to say. Similarly, if Brooke saw his job on the line over personal differences, it was natural that he would deny 'hating' his minister of defence, although the protestation of 'love' also rings hollow. To lose one's job over the Bay of Bengal (India–Burma–Malaya–Japan) strategy versus the south-west Pacific alternative (Australia–Philippines–Formosa–Japan) might be understandable; yet to lose it over a personality clash, especially after serving together for so long, would have been absurd. As it was, Churchill did not push the issue to the point of resignations or sackings, although neither did he ditch the concept altogether.

Despite potentially needing Marshall's support over Far Eastern policy, Brooke was not about to defer to him when it came to the Mediterranean. 'As far as Anvil is concerned I am giving up hope of getting Marshall to understand what the situation is in Italy,' he wrote to Dill. 'It has taken two months arguing with him for him to see that the situation in Italy now is what could be predicted some time ago.' He considered that staging Anvil at the same time as Overlord would be 'impossible'.[49] Brooke believed that the ten divisions Marshall now wanted for Anvil would not be enough 'if you want to *hold* forces opposite us in Central Italy'. Brooke ended the letter 'with best love to you both, Yours ever, Alan', a particularly affectionate signing-off from this somewhat emotionally buttoned-up Ulsterman and now field marshal. In the Far East, therefore, Brooke sought Marshall's help against Churchill over Culverin, while in the Mediterranean he sought Churchill's help against Marshall over Anvil. Meanwhile, in Washington, Marshall complained to Dill that Roosevelt was being stubborn over the issue of unconditional surrender; he wanted to find a definition of victory that would encourage enemy forces to surrender, but he told Lord Halifax that he was 'up against an obstinate Dutchman [that is, Roosevelt] who had brought the phrase out and didn't like to go back on it'.[50]

'So many interesting things are happening', wrote Admiral Cunningham on 1 April 1944, 'that I think it behoves me to keep a diary.' In fine handwriting, his first entry was about how the Joint Chiefs of Staff were being 'quite inflexible' over Anvil. Yet another public servant was now keeping a daily journal, which has never been published but is invaluable in flooding light upon the deliberations of everyone taking part in the higher direction of the western part of the war. Three days into keeping it, when the War Cabinet was informed that there could be as many as 160,000 civilian casualties as a result of bombing the French railway network prior to Overlord, Cunningham noted, 'Considerable sob stuff about children with legs blown off and blinded old ladies but nothing about the saving of risk to our young soldiers landing on a hostile shore. It is of course intended to issue warnings beforehand.'

April Fool's Day 1944 also saw Dill warn Brooke that the Joint Chiefs of Staff, apropos the reinforcements being sent to Italy, were 'shocked and pained to find out . . . how gaily we proposed to accept their legacy while disregarding the terms of their will'. Among themselves the US Planners drew up a memorandum for General Handy entitled 'What Shall We Do about Anvil?' which argued that, without the operation, the US would be 'committed to a costly, unremunerative, inching advance in Italy', which would be politically unpopular in France and at home and might end with American troops being used as occupation forces in Austria, Hungary and southern Germany.[51]

From the adjective 'unremunerative' it is clear that a certain degree of cold-bloodedness was needed at the senior Staff level in order simply to continue to do the job. 'If you let yourself get all involved in the personalities and cry that "here's a poor man gonna get killed", you'll lose your country,' recalled the OPD Planner Paul Caraway in his unpublished memoirs:

As far as we were concerned, when we said that we had ten divisions that we were going to put into this operation, those ten divisions weren't a hundred and fifty thousand people, they were 150,000 units, ones, entities, and we calculated with a completely passionless arithmetic as to what those 150,000 units pack. We expected to get back the maximum number possible, and we hoped the commanders would do all right, but what we had to have were results.[52]

Caraway's brother was in the forefront of every major attack of the 28th Division, but, as Caraway concluded after the war, 'Of course you do it this way.'

From Cunningham's journals it is evident that the Chiefs of Staff were looking towards the post-war situation, with a suspicious eye towards Russia, almost before any other British government agency or institution. Over the future of Middle Eastern oil supplies, for example, he noted in early April 1944 that they were approving the idea of American involvement there, 'so as to have USA support should Russia in post-war days cast sheep's eyes at Iranian or Iraqi oilfields'. This showed impressive foresight, considering that parts of the Crimea were still under German occupation at the time.

On Easter Monday, 10 April, Dill wrote to Marshall to 'thank you for all the understanding consideration which you have shown us, and me in particular, in our difficult negotiations concerning the Anvil–Overlord disagreement. It has made me feel once more that honest disagreements are of relatively small importance so long as we are completely honest and frank in all our dealings.'[53] This was not how Brooke felt. Only two days earlier the CIGS had told his diary that the Joint Chiefs of Staff 'have at last agreed to our policy, but withdrawn their offer of landing craft from the Pacific!! This is typical of their methods of running strategy.' Brooke believed that the Americans were using the landing craft 'as bargaining counters' in trying to pursue a Pacific over a Mediterranean strategy. This was unwarranted; in fact, the Americans were constantly using landing craft in the Pacific for precisely the purpose for which they had been built.

Two days later, Churchill wired Marshall via Dill, saying that he was convinced that the decision to implement Anvil could not be taken until the Anzio beachhead had been linked up to the British Eighth and American Fifth Armies in Italy and the initial results of Overlord evaluated. Marshall replied the next day that they appeared 'to be agreed in principle but quite evidently not as to method'. He argued that in order to keep options open when the time came, preparations for Anvil needed to be made immediately, even though they 'may be at the partial expense of future operations in Italy', and if Anvil turned out to be the wrong operation, Maitland Wilson would therefore always have 'an amphibious force available to carry out another and less difficult amphibious operation'. He added that the 'momentum' of US operations in the Pacific meant that the forces there could not be 'hamstrung'.[54]

Churchill then proposed to Roosevelt that a joint telegram, to be signed 'Roosevelt–Churchill', be sent to Stalin informing him of the agreed date of Overlord, 1 June 1944. A second paragraph should promise a 'heavy offensive which we shall launch in Italy with all out strength in mid-May'. However, as Marshall told Roosevelt, Combined Chiefs of Staff representatives in Moscow had already informed the Russians about Overlord, and Churchill's second paragraph might be taken in London to infer the cancellation of Anvil. He therefore advised the President not to authorize the sending of the telegram, and suggested another, somewhat anodyne, draft to Churchill, which was sent off unchanged by Roosevelt.

By then the British Chiefs believed that, in Cunningham's words, 'the only thing to do' was to 'abandon Anvil'.[55] On 15 April, as Brooke 'rattled through the business in great style', they drafted a new directive to Maitland Wilson 'cutting out Anvil' altogether. The next day Churchill wrote to Marshall regretting that landing craft could not be diverted from the Pacific to the Mediterranean, as there was no definite date commitment to Anvil. He did not want the landing craft to come from Italy, as he could not 'agree beforehand to starve a battle or have it break off just at the moment when success, after long efforts and heavy losses, may be in view'.[56]

The British and Americans knew that without extra landing craft there could be no two-divisional 'harvest bug' move to break the Italian stalemate further north, and probably no Anvil either. 'Dill tells me that you had expected me to support Anvil more vigorously in view of my enthusiasm for it when it was first proposed by you at Teheran,' Churchill told Marshall. Yet that was before the Allied attack got bogged down south of Rome. According to the Prime Minister, the Germans were committing to Italy the very divisions that Anvil had been designed to divert from Overlord. It was a difficult thesis to prove, and his last paragraph was classically Churchillian, all about how 'We must throw our hearts into this battle', in Italy, 'and make it like Overlord an all-out conquer or die'. It was just the kind of language that had no purchase with George Marshall.

Simultaneously, the British Chiefs were telling Marshall via the Joint Staff Mission that 'we cannot possibly agree, here and now, that preparations for an Anvil should have priority over the continuation of the battle in Italy,' and this would be true even after the Anzio beachhead had joined up with the main battle line. Without Pacific landing craft, they stated, far too precipitately and didactically as it turned out, 'the possibility of Anvil, as a supporting operation to Overlord, is terminated'.[57] Like the British words 'demand' and 'tabled', 'terminated' did not go down well with the Americans. It was always dangerous to present a man in charge of a six-million-man army with a fait accompli, but by 18 April Marshall had to conclude, as he told Churchill, that 'Since Eisenhower's assault is not to be supported by a landing in southern France, every possible deceptive effort ... will have to be utilized to hold the German divisions in southern France during the critical days of Overlord.' Marshall seemed to be admitting that Anvil was moribund; but was he?

The next day, Churchill once again expressed his severe doubts about Overlord itself, minuting to the Foreign Office: 'This battle has been forced upon us by the Russians and the United States military authorities.' His lack of enthusiasm was manifest, yet although he quoted from other parts of the minute in his war memoirs, that sentence and others like it were excised; indeed it is next to impossible for any reader of *Closing the Ring* to spot the slightest Churchillian doubt about the success of Overlord six weeks before it was launched.[58]

On 22 April Churchill attempted to set up a line of communication with Marshall separate from both Roosevelt and Brooke, writing to Dill that the Chiefs of Staff 'did not much like' Marshall's last cable, which had refused the landing craft. 'Let me know if he is offended by my corresponding through you with him,' he went on. 'I thought that as the

President was away he would readily understand my difficulties. If he does not like it, you may assure him that it shall not happen again. I will send direct to the President, who will soon be coming back. This will not mean any personal cooling-off in my relations with the Senior American Officer.'[59]

Dill replied that he could see little to dislike in Marshall's landing-craft cable and that he felt it 'wrong for you to raise direct with Marshall military questions demanding an answer when such an answer can only be given by the United States Chiefs of Staff'. Problems would arise when King discovered what was going on, yet since 'Marshall often wants to know how you personally are thinking,' Dill suggested that Churchill use him as the go-between. 'The President, as you know,' Dill ended, 'is not militarily minded and you will, in my view, gain little by referring purely military questions to him.'[60] This was an astonishing remark to make at that stage of a world war about one of America's greatest wartime presidents, but it was essentially true. Without telling Brooke, therefore, Dill had consented to be Churchill's go-between with Marshall behind Brooke's back, while he was also Brooke's long-term go-between with Marshall behind Churchill's.

Churchill would occasionally address questions to British Staff officers besides Brooke, but they automatically sent the answer to the CIGS first for approval. Brooke recalled that Churchill had once asked him: 'How is it that whenever I write to any officer of the General Staff at the War Office I get a reply from you?' Brooke told him that he 'should prefer him to address such minutes direct to me, but that even if he chose to ignore the chain of responsibility he would still get replies from me!'[61] The CIGS was a sufficiently experienced Whitehall departmental warrior not to lose control over these all-important lines of communication.

At the London Conference of Dominion prime ministers that opened on 1 May, Churchill made the same joke that he had on several other occasions to different audiences. The Americans, he said, 'all along said we were leading them up the garden path in the Mediterranean', which, while true, 'had provided them with nourishing vegetables and fruits. Nevertheless, the Americans had remained very suspicious.'[62] This had been funny the previous year when Tunisia surrendered with nearly a quarter of a million prisoners taken, and also when Sicily had fallen in the space of five weeks, but the punchline had worn thin by 1944, when

Salerno, Monte Cassino and Anzio had proved no garden path and it was difficult to see what strategic fruits were still waiting to be plucked after Rome.

Staying at Chequers on the weekend of 6 to 8 May, to meet William Mackenzie King and John Curtin, Brooke was shown to a small study in which the secretaries worked, where Churchill told him in confidence that 'Roosevelt was not well and that he was no longer the man that he had been, this he said also applied to himself.' Churchill added that he could still sleep well, eat well, 'and especially drink well!' but that he no longer jumped out of bed the way he used to, and felt as if he would be 'quite content to spend the whole day' there. This was the first time that Brooke had ever heard him 'admit that he was beginning to fail'.[63] For all the adrenalin that pumped through the amazing, bull-like constitution of the Prime Minister, he was now in his seventieth year and had suffered pneumonia among other illnesses (though not the full-scale heart attack that is sometimes attributed to him in December 1941). Yet whereas the sixty-two-year-old Roosevelt was dead in under a year, Churchill had twenty more to live.

Thursday 11 May was to witness Harold Alexander's and Mark Clark's Diadem offensive, which finally broke through the Gustav Line – Monte Cassino fell to the Poles a week later – linked up with the Anzio beachhead and captured Rome on 4 June, despite a highly skilled retreat by the Germans that preserved most of their forces. Marshall put much of the credit for this down to Clark's brand new 85th and 88th Divisions, which he thought proved that the US Army was ready for Overlord. There would now be no opportunity for German redeployment from Italy to France to deal with Overlord, so one of the main objects of the Allied campaign had been fulfilled. Indeed, late May brought the news that the Germans were taking four divisions from western Europe to reinforce Italy. 'Whatever happens,' exulted Cunningham, 'the battle has fulfilled its purpose of keeping the Italian divisions' – by which he meant the German divisions based in Italy – 'away from the Overlord battle.'[64] It was not now strategically necessary for Churchill and Brooke to insist upon another, painful, costly, drawn-out mountainous advance up to the Pisa–Rimini Line much further north, yet this is what they did, partly because it was a British-led operation.

On 15 May almost the entire upper echelon of the Allied High

Command except the Joint Chiefs met at St Paul's School in Hammersmith in west London in order to be briefed on the Normandy landings by Eisenhower, Montgomery (who had once been a pupil), the naval Commander-in-Chief of the Allied Expeditionary Force Admiral Sir Bertram Ramsay, the Commander of the US Strategic Air Forces in Europe Carl 'Tooey' Spaatz (who made a bad impression by reading his speech), Sir Arthur ('Bert' or 'Bomber') Harris of Bomber Command, Sholto Douglas and three others. The King also spoke briefly, before presenting the American General Omar Bradley with the insignia of a Knight Commander of the Order of the Bath. In his speech, Churchill ought to have chosen a better form of words than 'Gentlemen, I am hardening towards this enterprise,' which let everyone know, in Eisenhower's view, that the Prime Minister 'had long doubted its feasibility' and had only finally, 'at this late date, come to believe with the rest of us that this was their true course of action in order to achieve the victory'.

In fact the preparations for the Normandy landings, as Eisenhower knew better than anyone, had been progressing with Churchill's active political support for several years. All that was in doubt was the date of their launch. Millions of maps had been printed; thousands of aerial photographs compiled; millions of rounds of ammunition had been stockpiled; hundreds of miles of roads had been built; 6,250 pounds of sweets, 12,500 pounds of biscuits and one hundred thousand packets of chewing gum had been distributed; sixty days' supply of poison gas was prepared for retaliatory use; 25 square miles of west Devon were evacuated of its civilian population for training; mass rehearsals were conducted with live ammunition; vast encampments were built all over southern England with their own water supplies, field bakeries, post offices and camouflages; immense ingenuity and inventiveness were directed towards making the assault a success, including artificial harbours and underwater petrol pipelines.[65] Although the Americans brought over a great deal in convoys – necessitating forty ships per armoured division – the British provided a huge amount in terms of logistics too, which Churchill never begrudged, but rather drove through with insistence on administrative efficiency and vigour.

In a War Cabinet discussion on Monday 22 May, regarding the request of a Yugoslavian general to be parachuted back into his homeland in order to make contact with Marshal Tito, Churchill joked to the

War Cabinet: 'I'm not sure I should make a good landing by parachute; I'd break like an egg.' (The mental image of Humpty Dumpty is hard to avoid in this context.) He then reported on Italy: 'The battle is very heavy. Alex is pleased. The French have made great advances on the left flank. The [Allied] armies are hinging north of Cassino . . . The next [few] hours may produce very remarkable results. There have been seventeen thousand total [Allied] casualties.' Brooke then listed the advances made since the previous Monday, starting with a night attack across the River Rapido. Churchill was impressed by the way that Germany was being taken on in Italy 'by four separate countries', Britain and her Dominions, America, France and Poland, and insisted on 'publicity for Britain in communiqués', a regular demand of his. Otherwise he felt that although they had a quarter of the troops engaged, it 'looks as if the British were laggards in the show'.[66]

Two days before that meeting, on Saturday 20 May 1944, Admiral Cunningham had been driving from a football match to his country home in Bishop's Waltham near Southampton in Hampshire when he knocked a postman off his bicycle on the Tolworth roundabout; the victim died two days later. Cunningham employed characteristically nautical terms to describe to his journal how the man had 'turned sharp across my bows and I caught him on the back wheel. The car turned 180 degrees and charged up on to the pavement stern first . . . the Bentley nearly capsized.'[67] The coroner's verdict of accidental death at the inquest, the admiral recorded, was 'Very satisfactory for me but a tragedy for the poor widow.' Brooke and Cunningham now had something somewhat macabre in common, right down to the make of car they had both been driving. (The two men also died within five days of one another in June 1963, Cunningham in the back of a taxi taking him from the Athenæum Club to Waterloo Station.)

The human cost of the Italian campaign was brought home heavily to George Marshall when he arrived at his office on the morning of Tuesday 30 May, to find a personal radio message from Mark Clark informing him that Allen Brown, his wife Katherine's twenty-seven-year-old son, had been killed in action the previous day near Compoleone, south of Rome. Having no children of his own, Marshall had been a devoted stepfather, and he was utterly devastated. Brown had enlisted as a private and had progressed on his own merits to become a second lieutenant in the 1st Armored Division, the rank he held when he opened his tank turret to direct a group of Italian refugees down a road towards safety, just as a German sniper struck. Only the day before, the Marshalls had received a letter from him, looking forward confidently to the end of the war.

'Allen was the apple of his eye,' Frank McCarthy recalled of his boss, who as well as grieving with Katherine talked to friends who had also lost sons in the conflict, including Lord Halifax and Harry Hopkins.[68] Their only possible consolation was that within a week the beginning of the liberation of western Europe would be at hand. 'This is a distressing message to send,' Marshall wrote, before giving the circumstances of Allen's death to his widow Margaret.[69] The couple had a two-year-old son. Allen Tupper Brown was buried in the beautifully kept Sicily–Rome American Cemetery and Memorial just outside the town of Nettuno, 38 miles south of Rome. To have lost his beloved stepson was painful enough for George Marshall, but to have lost him to a campaign that he had never believed would bring victory significantly closer must have made it more terrible still.

18

D-Day and Dragoon:
'This world and then the fireworks!'
May–August 1944

A democracy cannot fight a Seven Years' War.

General George C. Marshall, 1957[1]

'There have been recently disquieting signs of a possible divergence of policy between ourselves and the Russians in regard to the Balkan countries and in particular Greece,' Churchill wrote to Roosevelt on 31 May 1944. 'We therefore suggested to the Soviet Ambassador here that we should agree between ourselves as a practical matter that the Soviet government would take the lead in Roumanian affairs, and ourselves in Greece. I hope that you would give this proposal your blessing. We do not, of course, wish to carve up the Balkans into spheres of influence, and in agreeing to such an agreement we should make it clear that it applied only to wartime conditions.'[2] The last sentence was superfluous: of course such an agreement would be a carve-up and equally obviously it would survive into the post-war world.

As might have been expected, Roosevelt replied negatively, saying that such a deal would only 'result in the persistence of difficulties between' Britain and the Soviets and instead they should 'establish consultative machinery and restrain the tendency towards the development of exclusive spheres'. The President was not about to challenge the Red Army in eastern Europe. Before the development of the atomic bomb – the threat of the use of which was anyway utterly unthinkable against their courageous ally – there was nothing much the Western Allies could do about the Red Army, which mobilized twenty-five million troops during the Great Patriotic War. In the feverish 'Red Scare' of the early Cold War, Roosevelt's supposedly appeasing policy towards the USSR was made much of politically by the Republican right and Senator Joseph McCarthy, but there was no genuine alternative, and indubitably not one that would have been acceptable to domestic political opinion in the West.

Churchill shared Roosevelt's belief that it was possible to win the friendship of Stalin, and that it was worth something once gained. Yet on 5 June 1944 the Yugoslavian Communist statesman Milovan Djilas met a 'lively, almost restless' Stalin at his dacha outside Moscow. Stalin left Djilas – and, through him, the Yugoslavian partisan leader Marshal Tito – in absolutely no doubt about the true level of trust and warmth he felt for his Western Allies. 'Perhaps you think that just because we are allies of the English we have forgotten who they are and who Churchill is?' Stalin asked. 'There's nothing they like better than to trick their allies ... And Churchill? Churchill is the kind of man who will pick your pocket of a kopeck, if you didn't watch him. Yes, pick your pocket of a kopeck! By God, pick your pocket of a kopeck! And Roosevelt? Roosevelt is not like that. He dips in his hand only for bigger coins. But Churchill? Churchill – will do it for a kopeck.'[3] Stalin then ridiculed Overlord, predicting that it would all be called off if there was fog in the English Channel. 'Maybe they'll meet with some Germans!' he continued, accusing the Allies of cowardice.

Yet, far from cowardly, British strategy had been like that of the matador in a bullfight, and now was the moment for the sword to be thrust between the bull's shoulderblades. 'It is very hard to believe that in a few hours the cross Channel invasion starts!' wrote Brooke on 5 June. 'I am very uneasy about the whole operation. At the best it will fall so very very far short of the expectation of the bulk of the people, namely all those who know nothing of its difficulties. At the worst it may well be the most ghastly disaster of the whole war. I wish to God it were safely over.' Churchill reacted very differently; after a 'good lunch and as usual lots of wine' at Downing Street, Cunningham wrote, 'PM very worked up about Overlord and really in almost a hysterical state ... He really is an incorrigible optimist. I always thought I was unduly so but he easily outstrips me.'[4] Perhaps Churchill's optimism hid a deeper fear, and was primarily adopted to put heart into those around him, part of the duty of a leader on the eve of so great an undertaking as the Normandy landings.

The Prime Minister was not quite so optimistic at the evening War Cabinet. After extolling Alexander's 'brilliant campaign' to capture Rome – which had finally fallen the day before after nine months – he said of Overlord that the 'Danger of this operation is very great during the next thirty or forty days.' Nevertheless he did think that 'We will get ashore and establish [a] bridgehead.'[5] Over Rome, the Secretary for India Leo Amery recorded in his diary that 'Winston expressed himself very disappointed that Brooke could not assure him that we could overtake and destroy all the retreating Germans.'

Most of the rest of the meeting was spent discussing de Gaulle, who had refused to broadcast to the French people on their liberation unless certain political conditions were met by Eisenhower. 'I'm nearly at the end of my tether,' Churchill said, although he accepted that 'it doesn't make the slightest difference to the outcome.' When Eden tried to explain de Gaulle's objections, the Prime Minister interjected: 'De Gaulle refuses to participate. It is an odious example of his malice . . . he has no regard for common causes – I may have to exhibit him in his true light, as a false and puffed up personality.'[6] Cunningham summed it all up as: 'Prima Donna de Gaulle making a nuisance of himself.'

The first time that Major Eisenhower met Harry Butcher, at the home of Butcher's brother's landlord in Chevy Chase in 1927, Ike had performed a strange party trick: 'He stood stiffly erect, slowly fell forward without moving a muscle, but, at the last instant just before it seemed as if he would break his nose on the floor, his strong hands and muscular arms quickly broke the fall.' Now, seventeen years later, would he get a bloody nose in Normandy? He feared defeat enough to write a resignation letter to Roosevelt beforehand, which turned up in one of his pockets some time afterwards.

As they went to bed that night, Churchill said to Clementine: 'Do you realize that by the time you wake up in the morning, twenty thousand men might have been killed?' The actual figure for D-Day was 4,570 Allied soldiers killed, comprising 2,500 Americans, 1,641 Britons, 359 Canadians, 37 Norwegians, 19 Free French, 12 Australians and 2 New Zealanders. Afterwards, however, the numbers stayed high, and in the thirty-five days after 6 June Allied casualties – that is, killed and wounded – ran at an average of 1,800 per day, as compared to 2,121 per day during the battle of Passchendaele.

It is not known whether Eisenhower on the morning of D-Day got out of bed crying, 'This world and then the fireworks!', as Butcher recorded he used to do in the mornings, but he was called early by

Admiral Ramsay with good news about the naval side of the landings and shortly after 6.40 a.m. there was more from the RAF's Trafford Leigh-Mallory about the air situation, with only twenty-one of the 850 American C-47s and eight out of the four hundred British planes missing, and only four gliders unaccounted for. Most 'Hun' night-fighters, Leigh-Mallory reported, were over the Pas de Calais and only three had flown over the beaches themselves.[7]

On 6 June 1944 there were twenty-eight German divisions stationed in south-eastern Europe, twenty-eight in Italy (of which only twenty-three had an effective manpower strength), eighteen in Norway and Denmark, and fifty-nine in France and the Low Countries.[8] Some historians quote as low a figure for France and the Low Countries as fifty divisions, with the confusion arising from the classification of some of the divisions as 'reforming' and 'refitting' at the time of the invasion, including at least half a dozen in the south of France.

Everything is relative, and in the words of the British Official History, 'in the months preceding Overlord the number of German divisions rose in France and the Low Countries at times when they [also] rose slightly in Italy and South-east Europe'. In January 1944, there were twenty-eight panzer divisions facing the Russians and eight in the rest of Europe, whereas by mid-June these figures had changed to eighteen and fifteen respectively.[9] Hitler was thus reinforcing both France and Italy at the expense of his Eastern Front. Yet that does not detract from Brooke's argument that, if there were no active Italian front, there would have been even more German divisions in France.

Five infantry divisions, three airborne divisions and three armoured brigades landed on and behind the five Normandy beaches on D-Day itself, which were built up to twenty-four divisions by D+30 days and more than thirty divisions by D+60. Within immediate range were six German infantry divisions and one panzer division, but within 200 miles there were a further twelve infantry and two panzer divisions, making twenty-one German divisions in all, many of them high-quality, veteran units.[10] D-Day thus needed all the diversions it could possibly get. (It also needed all the ingenuity it could get, but probably not the idea that Admiral King suggested to the Joint Chiefs of Staff for large numbers of American jackrabbits to be swept up, taken over to Normandy and released to bound ahead of the GIs, tripping the wires and letting off the mines and booby-trap grenades. The idea was duly noted in the records.)[11]

There have been many, such as Albert Wedemeyer, who have argued that Overlord would have been better undertaken in 1943 than 1944. Interviewed by SOOHP, General Handy recalled the time that Wedemeyer told Churchill that it 'might have shortened the war by a year':

The Old Man threw a complete conniption fit. But a good case could be made for Al's thesis. We took the scarcest thing and used it in the most ineffective way. We moved all our troops to Africa and we had to move them all back with the thing that was shortest of all, shipping. Now, if we had concentrated and continued to build up in the UK, you can easily picture we might have been ready a year ahead.[12]

Yet Handy himself went to Omaha Beach along with Rear-Admiral Cooke, and described it as 'carnage'. He remembered thirty years later: 'Well, hell, that was the one place I got close to the war . . . I tell you, we hung on there by our eyelashes for hours. It was God-awful . . . it was nip and tuck . . . Afterwards you lose sight of how close it is if it's a success.' He himself was losing sight of the likelihood of success in believing that the invasion could have taken place a year earlier.

'If you see fighter aircraft over you,' Eisenhower told his men just before D-Day, 'they will be ours.' He was right: Allied planes made 13,688 sorties that day, to the Germans' 319, but it was not a promise the Supreme Commander could have made his men before 1944. It is true that up to two million slave-labourers had built the Atlantic Wall since 1942, but set against that are the factors that the Germans could have concentrated forces faster against the beachhead in 1943 than in 1944, when bombing had ruined many supply lines; American troops coming to Britain would have had to face the U-boats in 1942 and 1943, when the menace was at its height; without Torch and Husky the Germans might have taken the offensive in the Mediterranean – as Rommel begged Hitler to do – perhaps with Malta falling, the Middle East being overrun, Turkey brought into the war on the Axis side, and Russia left vulnerable through the Black Sea.[13] Hundreds of thousands more Germans had been killed on the Eastern Front by 1944 than was so in 1942, and above all, over the question of morale, acute German officers recognized by June 1944 that the war was likely to be lost in a way that they simply did not before the battle of Kursk.

What is certain is that the invasion could not have been delayed much beyond the summer of 1944, for London was coming under devastating nightly attacks from both V-1 flying bombs and V-2 rockets. For all the deliberations of the Government's special anti-V weapon 'Crossbow' Committee, one of the few defences against the V-1s was to fly close to them and tip their wings – an immensely risky and difficult manoeuvre if one was lucky enough to intercept them at all – and the only response to the V-2s was to destroy their very well-camouflaged launching sites, preferably from the ground.

In 1956 Marshall made a terrible admission to Pogue about the lack of proper intelligence before D-Day. 'Don't quote this,' he told his biographer, but 'We didn't know we were going to hit such rough country . . . G-2 let me down every time in everything. They never told me what I needed to know. They didn't tell me about the hedgerows, and it was not until later, after much bloodshed, that we were able to deal with them.'[14] Later in the same interview, after Marshall had said, 'We had to pay in blood for our lack of knowledge,' he repeated: 'Don't print that.' It is traditionally an invocation, along with 'Burn this letter,' that automatically guarantees careful preservation and extensive quotation. The admission that Marshall was not warned about the *bocages* – the deep, thick, ancient Norman hedgerows that gave the German defenders such fine defensive cover – is a serious one, and a significant failure of US Military Intelligence (G-2). Brooke knew all about them because he had retreated over precisely that ground to evacuate from Cherbourg in June 1940, but his warnings were largely disregarded as yet another excuse for not invading.

The day after D-Day, Alexander reported that if he were left with his twenty-seven divisions in Italy, and not lose any to Anvil, he could break through the Apennines into the Po Valley, take eighteen divisions north of Venice and force the Ljubljana Gap between Italy and northern Yugoslavia. Once there, he stated in his memoirs, 'the way led to Vienna, an object of great political and psychological value'. This prospect appealed to Churchill and Clark, but not to very many others. Brooke told Churchill that with Alpine topography and winter weather, 'even on Alex's optimistic reckoning . . . we should have three enemies instead of one'.[15]

Marshall vociferously opposed forcing the Ljubljana Gap, arguing that Eisenhower needed the southern French ports so that he could deploy on a much wider front, and that the Germans would merely withdraw from north Italy to the Alps under Alexander's attack, which could then be held with far smaller forces. Alexander forever thereafter believed that his 'lost' seven divisions – about which he lamented as much as did the Emperor Augustus for his lost legions of the Teutoburgerwald in AD 9 – would have 'broken the Gothic Line in August, crossed the Po in September, captured Venice and Trieste in October' and so on. Since they had been withdrawn for the Overlord assault, pursuant to the agreement between Brooke and Marshall, we shall never know what they might have achieved elsewhere, but Anvil could not have been a success without them.

Churchill's dreams of British Commonwealth forces planting the Union Jack over Schönbrunn and the Hofburg before the Russians arrived in Vienna was ended by Brooke, who knew Marshall's view of it. There would still be plenty of teeth-gnashing before Churchill relinquished his project, which he clung to with the tenacity he had also shown over the operations in northern Norway, Sumatra and the Andaman Islands. Yet Roosevelt had promised Stalin Operation Anvil at Teheran, and Marshall meant to honour it.

At 7.40 p.m. on Friday 9 June Brooke and Cunningham went to Euston Station to meet the train bringing the American Joint Chiefs of Staff from Holyhead. If swift strategic decisions needed to be taken – including an evacuation from Normandy, though no one mentioned that – the Combined Chiefs would be able to go into continuous session. 'Marshal [sic] as charming as ever and King as saturnine,' recorded Cunningham. 'I can't bring myself to like that man.'[16] It is interesting that even that late in the war and after a long stay in Washington sitting on the Combined Chiefs of Staff with him, Cunningham could not spell Marshall's name correctly, although he also spelt words such as 'Bethnell Green', 'cardenal', 'gardiner' or 'gardner' and so on. The use of the word 'saturnine', meaning 'gloomy', was a good one for King, however.

The next day the Combined Chiefs met to discuss Normandy and Italy, before Marshall went to Chequers in the evening and Cunningham went home to the country for supper with Admiral Ramsay, where he opened a bottle of Turkish wine 'to celebrate the invasion', a curious way of doing it. At their Combined Chiefs of Staff meeting on Sunday 11 June, Marshall called for an early Anvil and 'advancing as much as possible the target date of amphibious operations in the Mediterranean'.[17]

Handy recalled discussing Anvil with Marshall and concluding: 'We were a damn sight better strategists than we ever had any idea we were.'[18] But were they? Donnelly considered the operation 'a big success' because the freeing of Marseilles gave Eisenhower another big harbour to help support the operations on the Western Front. Yet it was really Antwerp that they needed, rather than a port 480 miles south of Paris. In 1958, Brooke said of Marshall and Anvil: 'I was in disagreement with him because I was rather frightened that if you took many forces from Italy it took some time from the moment you removed them from Italy to the moment they attacked in France ... and what we should have gained on the swings we should lose on the roundabouts ... It was a matter of balancing the time of it.'[19]

Ian Jacob used a different argument in his discussions with Chester Wilmot in the spring of 1948, suggesting that the Americans:

could never appreciate the importance of threats. Our idea was always to use our sea power so as to threaten and to bluff ... If you did not show your hand you forced Hitler to protect himself against all possible threats and thus you gained much wider dispersion of his forces. The Americans' only idea of winning was by fighting. They never realized that strategy is successful or not in proportion to the amount of actual fighting which it involves. The less fighting, the more successful the strategy. This was an idea the Americans could never comprehend; they always wanted bigger and better battles.[20]

This – the strategy of Sun Tzu – might have been a sophisticated enough argument up to 6 June 1944, but from that date victory over Nazism certainly was a question of fighting bigger and better battles, and troops in France and Germany were of far more use than those going up the Apennines at slow speed and great cost.

Cunningham recorded that, at a Combined Chiefs of Staff meeting on Anvil at the American HQ at Stanwell Place in Staines, Surrey, on 11 June, Marshall 'drew a vision before our eyes of an airborne landing of five divisions in connection with a seaborne operation to seize a good port', probably either Marseilles or Toulon. They also considered the three alternatives for Alexander's army once it had broken through the Gothic Line, namely an amphibious operation in the Loire area or the Bay of Biscay, one in the south of France preferably in the Sète area, or one in the north-eastern Adriatic in conjunction with an advance from the Pisa–Rimini Line. Cunningham wrote after a 6.30 p.m. Cabinet the next day that Churchill, 'looking a bit exuberant from his trip to the beachhead, was a bit childish at times'.

'From the time the Normandy landing was achieved,' argued Liddell Hart, Brooke and Churchill 'ceased to have any important influence on the course of the war. Both strategically and politically, American influence became overwhelmingly predominant, and dictated the Allies' course ... [Churchill] had in effect become, as he earlier proclaimed himself, merely the American president's "lieutenant".'[21] Liddell Hart put it in his typically provocative way, and Churchill's self-designation as Roosevelt's lieutenant had only been used at Torch because he had felt that the Vichy French would resist Britons more than Americans. Nonetheless it was true that in all the great strategic controversies of the rest of the war – and there were many hard-fought ones still to come – it is hard to spot any that the British won. The preponderance of American over Commonwealth troops has already been noted, and once the Americans were ashore they no longer needed to defer to the British in the way they had when they required the United Kingdom as their launch-pad.

At a dinner for the King and the American Chiefs of Staff at Downing Street on 14 June, Churchill and Brooke had a long argument about the amount of equipment and supplies an invading army needed in its wake, with 'Winston maintaining in his rhetorical fashion that the progress of any army could only be delayed by the importation of dental chairs and units of the YMCA'. Brooke stubbornly argued that 'no army could fight, let alone progress, unless it had adequate supplies of ammunition.' Marshall and the others listened politely until 2 a.m., while Smuts and Attlee, sitting on either side of Sir Alan Lascelles, 'slept unashamedly'.[22]

On the evening of 17 June, Marshall and Arnold arrived in Naples, flying over the Salerno and Anzio beaches at a height of 300 feet on their way there. Marshall visited his stepson's grave the next day, and wrote to tell Margaret that 'Allen's plot is on the main pathway through the cemetery, a short distance beyond the flagpole.' He spoke to the men in Allen's unit who were with him when he was killed, before going on to Mark Clark's Fifth Army HQ.

By doleful coincidence, Brooke was also brought face to face with the personal cost of war that same day when a V-1 flying-bomb hit the roof of the Guards' Chapel on Birdcage Walk during the Waterloo Day morning service. Several of Brooke's friends and colleagues were among the 122 people who perished, including his great friend Ivan Cobbold, with whom he had been fishing on the River Dee six weeks earlier.[23] Brooke described how the 'ghastly blow' was made even worse for him by one of those macabre coincidences of wartime; his military assistant came in to tell him about the tragedy at the precise moment that he was opening a lunch invitation from the dead man.

Churchill had spent months trying to organize the first conference since Cairo, but as there were no crucial strategic issues or matériel shortages disrupting Allied planning, Roosevelt had kept putting it off. He actively did not want to meet Churchill during the Anvil dispute, and feared that any meeting might make Stalin suspicious. With the Allies victorious everywhere except China, and both British and American intelligence services predicting victory by Christmas, another meeting to plot the European endgame could not long be delayed; however, the Americans kept avoiding the issue.

Violent storms between 19 and 24 June swept away the American Mulberry harbour at Arromanches, and came just at the same time that Hitler reinforced Kesselring in Italy, thus stalling the Allied advance towards the Gothic Line. Eisenhower pressed Marshall harder than ever for a large-scale Anvil that would bring more Allied troops ashore in southern France and, it was hoped, divert more Germans away from Normandy, where Cherbourg and Caen had yet to fall. 'Wandering off overland via Trieste to Ljubljana repeat Ljubljana', wrote Eisenhower, 'is to indulge in conjecture to an unwarranted degree at the present time.'[24] It is hard to know whether the repetition of Ljubljana was intended as emphasis of its strategic absurdity or to lessen the chances of mistransmission of an unusual word, or possibly both.

Mountbatten's chief of staff Lieutenant-General Henry Pownall – yet another diarist – alluded to 'a considerable breeze' (that is, row) between Brooke and Marshall on 19 June over General Stilwell, a friend of Marshall whom Brooke despised.[25] Yet these were minor disagreements compared to the one over Operation Anvil that was to burst into flames the next day. On Tuesday 20 June Admiral King, without consulting the Admiralty, the Combined Chiefs or either of the Supreme Commanders Eisenhower and Maitland Wilson, ordered three battleships, two cruisers and no fewer than twenty-six destroyers to leave the Overlord flotilla in the Channel and sail to the Mediterranean as part of the Anvil invasion force, even though there was no invasion date yet set. Churchill immediately telegraphed Roosevelt to say he was 'much concerned' about this completely unilateral action; indeed so exercised was he that Cunningham recorded that 'we had to curb him a bit'.[26] Eisenhower countermanded King's order, so the naval force was able to stay put until Cherbourg had fallen a week later.

At a 10.30 p.m. British Staff Conference on 21 June with Churchill, Eden, Macmillan and the three Chiefs present there was 'a rambling discussion about the merits of advancing right up through Italy to the Italian Alps as opposed to landing a few divisions in the South of France', during which Cunningham could 'see the advantage of landing the French divisions in their own country and letting them rouse France'. Eden considered that 'Brookie appeared to be making pretty heavy weather of differences that cannot arise for many moves ahead, if ever.' He also believed that 'the failure to have minds meet' over Anvil was in large part down to Dill's absence.[27] Recovering from a serious bout of illness in Washington, the field marshal wasn't present to oil the wheels between Americans and British, although by that stage Cunningham believed that Dill had completely 'gone native' in America.

On 23 June Eisenhower – based at his headquarters in Bushy Park in the London suburbs – again demanded a strong Anvil, while the British were doubting the necessity of having one at all. 'France is the decisive theatre,' he told the Combined Chiefs of Staff, saying that they 'took this decision long ago. In my view the resources of Great Britain and the United States will not allow us to maintain two major theatres in the European War, each with decisive missions.' At the time, though, Brooke was unable to see that the Italian campaign had largely achieved its purpose and was now no longer central to victory. Because he knew, through Ultra decrypts, that Hitler wished to defend the rest of the Italian peninsula, Brooke hoped that Alexander would continue to engage large German forces there. It was simply a strategy for attrition by then, however, whereas opportunities for outright victory were now opening up in France.[28]

The withdrawal of Alexander's French mountain divisions from Italy for Anvil caused Brooke and Churchill immense consternation. On 26 June Maitland Wilson said this must start within forty-eight hours in order to meet Anvil's target date of 15 August. At a Chiefs of Staff meeting at 11 o'clock that morning, the War Office Planners backed Anvil, but Brooke and Portal wanted to cancel it altogether and carry on fighting in northern Italy instead. 'I find the arguments so evenly balanced that I have difficulty making up my mind,' wavered Cunningham, 'so allowed myself to be guided by the other two whose arguments are sound enough if a little specious in certain directions.'[29]

After meeting Churchill that evening, a telegram to Marshall – who had got back to America via Italy on 21 June – was drafted in which the Chiefs of Staff 'took a firm line turning down Anvil and pressing for the completion of the North Italian campaign'. Cunningham felt that the British 'are in the position of the man in possession: the campaign is going on.'[30] Although 'the man in possession' is said to have nine-tenths of the law, Cunningham was wrong if he thought that possession of the Anvil divisions might be worth nine-tenths of grand strategy against a hardened opponent like Marshall.

The British telegram to Marshall should have been worded far better than the bald statement that, in considering the 15 August date, 'The withdrawal now of forces from Italy to achieve this target date is unacceptable to the British Chiefs of Staff' because it 'would hamstring General Alexander so that any further activity would be very modest. The adequacy of air resources for both Italy and Anvil is gravely doubted.' It was as terse as Brooke's speaking voice, but by that stage of the war such a tone was beginning to be counter-productive. For, as Marshall showed the next day, two could use the word 'unacceptable'.

'The British proposal to abandon Anvil and commit everything to Italy is unacceptable,' Marshall cabled Eisenhower on 27 June, adding: 'It is deplorable that the British and US disagree when time is pressing. The British statements concerning Italy are not sound or in keeping with the early end of the war ... There is no reason for discussing further except to delay a decision which must be made.' The Joint Chiefs of Staff official reply to the Chiefs of Staff was almost identical. Marshall meanwhile asked Handy to prepare a brief memorandum for Roosevelt of the key documents so far, 'taking care to include the urgent arguments put forward by Eisenhower for the support of Overlord'. Marshall wanted the President fully informed, and under no illusion about the necessity for Anvil in helping Eisenhower.

Anvil had been agreed to by Brooke and Churchill in Teheran and Cairo. Eisenhower wanted it as soon as possible. The Italian campaign was slowing down in central Italy. The Supreme Commander in the Mediterranean had said that if Anvil was to proceed – which he did not want it to – then the troops must be withdrawn within hours. The crisis moment had therefore come, and on 28 June Roosevelt came down firmly on Marshall's side, telling Churchill:

I really believe we should consolidate our operations and not scatter them. It seems to me that nothing could be worse at this time than a deadlock in the Combined Staff as to future course of action. You and I must prevent this and I think we should support the views of the Supreme Allied Commander. He is definitely for Anvil and wants action in the field by August 30th and preferably earlier. It is vital that we decide at once with our long agreed policy to make Overlord the decisive action.[31]

Roosevelt's evocation of Eisenhower meant that he and Marshall could present themselves as supporting the views of the commander in the field. How Churchill must have regretted unilaterally offering Roosevelt the decision as to who should be supreme Allied commander, getting so little in return.

Churchill was deeply influenced by the fact that the Mediterranean theatre was principally British, with a British supreme commander in Maitland Wilson, a commander-in-chief Allied Forces Italy in Alexander, British corps commanders, many British divisions, and heavy Royal Navy participation, although of course Mark Clark's US Fifth Army, the USAAF at Foggia and several other important American units also served there. Churchill now feared that this primary British theatre would be severely weakened by the American-dominated Anvil, and later wrote that this was 'The first important divergence on high strategy between ourselves and our American friends', implying that everything else had been merely a matter of timing.[32] This was a convenient post-war fiction, but Anvil certainly saw a major clash.

Before returning to his duties at the Allied Forces HQ in the Palace of Caserta, north of Naples, Harold Macmillan saw Churchill on the evening of 28 June after Roosevelt's reply had arrived. 'It was not only a brusque but even an offensive refusal to accept the British plan,' he wrote in his diary. 'It so enraged the P.M. that he thought of replying to the President in very strong terms.' After consideration, however, it was decided that the British Chiefs should reply formally to Marshall, saying that they could not change the advice that they were giving the

War Cabinet, 'to whom they had the duty of giving the best professional opinion which they could form'.[33]

Macmillan records Churchill as 'still exceedingly anxious' to cross the Pisa–Rimini Line, seize the Po Valley and open the possibility of an advance on Trieste in the spring. Yet such was the preponderance of American troops in the European campaign that, as Macmillan recognized,

we should have to give in if Eisenhower and Marshall insisted upon 'Anvil'. We can fight up to a point, we can leave on record for history to judge the reasoned statement of our views, and the historian will also see that the Americans have never answered any argument, never attempted to discuss or debate the points, but have merely given a flat negative and a slightly Shylock-like insistence upon what they conceive to be their bargain.

These embittered views were shared by, and possibly inspired by, the Prime Minister.

'The deadlock between our Chiefs of Staff raises most serious issues,' Churchill replied to Roosevelt. 'Our first wish is to help General Eisenhower in the most speedy and effective manner. But we do not think this necessarily involves the complete ruin of all our great affairs in the Mediterranean and we take it hard that this should be demanded of us ... I think the tone of the American Chiefs of Staff is arbitrary, and I certainly see no prospect of agreement on the present lines. What is to happen then?' In fact the 'arbitrary' tone had begun when Brooke had drafted the telegram about what was 'unacceptable'.

To try to persuade Roosevelt, Churchill attached a twelve-page cable setting out the advantages of a Balkan strategy against the long distances that any landing in Toulon would have to go before it could engage significant German forces. He ended: 'Let us not wreck one great campaign for the sake of winning another. Both can be won.' Brooke was 'weak as a cat' from influenza at the time, and the prospect of his being brought from his sickbed elicited the request from Marshall that 'on no account should we worry the Field Marshal'. The US Army Chief's own view was that as far as the clash over Anvil was concerned, 'there is a big part played by the Prime Minister in the present affair', although this was immediately and categorically – and untruthfully – denied by Hollis from the War Cabinet Office.

Roosevelt sent a thirteen-paragraph reply to Churchill the next day,

29 June, urging that the Combined Chiefs of Staff directive authorizing Anvil be issued immediately. 'My interest and hopes center on defeating the Germans in front of Eisenhower and driving on into Germany rather than limiting this action for the purpose of staging a full major effect in Italy,' he wrote, before saying specifically,

I cannot agree to the employment of United States troops into the Balkans . . . History will never forgive us if we lose precious time and lives in indecision and debate. My dear friend, I beg of you let us go ahead with our plan. Finally, for purely political considerations over here, I should never survive even a slight set-back to Overlord if it were known that fairly large forces had been diverted to the Balkans.

It was relatively rare for Roosevelt to plead in aid domestic political pressures over military operations, but he had an election to fight four months hence.

Churchill's reply on 1 July was anguished. Even though he began with the first person plural – 'We are deeply grieved by your telegram' – he soon slipped into more intimate vernacular, saying that this was 'the first major strategic and political error for which we two are responsible. At Teheran you emphasized to me the possibilities of a move eastward when Italy was conquered.' He claimed that 'No one involved in these discussions has ever thought of moving armies into the Balkans,' but stated that Istria and Trieste were strategically and politically important positions, 'which, as you saw yourself, very clearly might exercise pro-found and widespread reactions, especially now after the Russian advances'. Finally Churchill stated that:

If you still press upon us the directive of your Chiefs of Staff to withdraw so many of your forces from the Italian campaign and leave all our hopes there dashed to the ground, His Majesty's Government, on the advice of their Chiefs of Staff, must enter a solemn protest . . . It is with the greatest sorrow that I write to you in this sense. But I am sure that if we could have met, as I so frequently proposed, we should have reached a happy agreement.[34]

That is precisely what Marshall had feared, and was one of the reasons Churchill did not meet the President at all throughout the nine months between December 1943 and September 1944, despite having seen him thrice in seven months in 1943. Churchill's force of personality was blunted once it was translated on to printed telegraph slips or blared over transatlantic scrambler telephone lines, with gaps between each crackling transmission. He ended his telegram: 'I need scarcely say that we shall do our best to make a success of anything that is undertaken.' Privately, Lascelles noted that 'Winston is very bitter about it, and not so sure that he really likes FDR.'[35]

On Saturday 1 July, Roosevelt gave Marshall the necessary orders to proceed with Anvil, telling Churchill, 'I always think of my early geometry – a straight line is the shortest distance between two points.' His decision underlined the paramountcy of the United States by the summer of 1944. Yet it is worth asking objectively, with the historian's luxury of hindsight, who was right: was Anvil a useful employment of so many divisions so far from the main struggle hundreds of miles to the north? Churchill and Brooke might have been muscled into it by Roosevelt, Marshall, Eisenhower and – at one remove – Stalin, but they were right to resist, because the military logic for Anvil was questionable and its results limited. Although one hundred thousand Germans surrendered in that campaign, they were not all high-quality troops, and Patton had got to Dijon, 300 miles from the Normandy beaches, on 12 September, before the Anvil force arrived from the south.[36]

'Please accept my cordial greetings for Independence Day,' Brooke wrote to Marshall on Tuesday 4 July. 'The operations now proceeding in Europe show the complete understanding and mutual trust existing between the USA and the British Armies both fighting for the ideals which you are celebrating today.' It was a good-natured gesture after a bruising defeat, during which Brooke had on at least three occasions toned down telegrams from Churchill which if sent would have caused deep offence to Roosevelt and Marshall, who had now shown beyond doubt that they were in the political and military driving seats. 'The trouble is the PM can never give way gracefully,' observed Cunningham. 'He must always be right and if forced to give way gets vindictive and tries by almost any means to get his own back.'[37] Neither Marshall nor Brooke believed that 'complete understanding and mutual trust' had characterized their dealings over Anvil during the previous fortnight, but what was needed was some pleasant-sounding hypocrisy. At Cabinet that day, Churchill complained of 'Differences with Americans. History will approve the use that the Allies have made of the Mediterranean.'[38]

Churchill's bitterness over the Anvil defeat is evident from his minute to the Chiefs of Staff of 6 July. 'Let them take their seven divisions,' he wrote, 'let them monopolize all the landing craft,' before adding, 'I hope you realize that an intense impression must be made upon the Americans that we have been ill-treated and [are] furious . . . If we take this lying down, there will be no end to what will be put upon us. The Arnold–King–Marshall combination is one of the stupidest strategic teams ever seen.'[39] Churchill reproduced the first part of this acidic memorandum in his war memoirs, although not of course the final sentence.

That evening the British Chiefs underwent a 'frightful meeting with Winston' lasting four hours until 2 a.m., which Brooke described as 'quite the worst we have had with him'. Although he wrote that fairly often, this one – a meeting of the Defence Committee – was indeed the worst of the war. The Prime Minister was very tired as a result of his Commons speech about the V-1 threat, and emotional over his defeat by the Americans over Anvil, and according to Brooke 'he had tried to recuperate with drink. As a result he was in a maudlin, bad tempered, drunken mood, ready to take offence at anything, suspicious of everybody, and in a highly vindictive mood against the Americans. In fact so vindictive that his whole outlook on strategy was warped.'[40]

In the course of the meeting, Brooke accused Churchill of belittling Montgomery and Alexander, the former for being over-cautious – the key town of Caen in Montgomery's sector had still not fallen – and the latter for not having simply outflanked Monte Cassino back in May. A discussion on the Far East degenerated into a row over all the old subjects, especially Culverin, and Churchill wound up by falling out with Attlee and 'having a real good row with him concerning the future of India!' The Chiefs withdrew from the meeting 'under cover of this smokescreen just on 2am, having accomplished nothing beyond losing our tempers and valuable sleep!!' After the war Brooke added that Churchill had been 'infuriated, and throughout the evening kept shoving his chin out, looking at me, and fuming at the accusation that he ran down his generals'.

Cunningham agreed with Brooke's assessment, lamenting that 'the PM was in no state to discuss anything. Very tired and too much alcohol . . . he was in a terrible mood. Rude and sarcastic.' The admiral hoped that Churchill's 'obstinacy and general rudeness may be the last flurry of the salmon before you get the gaff into him'.[41] It was not. Another diarist present that night, Anthony Eden, wrote that Churchill clearly hadn't read the strategy paper they were discussing 'and was perhaps rather tight'. He recorded an exchange that went:

BROOKE: 'If you would keep your confidence in your generals for even a few days, I think we should do better.'

CHURCHILL: 'When have I ever failed to support my generals?'

BROOKE: 'I have listened to you for two days on end undermining the Cabinet's confidence in Alexander until I felt I could stand no more. You asked me questions, I gave you answers, you didn't accept them and telegraphed to Alexander who gave the same answers.'[42]

There was more in the same strain, with 'Winston protesting vehemently'. Eden thought Churchill 'was clearly deeply hurt on his most sensitive spot, his knowledge of strategy and his relationship with his generals'. The Foreign Secretary, who regularly took Churchill's side against Brooke, nonetheless concluded: 'Altogether a deplorable evening which couldn't have happened a year ago. There is certainly a deterioration.'[43]

Alexander, who had considerable personal charm, had commanded the rearguard at Dunkirk and the retreat of the British Army in Burma before becoming C-in-C successively of the Middle East and of the 19th Army Group in North Africa. Churchill liked him personally, not least because he was a fellow painter and Old Harrovian, but he did abuse his generalship. Brooke and Cunningham, by contrast, abused Alexander only to their own diaries – as 'a void' and 'not a good general' respectively – and doubtless also to each other.

In early July, Sir John Dill suffered a mild heart attack, and Marshall arranged that he be taken to the military hospital in White Sulphur Springs, West Virginia, for a proper rest-cure. He had previously suffered rectal bleeding, and his haemorrhoids were removed by injection, causing severe anaemia which was only recognized late. Writing from hospital to Brooke about a problem that had arisen over Stilwell, Dill said: 'It is odd how that charming person Marshall can fly off the handle and be so infernally rude. Also he gets fixed ideas about things and people which it is almost impossible to alter. I am so very sorry that I was not there when the Anvil question came up, but I fancy he was pretty fixed on Anvil and most likely impossible to move.' It was a new and rare side to Marshall, and so Dill's 'oil-can' had been needed more than ever, but tragically it turned out that his illness was mortal.

On 10 July, the day after Caen finally fell, Churchill reported to the War Cabinet that Overlord's 'daily discharge' in Normandy amounted to twenty-five thousand men, seven thousand vehicles and 30,000 tons of stores. He was concerned about German soldiers who were captured shortly after setting delayed-action mines that killed Allied troops, and suggested that they should be warned that they would be personally 'held responsible' if they did not reveal where these booby-traps were.

After Brooke had reported on the situation in Normandy, Churchill said that the fighting there had drawn in German reserves, and that twenty-nine Allied divisions (fifteen American, fourteen British Commonwealth) were presently engaging twenty-three German. Altogether British and Canadian casualties came to twenty-six thousand out of a total of sixty-four thousand Allied, and 51,393 prisoners had been taken. Nonetheless, Churchill emphasized that ministers should 'Not encourage people to expect war to end this year. No right to count on it . . . Don't minimize what we [must] do.'

After the fall of Caen, the Germans continued to put up strong resistance at Saint-Lô, which did not fall to the US XIX Corps until 17 July, after which the British and Canadians pushed south and east of Caen with the support of thousand-bomber raids. It was another month, however, on 18 August, before the Falaise gap was closed by the Allies, trapping the Germans south of the Normandy battlefield. After that it was only a matter of a week before Paris was liberated.

Although it had been left pending since the furious exchange of memoranda in late March, when neither side accepted defeat but wished to avoid resignations and sackings on the issue, Far Eastern strategy could not be long ignored. The total dichotomy in thinking between Churchill and the Chiefs of Staff over defeating Japan was once again laid bare at a Defence Committee meeting on 14 July, after which Cunningham wrote that they had been 'treated to the same old monologue' about how much better it was to take the tip of Sumatra and then the Malay Peninsula and finally Singapore than to join with the Americans to fight Japan in the south-west Pacific. The First Sea Lord assumed that the politicians were 'obviously frightened of the Americans laying down the law' about what was to happen 'to the various islands, forts and other territories' once Japan was defeated.

The way to ensure the Empire's proper treatment in the final peace treaty, thought Cunningham, was to stick closely to the Americans in their long campaign from Australia to the Japanese mainland, but

Churchill, Attlee and Eden 'will not lift a finger to get a force to the Pacific; they prefer to hang about the outside and recapture our own rubber trees.'[44] It was during that meeting that Attlee passed Eden a note saying of Churchill's chairmanship: 'Two hours of wishful thinking.' Cunningham was wrong about the Foreign Secretary supporting Culverin, since Eden actually felt that because Sumatra 'is remote from the centre of conflict we shall be regarded by Americans as having played virtually no part in defeat of Japan'.[45]

On 16 July Marshall wrote to Lieutenant-General Jacob L. Devers, the commander of Anvil, that 'If the forces in Italy bog down on the Pisa–Rimini Line, we should not long delay putting the Fifth Army divisions into the fight in southern France. I hope that Alexander will quickly get into the Po Valley. Then the Fifth Army, or portions thereof, could be moved into France, possibly some of it moving overland . . . The important thing is that we push Anvil to the utmost as the main effort in the Mediterranean.'[46] Here was another row waiting to happen, for on the same day in London Cunningham recorded a Chiefs of Staff meeting at which Churchill was 'in a very sweet and chastened mood probably remembering last Thursday' – when he had been drunk – and there was 'Much discussion on whether it was wise to try to bind the American Chiefs of Staff to leave the present forces less the Anvil contingent in Italy to enable Alexander to plan his forthcoming campaign to carry the Pisa–Rimini Line on a firm basis'. The British Chiefs were 'rather against it' as it might put 'bad ideas' of using the Fifth Army in Italy 'as a reserve for Anvil'. Yet from Marshall's letter it was clear that they had indeed already started to think that way.

Casualty numbers from the V-1 campaign given to the War Cabinet on 24 July were 'thirty thousand-odd of whom four thousand odd killed'. Children who had returned to the cities from the countryside were evacuated all over again, and despite the success in Normandy, national morale suffered. The nightly bombing, with its effects of sleeplessness and general strain, needs to be borne in mind when considering the ill-tempered mood that often descended on crisis meetings during this stressful time. The Cabinet War Rooms began to be used regularly again for almost the first time since the Blitz, and their stuffy atmosphere – especially with Churchill's cigars, Bevin's cigarettes, Attlee's pipe and so on – cannot have helped.

In the east, the Russians crossed into Poland on 23 July. It was a genuine cause for celebration, because few guessed at the time that they would stay there for forty-five years. Nevertheless, at a Chiefs of Staff meeting on 26 July centred on British post-war security it was assumed that Germany needed to be included 'in the Western powers organisation' and 'It was generally agreed that Russia would be the only danger in the foreseeable future.'[47] Churchill agreed, telling Charles Moran, 'Good God, can't you see that the Russians are spreading across Europe like a tide; they have invaded Poland, and there is nothing to prevent them marching into Turkey and Greece!' Operation Anvil he described as 'Sheer folly', lamenting that the 'ten divisions could have been landed in the Balkans . . . but the Americans would not listen to him.' Moran concluded that Churchill was 'distraught, but you cannot get him down for long.' He sat up in bed as his speech quickened and he expounded on how 'Alex might be able to solve this problem by breaking into the Balkans. Our troops are already in the outskirts of Florence. They would soon be in the valley of the Po.' Churchill's promise to Roosevelt on 1 July that 'No one involved in these discussions has ever thought of moving armies into the Balkans' therefore was obviously completely misleading.

In common with other operations, such as Roundup/Overlord and Super-Gymnast/Torch, for security reasons Operation Anvil was renamed Dragoon just before it took place. The joke went around that the word was chosen because Churchill felt he had been dragooned into it. Writing to General Maitland Wilson ('My dear Jumbo') on 2 August, Brooke blamed Alexander for the failure to persuade the Americans: 'Alex's wild talk about his advance on Vienna killed all our arguments dead!' Brooke then dilated very perceptively upon the loss of British influence in Washington. Dill's illness 'had a great deal to say to it', he thought, but there 'is more to it than that'. The Americans 'now feel that they possess the major forces at sea, on land and in the air', and with 'all the vast financial and industrial advantages which they had enjoyed from the start' they were in a far stronger position. 'In addition they now look upon themselves no longer as the apprentices at war, but on the contrary as full blown professionals. As a result of all this they are determined to have an ever increasing share in the running of the war in all its aspects. I can assure you that we are watching these unpleasant new developments very carefully.'[48] Brooke allowed himself a rare bout of over-optimism when he told Maitland Wilson that now the Normandy operation had broken through into Brittany, 'It becomes more evident every day that the Bosche is beat on all fronts. It is only a matter of how many more months he can last. I certainly don't see him lasting another winter.'

As if to underline Brooke's earlier point, that same day the Director of the US Bureau of the Budget, Harold Smith, wrote to Roosevelt to report: 'It is highly probable that by the end of September we shall have a total army strength in excess of eight million.' With private soldiers (GIs) in the US Army receiving $50 per month, corporals $66, sergeants $78 and master-sergeants $138, Smith's financial concerns were legitimate, even for an economy as strong as that of the United States. So on 15 August the President asked Marshall for a memo 'on this matter of over-strength of the Army'. A week later Marshall replied in detail, admitting that the Army was 5 per cent over-strength, at 8.05 million, but the reason was a deliberate recent increase of 150,000 to meet the expected casualties in the Overlord and Anvil landings.

At a Chiefs of Staff meeting in London at 10 p.m. on 4 August there was some discussion about shifting Dragoon from the south of France to Brittany, which Churchill said Eisenhower had already recommended to Marshall. The Chiefs of Staff therefore telegraphed Washington supporting the change, while warning Maitland Wilson 'that it might come off. All this due to the spectacular advances of the US armoured forces in Brittany.'[49] Yet the very next day, when Churchill and Cunningham flew to the Cherbourg peninsula but were unable to land owing to fog, they stopped instead at Eisenhower's headquarters at Sharpener Camp on Thorney Island, a peninsula in Chichester Harbour, West Sussex. Over an 'excellent' lunch, they discussed moving Dragoon across to Brittany, and to Cunningham's surprise found Eisenhower 'dead against it and had never sent a message putting it forward'.[50] Cunningham believed it was 'very apparent that the PM, knowingly or not', had 'bounced' the Chiefs of Staff into sending their telegram to the American Chiefs of Staff. Sure enough, two days later the Joint Chiefs of Staff turned down the Brittany idea out of hand, and Cunningham blamed Churchill for being Machiavellian over what was probably only a genuine misunderstanding.

'What a drag on the wheel of war this man is,' Cunningham wrote four days later after having three meetings in one day at which Churchill tried yet again to push for the Sumatran operation. 'Everything is centralized in him with consequent indecision and waste of time before anything can be done.' Eden noted that Churchill 'generally seemed very tired and unwilling to address himself to the arguments', and as a result, 'Brookie became snappy at times, which didn't help much.'[51] What earlier in the war had been a creative tension between Churchill and Brooke was fast becoming simply mutual friction.

The next day Cunningham was furious that Churchill had ordered Ismay not to circulate a paper on Far East strategy before their 10.30 p.m. meeting, even though it had been ready at 4 p.m. 'Thus we are governed!!' he wrote afterwards. 'I presume he himself has such a crooked mind that he is suspicious of the Chiefs of Staff.' Part of the problem was the absurd times that these meetings were scheduled: ill-temper could almost be guaranteed at meetings that night after night began at 10.30 p.m. and which could often go on for three or four hours. Cunningham wrote – admittedly after a long, busy and very trying day – that the meeting had been 'A breeze at the start' when Brooke asked for time to consider the Prime Minister's new paper. Its first four paragraphs were devoted to the subjects on which they were to approach the Joint Chiefs of Staff and 'the way it was to be done, and the fifth arranging to double-cross them', added the irascible First Sea Lord. 'I often wonder how we expect the US Chiefs of Staff to have any respect for us. We allow our opinions to be overridden and ourselves persuaded against our own common sense at every turn.'[52]

On Friday 11 August, the Chiefs of Staff were warned by Ismay in a closed session that Churchill 'was just raving last night and absolutely unbalanced. He cannot get over having not had his own way over Anvil.' To Cunningham's surprise it was the normally placating Charles Portal 'who suggested that we must have a showdown with him before long if he went on as he is now'. Cunningham agreed, particularly disliking the way Churchill was attempting 'to dictate' to the Chiefs of Staff what they should say to the Joint Chiefs of Staff. They decided to hold the issue off for twenty-four hours to let the Prime Minister 'recover his balance a bit', especially as he was off to Algiers and Italy for a fortnight anyhow. They also concluded that 'after their exhausting week with the PM' they would all three go off for the weekend a day early. With the Vice-Chiefs taking the Saturday morning meeting instead, Cunningham went down to Hampshire and repaired his lawn-mower.[53]

*

On Tuesday 15 August 1944 – Napoleon's 175th birthday – Operation Dragoon (formerly Anvil) was launched, with seventy-seven thousand troops of the US Seventh Army landing along the French Riviera coast between Cannes and Cavalaire, while an American and British airborne force of nine thousand men landed a few miles inland. The Free French also took part, and casualties amounted to only 520 of the seaborne force, of whom fewer than one hundred were killed. Ten weeks after D-Day, the reserves of German Army Group G had long before moved northwards.

Both Toulon and Marseilles fell on 28 August, Lyons on 2 September, and ten days after that contact was made with Patton's Third Army at Dijon. With one hundred thousand German prisoners taken, the south of France liberated, 300 miles covered in a month and good supply bases in Toulon and Marseilles secured, Dragoon was, on its own terms, a military success, although the Germans were already withdrawing from southern France in any case. Considering the mutual suspicion and sustained anger that its planning caused in the highest Allied counsels of the war for so long, Dragoon today seems like a typhoon in a teacup, yet on 2 November 1944 Churchill told Bedell Smith and Cunningham that 'History would pronounce on the Dragoon operation,' implying that the verdict of Clio would be unfavourable. Cunningham suggested that it would depend on who wrote the history, to which Churchill replied that 'he intended to have a hand in that'.[54]

19

Octagon and Tolstoy: 'It takes little to rouse his vengeful temper'

August–December 1944

The personal relations between Roosevelt and Churchill illustrated a real alliance of interests; the personal relations between Roosevelt and Stalin concealed a real opposition of interests.   Hugh Trevor-Roper to Bernard Berenson, 1948[1]

The loss of General Alphonse Juin's French Expeditionary Corps, mountain-trained units intended to break the Gothic Line, to Dragoon effectively forced Alexander to winter in the Apennines, fighting a campaign of attrition with what he had left. At that point the Germans ought strategically to have withdrawn from the Po to the Alps, thus freeing up divisions to fight in France, but the Führer's no-withdrawal obsession precluded that. It is small wonder that Churchill was actually relieved that Hitler survived the 20 July Bomb Plot, telling the House of Commons in September that 'It would be most unfortunate if the Allies were to be deprived in the closing stages of the struggle of that form of warlike genius that Corporal Schicklgruber has so notably contributed to our victory.'

As the war progressed, such political factors as the post-war political configuration of Europe weighed more heavily with the politicians, but not with the generals whose sole job was to see every campaign in terms of what would bring victory soonest with the minimum loss of Allied life. 'By and large we were not influenced by political factors in making our military decisions,' claimed Portal. Churchill and Eden were, of course, but, as he rightly pointed out: 'That was their job.' Over such issues as Operation Torch, the Bay of Bengal strategy and the liberation of Vienna and Berlin – all of which had deeply political overtones – the Chiefs of Staff resolutely considered only the military implications. That does not mean that they were not also keenly aware of the potential dangers that arose from a vast Red Army pushing towards the heart of Europe.

The next issue to divide the Americans and British was over what kind of supreme commander Eisenhower should be. Was he to function as essentially a non-executive chairman of the board, overseeing but not interfering much in the activities of his Army commanders, and calming the prima-donna tendencies of his lieutenants such as Bernard Montgomery (21st Army Group), George Patton (Third Army) and Omar Bradley (12th Army Group)? Or was he going to exercise direct daily military control over the immense scope of the war in France, the Low Countries, north-west Europe and Germany, at least as much as any individual reasonably could? In very crude outline, Churchill and especially Brooke wanted the former, while Roosevelt and especially Marshall expected and demanded the latter.

That question then led to the next: what kind of front would the Allies choose in the drive to the Rhine and beyond? Would it be a 'broad' one that comprehensively forced the Germans back towards the Fatherland, with two major advances on wide fronts north and south of the Ardennes, or would the attack instead be on 'narrow' fronts, spearheaded by several faster thrusts to try to capture important targets deep within Germany, possibly even including Berlin before the Red Army reached it? Here again, roughly speaking, Roosevelt and Marshall supported Eisenhower's inclination for the former, while Brooke and Churchill tended to opt for Montgomery's and Patton's preference for the latter.

On the question of what the armies in Italy under Alexander and Clark would do once Lucian Truscott's Fifth Army and Sir Richard McCreery's Eighth Army broke through the Gothic Line, the Americans strongly deprecated proposed moves towards Trieste, Istria, the Ljubljana Gap, Vienna and the Balkans. Over Far East strategy, the British – with Churchill now generally persuaded and included – wanted the Royal Navy to join in the reconquest of the south-west Pacific alongside the US Navy, but it was (rightly) feared that Admiral King wanted to spurn this help. All these grand-strategy issues, and more localized ones as they arose, gave Roosevelt, Churchill, Marshall and Brooke ample incentive to continue dancing their complicated, fast-moving, intimately interlocking minuet.

On 18 August, Lord Halifax and Alec Cadogan visited Marshall at the Pentagon. Halifax noted that Marshall was 'quite optimistic' about the campaign in north-west France and showed them maps of how advance elements of the Allied forces were only about 12 miles from Paris. Marshall said he believed that, since the Germans were exhausted and starved of oil and resources, Eisenhower should adopt a broad front, taking local setbacks as they came. At a meeting between Montgomery and Eisenhower five days later, Monty told Ike that in his opinion the Supreme Commander should stay aloof from the land battle, and that his 21st Army Group should make a single bold thrust along a narrow front straight to Berlin.

That meeting seems to have had precisely the opposite effect of what Montgomery hoped for, because Eisenhower very soon afterwards set up his own Command HQ in France. At a Chiefs of Staff meeting at this time, Cunningham sought to moderate criticism of the Americans, pointing out that 'We should do exactly the same if we had two-thirds of the troops in the field.' Employing a phrase that should have expired in 1914, he told his journal: 'I am full of hope that the war will be over by Christmas.'[2] It was ominous, but an appreciation that the British Joint Intelligence Committee encouraged too, with their generally over-optimistic estimates of Allied strengths and German weakness.

On 29 August Churchill sent Roosevelt a telegram about the Mediterranean in which the final paragraph once again brought up their Teheran conversation. It ended, 'I am sure that the arrival of a powerful army in Trieste and Istria in four or five weeks would have an effect far outside purely military values.' Although the condition of Hungary could not be predicted, he believed that having troops there would leave the Western Allies 'in a position to take full advantage of any great new situation'.[3] Roosevelt passed this on to Marshall, who asked McNarney and Handy to work on a draft reply that covered Italy in full but deliberately bypassed Istria completely. Churchill cannot have failed to mark the implications.

On a visit to Montgomery, Brooke satisfied himself that the Allied armies in the north were strong enough to destroy the German forces ranged against them. Montgomery's mission was to take the British and Canadian armies up the coast, while Bradley commanded the nine-division 12th Army Group on his right. Patton's Third Army, part of the 12th Army Group, was to make for the German border south of the Ardennes mountain region, through which the Germans had attacked in 1940. The broad-front strategy had prevailed for the moment, though Brooke was far from happy with it.

At a lunch party at Downing Street on 30 August – the day that the Red Army entered Bucharest and the seat of French government was transferred from Algiers to Paris – Cunningham was asked by Lord Camrose about Eisenhower taking over the day-to-day command of the ground forces in two days' time. He defended it as 'expected' and 'the correct procedure', and then overheard Camrose telling Clementine Churchill that it would be 'most popular' if Monty were made a field marshal. 'I trust it will come to nothing,' wrote Cunningham, very prematurely, as two days later it did. The reason seems partly political; Churchill, wearing 'a sumptuous pale-blue dressing gown of oriental design' in his Annexe, told Lascelles that the promotion 'will put the changes in command in their proper perspective'. With a full-scale conference about to take place in Quebec, the Prime Minister wanted Montgomery's rank to be at least notionally superior to that of the senior American generals, even including Marshall. 'The Americans would not like it,' remarked Cunningham. 'I don't much myself.'[4]

On 3 September the British Second Army liberated Brussels, and the next day Antwerp was taken and flying-bomb sites in the Pas de Calais started to be destroyed. With the Americans having liberated both Rome and Paris – although for political reasons the Free French were allowed to take the fore in the latter – victory in the west seemed within the Allies' grasp. Now even Roosevelt and Marshall accepted that the time had come for another conference.

Churchill seemed in a good mood at 11 a.m. on Tuesday 5 September in the saloon car of his train on the way up to the Clyde, where he was going to board the *Queen Mary*. Cunningham reckoned that 'If he keeps up his present attitude things should go well in Quebec and it will be what the Americans called "a love feast". But it takes little to rouse his vengeful temper and he will do anything then to get the better of our allies.'[5] At 5 p.m. Churchill called Cunningham back to the saloon to say that there was a rumour that Germany had capitulated, and what if, two days out to sea, it proved to be true? 'The only thing to do was to turn the ship round and come back,' replied the First Sea Lord. On board the great liner, which sailed from Greenock for Halifax, Nova Scotia, that night, were Winston and Clementine Churchill, Brooke, Portal, Cunningham, Leathers, Cherwell, Ismay, Hollis and Colville.

As the Prime Minister crossed the Atlantic, the President was attending a meeting at the White House to discuss Henry Morgenthau's extraordinary plan to deindustrialize post-war Germany. 'There is no reason why Germany couldn't go back to 1810,' expounded Roosevelt at some length, 'where they would be perfectly comfortable but wouldn't have any luxury.' In fact 1810 saw the German Confederation dominated by Napoleon, with Prussia still seething with revanchism after her humiliation at Jena–Auerstadt, but the general point was made. Roosevelt's reverie of a defanged Teutonic rural idyll was darkened only by the idea that Britain might be the ultimate beneficiary of the lack of competition from German iron and steel manufacturers. Discussing the Saar and the Ruhr, with Hopkins and Morgenthau arguing on one side and Stimson and Hull on the other, it was 'a very unsatisfactory meeting', as Roosevelt worried 'that the English would have the advantage of the steel business if the Ruhr were closed' and consequently 'he had the idea that this thing was good for England.'[6] The assumption that that could therefore not also be good for the United States shows how far Roosevelt's thinking had come since the Riviera and Arcadia conferences.

On Friday 8 September the first V-2 rocket-propelled bombs fell on Britain, and Churchill held a wide-ranging Staff Conference on the liner in which, according to Cunningham, 'he was in his worst mood. Accusing the Chiefs of Staff of ganging up against him and keeping papers from him and so on.' He refused to accept that after Kesselring was defeated, Italy 'becomes a secondary front and that the real work is on the Russian and Western fronts', even though that had effectively been true ever since D-Day. Churchill still hankered after an amphibious operation against Istria, even though the Chiefs of Staff thought it 'of no military consequence and so on and so on'.

Churchill's true animus was against the Americans, however, 'who he accuses of doing the most awful things against the British. There is no question he is not well and is feeling this hot sticky weather,' thought Cunningham. The liner was in the Gulf Stream where the water was 72 degrees Fahrenheit, and when Churchill tried to persuade Commodore James Bissett to change course towards cooler climes, Cunningham had to go with the commodore – who had first gone to sea in 1898 – to talk the Prime Minister out of it. 'I am afraid that he is very definitely ill and doubtful how much longer he will last,' wrote Brooke the next day. 'The tragedy is that in his present condition he may well do untold harm!' Ten years after Brooke wrote that, Churchill was Prime Minister again, and twenty years later had outlived both Cunningham and Brooke.

The trip on the *Queen Mary* witnessed another particularly low point in Churchill's relationship with the British Chiefs. On 10 September, just before the ship docked at Halifax, Brooke complained – in perhaps his most oft-quoted and notorious diary entry – that at their noon meeting on Culverin that day Churchill:

knows no details, has only got half the picture in his mind, talks absurdities and makes my blood boil to listen to his nonsense. I find it hard to remain civil. And the wonderful thing is that ¾ of the population of the world imagine that Winston Churchill is one of the Strategists of History, a second Marlborough, and the other ¼ have no conception what a public menace he is and has been throughout this war! . . . Without him England was lost for a certainty, with him England has been on the verge of disaster time and again.[7]

Redeeming Churchill (and himself) somewhat was Brooke's post-war comment that these remarks were written 'at a moment of exasperation'. A married couple who felt like that about each other could always get divorced, but that route wasn't open to Churchill and Brooke, although once the war was over they saw little of each other by choice.

When the Churchills' train arrived at Quebec at 10 a.m. on Monday 11 September, Franklin and Eleanor Roosevelt were just disembarking from theirs. The High Command again had suites on the fifteenth floor of the Château Frontenac, airy in the heat and with fabulous views overlooking the river. The place cannot have had happy memories for Brooke, however, since it had not been far off, on the Citadel terrace, that Churchill had told him that he would not after all be commanding Overlord, which had since been such a success. Cunningham noted that the American Chiefs of Staff seemed 'in good form and very friendly'. They certainly arrived mob-handed, their military delegation alone numbering 125.

At lunch at the Citadel that day with the Roosevelts, Mackenzie King and the Governor-General of Canada, the Earl of Athlone (the King's uncle, who was married to Queen Victoria's granddaughter Princess Alice), Churchill told the President that he was 'the head of the strongest military Power today, speaking of air, sea and land'. Roosevelt replied that it was 'hard for him to realize that, as he did not like it himself. He could not feel that way.'[8] Perhaps in order to equalize their relative worth, Churchill went on to say 'quite frankly that if Britain had not fought as she did at the start, while others were getting under way, America would have had to fight for her existence. If Hitler had got into Britain and some Quisling government had given them possession of the British Navy, along with what they had of the French fleet, nothing would have saved this continent,' especially with Japan preparing to strike. According to Mackenzie King's notes, 'The President was inclined to agree with him that they could not have got ready in time.'[9] Churchill's message was clear, and was not disputed by Roosevelt: the Americans might be providing the men, money and matériel today, but four years earlier the British had provided time: an equally important element for the defence of Western civilization.

The Second Quebec Conference (codenamed Octagon) witnessed the 172nd to 176th Combined Chiefs of Staff meetings back in the Salon Rose between Tuesday 12 and Saturday 16 September, as well as two plenary sessions at the Citadel on the 13th and 16th. The British and American Chiefs of Staff had not met since the Americans had come to London four days after D-Day in June. At the first Combined Chiefs meeting at noon on 12 September, Brooke agreed with the optimistic assessment of the Joint Intelligence Committee that Germany was crumbling, and in those circumstances he saw 'great advantages in a right swing at Trieste and an advance from there to Vienna'.[10] Although German and Austrian resistance would mean not getting there till after the winter, he believed that seizure of the Istrian peninsula 'not only had a military value, but also a political value in view of the Russian advances in the Balkans'.[11]

How different was this – albeit short-lived – stance from the one that Brooke later claimed to have adopted. 'We had no plans for Vienna,' he stated in *Triumph in the West*, 'nor did I ever look at this operation as becoming possible.'[12] Yet the minutes of the conference (which Brooke approved at the time) and the contemporary diaries of many of its participants are incontrovertible. At Octagon Brooke supported the Vienna strategy, and for manifestly political reasons, in complete contrast to his stated stance of not allowing such considerations to sway his judgement.

Had he been persuaded by Churchill, or Alexander, or his own anti-Bolshevism? We cannot know, and he soon afterwards changed his mind anyhow and thereafter denied he was 'ever' tempted by the Vienna option. For all that he is well represented by the impressive statue of him outside the Ministry of Defence in Whitehall, which rightly proclaims him a 'Master of Strategy' on its pedestal, Brooke was not above altering his point of view. The plan he presented to the Americans at Octagon was substantially the same one that he had decried when Alexander first mentioned it, and had argued against with Churchill. Yet by September 1944 he was seemingly all in favour. Unlike their statues, human beings – even Field Marshal Sir Alan Brooke – are not made of bronze and granite.

Marshall told the British that he did not intend to weaken the Fifth Army in Italy by reinforcing the Seventh Army in southern France. (There were already four hundred thousand Allied troops taking part in Dragoon, after all.) He also promised to hold landing craft in readiness for a possible British amphibious assault in Istria, in contrast to everything he had previously said. That same meeting agreed to Eisenhower's proposal to consider Montgomery's northern approach – from Holland to the Ruhr – into Germany. Cunningham therefore found the American Chiefs 'in a most accommodating mood and had no disputes', and Charles Donnelly described the 'tone' at Octagon as 'much more relaxed and agreeable than had been the case in previous meetings. It was no longer a question of if the Axis could be beaten, but when.'[13] At 6.30 that evening, Cunningham found Churchill 'in a mood of sweet reasonableness'. Was everyone going to be on his best behaviour throughout the conference? No; Admiral King had yet to speak.

Churchill told Colville that he feared Roosevelt had got 'very frail' since he had seen him last. Colville later said of Roosevelt – to whom he was introduced for the first time at Quebec – 'I heard him say nothing impressive or even memorable and his eyes seemed glazed.' At his talks with Roosevelt at the Citadel, Churchill warned about 'the rapid encroachment of the Russians into the Balkans and the consequent dangerous spread of Russian influence in the area', something at which the American side took relatively little alarm. All that the conference minutes state is: 'Balkans: Operations of our air forces and commando-type operations continue.' Cunningham told Pogue in February 1947 that, at least up until the Yalta Conference, Churchill 'thought he could help Russia by going into the Balkans or into Austria'. Cunningham did 'not believe that he was motivated at that time by fear of Russia'.[14] The truth was different.

Ian Jacob believed that Marshall and the American Chiefs 'looked upon the Balkans as a political jungle and they weren't going to have their troops in there. They regarded the whole war in Europe merely as a problem for a fire brigade. The fire was in Germany, therefore you sent the fire brigade by the shortest road into Germany ... To them it was as simple as that.'[15] If the British wished to get entangled in Balkan intrigues and struggles, Marshall seemed to be saying, he might provide some landing craft but would otherwise leave it entirely up to them.

'It was a dazzling idea, this grand project of reaching Vienna before our Russian allies,' wrote General Alexander in his memoirs, 'and we discussed it informally at my headquarters.' Yet taking the route to Vienna along the so-called Ljubljana Gap involved horrendous difficulties. The 'Gap' was a col 2,000 feet high and 30 miles wide leading to the Save Valley. Between the Save and Vienna is the Karawanken mountain range, with 6,000-foot peaks through which only two roads descended into the Klagenfurt valley. After that there were 200 miles of roads through yet more narrow valleys. 'The powers of recovery of the German forces were a matter of record,' points out Sir Michael Howard. 'They would be falling back along their own lines of communication; at the Ljubljana Gap they would have had a front to defend about one-quarter of the length of the Pisa–Rimini Line ... Finally, the distance from Rome to Vienna is some six hundred miles – about three times the distance from Naples to Rome which it had taken the Allies six months to cover.'[16] It seems surprising that a strategist of Brooke's eminence could ever have proposed such a scheme to the Combined Chiefs of Staff, though not that he should later have denied doing so. He had sarcastically criticized Churchill's Jupiter plan for proposing to 'advance victoriously over one mountain range after another' in northern Norway, yet that is roughly what he himself now advocated in the push to Vienna.

Rear-Admiral Morison explained that the Ljubljana Gap, 'narrow, tortuous, dominated by mountain peaks, would have been a tactical cul-de-sac'.[17] A railway that ran through a large number of tunnels could have been easily destroyed, while the two-lane road could have supported two divisions at most. Furthermore, if it turned into a race to Vienna and Budapest, the Russians would comfortably have won it from the north-east. Even with a Trieste landing taking place in September at the earliest, the Western Allies had run out of time, as the Russians were already in Bucharest.

*

At the first plenary meeting of the conference, held at 11.45 a.m. on Wednesday 13 September, Churchill opened the proceedings with an overview of everything that had happened since they had last all met in Cairo:

Although the British Empire had now entered the sixth year of the war, it was still keeping its end up with an overall population, including the British Dominions and Colonies, of only seventy million white people. The British Empire effort in Europe, counted in terms of divisions in the field, was about equal to that of the United States. This was as it should be. He was proud that the British Empire could claim equal partnership with their great ally, the United States, whom he regarded as the greatest military Power in the world. The British Empire had now reached its peak, whereas that of their ally was ever-increasing.[18]

While the subtle reference to how much longer the Empire had been fighting the Axis was reasonable, it was somewhat disingenuous of Churchill only to count the white population of the Empire and using that in contrast to the much larger population of the United States. At two-and-a-half million men, the Indian Army was the largest volunteer force in the history of mankind; it had soldiers fighting in almost every theatre, lost eighty-seven thousand of them during the conflict and won thirty Victoria Crosses. It surely deserved to be taken into account.

A 'tentative programme and time-table' had been drafted by the British on the *Queen Mary*, ostensibly 'to save time', which was agreed by the Combined Chiefs of Staff.[19] Attempts to control the agenda had long been a British manoeuvre, and on three of the proposed four days of meetings some aspect or other of 'British participation in the Pacific in the war against Japan' found its way on to the schedule. At the first plenary session, Churchill expounded his Far East strategy, declaring that 'He had always advocated an advance across the Bay of Bengal and operations to recover Singapore, the loss of which had been a grievous and shameful blow to British prestige which must be avenged. It would not be good enough for Singapore to be returned to us at the peace table. We should recover it in battle.' He also found time to be gracious about Dragoon, congratulating the Joint Chiefs of Staff on the operation 'which had produced the most gratifying results'. Indeed, Churchill was exuberant, observing that in general 'everything we had touched had turned to gold, and during the last seven weeks there had been an unbroken run of military successes'.[20]

In his answer, Roosevelt said he believed that the enemy would soon retire to the Alps and the right bank of the Rhine, but 'The Germans could not be counted out and one more big battle would have to be fought.' This was perceptive, and a full three months later the battle of the Bulge was to prove him right. The President did not agree with Churchill over Singapore, however, which he thought it possible to bypass, since the fortress 'may be very strong and he was opposed to going up against strong positions.' The Americans had been very successful at 'island-hopping' in the Pacific, leaving stranded Japanese garrisons in the rear for, as Roosevelt put it, 'mopping up later'.

Churchill disagreed with Roosevelt, arguing that 'there would undoubtedly be a large force of Japanese in the Malay Peninsula and it would help the American operations in the Pacific if we could bring these forces to action and destroy them in addition to achieving the great prize of the recapture of Singapore.' In reply, Roosevelt 'referred to the almost fanatical Japanese tenacity', especially at Saipan, where he said 'not only the soldiers but also the civilians had committed suicide rather than be taken.'[21] In fact the civilian families who leapt from the 220-foot cliff at Marpi Point on Saipan in the Mariana Islands on 9 July 1944 did so under compulsion from the Japanese military, but more than seven thousand Japanese soldiers committed suicide there too, so Roosevelt was right about their fanaticism in general terms.

Octagon's major point of contention then arose when Churchill asked what plans the Americans had for employing the Royal Navy in the Pacific after the collapse of Germany. The British wanted to be seen to be taking an active part in the victory over Japan, wanted the return of their many bases and possessions, and wanted the prestige in Australasia of having helped liberate the Far East. This could be done only by active Royal Navy involvement in the final victorious campaign. Although Roosevelt said he wanted 'to use it in any way possible', at that point Admiral King – who didn't want it used at all – said that a paper had been prepared for the Combined Chiefs of Staff. It sounded like a delaying tactic to avoid a decision being taken in the plenary session before the two heads of government, and Churchill asked 'if it would not be better to employ the new British ships in place of battle-worn vessels of the United States'. King reiterated that 'the matter was under investigation'. Churchill then bluntly said that 'the offer had been made and asked if it was accepted'. At this point Roosevelt stepped in and said categorically: 'It is.' Cunningham recalled FDR as replying: 'No sooner offered than accepted.'[22] King 'glowered' at his commander-in-chief's intervention, but Churchill's insistence had yielded the result the British wanted. These meetings were not simply for polite mutual congratulation after all, but the struggle was not over.

Donnelly thought that 'Probably the chief reason for King's dissent was that the US Navy . . . were smelling a not-too-far-off victory over Japan, a victory they were loath to share with an eleventh-hour entry.' King also recognized that the active involvement of the Royal Navy in the south-west and central Pacific might entail the United States giving up bases that she would find it difficult to get back from Britain afterwards. (The US took leases on fourteen Atlantic and Caribbean bases for ninety-nine years in 1940, some of which she still uses to this day.) In his memoirs Cunningham recalled that the Chiefs of Staff had been pleasantly surprised when Churchill offered the British Fleet to operate alongside the US Navy, as he had hitherto wanted to use it exclusively against Singapore, Malaya and Borneo; it looked like a step back from the Bay of Bengal strategy.

The next meeting of the Combined Chiefs of Staff, at 10 a.m. on Thursday 14 September, once again in the Salon Rose conference room of the Château Frontenac, was climactic. Far from Octagon being a 'love feast', it saw the most aggressive Staff clashes of the war so far. The Joint Chiefs of Staff had circulated a paper suggesting that the British Fleet 'should be on the western flank of the advance in the south-west Pacific', which would have accorded it a minor role at best. So Brooke started off by saying that the British Chiefs were 'disturbed' by the statement, adding that he 'realized that this paper had been written before the plenary session on the previous day. He felt that it did not entirely coincide with the proposal put forward at that conference and approved by the President.' Brooke then emphasized that 'For political reasons it was essential that the British Fleet should take part in the main operations against Japan.' British prestige in the Far East was intimately bound up with being in at the kill.

'It might be that the British Fleet would be used initially in the Bay of Bengal and thereafter as required by the existing situation,' answered Admiral Leahy, a classic delaying compromise, but Cunningham pointed out that the main Fleet 'would not be required in the Bay of Bengal since there were already more British forces there than required'. He went on to say that a supply Fleet Train operating out of Australia, consisting of repair boats, ammunition transports, tankers, store-vessels, salvage craft and floating hospitals, would mean that a force of four battleships, six large carriers and twenty light Fleet carriers could operate unassisted for several months. The British Chiefs of Staff therefore wished to see the Fleet operate in the main battle theatre against Japan, which was in the central Pacific rather than the Indian Ocean.

Admiral King immediately weighed in, complaining that 'at the plenary meeting no specific mention of the central Pacific had been made', to which Brooke answered that 'The emphasis had been laid on the use of the British Fleet in the main effort against Japan,' which implied the central Pacific. King replied that 'he was in no position now to commit himself as to where the British Fleet should be employed.' Portal then quoted from an earlier CCS document agreed to by the Joint Chiefs of Staff, which stated in paragraph nine that the British Fleet should indeed be used 'in the main operations against Japan'.[23] Cunningham 'stressed that the British Chiefs of Staff did not wish the British Fleet merely to take part in mopping-up operations in areas falling into our hands'. Leahy replied that 'he felt that the actual operations in which the British Fleet would take part would have to be decided in the future,' suggesting the reconquest of Singapore instead.

Portal was not about to be fobbed off with that, reminding the Combined Chiefs of Staff that Churchill had only the previous day offered the British Fleet for use 'in the main operations against Japan'. At this point King said 'that it was of course essential to have sufficient forces for the war against Japan. He was not, however, prepared to accept a British Fleet which he could not employ or support.' This was despite Cunningham's statement that it could support itself out of Australia with its large Fleet Train. 'It would be entirely unacceptable for the British main Fleet to be employed for political reasons in the Pacific and thus necessitate withdrawal of some of the US Fleet,' King insisted. At some point in these discussions, he went so far as to describe the Royal Navy as a 'liability'.

Portal thereupon 'reminded' Admiral King that the Prime Minister had suggested that certain of the newer British capital ships should be substituted for certain of the older American ships. Cunningham added that, the very day before, Roosevelt and Churchill had agreed 'that it

was essential for British forces to take a leading part in the main operations against Japan'. Whereupon, astonishingly, King 'said that it was
not his recollection that the President had agreed to this', and anyhow
'He could not accept that a view expressed by the Prime Minister should
be regarded as a directive to the Combined Chiefs of Staff.'[24]

Portal – repeating himself almost word for word for the third consecutive time, a sure signal that the talks were getting deadlocked and
ill-tempered – said 'that the Prime Minister felt it essential that it should
be placed on record that he wished the British Fleet to play a major role
in the operations against Japan'. Brooke added that 'as he remembered it, the offer was no sooner made than accepted by the President.'
His use of Roosevelt's exact phrase was telling. King sensibly changed
tack at this, and asked for specific British proposals, whereupon Portal
went back to quoting from paragraph nine of the CCS document that
related to the Royal Navy playing 'a full and early part' in 'the main
operations' against Japan. Leahy accepted that, with the caveat that
they could 'not say exactly where the Fleet could be employed at this
moment'.

King, however, would not let the matter rest there, but asserted 'that
the question of the British proposal for the use of the main Fleet would
have to be referred to the President before it could be accepted'. Cunningham repeated Brooke's point that it already had been, to which King,
trying to widen the argument and bring in Arnold, replied 'that the
Prime Minister had also referred to the use of British air power in
the Pacific'. Arnold said that the amount of British air power would
depend on the development of suitable facilities, to which Portal added
that he would put forward proposals for 'air facilities available in the
bases in the Pacific so that the British could play their part'.

Marshall then spoke for the first time, suggesting that 'the best method
would be a statement of numbers of aircraft and dates at which they
would be available.' Arnold agreed. Brooke stated that since the Combined Chiefs had accepted the principle of the British Fleet operating in
the central Pacific – which King by then most certainly had not – British
land forces 'could only arrive at a later date'. King asked whether 'it
was intended to use the British Fleet only in the main operations and to
make no contribution to a Task Force in the south-west Pacific'. He and
Marshall then disagreed over the Task Force proposal, and in King's
authorized biographer's view, 'they nearly had words.' After King had

criticized Marshall, Leahy told him: 'I don't think we should wash our linen in public.'[25]

For the British this could hardly have played out better, with an open US Army-versus-Navy spat, in which the US Navy representatives were themselves split. Brooke stoked up the situation by repeating that 'The British Fleet could of course play a part in operations in the south-west Pacific if they were required.' Finally the Combined Chiefs 'Agreed that the British Fleet should participate in the main operations against Japan in the Pacific', in a manner both 'balanced and self-supporting'.[26] This meant that it could expect no logistical support from Admiral King's US Navy, which anyone present could probably have guessed anyway.

'King made an ass of himself,' recorded Cunningham in his diary, 'and having the rest of the US Chiefs of Staff against him had to give way to the fact that the British Fleet would operate in the central Pacific, but with such bad grace.' Arnold recounted the way that 'King hotly refused to have anything to do with it. All Hell broke loose! Admiral King could not agree that there was a place for the British Navy in the Pacific, except for a very small force. The American Navy had carried the war all the way from Honolulu to the west and it would carry it on to Japan!'[27] In the end the issue was never resolved because the Japanese surrendered only three months after Victory in Europe Day, although Task Force 57, the British carriers under Vice-Admiral Sir Bernard Rawlings, did play a part in neutralizing Japanese airfields in the Sakishima Gunto islands, 250 miles south-west of Okinawa, and by frequently attacking Japanese forces on Formosa.

Arnold's own role had been almost as unhelpful as King's, however, and when later Churchill said to him, 'With all your wealth of aerodromes, you would not deny me the mere pittance of a few for my heavy bombers, would you?', Arnold replied that the Super Fortresses had started to move in to Guam, Saipan, Tinian and Iwo Jima, and it would take a Combined Chiefs of Staff decision to reverse that and replace them with Lancasters. This was all a ridiculous dispute about pride, prestige and post-war positioning. Of course the Americans should have grasped the chance to broaden the load, perhaps saving American lives in the process. Instead, as King's biographer accepts, 'Not only was King unwilling to share in the glory, but, most galling of all, he called the Royal Navy a liability.'[28]

Recalling the row, Brooke commented that King had 'lost his temper entirely'. Portal later remembered 'blunt speeches and some frayed tempers', while Arnold's biographer rightly described that meeting as 'one of their most emotional and acrimonious confrontations during the war'.[29] King's own semi-autobiographical account completely failed to mention it, but as Arnold's biographer asserts, the book 'made convenient omissions where his habitual bad manners were concerned'. On that occasion it seems to have been Marshall who calmed the situation, and the minutes recorded the very obvious compromise – but arguably also commonsensical – solution, 'that the method of the employment of the British Fleet in the main operations in the Pacific would be decided from time to time in accordance with the prevailing circumstances'.[30] This could be taken to mean absolutely anything or everything, which was what was fully intended.

The open disagreement between Marshall and King over the southwest Pacific Task Force – washing their dirty linen in public, as Leahy had called it – could not have happened among the British Chiefs of Staff, who strictly adhered to collective responsibility. Once decisions were taken internally, they were presented as unanimous to the outside world, whether it was to Churchill, the War Cabinet or the Americans. It was the secret of Brooke's power, and he knew it, acknowledging it in generous terms when his two colleagues retired after the war. As is clear from Cunningham's diary, like any independently minded man he sometimes disagreed with his colleagues, but he stuck to the ethos of the Committee that he had joined relatively late on in the war. Over Anvil, for example, he wrote in his autobiography: 'On the whole I was neutral, though when it came to a collective decision I was at one with my colleagues.'[31]

On the evening of the row, after a vast dinner, Colville recalled that 'there was a shockingly bad film chosen by the President. The PM walked out halfway through which, on the merits of the film, was understandable, but which seemed bad manners to the President.' At the next Combined Chiefs of Staff meeting, at 10.30 a.m. on Friday 15 September, 'Everything went sweetly,' with King 'more or less resigned' to having the Royal Navy in the central and south-west Pacific after victory in Europe. It was not so sweet at the 6 p.m. Staff Conference with Churchill, however. The Prime Minister tried to amend the Combined Chiefs' final report until it was pointed out that it had been agreed with the Joint Chiefs of Staff and so was not susceptible to change without their permission. 'He was just at his worst and Brooke was very patient with him,' wrote Cunningham. 'Looked likely to wreck all the good that had been done. Finally we left him as he had to see General Marshall but we briefed Eden to talk to him.' This, too, conformed to the pattern of earlier conferences: a last-minute threat by the Prime Minister to destabilize the whole network of agreements was seen off by Brooke at the eleventh hour.

That same day Roosevelt and Churchill, amazingly enough, initialled the Morgenthau Plan, which said that Germany needed to be turned 'into a country primarily agricultural and pastoral in character'. Brooke was fundamentally opposed, already seeing Germany as a future 'ally to meet the Russian threat of twenty-five years hence'.[32] Considering that twenty-five years after writing that in 1944, West Germany was an integral part of NATO and Russia had just crushed the Prague Spring, Brooke was more acute than either Roosevelt or Churchill at the time. Once Churchill had properly examined the plan, which amounted to an agricultural Treaty of Versailles and would have hardly allowed a fraction of Germany's population to survive on her own territory, he rightly denounced it as 'unnatural, unchristian and unnecessary'.

The final Combined Chiefs of Staff session at 11 a.m. on Saturday 16 September 'went very happily', as did the final plenary session and press conference at the Citadel. The British Chiefs of Staff had plans to go off fishing together that afternoon, but at 2.30 p.m. a message arrived to say that Churchill wanted to see them that night to discuss operations in Burma. Portal wrote a note pointing out all the arrangements he would be wrecking, and the Prime Minister cancelled the meeting. 'He really is a most selfish and impossible man to work with,' wrote Cunningham of the incident.[33] This was undeserved: Churchill was prime minister and minister of defence during a world war and he wanted to discuss a major theatre of operations with his Chiefs of Staff, while they wanted to go off fishing. As he had been the one to give way, it was hardly the moment to accuse him of selfishness.

Being a keen fisherman was almost a precondition of entry to the Chiefs of Staff Committee in the Second World War, on both sides of the Atlantic. After a Chiefs of Staff meeting in April 1944, Cunningham had gone off to the Army & Navy Stores in London to buy tackle, and there he found Portal buying waders. A fortnight later Brooke presented him during another Chiefs of Staff meeting with an all-purpose dry fly. After Octagon the three men flew off to fish three lakes in canoes, and had 'good sport'. At the same time, Arnold and Marshall went on a ten-day fishing break in the High Sierras, riding out together from Bishop, California. At altitudes of 10,000 feet, 'in the middle of an excellent fishing ground', they were kept in touch with Washington 2,400 miles away by radio. (On one occasion top-secret papers in a securely locked pouch were misdropped by an Army courier plane 2 miles away from the camp. 'To say there was confusion, apprehension and concern is putting it mildly,' recalled Arnold.)

After fishing, the British Chiefs made their way back via New York, where Cunningham stayed at the Knickerbocker Club on Fifth Avenue. He tried to buy fishing tackle at Abercrombie & Fitch, and noticed in a department store that his devoted secretary, Captain A. P. Shaw, 'appeared much interested in buying undies . . . for a lady who I sensed was not Mollie Shaw'.

Roosevelt could not attend the Second Moscow Conference (codenamed Tolstoy) because of the presidential elections, and it was at this conference that Churchill concluded the notorious 'percentages' deal with Stalin. The Prime Minister had already suggested an arrangement of this kind to Roosevelt on 31 May, but had been turned down. Nonetheless, he went ahead and offered the Soviets paramountcy in Roumania if Britain were given a free hand to put down the Communist insurgency in Greece. Stalin had probably already decided not to intervene there in any event, and so placed a big blue tick on what Churchill, with some understatement, later called this 'naughty document'.

Realpolitik in eastern Europe hardly came more blatant and brutal than in Soviet policy towards Poland. On 2 October, discussing the way that Stalin had cold bloodedly refused clearance for aircraft to help Warsaw during the recent uprising there, Eden told the War Cabinet that the Polish President and Commander-in-Chief had been 'unhelpful' in the publicity they had given to Russia's actions. Churchill then spoke of 'These heroic people dogged by their maladroitness in political affairs for three hundred years.'[34] It is one of the less attractive aspects of British policy-making in this period that the Government constantly gave the Russians leeway over the Poles, even to the point of declining to recognize that Stalin, and not the Nazis, had committed the Katyń

Massacre of Polish officers in 1940 despite overwhelming proof to that effect.

Averell Harriman and John R. Deane, head of the US military mission to Russia, were invited to the military conferences with Stalin, and were amused by the way that Brooke gave excellent presentations of Eisenhower's plans, which were then subjected to constant interruptions by Churchill, who would 'leap from his seat and stride to the map in order to emphasize the magnitude or difficulties of certain phases of the Anglo-American operations'.[35] It is doubtful that Brooke took much solace from the fact that General A. E. Antonov, the Red Army Deputy Chief of Staff, was subjected to much the same kind of behaviour by Stalin. Deane also recalled with embarrassment a meeting on the evening of 14 October in the large conference room outside Stalin's office, in which Stalin asked him how many divisions the Japanese had, and he did not know. 'The day went to the British,' he admitted, after Brooke 'quickly thumbed through his papers and came up with the right answer'.

Because Roosevelt and Marshall were not present, the Second Moscow Conference was not able to resolve major issues in eastern Europe, and when Churchill did complete his percentages deal with Stalin, it was not ratified by the Americans, insofar as he even explained it to them. Churchill went on to claim, rightly, that Britain had nonetheless saved Greece from 'the flood of Bolshevism'. Speaking to Leo Amery soon after returning from Russia, Brooke said that 'the change even in two years away from proletarian Communism to uniforms, decorations, rigid class distinctions, etc and towards old fashioned nationalism is very marked. There is fearful squalor behind the façade.'[36]

In early October, Churchill having effectively been given the go-ahead by Stalin, Lieutenant-General Ronald Scobie's III Corps began to occupy Athens in Operation Manna, at least once Generaloberst Alexander Löhr's Army Group E evacuated Greece in order to avoid being cut off by Soviet forces. Rather like the Suez Crisis of 1956, this could hardly have come at a worse time in the American electoral cycle. As the Germans had started to withdraw from Greece, Churchill cabled Roosevelt to warn him that the Communist-dominated EAM (National Liberation Front) and its military wing ELAS (National Popular Liberation Army) would soon fill the power vacuum in Greece and crush all opponents unless Athens were swiftly occupied by the Allies.

At the time, Roosevelt replied that he had no objections to this.

When Scobie arrived, however, fighting broke out between ELAS and supporters of King George II of the Hellenes, and there was much criticism in the American press about Limey attempts to impose a reactionary, monarchical regime on freedom-loving Greek republicans. (In fact the King favoured a liberal democratic constitution, and most of the Greek republicans were pro-Soviet Communists.) Stalin, under the terms of the 'naughty document', did his bare minimum to support ELAS and EAM, but Roosevelt failed to say a word in Britain's favour. Churchill said he understood the President's difficulty, but he privately resented the complete lack of moral support afforded him on that occasion. Marshall utterly opposed any American involvement in Greece to assist Britain (although when he became secretary of state he supported US intervention there).

On Tuesday 7 November 1944 Franklin D. Roosevelt was re-elected with 25.6 million votes, against 22.0 million cast for the Republican candidate Thomas E. Dewey. Marshall congratulated him 'with great respect and compete loyalty to your leadership'. Yet neither Roosevelt nor Marshall altered his stand to a more supportive one over Greece, and on 10 December it was discovered that Admiral King – never one to let an opportunity to discomfit the British go by – had even ordered the American landing craft in the Mediterranean to cease taking British troops and supplies to Greece. This was later quietly countermanded by Harry Hopkins, acting on Roosevelt's behalf.[37] There were unmistakable echoes of the incident six months previously when King had unilaterally tried to move the flotilla from the Channel without consulting the Admiralty, though at the next day's Chiefs of Staff meeting Brooke reported that 'Hopkins had asked the PM that there should be no recriminations and the PM had agreed.'

Marshall later recalled: 'We were very much afraid that Mr Churchill's interest in matters near Athens and in Greece would finally get us involved in that fighting, and we were keeping out of it in every way we possibly could.' On 13 December, Roosevelt cabled Churchill to say that 'the traditional policies of the US' meant that as head of state he had to be 'responsive to the state of public feeling' against Britain on the Greek issue, and he concluded, 'I don't need to tell you how much I dislike this state of affairs as between you and me.' Churchill replied generously: 'I have felt it much that you were unable to give a word of explanation for your action, but I understand your difficulties.' The new burden of combating Communism in south-eastern Europe therefore looked as if it would be carried entirely by the British.

On 2 November, Walter Bedell Smith flew to London for lunch with Churchill and the Chiefs of Staff. 'Some good talk', recorded Cunningham, who learnt there of the huge attacks planned by Eisenhower along the whole Allied line, with Patton starting off with nine divisions in three days' time and Bradley and Montgomery leading the main attack towards the Cologne–Ruhr area. 'Brooke obviously does not think the main attack has enough weight behind it and looks on Patton's attack as too great a diversion of strength,' noted Cunningham. 'However these Americans are often right!!' For all the regular confluence of views between the CIGS and the First Sea Lord, that was simply not a sentence that could have been written by the former. Churchill joked that in the discussions with the Americans over whether Alexander should be given two extra divisions, the opinion at Eisenhower's headquarters was 'that it would be cheap at the price provided Monty accompanied the two to Italy and Alex came in his place'.

On 4 November, Field Marshal Sir John Dill died in Washington. 'We mourn with you the passing of a great and wise soldier, and a great gentleman,' Marshall wrote to Brooke as part of a long and heartfelt letter that concluded: 'His task in this war has been well done.' By a special Act of Congress, Dill was buried at Arlington Cemetery, the only non-American to have been accorded that honour. There is a very fine equestrian statue of him there, cast with such attention to detail that it is even possible to make out the rosettes on three of his campaign medals.

Marshall read the lesson at the funeral. Another of the mourners, a cousin of the general, noted afterwards: 'I have never seen so many men so visibly shaken by sadness. Marshall's face was truly stricken . . . It was a remarkable and noble affair.'[38] Giving Marshall a silver tea service to remember her husband by, Dill's widow Nancy wrote to say: 'He really loved you, George, and your mutual affection meant a great deal to him – he always trusted you implicitly.' Marshall's reply mentioned 'the intimate bond between Dill and myself'. Under the impression that Dill was about to be recalled to London for being too pro-American, Marshall had organized for a number of American honorary degrees to be bestowed on him, to convince Churchill of his prestige.

Charles Donnelly spotted that Dill's death was 'a blow' because his assistance had been 'immeasurably valuable' transatlantically 'in calming bitter arguments which so often arose between Sir Alan Brooke and Marshall'. That job could not really be filled by Jumbo Maitland Wilson, who now replaced Dill in Washington, with Alexander taking over as supreme commander in the Mediterranean, 'a post for which he is totally unfitted' in Cunningham's view, because he was 'completely stupid'.[39]

Lord Halifax told Churchill, 'very much off the record', that Marshall had suggested Ismay would be the best successor for Dill. 'The latter's close contact with you and knowledge of your thought would, in Marshall's view, be of great value to the partnership between us.'[40] Halifax had warned Marshall that Churchill would probably be unwilling to lose Ismay, and he was right. It nevertheless shows the emphasis that Marshall placed on finding someone who was close to Churchill for this most sensitive of posts. Maitland Wilson was not a Churchill confidant; the rows that developed later would not have been avoided had Dill survived or had Ismay succeeded, but their intensity might have been lessened in either case. Certainly Marshall doubted whether Churchill – who had twice turned Dill down for a peerage – and even Brooke 'fully realize the loss you have suffered'.

At Octagon the Combined Chiefs of Staff had approved Eisenhower's plan to mount a major northern effort encircling the Ruhr, with secondary attacks towards Bonn and Strasbourg, yet this strategy satisfied neither Montgomery, who wanted to go straight to the Ruhr, nor Patton, who wanted to go to Berlin. Brooke, who was not insensible to Montgomery's vanity and attention-seeking, nonetheless protected the new field marshal from a prime minister who he believed was jealous of Monty's popularity with the public, and also from those Americans who had long ago spotted an anti-American prima donna.

'I first got to know Brooke ("Brookie" as he has always been to me) in 1926 when I went as an instructor to the Staff College, Camberley,' recalled Montgomery in his memoirs; 'he was already there, as instructor in artillery. I quickly spotted that he was a man of outstanding character and ability, and my liking and respect for him can be said to have begun then.' With uncharacteristic modesty, Montgomery also wrote, 'He was well tuned to my shortcomings and often administered a back-hander, sometimes verbally and sometimes in writing; in neither case could they ever be misunderstood!' Overall, Montgomery's 'feeling was that in strategic matters Brookie was generally right and Marshall wrong'.[41]

Brooke was in many ways Montgomery's mentor, and so when on 22 November 'Monty' wrote to 'My dear Brookie' from 21st Army Group, he was willing to listen sympathetically. Montgomery complained that Eisenhower seemed 'to have a curious idea that every Army Command must have an equal and fair share of the battle', and he pointed out that two army groups were involved north of the Ardennes and two south of it, with Omar Bradley's command split between them. Montgomery proposed that since 'Ike seems determined to show that he is a great general in the field,' the theatre should be divided 'naturally into two fronts – one north of the Ardennes and one south', with him commanding the northern and Bradley the southern sectors; meanwhile 'Ike should command the two fronts, from a suitable Tactical HQ.'[42]

Brooke approved of this. It is not necessary to entertain the belief of Eisenhower's son John that Brooke's 'zealous' support for Monty against his father was 'intensified . . . by his own personal disappointment' over the post of supreme commander, a serious charge for which he gives no supporting evidence.[43] But before Brooke could genuinely help at all, let alone zealously, later that month Montgomery wrote to Eisenhower asking for some of Patton's troops in 12th Army Group to be moved north of the Ardennes and calling the current campaign a 'failure', something to which Ike understandably took great exception. If Patton, Clark and Bradley – not to mention some Britons such as Cunningham – hadn't existed, then Montgomery would have been his own worst enemy. Since they did, he was spoilt for choice.

'Monty wants to command a northern group of armies and the Americans are always suspicious of his motives,' concluded Cunningham after a Chiefs of Staff meeting on 4 December, 'The fact is we are stuck.' Three days later Cunningham and Portal and their wives dined at Brooke's flat; after dinner he showed them his bird films, some made in his garden and one on a trip to the Faroe Islands, which Cunningham claimed to his diary were 'Most interesting.'

On 29 November Churchill made clear his objections to the early liberation of the Channel Islands, telling the War Cabinet that while the twenty-eight thousand Germans there 'can't get away', if they surrendered Britain would 'have to feed them'. The islands therefore remained under German occupation for the eleven months between D-Day and V-E Day.[44]

On 12 December 1944 Churchill and the Chiefs of Staff convened in the Map Room in the Annexe after the 5.30 p.m. Cabinet to meet Eisenhower and Tedder and hear their plans for an attack by simultaneous thrusts across the Rhine, one north of the Ruhr and the other from Mainz–Frankfurt towards Kassel. Brooke 'vehemently argued' for Montgomery's idea of 'a really strong thrust north of the Ruhr' eastwards, with the Frankfurt area to be merely subsidiary. 'Ike was good and kept an even keel,' thought Cunningham. 'He was obviously impressed by the CIGS's arguments but refused to commit himself.'[45] At dinner, which went on until 1.30 a.m., Eisenhower said that he had recently rejected an offer of half a million dollars by a newspaper to allow articles that he had not written to be published under his name.

The meeting in the Map Room exposed the disagreement over strategy between Eisenhower and Brooke. As in the past, Eisenhower turned to his own mentor, Marshall, writing the next day to say that 'Field Marshal Brooke seemed disturbed by what he calls our "dispersion" of the past weeks of this campaign.' He added that until the floods in the lower Rhine valley he had tried to give everything he could to the northern thrust but these had naturally foiled it. At the War Cabinet on 13 December, Brooke set out Eisenhower's suggested plan and 'also gave his reasons for not accepting it and stressed the necessity of making the punch north of the Ruhr strong enough not to be held up at the expense of attacks elsewhere'. Before a full-scale Anglo-American argument could develop between Marshall and Brooke over Eisenhower's strategy, however, Hitler intervened. Before dawn on Saturday 16 December 1944, the Germans unleashed one of the greatest surprise offensives in the history of modern warfare.

Operation Herbstnebel (Autumn Mist) flung no fewer than twenty-four divisions – the last of Germany's reserves – into an all-out attempt to split the United States and the British Commonwealth armies and recapture Antwerp. The Führer chose to attack through the same semi-mountainous and wooded Ardennes region on the east Belgian–German–north Luxembourg borders where four-and-a-half years earlier his armour had broken through to deliver him the Fall of France.

20

Autumn Mist: 'We have been having a bit of a party out here!!'

December 1944–February 1945

The time has now come when the German Army must rise again and strike.
Field Marshal Gerd von Rundstedt, 16 December 1944

The Ardennes Offensive, also known as the battle of the Bulge from the 55-mile-deep protuberance that it created in the Allied lines, was Hitler's final chance to split the Allied armies by taking Antwerp, and then to defeat them in detail. By attempting to cross the River Meuse and strike at the hinge between his foes, Hitler's last gamble was remarkably similar to that of Napoleon during the Hundred Days. Although the battle lasted forty-four days and was the largest land battle in American history, Hitler had no more success than the Emperor.

As with so many of his coups in the past, Hitler chose a Saturday to unleash his surprise stroke, and like them it met with startling initial success. Eisenhower had left the defences relatively thin in the Ardennes Forest in order to concentrate on seemingly more profitable areas to its north and south, so two hundred thousand Germans, commanded by Field Marshal Gerd von Rundstedt, were able to attack eighty-three thousand Americans across a 60-mile front.

Allied intelligence and aerial reconnaissance had not spotted the vast congregation of German armour from Army Group B moving through the dense forests. German security was excellent: radio and telephone transmissions concerning the attack were *verboten*, only land-lines or messengers were permitted; troop movements took place at night or in bad weather and corps commanders were not given their assignments until days beforehand. On the night of the attack itself, artificial moonlight was created by bouncing searchlight beams off low clouds. Furthermore, a special unit of two thousand men of Panzer Brigade 156 – including 150 English-speakers – were dressed in American uniforms to increase the confusion. The German capacity for counter-attack in the Ardennes as late as the winter of 1944 – undertaken through deep snow in the worst Belgian winter in living memory – must give pause to those who believe a 1942 or 1943 Overlord would have fared better.

On 18 December north Burma was finally cleared of Japanese, and two days later Roosevelt bestowed the ultimate commission of general of the Army on Marshall, raising him to the same rank as his former commander, 'Black Jack' Pershing, and awarding him a fifth star. It was a fitting recognition of the way that Marshall had overseen the creation of a vast army virtually from scratch, even if the promotion unfortunately fell on the day before that Army's worst humiliation of the war against Germany: on 19 December, nearly eight thousand men of the US 106th Infantry Division surrendered to the Fifth Panzer Army in the Ardennes. Elsewhere, however, and especially at Bastogne and Saint-Vith, fierce American resistance slowed the German juggernaut and threw its well-laid plans – which depended on capturing US fuel dumps by specific dates – badly out of kilter. That day Churchill told Cunningham that he 'preferred a tortoise with its head out even if it looked like biting him'.[1]

Eisenhower's reactions were commendably fast, and by midnight on the second day of the offensive the 82nd and 101st Airborne Divisions were on the move; in all a quarter of a million men and fifty thousand vehicles were detailed to destroy it. Patton was ordered to wheel virtually his whole army of six divisions sharply to the left and hit the offensive from the south. Because the Germans had split the 12th Army Group north and south of the Bulge, destroying much of their communications, Eisenhower temporarily transferred the US First and Ninth Armies to the north under Montgomery's command, something for which Omar Bradley never really forgave him.[2] 'Brad was absolutely livid,' recalled Air Marshal Sir Arthur Coningham. 'Walked up and down and cursed Monty. Was startling to see Brad like this. Because of his personal loyalty to Ike, Brad stuck out the show.'[3] A jingoistic and misinformed British press portrayed the move as the heroic Monty being called upon to save the day from a hapless American, an interpretation which Montgomery ought to have done more to dispel the moment he could. Instead he gave a press interview at his headquarters in which, although he praised the ordinary GI, he failed to give proper credit to the American High Command and seemed to hog the glory for closing the Bulge.

Operation Herbstnebel ran out of momentum and especially petrol by Christmas Eve, and on 3 January 1945 the First Army counter-attacked from the north, linking up with the Third Army two weeks later. 'We have been having a bit of a party out here!!' Montgomery wrote to Portal.[4] (As if misunderstanding such gung-ho spirit, the published diaries of General Hap Arnold attribute to Colonel Frederick W. Casfie, the son of one of Arnold's West Point classmates killed in combat that Christmas Eve, the posthumous award of 'the Medal of Humor'.) To put the Ardennes Offensive in context, however, the Joint Intelligence Committee estimated that the Germans had 105 divisions on the Western and Italian fronts at the time, but 149 on the Eastern. The battle of the Bulge, for all the potential danger it posed in the west, was only half the size of the battle of Kursk, for example.

Early in January, Churchill and Brooke visited Eisenhower's Supreme Headquarters Allied Expeditionary Force (SHAEF) at the Trianon Palace hotel in Versailles. Cunningham later flew over to attend the funeral of his friend Admiral Sir Bertram Ramsay – the mastermind behind the naval side of Dunkirk and D-Day who had died in a plane crash on 2 January – and also visited SHAEF. He found that 'Montgomery's patronising talk to the Press had done little to improve matters. Ike and the American generals were all up in arms and the tone of the British Press was making it quite impossible to put American troops under Montgomery's command.'[5] That night, Brooke and Bracken made 'unfavourable comment' about Montgomery's interview, even though he had extravagantly praised the courage of the American fighting man.

The battle was not finally won until Sunday 28 January, when the last of the Germans were cleared from the Bulge. They had got tantalizingly close to the Meuse but never quite reached it. They incurred one hundred thousand casualties, the Americans eighty-one thousand, including nineteen thousand killed and fifteen thousand captured. The British, who fought only at the tip of the Bulge, suffered 1,400 casualties. Both sides lost around eight hundred tanks, which by that stage in the war the Germans could no longer afford. 'Any army can go through your force if they are willing to risk losses or if they are willing to weaken their own front so that they can't prevent a counter-attack,' explained Walter Bedell Smith, Eisenhower's chief of staff, to Pogue after the war. 'Germany did both those things. The result was that the Germans used up most of their armor and had practically nothing to oppose us later on.'[6] Bedell Smith provided Cunningham with some of the Wehrmacht's champagne to toast the New Year, with no one enquiring too closely where it had originally come from.

Captain Cyril Falls, the military correspondent of *The Times*, had written to Churchill on 21 December 1944 – 'Now Rundstedt has shown us how an offensive should be conducted' – a letter critical of both Alexander and Eisenhower, and which suggested that political considerations had governed the senior Anglo-American military appointments. In his reply, Churchill praised Alexander's command in the Western Desert and said that Rundstedt's offensive would probably shorten the war. 'All these Commands have been made on their merits,' he assured Falls, 'but the British and Americans have to be represented to a very large extent in accordance with the forces employed in the different theatres,' and 'in all these questions the United States forces are already between two and three times as numerous as ours.' Churchill concurred in Falls' praise for Brooke, saying that 'had the Command in North-West Europe fallen to the British instead of the United States, he was already chosen as its commander. However in his present great situation he is able to exert an immense influence over all the theatres of war.'[7] Of course Churchill had in effect admitted that Brooke could not have had the command whatever his merits, because of political considerations, thereby substantially confirming what Falls had alleged.

Leo Amery recalled that in late January 1945 Churchill asked Brooke how many divisions the Russians had, and 'when he said five hundred one could feel the shudder going through the Cabinet.'[8] Russian divisions tended to be smaller and many were under-strength and under-equipped at that time, but nonetheless the USSR mobilized more troops in the war than Germany, Britain and Italy combined. It was in order to keep this vast Russian strength within certain European confines, and to prevent chaos during the death-throes of the Third Reich, that a second Big Three conference was slated to take place at Yalta, 345 miles southeast of Sevastopol in the Crimea in mid-February. Stalin had claimed to Roosevelt and Churchill that his doctors had advised him not to leave the Soviet Union, even though Roosevelt was far more ill than he. With a sublime disregard for historians' convenience, this gathering was also codenamed Argonaut, the same name as the Second Washington Conference of June 1942. Churchill wanted to see Roosevelt and Marshall before Yalta, in order – rather as at Cairo – to try to agree common ground before meeting the Soviets. The place he chose was the brave little Mediterranean island of Malta. 'If you do not wish to spend more than one night at Malta,' he wrote to an evidently reluctant president, 'it could surely be arranged that both our Chiefs of Staffs should arrive there say a couple of days before us and have their preliminary discussions.' So little did Roosevelt want to collude with the British that Harriman even asked Stalin not to tell Churchill about the conference arrangements for Yalta until the last moment.[9]

In his answer to Churchill's request about Malta, Roosevelt said that 'in view of the time available to me for this journey it will not be possible for us to meet your suggestion and have a British–American staff meeting at Malta before proceeding to Argonaut. I do not think that by not having a meeting any time will be lost at Argonaut.' A correspondence that had begun as mutually affectionate billets-doux and jokes had progressed to cold notes of mutual suspicion. In the event, another, very short conference was scheduled to take place on Malta, codenamed Cricket.

At the War Cabinet of 8 January 1945, Churchill asked of the celebrations for soldiers who were returning on leave from the Middle East: 'Why not brass bands?'[10] That was to be the last time that the War Cabinet met at Downing Street for some weeks. The V-2 campaign meant that the next day it convened underground in the Cabinet War Rooms, then in the Map Room in the No. 10 Annexe, then back in Downing Street for one meeting only, but not there again until 3 April. 'The Angel of Death is abroad in the land,' said Churchill of the V-2s, slightly misquoting John Bright's philippic against the Crimean War, 'only you can't hear the flutter of his wings.' The atmosphere returned to that of the 1940 Blitz, with decisions taken in emergency conditions underground. 'It got pretty stale down there sometimes,' recalled Paul Caraway, 'but that was bearable.'

On 9 January, Brooke considered 'a new scheme' of Churchill's under which Air Chief Marshal Sir Arthur Tedder, Eisenhower's deputy supreme commander, was to get a post at the Air Ministry and Alexander would leave Italy and take his place. Churchill wanted a high-profile Briton to be co-supreme commander in all but name. 'This presupposed that divisions were withdrawn from Italy and the campaign there died away to a defensive one,' thought Cunningham, who told Brooke that 'the Americans would take it as an insult and think that Alex was being sent to hold Ike's hand.'[11] Another idea was simply to swap Tedder and Alexander around.

On 12 January Eisenhower wrote to Marshall to say that he didn't oppose that concept, but Marshall thought it seemed like an admission of failure in the Ardennes. Brooke liked the idea of interposing Alexander – whom he did not rate very highly as a strategist – in between Eisenhower (whom he rated even lower) and the various army group, army and corps commanders. A plan to have Montgomery as Eisenhower's land commander had briefly cropped up in the autumn, largely promoted by Montgomery himself. Now Churchill and Brooke were hoping to revive it for Alexander. On 10 January, Roosevelt sent Churchill a Joint Resolution of Congress awarding Dill a posthumous Distinguished Service Medal and donating $25,000 to erect the equestrian memorial in Arlington Cemetery. The President wrote that it was 'evidence of a very wholesome state of mind in the midst of the bickerings that are inevitable at this stage of the war'.[12]

On 28 January, the day of the final victory in the battle of the Bulge, Marshall met Eisenhower near Marseilles, partly in order to demonstrate his continued confidence in him in the face of British pressure effectively to relieve him of the day-to-day control of the ground forces. It allowed the Army Chief of Staff to hear about the Supreme Commander's future strategy, and thus arm himself against any British criticism at the Cricket Conference to which he was headed. The trip gave Marshall an unexpected chance to learn about the geography of the Balkans, in somewhat surprising circumstances. Staying with a Polish-born American liaison official in a villa outside Marseilles, Marshall had his hair cut by a barber who it transpired came from the crest of the Ljubljana Gap. The butler, who did the translation, also turned out to have been born and bred in the valley below, 'So they spent about an hour and a half educating me as to the country,' recalled Marshall. When he met the British soon afterwards in Malta, 'they were astonished' at his seemingly intimate knowledge of the geography of the region, and assumed that he must have once spent a summer there before the war.[13] It stood him in very good stead. If Marshall was unwilling to commit American troops to Leros, Rhodes or Athens, he had always looked even less favourably on a political scheme to send armies into the Balkans. 'The only thing the British hadn't put in was trying to get to the North Pole,' he once commented to Pogue about his ally's supposed fondness for indirect 'sideshow' operations. His problem, as ever, was with the President. 'I was frankly fearful of Mr Roosevelt introducing political methods, of which he was a genius, into a military thing which had to be on a fixed basis,' he said in 1957. 'This was particularly so in regard to the Balkan states and the now-termed satellite states. You can't treat military factors in the way you do political factors. It's quite a different affair.' Marshall felt that his brief was not to save eastern Europe from Communism but instead to win the war in the shortest possible time and with the fewest possible Allied lives lost. After the Iron Curtain descended he was severely criticized for this, but at the time he did his military duty, leaving the political consideration to the politicians.

Eisenhower set out his plans for the double-envelopment of the Ruhr, and explained in detail his view of the European endgame, so that by the time Marshall left Marseilles on 29 January (he arrived in Malta the next day), he was well prepared for yet another showdown with the British over Eisenhower's role and his strategy, whether they wanted one or not.

At the War Cabinet that same day, after hearing about the Wehrmacht withdrawing back to the Fatherland from Norway, Churchill likened Germany without military reserves to 'Living in the middle of a spider's web and not having a spider'.[14] He then left for Northolt aerodrome and Malta, where he arrived with a temperature and went straight to bed that afternoon on board the cruiser HMS *Orion*. Brooke and Cunningham had already left that morning with Jacob, who was 'nearly knocked [out]' in the unpressurized cabin at an altitude of 12,000 feet and had to be given oxygen. They were met at the airport by Admiral Sir James Somerville, the head of the British naval delegation in Washington, and driven to Admiralty House where they were guests of the man who had succeeded Sir Andrew Cunningham as commander-in-chief Mediterranean Fleet, rather confusingly called Vice-Admiral Sir John Cunningham. Somerville had a lady flag lieutenant, noted Andrew Cunningham, 'and the things he says to the poor girl are quite scandalous'.[15]

The first two Combined Chiefs of Staff meetings of the Malta Conference – the 182nd and 183rd of the war – took place on the morning and afternoon of Tuesday 30 January in the Montgomery House in Floriana, a suburb of Valletta, where the American delegation was staying. Procedure was settled quickly and Andrew Cunningham felt 'We have no serious differences,' but then he had also said that at Octagon on the eve of the discussion on the employment of the Royal Navy in the Pacific.

Marshall outlined Eisenhower's plan for the main Allied attacks to be undertaken on a broad front north of the Ruhr by Montgomery's 21st Army Group with a second attack between Frankfurt and Kassel by Bradley's 12th Army Group. He 'considered it essential that there should be more than one possible line of advance', with the majority of the reinforcements then fed into whichever seemed to be doing best.[16] The British worried that shifting the 15th Air Force from the Mediterranean to Eisenhower might damage their position in Italy, but Marshall said that it needed to be employed wherever the weather proved advantageous, and that the move wasn't permanent.

At 2.30 p.m. the next day, 31 January, Andrew Cunningham spotted 'Some differences over the Western front strategy and also minor ones over the Mediterranean strategy'. However, 'Much was deferred as we pushed a number of new minutes at one another.' That evening John Cunningham hosted a dinner at Admiralty House for the American and British Chiefs of Staff, as well as Maitland Wilson, Admiral Stark and many others, at which M. Bellizzi's twenty-piece band 'played splendidly' until midnight. Everyone was very complimentary to Andrew Cunningham about the furniture and fittings of Admiralty House, which had been chosen by his wife Nona when they were stationed there from 1939 to 1942. 'It was like old times,' the ex-Commander-in-Chief reflected. Agreement was reached over issues as wide ranging as the priorities of the combined bombing offensive, the danger posed by German jet-fighters against Allied piston-engined planes, the allocation of resources between India, Burma and China, co-ordination with the Soviets (especially over bombing, from which was to come the destruction of Dresden a fortnight later), and the U-boat threat in British waters too shallow for the ASDIC underwater sonar device to work effectively. Furthermore, the Ljubljana Gap concept was effectively killed off – with the help of Brooke, who had by then had time to examine the operation more closely – and the British were also persuaded to go on the defensive in Italy and move five divisions from there to fight under Eisenhower.

Yet Thursday 1 February witnessed what have been described as 'the most violent disagreements and disputes of the war.' Churchill wrote to

Clementine from HMS *Orion* that he had had luncheon alone with Admiral King the day before and Marshall that day, and could report: 'Both are in great form and all the conversations at the Conference have been most friendly and agreeable.'[17] But at the 2.30 p.m. Combined Chiefs of Staff meeting that same afternoon, coming straight from lunch with Churchill, Marshall called for an off-the-record session, and there he told Brooke exactly what he thought of his demand that Montgomery be given more American troops to effect the Rhine crossing, and much more besides. As Andrew Cunningham recalled it, Marshall:

let off rather a tirade about trying to fight the Western front battle by committee and also about the constant pressure maintained by Montgomery and the Prime Minister on Eisenhower. He said some pretty straight things about Montgomery, allowing personal feelings to enter into things. Brooke was not too good and we only noted Eisenhower's submitted plan although both Portal and myself and King would have liked, under conditions, to note it with approval. Marshal [*sic*] practically made the question one of confidence in Eisenhower.[18]

Of course Marshall privately included Brooke himself in the list of those Britons exercising improper pressure on Eisenhower, even if he did not say so to his face. It had, after all, been Brooke who had on 12 December criticized Eisenhower's strategy most 'strenuously' in the Annexe Map Room.

Quite what 'pretty straight things' Marshall said about Montgomery on that occasion we do not know, but a flavour of them might be taken from his strictly 'off-the-record' remarks to four Pentagon historians in 1949, to whom he said: 'The Eighth Army had committed about every mistake in the book. It was no model campaign. The pursuit of Rommel across the Desert was slow. The British even laid a minefield in front of them which benefited the Germans more than it did the British ... Montgomery left something to be desired as a field commander.'[19] He added that El Alamein 'gave a great boost to morale but was blown up out of all proportion to its importance' and that Montgomery's performance in north-west Europe was no better. If these were the kind of remarks Marshall made about Montgomery at Malta in 1945, as well as in Room 2E844 of the Pentagon four years later, it is understandable that the atmosphere became, in his own words, 'very acidic'.

Brooke stuck up for his protégé as best he could, and when Marshall asked him to approve Eisenhower's plan for the Western Front, he simply refused to do so, agreeing only to 'take note' of it. After the war Brooke wrote that 'through force of circumstances' he had to accept the plan, because they 'were dealing with a force that was predominantly American, and it was therefore natural that they should wish to have the major share in its handling.'[20] Marshall had protected his own protégé, Eisenhower, and the broad-front strategy of advance to the Rhine and Elbe, even suggesting to Ike that he threaten to resign on the issue.

Asked after the war to comment on what had happened at Malta, Marshall said that the session had been 'a very hot one. We had great difficulty in reaching a general decision.' He explained that 'Montgomery wanted certain troops and a lead in the crossing of the Rhine,' troops which had largely to come from the Americans. Eisenhower had given him practically all he asked for, but it had not satisfied him. 'It was getting to be a quite serious political matter,' and Marshall and Roosevelt 'had a hard time beating it off'. Marshall also recalled telling Brooke that he and Roosevelt 'hardly ever saw' Eisenhower, who was 'under the guns from Mr Churchill almost twice a day at times and very, very frequently all the time'. It is evident that Marshall personally blamed Brooke too, telling the Pentagon historians that 'the real influence' being brought to bear on Eisenhower 'was the direct influence of Churchill and Alan Brooke. They were seeing him every week, and not going through the Combined Chiefs of Staff. We here in Washington were playing according to the rules.' He claimed of the prestigious Rhine crossing planned in March (codenamed Plunder) that the British 'were trying to restrict this thing so as Bradley couldn't advance on the Rhine', and that 'They were all afraid of Patton getting loose down there.'

On other occasions Marshall referred to that meeting as 'terrible', which it clearly was.[21] Deep and long-held American suspicions had come out into the open, and were very fully aired. 'It was rough,' said Marshall; 'these sessions of lively arguments came up – and they were lively and they were very frank – but we always came to a harmonious conclusion.' With his immense sense of fairness and objectivity, Marshall pointed out that 'We Americans must keep in mind that the British . . . gave supreme command to Eisenhower in Africa when we had very few troops there and they had the dominant armies. They gave the supreme command, and reiterated it, to General Eisenhower, when General Montgomery's famous Eighth Army came up along the northern rim of

Africa,' and even though 'he was outranked.' It was true; Eisenhower had only received his fourth star on 11 February 1943.

Marshall might well have also been venting irritation at the meeting provoked by what he later described as 'This patronizing attitude towards American troops' that he thought was 'rather widespread in English circles', instancing a time when the King 'started telling me how fine it was to have Eisenhower in nominal command with Montgomery at his side, etc'. Marshall diplomatically confined himself to replying: 'That's very interesting, Your Majesty.'[22] He was less restrained when General Alexander remarked to him, 'Of course your American troops are basically trained,' replying tartly: 'Yes, American troops start out and make every possible mistake, but after the first time they do not repeat these mistakes. The British troops start out in the same way and continue making the same mistakes over and over, for a year.'[23]

Walter Bedell Smith was present at the closed session of 1 February, and he later recalled that Brooke was 'upset' about Eisenhower's plan for the Rhine crossing and wanted 'a directive to Ike which would require him to give a certain amount of troops to Montgomery'. Bedell Smith told Brooke that that would amount to a vote of no confidence in Ike, who would offer his resignation, upon which Brooke 'disavowed any intention of getting rid of him'. It was hardly a ringing endorsement, nonetheless, and echoes other criticisms, such as that of the splendidly named Colonel C. H. Bonesteel III, who was in the Planning Department of 12th Army Group, and who said that Ike 'Never really commanded. He was an arbiter or tribunal between services.'

A week after the row, Bedell Smith wrote to Marshall's deputy chief of staff Tom Handy about a:

bitter argument with Field Marshal Brooke who wished to revise Ike's directive in such a way that he could hardly move a division except north of the Ruhr. I had a couple of long talks with him after we got back to our rooms, and I give him credit for complete honesty in this matter, a tribute I have never paid him before; but he is stubborn as Hell, and stood out until finally GCM called a closed conference at the end of one of the sessions, spoke his mind as only he can do, for about fifteen minutes, and, as a result, the matter was dropped.

Bedell Smith added that 'it would have been criminal' if Eisenhower had 'staked everything on one narrow thrust north of the Ruhr'.

After the war, Eisenhower signally failed to repay the support that

Marshall had shown him at this crisis moment at Malta, and generally in having promoted him from lieutenant-colonel to four-star general in the less than two years between March 1941 and February 1943, and to five-star General of the Army on 20 December 1944 (only two days after Marshall himself). When in October 1952 Marshall came under violent criticism from the Wisconsin Senator Joseph McCarthy for having let China 'fall' to the Communists when secretary of state, Eisenhower excised a paragraph of one of his election speeches in Milwaukee that described Marshall as 'dedicated with singular selflessness and the profoundest patriotism to the service of America'.[24] As President John F. Kennedy later put it: 'No man is less loyal to his friends than Eisenhower. He is a terribly cold man. All his golfing pals are rich men he has met since 1945.'[25]

The morning after the 'terrible' meeting, on Friday 2 February, Brooke learnt that his aide-de-camp, flatmate and friend Captain Barney Charlesworth had been killed in a plane crash near Pantelleria. 'He was always cheerful and in good humour no matter how unpleasant situations were,' wrote Brooke of the loss, which came only eight months after the death of Ivan Cobbold and three months after that of Dill. Brooke was desperately sad for Barney's wife Diana, and he found it hard during the day to keep his thoughts on the vital business at hand, 'and not let them wander off to Barney'. It is sometimes easy to forget, when dealing with these giants of mankind's greatest war, that they were subject to ordinary human emotions too.

After a truncated, deadlocked Combined Chiefs of Staff meeting at Montgomery House, the fourteen-month-old heavy cruiser USS *Quincy* sailed into the Grand Harbour at Valletta, with Roosevelt and Leahy on board. The Joint Chiefs of Staff reported to the President at 4.30 p.m. and at 6 p.m. they were joined by Churchill and the British Chiefs of Staff. No changes at SHAEF or alterations of strategy were requested by Brooke at that meeting, in the certain knowledge that they would be turned down. Indeed at Malta, where Roosevelt pointedly stayed for less than twenty-four hours, the President deftly warded off British attempts to discuss anything of importance at any length.

Churchill did try to use the opportunity to argue that 'we should occupy as much of Austria as possible, as it was undesirable that more of Western Europe than necessary should be occupied by the Russians,' to which Roosevelt reacted neither positively nor negatively.[26] This was doubtless a relief to Marshall. Since Brooke by then anyway believed that the Ljubljana Gap concept was unworkable, and Churchill could not interest the Americans in it, and especially not Marshall, it died rather as the Prime Minister once claimed that Sledgehammer had, as a victim of Darwinian forces. Churchill and Roosevelt were not to speak privately again for another three days, and so next to nothing was discussed about how to deal with the Soviets at Yalta. Churchill fully recognized this increasing lack of influence with Roosevelt, telling Charles Moran in October 1951 that he had once 'had great influence over the President' but this had ended 'about three months before Yalta; then he ceased to answer my letters.'[27] In all, Churchill wrote 201 more letters and telegrams to Roosevelt than he received from him.

'We left Malta in the darkness,' recalled Donnelly thirty-four years later, 'like migrating swans.' Because of the short, 5,000-feet runway at Luqa airfield, the four-engined C-54s took off from 1.50 a.m. with flaps all the way down, fuel mixture at 'full rich' and throttles wide open, one plane every ten minutes. The night was dark and cloudy and, as the planes roared down the runway past the operations office, 'eerie bluish flames poured out of the exhausts and the noise was deafening'.[28]

Marshall arrived at Saki airfield in the Crimea, south-east of Tarkhan-hut Cape, in the early hours, to find large tents with tables full of vodka, caviar and Russian wine for breakfast. He took one look and merely said, 'Let's get going,' leaving the banquet for others. They were driven to their quarters in the conference meeting place, the murdered Romanov family's former holiday residence, the Livadia Palace. At 50 to 100 yard intervals along the 90 miles of the overland route across the peninsula, male and female Russian soldiers stood to attention, saluting each car as it passed. The three British Chiefs of Staff landed at 9.30 a.m. and, after breakfasting in one of the tents, went off in one car to the Vorontsov Villa overlooking the Black Sea at Alupka, 12 miles from Yalta. 'A sort of Scottish baronial Moorish mixture', opined Cunningham. 'The place is very crowded, bathrooms few and far between.'

Brooke, Cunningham and Leathers shared one bathroom, and the Chiefs of Staff office was situated in the library. Cunningham was delighted to find a history of Hampshire that described the ruins of the medieval Waltham Palace, on whose site he lived in a modern house. It was believed that the villa – built by one of the British Empire's greatest architects, Edward Blore, in 1837 – had been given to Field Marshal Erich von Manstein, commander of Army Group South, by Hitler as a reward for taking the Crimea, which was why it was more or less intact.

If the British thought themselves somewhat crowded at Yalta – which they pronounced to rhyme with 'Malta' – the Americans at the Livadia Palace had sixteen colonels sharing one room.[29] Donnelly and six other Planners were accommodated in the Romanov children's classroom on the second floor; although none of the seven snored, he recalled that there were bedbugs, that 'bathing and toilet facilities were very primitive' and that only FDR had a private bathroom. Just as the bedbugs were impervious to sprays, according to Harry Hopkins' son Robert, so too did electronic bugs abound at Yalta, as the secret policemen of the NKVD listened in to delegates' conversations.

George Marshall did not have to share a room: he had the Tsar and Tsarina's bedroom on the second floor all to himself. Space considerations apart, great efforts had been made to accommodate the American delegation comfortably: accoutrements from Moscow's luxurious Metropol Hotel had been moved en masse to the Livadia – even the maids' uniforms had 'M' on them – and as the conference opened a Soviet briefing paper entitled 'Notes on the Crimea' was distributed to the American delegation tracing the area's history and geography. A slightly contentious account of the Crimean War put the Allied capture of Sevastopol entirely down to the French, and of course did not mention that the Russians lost that war. Of the Livadia Palace it said: 'General Marshall is occupying the Imperial bedroom and Admiral King the Tsarina's boudoir,' but without referring to the circumstances of the change of ownership.

Downstairs, Roosevelt and his daughter Anna Boettinger occupied rooms close to the plenary meeting room on the ground floor, with Harriman and the secret service detail stationed near by. (Security was continuously tight. When Roosevelt's secret service contingent, Churchill's military protection unit and Stalin's bodyguards with short repeating rifles slung across their chests all crowded into the same room before Big Three meetings, Ed Hull 'couldn't help but feel that if someone had set off a firecracker all hell would have broken loose'.)[30]

There was some entertainment in the evenings for the US Chiefs of Staff, including a preview of the movie *National Velvet* starring Mickey

Rooney and Elizabeth Taylor. The British were entranced by unrationed breakfasts that featured caviar, salmon, tangerines, butter and even Stalin's favourite naturally aerated Georgian mineral water, Borzhoni.

Stewart Crawford, Portal's private secretary, agreed about the idyllic atmosphere and remarked upon 'The splendid scenery along the coast near Yalta. The pleasant cypress-studded slopes covered with villas from the Tsarist days . . . wonderful colouring in light neutral tints'.[31] Yalta's colouring was indeed light and neutral, but the results of the conference were anything but. Few suspected it at the time, and it was perhaps inescapable, but the decisions taken there were to reduce the eastern half of the Continent of Europe to a state of modern vassalage that was to last nearly half a century.

21

Yalta Requiem: 'They were ending the war in no friendly spirit'
February–May 1945

States which have no overseas colonies or possessions are capable of rising to moods of great elevation and detachment about the affairs of those who have.
Winston Churchill, February 1942[1]

Although the Yalta Conference had many issues to discuss which were of lasting consequence to the post-war world – including the future of eastern Europe (especially Poland), reparations from and the partition of Germany, the founding of the world organization called the United Nations, Anglo-American representation on the Control Commissions of Roumania and Bulgaria, the future of Iran and China, and the timing of Russia's declaration of war against Japan – these mostly concerned politics more than grand strategy. The disputes between various politicians and diplomats were in the spotlight at Yalta, unlike earlier conferences where it was squarely trained on Marshall and Brooke. The military questions that remained to be settled were mostly administrative, such as over POWs and the demarcation of areas to be bombed. Even some political questions that had military overtones, such as Roosevelt's decision to let Russia take Japan's Kurile Islands, were not discussed with the military advisers, possibly, in that particular case, because they would have recommended very strongly against it.

During the wide-ranging political discussions at Yalta, Marshall and the other Chiefs rarely said a word, but stuck closely to their military briefs. After the war Marshall was unfairly accused by the McCarthyites of playing a major political role in appeasing the Soviets at Yalta, an accusation he understandably resented. He had not even been present at many of the political meetings and dinners, and had concentrated solely on the military issues. The extensive documents recording these discussions – principally the relevant Foreign Relations of the United

States volume of official papers – fully bear Marshall out over this, although they were not published in time to exonerate him.

Marshall was well aware that politicians had to look at issues through a different prism from soldiers. 'Of course, Mr Churchill and the President were the dominant factors in all arrangements and all guidance,' he said. 'And they were the great political leaders of their countries, but they were also the military leaders and it was quite a delicate issue back and forth, particularly in matters like the Mediterranean, the soft underbelly of Europe, the Balkan states, the march on Berlin, and things of that sort.' Marshall insisted to Pogue that, other than the shortage of landing craft, there was nothing that the Joint Chiefs of Staff discussed more than political factors, at least among themselves. 'But we were careful, exceedingly careful, never to discuss them with the British . . . because we were not in any way putting our neck out as to political factors, which were the business of the Head of State.'[2]

When the conference opened, on Sunday 4 February 1945, the Red Army had crossed the River Oder and were only 40 miles from Berlin. Under orders to seize as much territory as possible in the last moments before the Big Three met, they were badly over-extended and unable to commence their next, albeit final, assault for another ten weeks, although of course the Western Allies did not know that. It was hardly surprising, therefore, with the battle of the Bulge won only a week earlier, that senior American diplomats advised Roosevelt to ratify the Occupation zones of Germany as soon as possible.[3]

Churchill's comments to Colville, Moran and others about Roosevelt's state of health the previous year proved prescient. Marshall said the President 'looked very, very tired' at Yalta; 'I was quite shocked by his looks.'[4] Ismay thought that FDR 'was more than half gaga' there.[5] Of course there might well be plenty of hindsight in the almost unanimous testimony about how Roosevelt appeared, but it was pretty unvarying. Donnelly found him 'gaunt, his eyes sunken deep in his lined face; he looked very tired and ill, as though he were existing on pure iron determination to see the war to the end.'[6] Hull agreed: 'The President looked dreadful when he was wheeled into the room – sagging jaw, drooping shoulders. He appeared almost oblivious of his surroundings and of his guests. After several strong martinis, however, he seemed to come to life.'[7] Brooke's interpreter Hugh Lunghi recalls that he watched FDR's plane touch down at Saki and saw the President, 'waxen cheeked, looking ghastly, his familiar black naval cloak over his shoulders, hat-brim turned up in front, being helped into a jeep'. Stewart Crawford wrote three months after the conference that Roosevelt had looked 'half dead with grey sunken cheeks and little spark of vitality'.[8] Nonetheless, Admiral Emory S. Land, the chairman of the US Maritime Commission, told Sir Alan Lascelles that Roosevelt was not so ill at Yalta as the photos of him there might suggest, but was merely 'having trouble with his dentures', which had 'affected his speech and caused his face to fall in unduly'. In the unlikely event that spin-doctors ever find themselves in need of a patron saint, they should choose Admiral Emory S. Land.

Between Sunday 4 and Sunday 11 February there were eight plenary sessions, five Joint Chiefs of Staff meetings, three Combined Chiefs of Staff meetings, two bilateral meetings between the US and Soviet Chiefs of Staff and numerous other convocations of smaller, more specialized groups. The Combined Chiefs met at Stalin's HQ, the Yusopov Villa at Koreis, 6 miles from the Livadia Palace. By the end of the conference all the issues facing the Allies had been fully talked through; for all that historians still debate the outcomes of Yalta, few deny the sheer reach of the discussions there.

Although victory over Germany was no longer in doubt, the nature of the post-war world order certainly was. The question of Poland took up much time, with the Soviet Union supporting the claims of her Polish puppet government, the Lublin Committee, for territorial compensation against Germany, by moving the new Polish–German border to the Oder–Neisse line, which the Western Allies accepted. Churchill and Roosevelt did manage to win an agreement for the Polish Government to be formed 'on a broader democratic basis', including members of the London-based Polish government-in-exile, one of a number of Soviet promises that was not subsequently acted upon. Yalta confirmed the unconditional-surrender policy, mapped out zones of control in Germany for each of the Big Three powers, and established an Allied Control Council to administer national policies for the country as a whole. This Council would consider issues such as reparations and the punishment of war crimes that the conference failed to agree upon.

As the price for Stalin agreeing to declare war on Japan three months after the German surrender, the Western Allies secretly promised that Russia could have the territories she had lost in the Russo-Japanese War of 1904–5, but she was also offered further concessions concerning Port Arthur, Mongolia and Manchurian railways that were strictly speaking in China's gift rather than theirs, and the Chinese were not present at Yalta. Equally secret were the decisions over the size, extent and relative voting strengths in the General Assembly and Security Council of the United Nations. The number of seats each great power would have, and the veto rights, were the subject of much political horse-trading.

Agreements were also made at Yalta over transfers of prisoners-of-war from West to East, which became highly controversial once it was recognized that thousands of non-Soviet and non-Yugoslav citizens who had fought for Hitler, such as the Cossacks, had been forcibly returned to Stalin and Tito by the British and subsequently murdered. Because these were political rather than strictly military questions, however, politicians and diplomats dealt with them rather than Marshall, Brooke and their respective Staffs. There is no indication that Churchill and Roosevelt differed over the (misnamed) 'repatriation' policy concerning the Cossacks.

Roosevelt conferred with Marshall and others prior to the opening plenary session at the Livadia. On the question of direct day-to-day liaison between Eisenhower and the Red Army, Marshall said that 'the difficulty had been, not with the Russians but with the British who wish to effect the liaison through the Combined Chiefs of Staff', and he argued that with the Russians so far inside East Prussia 'there was not time enough' to go through that process. Since there had been time enough to go through it during Overlord itself, and indeed the Americans had crossed the Atlantic so that they could do just that, this sounds suspiciously like an *ex post facto* rationalization for the fact that the Americans wanted direct bilateral military relations with the Soviets that excluded the British. Realpolitik demanded nothing less by that stage in the war. Power had shifted. Financially and economically Britain was close to bankruptcy. 'To put it crudely,' as one economic historian has, 'in the end the net aid the United Kingdom received [from the United States and Canada] amounted to the equivalent of at least one full year of its own peak total war effort.'[9] Canada was pliable and supportive as ever, but the Americans could no longer be cajoled. If they wished to meet the Russian Chiefs of Staff bilaterally, Combined Chiefs of Staff rules would not be invoked to prevent them.

At the first plenary meeting at the Livadia Palace at 5 p.m. on

4 February, Roosevelt was asked by Stalin to take the chair. Marshall then summarized the post-Bulge situation in the west, stating that the Rhine would be crossed soon after 1 March, that 75,000 tons of supplies were coming through Antwerp daily, and that area bombing was destroying German capacity to fight back, having reduced German oil production to 20 per cent of its original capacity.[10] Although Cunningham thought Marshall 'went rather beyond his brief' in covering British air and naval matters, it seemed to impress Stalin.

The Russian dictator then said that, because of the Ardennes Offensive, the Soviets had started their winter offensive earlier than intended, and had done so in comradely duty and not because they were asked to by their allies. 'The President, who is undoubtedly in bad shape and finding difficulty in concentrating,' noted Cunningham, 'did not rise to the occasion but the PM did brilliantly . . . Stalin was good and clear in his points, the PM also very good but the President does not appear to know what he is talking about.'[11]

Other than this first plenary session, Marshall did not attend the Big Three meetings. He was present with Roosevelt among dinner guests on 5 February and at a Combined Chiefs of Staff meeting four days later with Churchill, and then finally on the day that the famous photographs were taken in the courtyard of the Livadia (which is entirely unchanged today). 'My God! How tired I am of it all!' wrote Brooke in characteristically peppery mood. His loathing of official banquets and especially the speeches made at them – 'insincere, slimy sort of slush!' – was very evident throughout. On 6 February he summarized the Burmese campaigns for the Russians, and Marshall reported that 'in the face of unparalleled difficulties' 44,000 tons of supplies had been flown over the Himalayas the previous month, which he described, somewhat hyperbolically, as 'the accomplishment of the greatest feat in all history', and beside which he said inter-Staff co-operation 'should be relatively easy'.[12] One problem frequently encountered was the reluctance of even high-ranking Russian military officers to commit themselves to anything, however minor, until it had been referred back to Stalin; the hitherto short life-expectancies of marshals of the Soviet Union made that a sensible precaution.

On Wednesday 7 February there was a tour of the city of Sevastopol, where hardly a dwelling had been left intact by the long siege in 1941–2 and subsequent fighting in 1944, and there were three stops for Crimean

War battlefields. Lunch with the Soviet admiral commanding the Sevastopol naval base consisted of 'many dishes of stale fish and only vodka to drink'.[13] Although Balaklava mattered much to men like Churchill and Brooke who had grown up with Tennyson's poem, the Prime Minister complained that the local Russian guides had shown 'no sort of feeling' there. 'Either they thought they had won the battle or they had never heard of it.'[14] Shortly afterwards the Countess of Ranfurly in Cairo received a letter from a member of the British delegation saying: 'I wish you could have seen Sir Alan Brooke, with a school history book in one hand, explaining the battle of Balaclava to an audience of field marshals. We stood on a little ridge on the end of that famous battlefield where the Charge of the Light Brigade took place. All around us were the twisted remains of German anti-tank guns.'[15]

Because there were now, for the very first time since Brooke and Marshall first met, no significant strategic differences between the two men – or at least none that Brooke could do anything about – relations ran smoothly. At the noon Combined Chiefs meeting on Thursday 8 February, Cunningham, quoting Byron's *Childe Harold* on the Duchess of Richmond's eve-of-Waterloo ball, reported that 'everything went as merrily as a marriage bell', and less poetically that there was 'complete agreement on all matters on the agenda'. In the past this had been the harbinger for a furious bust-up, but not this time. Even Admiral King defended the British, at least after his own fashion. At the first tripartite military meeting, the Commander-in-Chief of the Red Navy, Admiral Nikolai Kuznetsov, had criticized the Royal Navy for its lack of support in convoying supplies to northern Russia, and Admiral King asked him, 'rather sharply', what the Soviet Navy had achieved against the German Navy. He 'received an ambiguous reply'. (That was because during the first six months after Barbarossa the entire Russian Navy had managed to sink only one German cargo ship, the 3,700-ton *Baltenland*. The 3.6 million tons of enemy shipping Soviet propaganda claimed to have sunk between 1941 and 1945 is estimated to be around twelve times the genuine figure.)[16]

Even when King informed Cunningham that he did not intend to assign the British Pacific Fleet to the first phase of Operation Iceberg – the capture of the Ryukyu Islands south of Japan – claiming that he was 'uncertain of what MacArthur is going to do', Cunningham concluded: 'I doubt he is up to his usual game of trying to keep us out of it.' Although the next day Cunningham professed himself 'rather disappointed' at missing Iceberg, in the event the Royal Navy was fortunate to have escaped involvement in its terrible first phase, which comprised the capture of the island of Okinawa and which led to 7,374 American soldiers and 4,907 sailors being killed, and 31,807 soldiers and 4,874 sailors wounded, 36 ships being sunk and 368 damaged, and 763 aircraft lost – the highest number of American casualties in any single campaign of the war against Japan.

When they got back for lunch at the Vorontsov Villa after that morning's Combined Chiefs of Staff meeting, the British Chiefs found that a group from the Foreign Office had 'pinched' their luncheon room. When they tried to walk in anyway, a young official came up and told them 'to hold off'. No one ever talked like that to the Chief of the Imperial General Staff, and 'He got properly set about' by Brooke, much to Cunningham's amusement.

An indication of how strained the Roosevelt–Churchill relationship had become by the time of Yalta might be gleaned from an incident that took place one afternoon. When the Prime Minister arrived at the fifth plenary session on Thursday 8 February, ten minutes before the 4 p.m. meeting, he 'seemed a bit surprised' to hear that Stalin and Roosevelt had already been conferring for twenty minutes. He was ushered on his own into the President's suite to wait until his delegation had shed their hats and coats. Edward Stettinius, the new US Secretary of State, suggested that Admiral Leahy tell the President that Churchill was waiting outside. The admiral dutifully delivered the message and was back in a few moments. 'What did the President say?' asked Stettinius. 'Let him wait,' answered the blunt Leahy, indifferent to the presence of the Prime Minister's delegation. Stettinius was embarrassed, Hopkins seemed amused. Stettinius then apologized to Eden, who did not seem at all put out, assuring the Secretary of State that he 'thoroughly understood the whims of their masters and not to be upset about it'.[17] Perhaps Eden ought to have taken this act of casual rudeness on the part of the President as yet another sign of his willingness to appear somewhere between grandly nonchalant and simply offhand towards Churchill.

With Roosevelt's permission the Joint Chiefs of Staff held joint Staff sessions with the Russian Chiefs of Staff on 8 and 9 February, on the same days that the Combined Chiefs held their 187th and 188th sessions. These showed the shape of things to come. Of the Americans' argument that they did not want the world to break down into 'spheres of influence', Lord Halifax – who did not attend Yalta – wrote to his Foreign Office friend Charles Peake, Eisenhower's political adviser at SHAEF, that it seemed 'astonishing nonsense', because they were 'altogether ignoring that they have the biggest of all through the Monroe Doctrine. If any people have the gift of ignoring the beam in their own eye, it is surely them. But I have no doubt they think just the same about us.' Over such questions as India, Singapore and Hong Kong, they did indeed.

On Friday 9 February, after Churchill and Roosevelt had accepted the Combined Chiefs' final report at their plenary session at noon – for which Roosevelt was 'over half an hour late and not in good shape' – the photographs were taken in the courtyard at the Livadia.[18] Iconic though they are today, Cunningham thought them 'Very badly organized. Various people were fed into the picture at intervals behind the three great men,' including him.[19] For all the smiles in those photographs, the number of issues on which Churchill and Roosevelt found themselves ranged on opposite sides seemed to increase as victory neared, and by February 1945 these included the future of the Italian monarchy, the purchase of Argentine beef, civil aviation rights, Middle Eastern oil, Polish election supervision, Western involvement in the Balkans and the future of Greece.

Because he is usually accredited the victor at Yalta, it is sometimes forgotten that Stalin made a number of concessions there. He gave a firm date of entry into the Japanese war (three months after the German surrender); agreed to observe the provisions of the Atlantic Charter in eastern Europe by signing the Declaration of Liberated Europe, which affirmed 'the right of all peoples to choose the form of government under which they will live'; assented to France sitting on the Control Commission for Germany, and agreed that the USSR would join the new United Nations Organization, largely on Roosevelt's terms. Taken together these seemed significant, yet in reality they amounted to relatively little. The Soviet Union offered no written commitment that the United States could have the air and naval bases she needed in the war against Japan, and Stalin's commitment to the Japanese war arose largely out of his wish to be present at the division of spoils; the provisions of the Atlantic Charter and Declaration of Liberated Europe were never going to be seriously implemented in eastern Europe; France's zone in Germany had to be carved out of the Anglo-American zones, not Russia's; and it was objectively in Russia's interests to have a founding say in the United Nations.

Speaking in 1974, Ed Hull made the sensible but rarely heard argument that:

All that Yalta did was to recognize the facts of life as they existed and were being brought about . . . The only way we could have in any way influenced that in a different way was not to have put our main effort into France or the Low Countries but to put it into the Balkans . . . It might have meant that Bulgaria, Rumania, and possibly others of those Eastern European countries that are now Communist-dominated would have other type of control at present. But . . . it would also mean that all of Germany and probably a good portion of the Low Countries, Belgium, Holland and even France, might have Soviet influence over them rather than Western influence. To me there was no choice to make.[20]

The only way the Western Allies could have prevented the Soviet domination of eastern Europe was to have invaded the Continent in 1943, but that would have been to risk catastrophe in Normandy, and thus probable eventual Soviet domination of the entire Continent. Before criticizing Roosevelt and Churchill over the European endgame of 1945, it is important to recognize how limited were their options. When the Yalta Conference broke up, Brooke returned to London, Churchill left Saki airfield for Athens, Alexandria and Cairo, and Roosevelt and Marshall went back to Washington. The four of them were never to meet all together again.

Between noon on 21 June 1942 and the morning of 10 February 1945 – when Brooke said 'all the necessary goodbyes' – Western strategy-making between the four principals had brought the British and American armies to Africa, Sicily, Rome, Normandy, Paris and almost into the heart of Germany. In all they had met seven times – twice at Washington, at Casablanca, at Teheran, twice at Quebec, and at Yalta – and at these hard-fought meetings had hammered out a victorious strategy. There had been some individual defeats and disappointments in battle against the Axis, of course, but no campaign reversals. Above all, the contentious decision over the timing of the greatest amphibious assault in history had been justified by the only truly unanswerable criterion of warfare: success. A different quartet from

Roosevelt, Churchill, Marshall and Brooke might have taken different decisions, but it is unlikely that any would have significantly shortened the Second World War.

In Britain, a small number of Conservative MPs threatened to abstain during the parliamentary debates on the Yalta agreements, prompting Cunningham to write: 'One sympathises with the dissidents but they do not face facts. The most outstanding one being that Russia is in occupation of Poland and can do just what she likes there without us or the USA being able to stop her.'[21] The Commons debate on Yalta was won by the Government by 413 votes to nil with about thirty (mainly Tory) abstentions.

It is hard to be naive and cynical at the same time, but Roosevelt was both when it came to Stalin and the fate of the Poles. 'Of one thing I am certain,' he told the Polish Prime Minister-in-exile Stanisław Mikołajczyk, 'Stalin is not an imperialist.' To the former American Ambassador to France, William C. Bullitt, he also said: 'I have a hunch that Stalin doesn't want anything other than security for his country, and I think that if I give him everything I possibly can and ask for nothing in return, *noblesse oblige*, he won't try to annex anything and will work for a world of democracy and peace.' To the British diplomat Richard Law in late December 1944, the President said that 'he was not afraid of Communism as such. There are many varieties of Communism and not all of them are necessarily harmful.'[22]

Yet if Roosevelt was wrong about Stalin's intentions, to the point of believing that Soviet expansionism would no longer pose a serious global threat, then so too was Churchill, who told the War Cabinet at its first meeting after he arrived back in Britain from Yalta that it was:

Impossible to convey the true atmosphere of discussions between the [Big] Three. Stalin I'm sure means well to the world and Poland ... The military situation has undergone extraordinary change, in three weeks the Russian army crossed the River Vistula to the Oder. Stalin has offered the Polish people a free and more broadly based government to bring about an election; I cannot conceive any government has the right to be treated like that. Stalin about Poland said 'Russia has committed many sins about Poland – pacts and partitions – it is not the intention of the Soviet Government to do such things but to make amends.' President Roosevelt was very feeble – but when he showed he did not want a thing to be done, Stalin withdrew his request. Very important other matters were settled, including an agreement signed re: Japan on the basis that Russia gets back what she lost in the Russo-Japanese War subject to agreement with Chiang Kai-shek. Stalin had a very good feeling with the two Western democracies and wants to work quite easily with us. He was not jarred by the United States and us speaking the same language. My hopes lie in a single man, he will not embark on bad adventures. Re: Greece – Stalin was jocular . . . He does what he likes in Bulgaria, Roumania – and leaves us alone in Greece. He held his own people off; he made a bargain with them and they have a great desire to keep it. Russian troops have a wonderful bearing.[23]

The rest of Churchill's report had an almost what-I-did-in-my-holidays quality to it. 'Saw the Lion of Judah', he said of the Emperor Haile Selassie of Ethiopia, 'not good impression. I reminded him we liberated his country.' Then he met King Farouk of Egypt, but was much more impressed with Ibn Saud, the first king of Saudi Arabia, whom he thought a 'marvellous figure, splendid looking man, boasts of his virility and how often he attends to his harem. He must keep a card index.' Churchill was proud that although Ibn Saud's retinue came to Alexandria on an American destroyer, 'we sent them back on a cruiser'. There were sheep on board and a member of the entourage made coffee in the destroyer's magazine, which 'alarmed the Americans'. Churchill was amused by the way the retinue in Ibn Saud's hotel suite included an astrologer, a fortune-teller and slaves. He thought the meeting 'went off very well. I drank Mecca water – I'm not apt to taking it on occasions like that.' His party were presented with '£100 worth of perfume', as well as 'presents we have got to discuss with chancellor of the exchequer, diamonds and pearls'. Churchill 'Pleaded the case of the Jews', but Ibn Saud brought up the assassination the previous year of the British Minister in the Middle East by Stern Gang terrorists, saying: 'If they murder Lord Moyne, what do you expect them to do to the Arabs?'

The 'high spot' for Churchill had been Athens – his visit of 14–15 February codenamed Operation Freehold – where he had 'never seen such a mass of people, jammed together. Tremendous.' It would have been easy, so slowly did the crowds allow him to move, for someone to 'have a pot shot at you', after which there would have been a by-election, he joked, but nonetheless his antagonist Aneurin Bevan 'would not stand a chance'. The last comments were almost Pooterish considering the great events Churchill had been describing: 'Well, very enjoyable I must say.' He had brought back goldfish from Moscow to swim in the pond at Chartwell, and had 'Maintained one's own against the bugs – got it a bit in the gut at the banquet . . . Left Cairo at 2 a.m. this morning. I'm not the slightest bit tired.'[24]

Although the official Cabinet minutes reported over three-and-a-half pages, it might almost have been a different meeting altogether from that verbatim account we have thanks to Lawrence Burgis. Churchill's naivety about Stalin was still present in the official record – 'Premier Stalin had been sincere. He [Churchill] had a very great feeling that the Russians were anxious to work harmoniously with the two English-speaking democracies' and 'He was struck by the desire of the Russians to meet the President half way on points to which they thought he attached real importance' – yet there was unsurprisingly no mention of the agreement to allow Stalin to do 'what he likes' in Bulgaria or Roumania, let alone Roosevelt's feebleness or how Churchill's hopes for peace 'lie in a single man, he will not embark on bad adventures'. Three years later, at the time of the Berlin Blockade, these forecasts were looking myopic.

The official minutes did record Churchill saying that 'There was no question that the Russian Army was a formidable machine,' which of course explains why the best that Churchill and Roosevelt could have hoped for was Stalin's goodwill, since nothing else could dislodge the Red Army from Poland at that time.[25] Both the official minutes and Burgis' verbatim transcriptions nonetheless absolve Roosevelt from the criticism that he was the only Western leader who was naive about Stalin's real post-war intentions towards eastern Europe. In the historical discussion about Roosevelt's supposed naivety versus Churchill's supposed cynicism, the truth is more complex. Churchill was more naive than he liked later to maintain, but neither man's beliefs made the slightest difference when faced with a vast Red Army stationed squarely over Poland and East Prussia. (Churchill was also rather endearingly naive about his own popularity, telling the Cabinet that when he landed at RAF Lyneham that day, after his fourteen-hour flight from Cairo, 'I was very thirsty and we stopped at a railway hotel for a whiskey and soda, which was most welcome. But, do you know, they wouldn't let me pay for it!')[26]

*

The British plan for Alexander to replace Tedder as Eisenhower's deputy had been agreed to by Eisenhower himself, who nonetheless spotted the dangers of its being perceived as a British attempt to slip a land commander between him and his other senior officers. As he told Marshall on 20 February, 'Since Public Relations often cause me the biggest headaches, I wanted to make sure the CIGS clearly understood what might occur.' The whole idea was then downplayed by Brooke, who wrote a 'My dear Monty ... Yours ever Brookie' letter on 7 March saying he thought he had settled 'the Alex business' by having 'got in with Ike before the PM saw him and had long talk with him ... I told him to be quite frank with PM and tell him exactly what his fears were, and not allow himself to be overridden. As a result PM told me afterwards that he had had doubts as to the wisdom of the change.'[27]

Brooke advised Churchill to leave things as they were, and soon afterwards Churchill wrote to Roosevelt and Marshall (with a copy to Brooke) withdrawing the proposal due to 'the progress of the war', which was a pretty broad catch-all. Brooke also told Montgomery that since Tedder 'manages the whole air business on the Western front in Eisenhower's name' it wouldn't have worked anyway, and he wanted Alexander to occupy Austria after the surrender instead. Tedder later told Pogue that the whole move had originally been 'started by Monty, backed by Winston and Brooke', because he, Tedder, kept supporting Eisenhower over Montgomery. 'We didn't happen to support [Monty] against Ike,' Tedder told Pogue. 'He couldn't understand that we should be loyal to our commander.'[28] Whereas Tedder fought coalition warfare as it needed to be, Montgomery – and to an extent also Brooke – consistently saw it in terms of Anglo-American competition. Eisenhower's attitude was that he did not mind one officer calling another a 'useless sonofabitch' so long as the epithet did not include the word 'British' or 'American'.

Montgomery's reply to 'My dear Brookie' claimed that he was 'delighted that the Alexander business has been postponed: and I hope this will lead to a cancellation. The change would have upset matters, without any doubt. We are now on a very good wicket; Ike has learnt his lesson and he consults me before taking any action.'[29] Small wonder that the Americans thought of him as they did.

Brooke had already warned Montgomery that Churchill was 'determined to come out for the crossing of the Rhine and is now talking of going up in a tank!' Brooke thought that it would be safest to find 'some reasonably secure viewpoint (not too far back or there will be hell to pay)' from which the Prime Minister could watch what was happening. Montgomery replied that he would invite Churchill to stay with him in his camp, since that way he would 'be able to keep an eye on him and see that he goes only where he will bother no one'. When he went out later that month, Churchill assured Montgomery that his entire party would consist only of Brooke, his aide Tommy Thompson and Sawyers, 'four in all'. (Of course no one could have expected Churchill to see the Allies crossing the Rhine without his valet being present.)

On 15 March 1945 British troops reached the Rhine on a 10-mile front. Of the controversies over the so-called race to Berlin, Brooke later explained that the Occupation zones of Germany had been settled at Yalta, and that 'Russia having taken what you might call the major part in the land warfare certainly had to have an equal part to the other two, on the Eastern Front ... The advance into the country really had to coincide to a certain extent with what our final boundaries would be. That was what led to the stopping of the American advance at one point; they were going into territory that would eventually be occupied by Russia, they would lose men in doing so.'[30] As there was no point in doing that, there was no race to Berlin between Montgomery and Patton, or anyone else. Berlin was in the Soviet zone, and if the Allies had reached it first, they would simply have had to withdraw.

In a wide-ranging, ruminative message to Roosevelt two days later, Churchill reminisced about how their friendship was still 'the rock on which I build for the future of the world so long as I am one of the builders'. He said that he often thought back to 'those tremendous days when you devised Lend–Lease, when we met at Riviera [Placentia Bay], when you decided with my heartfelt agreement to launch the invasion of Africa, and when you comforted me for the loss of Tobruk by giving me the subsequent three hundred Shermans of Alamein fame'.[31] Churchill felt nostalgia for that nerve-wracking, but for him much happier, stage of the war when his relationship with Roosevelt was stronger and clearer than in the multi-nuanced later periods. Beneath this classic, slightly maudlin Churchillian reminiscence, there was a definite indication that it was Roosevelt rather than Churchill who had initiated the idea for Torch at Hyde Park in June 1942. Of course, occasionally Churchill did use such musings to try to bounce Roosevelt into action – such as his regular recollections of their 'Istria' conversation at Teheran to try to generate support for the Ljubljana Gap proposal – but there seemed to be no reason for this telegram, unless it was intended in the form of an elegy for a man who he by then suspected was dying. At about this time, Grace Tully sent a signed photo of FDR to Marshall, saying that the President 'was a little surprised to learn he hadn't already given one to him'. Perhaps Marshall suspected, too, that the President had less than a month to live.

The day that Churchill and Brooke set off to watch the Rhine crossing, Friday 23 March, Cunningham reported that the Chiefs of Staff that morning were 'much more interested in the American fishing bait I produced and which has concentrated blood in it and so bleeds, than in our proper business of which there was not much'.[32] With victories across all fronts, especially Patton's when he took one hundred thousand prisoners and 'smashed up' nine German divisions south-east of the Moselle river, there was now little over which Marshall and Brooke could cross swords.

Harry Butcher recorded that Eisenhower had taken 'special pleasure' from Brooke, 'who had once argued heatedly against the [Ruhr] plan', generously telling him, as the Rhine crossing was actually in progress, that Eisenhower had been 'right and that his current plans and operations are well calculated to meet the current situation'.[33] After Brooke had sent Eisenhower a similarly flattering telegram on US Army Day, Ike told Marshall that 'This was especially pleasing because of the past arguments we have had and to my mind shows there is a bigness about him that I have found lacking in a few people I have run into on this side of the water.'[34] Eisenhower felt that the victories west of the Rhine had made possible the bold advances of General Courtney Hodges' First and Patton's Third Armies towards Kassel. 'General Ike didn't wish to sound boastful,' wrote Butcher, 'but he was like a football coach whose team had just won a big victory and he couldn't help talking about the accomplishments of his players.' (When Butcher had first met Ike, in Washington, he was indeed coaching an army football team at Fort Benning.)

On 24 March General Sir Miles Dempsey's Second Army crossed the Rhine in Operation Plunder, watched by Churchill, Brooke and Eisenhower from a convenient hillock. The party saw Dakotas and gliders dropping parachutists and a Flying Fortress on fire. 'Several of the returning Dakotas were in trouble,' recalled one of the group, 'and three or four crashed before our eyes, bursting into flames as they struck the ground.' Nonetheless Churchill was not allowed to cross the Rhine during the battle, which left him 'glum and angry'. The man who had charged with the 21st Lancers nearly half a century before was now just too important to be risked with the 21st Army Group. 'I honestly believe that he would really have liked to be killed on the front at this moment of success,' wrote Brooke. 'He had often told me that the way to die is to pass out fighting when your blood is up and you feel nothing.' It was a relief for Brooke to get him home safely, and home was at last safer too: on 28 March the last of 1,050 V-2 rocket-bombs landed on Britain, having killed over 2,500 Britons (far fewer than the citizens of Antwerp, however, of whom thirty thousand had been killed by these weapons).

As it turned out, the British and Americans did have an opportunity for one final contretemps before the German collapse. On 29 March the British Chiefs of Staff received what Cunningham called 'Rather a disturbing telegram from Eisenhower direct to Stalin (a most improper procedure) in which he indicates he is shifting the axis of his main thrust to the south'. The British disapproved of this because they believed that denying the northern German ports to U-boats and the Dutch ports to E-boats and midget submarines was still vital at a time when shipping continued to be sunk. Churchill telephoned Eisenhower, who told him that the US Ninth Army was to be removed from Montgomery's command after it had linked up with the US First Army between Kassel and Paderborn, and placed under Bradley. Montgomery would therefore be left to cover the front that he had previously controlled with both the British Second Army and the US Ninth Army with the former force only, while General Henry Crerar's Canadian First Army 'mopped up on the left' – that is, northern – flank. 'Something curious has happened at Eisenhower's HQ,' concluded Cunningham. 'Perhaps the US generals have ganged up and insisted on their national army being under US command.'

Whatever the reason, the Chiefs of Staff sent a protest to the Joint Chiefs of Staff 'against this procedure and change of plan without any consultation with Combined Chiefs of Staff'.[35] Cunningham and Brooke then went off fishing for the weekend at Mountbatten's home, Broadlands in Hampshire, which Cunningham thought the 'No 1 beat on the

Test'. On Easter Sunday, however, they were summoned to Chequers to discuss Eisenhower's plan to strip Montgomery of half his command.

Churchill had sent Roosevelt a long telegram – copied to Brooke and specifically to Marshall via Maitland Wilson in Washington – in which he expressed his 'complete confidence' in Eisenhower, a classic precursor to criticism.[36] The telegram then described his doubts about the strategy relating to Berlin and the Elbe, saying he wanted to be certain that Marshall knew his thoughts. The response was prompt; in Cunningham's words, Marshall sent 'a very rough message in reply to ours'.

Brooke and Cunningham arrived at Chequers at 11.30 a.m. and went straight into a Staff Conference with Churchill, who was 'a bit savage' about the Joint Chiefs of Staff. He had drafted a telegram that Cunningham thought 'full of soft soap but with some pretty shrewd digs in it to the President'. He told them that Roosevelt 'was in a pretty bad way and only the last day or two had been writing his own telegrams'. After lunch – at which the Americans Bernard Baruch and Gil Winant were present – Portal and Hollis drafted a reply to Marshall's 'rough message'. Meanwhile Brooke settled into a chair in the library and read a book he had taken down from a shelf, entitled *The Theory and Practice of Prostitution*.[37] Churchill vetted the reply, upon which Cunningham 'tried to get away but Brooky and I were had by him for a stirrup cup in the library' and they didn't manage to leave until 5 p.m.

The next day Jumbo Wilson wired from Washington, having spoken to Marshall who had claimed that he 'could not really understand the fuss'. Marshall said he hoped the Anglo-American armies would reach the Berlin area by the end of April and he could not see much divergence between Churchill's proposals and Eisenhower's, 'since, with the right of the northern armies advancing on the line Hanover–Stendal, the left wing of the central group would be on the line Paderborn–Magdeburg, while the Fifteenth Army masked the Ruhr area'.[38] He added that he was not in favour of Eisenhower's central thrust going further than Leipzig. Meanwhile the Russians could have Dresden.

Tedder flew over to explain to the Chiefs of Staff that Eisenhower's original message had been sent in order to forestall Montgomery's orders for the advance. 'Monty has only himself to blame for the suspicion with which the Americans treat him,' concluded Cunningham. The 'only difference' in the plan, he thought, was that the American Ninth Army 'advances under Bradley's instead of Monty's orders'. There was a time when that difference would have led to a full-scale shouting match between Marshall and Brooke, but with the Joint Intelligence Committee now correctly predicting Germany's collapse in a few weeks, it hardly seemed worth the row. All passion spent, the British gave way gracefully.

Just before he died, Roosevelt had the satisfaction of receiving a report from Marshall stating that 'By about the end of April 1945, military operations on the continent of Europe will probably have reached the final stage – mopping up. The month of April will likely prove to be the transition period for Germany between organized resistance and utter defeat.'[39] Similarly, Marshall received an encomium from Churchill via Maitland Wilson: 'Pray further give him my warmest congratulations on the magnificent fighting and conduct of the American and Allied armies under General Eisenhower, and say what a joy it must be to him to see how the armies he called into being by his own genius have won immortal renown. He is the true "Organizer of Victory".'[40] Marshall replied modestly: 'Our greatest triumph really lies in the fact that we achieved the impossible, Allied military unity of action.'

Was it merely for the record's sake that Churchill sent Roosevelt a letter on 3 April seeming to suggest that they take Berlin, knowing perfectly well that there was no appetite in the General Staffs of either country to lose lives capturing a place they would later have to relinquish to the Russians under the terms of agreements already made? 'If they also take Berlin,' Churchill wrote almost rhetorically of the Russians, 'will not their impression that they have been the overwhelming contributor to the common victory be unduly printed in their minds, and may this not lead them into a mood which will raise grave and formidable difficulties in the future?' He did not explain what was 'undue' about the impression that the Soviets had been the 'overwhelming contributor', given that they had killed four out of every five Germans who died in combat. The Russian military dead of at least thirteen million was over twelve times that of the combined British and American, and their civilian dead (of around thirteen million) was a full two hundred times that of the Western Allies.

Roosevelt's curt reply to Churchill – 'I do not get the point' – ended with his 'regret that the phrasing of a formal discussion should have so disturbed you but I regret even more at the moment of a great victory we should become involved in such unfortunate reactions.'[41] Churchill could hardly have felt that it was worth while ripping up the various agreements made with the Russians over Occupation zoning in order to dash for Berlin. More likely he wished to put in writing that he was on the right side of the Cold War which he saw – earlier than anyone else except perhaps Brooke – was looming. Between Churchill's wildly over-optimistic report to the War Cabinet on returning from Yalta and this doleful telegram to Roosevelt only two months later, Stalin had given no indication that his promises of free and fair elections in eastern Europe had been genuine.

Of course Eisenhower also understood the political dimension involved in delineating where the military demarcation lines lay. As he wrote to Marshall on 7 April in a 'Personal, Eyes Only' message from his SHAEF HQ at Rheims, he thought his main thrust should be to the area including Leipzig, but with the left flank on the coast near Lübeck, which 'would prevent Russian occupation of any part of the Danish peninsula'. The Leipzig thrust also allowed him maximum flexibility and the opportunity to disrupt any Fortress Southern Germany concept that the more fanatical Nazis might conceive. Over the issue of whether he should have communicated directly with the Soviet High Command rather than through the Combined Chiefs of Staff, the Supreme Commander explained, 'Frankly, it did not cross my mind to confer in advance with the Combined Chiefs of Staff because I have assumed that I am held responsible for the effectiveness of military operations in this theater and it was a natural question to the head of the Russian forces to inquire as to the direction and timing of their next major thrust, and to outline my own intentions.'[42] Eisenhower added that the British on his Staff such as Tedder and Morgan agreed with his stance.

On Thursday 12 April, Churchill was discussing with Cunningham the somewhat prosaic subject of whether forty fishing trawlers might be released from anti-U-boat duty when the Prime Minister mentioned that the head of the SS, Heinrich Himmler, who was making desperate last-minute peace offers, 'appeared to be trying to show that he wasn't so bad as painted and PM said if it would save further expenditure of life he would be prepared to spare even Himmler'. Ever the seaman, Cunningham suggested that 'there were plenty of islands he could be sent to.'[43]

That morning, Franklin Delano Roosevelt died at Warm Springs, Georgia, of a massive cerebral haemorrhage. As with Dudley Pound and John Dill, the President's health could not take any more of the strain of fighting a global war. In a sense all three of them died for their countries, no less than any other combatant in that conflict. The Masters and Commanders had broken up; the quartet of power was over. Just as he had come to power in the same month as Adolf Hitler, so Roosevelt, his nemesis, also departed life in the same month. (Although Goebbels equated Roosevelt's death with that of the Tsarina Elisabeth in 1762, which split the enemies of Frederick the Great, according to his Luftwaffe aide Nicolaus von Below, the Führer took 'a more sober view devoid of optimism'.)[44]

The news of the President's death did not reach Churchill until midnight London time, but Jock Colville recorded him as 'very distressed' by it.[45] The Prime Minister 'toyed with the idea' of undergoing the still-dangerous journey of crossing the Atlantic in order to attend the funeral, but Lascelles opposed this 'firmly' on the King's behalf.[46] With decisions being taken in London concerning the fast-moving situation as the European war entered its denouement, it was pure romanticism even to consider making the journey for personal reasons, although it would have been a good opportunity to get to know the new President, Harry Truman. The fact that Eden and Attlee were both abroad at the time also militated against it. Nonetheless, by 1951 Churchill believed that missing the funeral was the biggest mistake he had made in the war, because hugely important decisions were being made for the rest of the war 'by a man I did not know', and he blamed Eden for the decision.[47]

Roosevelt's death was of course the first item on the agenda of the War Cabinet the next day, where Norman Brook recorded Churchill as describing it as a:

Profound shock. [A] Leap into the unknown. Truman's statement [said he] will keep present Cabinet and prosecute the war to the utmost against Germany and Japan. Truman will be [a] well man: FDR has been a sick man for months . . . Had thought of going to-day to funeral. But v[ery] private: in room at White House. Interment at Hyde Park. Relatives . . . only. Suggest A[nthony] E[den] sh[oul]d be present.[48]

Since Eden was not a relative, and anyway Roosevelt's funeral was not at all private, this seems misleading. But, despite the tension between the two Masters in the last year or so, there is no evidence to support the notion that Churchill's absence was 'because he felt the President had latterly become unsupportive', or that 'the emotional link was never as close as was commonly thought,' as some historians have suggested.[49] Roosevelt was laid to rest in a field on the Springwood estate close to his house on Sunday 15 April, under a large slab of Vermont marble. It was an indication of how professional Marshall had always wanted to keep his relations with the President that this was the very first time he had ever visited Hyde Park.

After Roosevelt's memorial service at St Paul's Cathedral on 17 April, Churchill was 'feverishly composing over the luncheon table his tribute to the President' which he was due to deliver in the House of Commons that afternoon. In the event the oration was delayed by an hour because an incoming Scottish Nationalist MP had chosen to bow to the Speaker without sponsors, which contravened a House of Commons resolution of 1688, and a debate and division had to take place. After the ridiculous came the sublime when Churchill, 'his voice thrilling with emotion', quoted from 'The Building of the Ship', the Longfellow poem that Wendell Willkie had brought over from Roosevelt in January 1941:

> Thou too, sail on, O Ship of State!
> Sail on, O Union, strong and great!
> Humanity with all its fears,
> With all the hopes of future years,
> Is hanging breathless on thy fate!

In the course of his address, Churchill said of Roosevelt:

What an enviable death his was! He had brought his country through the worst of its perils and the heaviest of its toils. Victory had cast its sure and steady beam upon him. In the days of peace he had broadened and stabilized the foundations of American life and union. In war he had raised the strength, might and glory of the great Republic to a height never attained by any nation in history.[50]

The next day a fit-looking Eisenhower visited Churchill's Map Room in the Annexe to discuss what would happen when the Anglo-American forces met the Soviet armies. Legally the Western Allies were on firm ground in occupying as much as they could reach before the Germans surrendered, as the zones already agreed at Yalta only came into operation then, but, as Cunningham recorded, 'Ike naturally does not wish to be faced with a situation in which some Russian general says on meeting that he proposes to advance to the limit of the Russian zone,' especially since Allied troops had already overrun much of the western part of the zone allotted to the Russians.[51] Churchill wanted to have a tactical zone fixed by the commanders in the field while operations were still in progress and only move into the Occupation zones later on. It was a recipe for friction with the Red Army, which the Americans considered completely unnecessary, given that they would have to relinquish the territory sooner or later anyhow.

Truman, who in all military matters understandably tended to defer to Marshall, followed the Joint Chiefs' line that it was best to adhere to the Yalta zoning arrangements whatever the legal or political circumstances. Brooke wanted Prague to be liberated by the Western Allies for the 'remarkable political advantages' that would accrue, but Marshall merely passed this information on to Eisenhower with the comment: 'Personally, and aside from all logistics, tactical, or strategic implications, I would be loath to hazard American lives for purely political purposes.'[52] In those places in the eastern zone where the Germans were still holding out it seemed common sense to let the Russians rather than the Western Allies fight them. Eisenhower replied, 'I shall *not* attempt any move I deem militarily unwise merely to gain a political prize unless I receive specific orders from the Combined Chiefs of Staff.' Eisenhower halted his troops at the Czech frontier, and when the Russians asked him to proceed no further he agreed, although reconnaissance elements of Patton's Third Army reached the vicinity of Prague, the furthest eastward progress of any Western army.

The Russians marked Hitler's fifty-sixth birthday on 20 April by reaching the outskirts of Berlin, and three days later the Allies finally got to the Po, on the same day that there was a junction between American and Russian forces at Torgau. There were still opportunities for minor spats between the British and Americans, and one 'royal row' between Admirals King and Somerville over bases for the British Pacific Fleet, in which Somerville turned down the offer of Brunei as ridiculously far from Japan. Yet these were as nothing compared to those of earlier years.

On the day that Adolf Hitler committed suicide, Monday 30 April 1945, Brooke came back from an unimpressive day's fishing on the Dee to hear Churchill discuss 'the foreign situation' at the 6 p.m. War Cabinet, at which 'He made a remark that though the Powers were at the end of their tether as regards fighting, they were ending the war in no friendly spirit. There was a tendency to quarrel.' Cunningham thought this 'Quite true. The French are very difficult and the Russians very suspicious and so difficult.'[53]

On Friday 4 May, the news came through that Montgomery had taken the unconditional surrender, at the hands of Admiral von Friedeburg and General Kinzel, of all German forces in Holland, north Germany, Schleswig Holstein and Denmark. That evening Churchill invited the Chiefs of Staff to No. 10 to celebrate, and, according to Ismay's account to Joan Astley that night, the Prime Minister had even 'with his own hands put out a tray of glasses and a drink'. When the Chiefs arrived, Churchill was on the telephone telling the King about his conversations with Montgomery and Eisenhower. Brooke recorded that the Prime Minister was 'evidently seriously affected by the fact that the war was to all intents and purposes over as far as Germany was concerned'. Churchill had tears in his eyes as he toasted each of the Chiefs in turn, and thanked them 'for all we had done in the war, and all the endless work we had put in "from El Alamein to where we are now"'.[54] That was the moment when Brooke, their chairman, ought to have reciprocated on behalf of the Chiefs of Staff Committee, and said at least a few words about Churchill's superb leadership during the struggle. 'It was a sad example of human imperceptiveness,' thought Astley, that none of the Chiefs thought to salute Churchill in a toast. Ismay was too modest in the presence of his seniors to do it himself. 'It is possible they were shy,' she concluded, 'it is certain that they were British, it is probable that they reacted as a committee, a body without a heart, and that each waited for the other to take the initiative. Whatever the reason it was an opportunity missed that the Grand Old Man, who had been the architect of the victory they were marking, did not receive a tribute from his three closest military advisers.'[55]

On Tuesday 8 May 1945 – Victory in Europe Day – Admiral of the Fleet Sir Andrew Cunningham, the First Sea Lord of the Royal Navy, summed up the 11 a.m. meeting of the British Chiefs of Staff Committee with the words: 'No important business.'[56] This must rank alongside 'Rien,' Louis XVI's famous diary entry for the day the Bastille fell. Others were less prosaic on that historic day; Marshall wrote to Churchill to say: 'I can bear personal witness to the grandeur of your leadership since the meeting in Newfoundland in 1941. I can never forget . . . the breadth of your vision . . . in effecting our combined plans.' Churchill replied:

'Under your guiding hand the mighty . . . formations which have swept across France and Germany were brought into being and perfected in an amazingly short space of time.'

On that momentous day in world history, vast, cheering crowds packed the streets of London, New York, Moscow, Paris and countless other Allied towns and cities in wild, all-night celebrations after five-and-a-half years of blood, toil, tears and sweat. In Washington, Secretary of War Stimson called Marshall into his office, where the leaders of the US General Staff had already gathered. Placing the general in the centre of the room he uttered a lengthy paean to 'the finest soldier I have ever known'. In MacArthur's recollection, Marshall 'responded with about two sentences and the thing was over'.[57] Meanwhile in London, Sir Charles Portal had a cup of tea at the Air Ministry, Sir Andrew Cunningham 'dined quietly at the flat' with his wife and son, while Sir Alan Brooke went 'back to the War Office to finish off work'.[58] They were busy men, and they still had a war to win.

Conclusion
The Riddles of the War

There is only one thing worse that fighting with allies, and that is fighting without them!   Winston Churchill to Sir Alan Brooke, 1 April 1945[1]

Harry Hopkins, speaking to Churchill's doctor Charles Moran at the Quadrant Conference in Quebec in August 1943, said that the Prime Minister had 'thrown in his hand' over the Second Front: 'Winston is no longer against Marshall's plan for landing on the coast of France.' Moran attributed Hopkins' evident bitterness over the long delays in the cross-Channel landings to the suspicion that Churchill's opposition to them had prolonged the war. 'Is Hopkins right?' he asked his diary. 'That must remain the riddle of the war.'[2] There are of course plenty of other riddles of the Second World War besides the ideal timing for D-Day, many of which, as I hope this book has shown, can be solved by viewing Anglo-American grand strategy through the invaluable prism of the interaction between the Masters and Commanders.

In considering the roles of Roosevelt, Churchill, Marshall and Brooke, it is important to remember that the decisions of Hitler and Stalin far more profoundly influenced the outcome than those of any Briton or American. After all, four out of every five Germans killed in combat between 1939 and 1945 died on the Eastern Front. Yet by sustaining Russia with massive amounts of aid, drawing off German strength from the Eastern Front by action in Africa, the Mediterranean and France, devastating German industry and cities through aerial bombardment, effectively protecting the USSR from Japan, and finally invading Germany itself from the west, the Western Allies also made an important contribution to victory.

When it comes to deciding which of the four American and British leaders influenced strategy the most, probably too much has been made of the Roosevelt–Churchill relationship. 'Each was personally fascinated not so much by the other', wrote Sir Isaiah Berlin, 'as by the idea of the other, and infected him by his own peculiar brand of high spirits.'[3] Yet historians too have been fascinated, and perhaps infected, by the idea of this friendship as being the ultimate lynchpin of the Western Alliance. In fact the realities of Realpolitik, often in the persons of Marshall and Brooke, constantly intruded on the relationship. When their countries' interests required Roosevelt and Churchill to be friends, they genuinely became so; when they needed to clash, they no less genuinely did that too. Yet unity of action was too great a prize to be jeopardized by lack of charm, especially from two of the most naturally engaging politicians of their era. (Roosevelt could be snappish about Churchill behind his back, and in December 1947, after being told that the late President had said of him that he had had one hundred ideas a day during the war, of which only four were good, the ex-premier told Moran: 'It is impertinent of Roosevelt to say this. It comes badly from a man who hadn't any ideas at all.')[4]

Roosevelt controlled his Administration just as completely as Churchill had ascendancy over the War Cabinet on matters strategic. Brooke won all the important debates within the British Chiefs of Staff Committee, whose members subsequently kept their disagreements entirely to themselves to protect their corporate strength. It is true that Marshall did not chair the American Joint Chiefs of Staff Committee, and had Admiral Ernest J. King as a constant irritant, but he was able to get his way there on all the major strategic issues. For all their domination over their own power bases and hinterlands, however, the Masters and Commanders as individuals never dominated each other. Their decisions were produced through hard-fought interaction using logical debate and compromise, over many months of constant and unimaginable stress that would have shattered lesser men. Above all, it must be emphasized that Churchill never once used his position as prime minister and minister of defence to overrule the Chiefs of Staff Committee, at least while Brooke sat on it.

Although Brooke must take immense credit for steering Churchill off his favoured but flawed operations such as Jupiter (northern Norway) and Culverin (northern Sumatra), it was a post-war fiction of Brooke's that he had not fully supported the Dodecanese and Ljubljana Gap schemes, at least at some stages of the policy-making process. It is clear from the records that he had, and his subsequent attempts to rewrite history are just as culpable as Churchill's.

For all the criticisms of Churchill in these pages, the obvious fact emerges that he was a genius, and the madcap schemes he occasionally came up with were merely the tiny portion of inevitable detritus that floated in the wash of his greatness. Had Britain been a dictatorship they would have been put into operation – as Hitler's were – but because she was a democracy they were blocked and eventually buried, usually by Sir Alan Brooke. For all the frustrations it caused him, Churchill preferred the democratic way of making war, and showed he had learnt the lessons of Gallipoli. The lack of a collegiate Chiefs of Staff system was one of the major reasons Germany lost the Second World War.

It might have surprised readers quite how often grand strategy was made as a result of last-minute compromises between Marshall and Brooke, just before they were due to present their final reports to Roosevelt and Churchill on the seven occasions they met together. When Dill was alive, these were facilitated by him; after his death in November 1944 it is noticeable how rows were not headed off, most spectacularly at the Cricket Conference at Malta. If it is true that Dill went somewhat native in Washington, it was nonetheless to his country's benefit.

Marshall did not envy the President's role in decision-making. 'It must be remembered the military responsibility in operations is very, very large, and it has with it a terrible measure of casualties,' he told Pogue. 'I know I was very careful to send Mr Roosevelt every few days a statement of our casualties, and it was done in a very effective way, graphically and . . . in colors, so it would be clear to him when he had only a moment or two to consider.' Marshall, quite as much as Brooke and everyone else on both Staffs, well understood that the statistics of killed, wounded and captured represented individual stories of tragedy, 'because you get hardened to these things and you have to be very careful to keep them always in the forefront of your mind'. These pressures simply did not exercise the Axis decision-makers; indeed by the end of the war Hitler privately stated that the German people had not in fact suffered enough.

The comparative capacities of democracies versus autocracies to wage war has been vigorously debated since the days of Thucydides, and the experience of the Second World War certainly contradicts the conclusions that Thucydides himself drew from the Peloponnesian War.

Because Nazi Germany was an autocracy, Hitler was able to impose a grand strategy on his generals that a few at the beginning, but many by the middle and almost all by the end, thought suicidal. Subservient subordinates such as Jodl and Keitel failed to ask searching questions, and few other German generals had the access or the courage to criticize their Führer's plans to his face, on the rare occasions that they were given the opportunity to be apprised of them beforehand. Flawed strategies, such as the 'no withdrawal' policies in Tunisia, Russia and Italy, were therefore not subjected to the kind of unsparing analysis that would undoubtedly have halted their adoption in a democracy. By complete contrast, the strategies of the Western Allies had to be exhaustively argued through the Planning Staff, General Staff, Chiefs of Staff and then Combined Chiefs of Staff levels, before they were even capable of being placed before the politicians, where they were debated in microscopic detail all over again. As we have seen, the British and American Chiefs of Staff spoke their minds without fear or favour, in a way that Hitler's lieutenants could not. Even Stalin, as the war progressed, gave more and more autonomy to the members of the Stavka (High Command) in Moscow, as well as to commanders in the field.

At the First Washington (Arcadia) Conference, Churchill, Roosevelt and Marshall simply overruled the objections of the crucially absent Brooke over the vital issues of unity of command and the Combined Chiefs of Staff system. At London (Modicum) in April 1942, Churchill and Brooke – to very different degrees – misled Marshall about the likelihood of an early attack in France. At the Second Washington (Argonaut) Conference in June 1942, Churchill and Roosevelt conspired behind Brooke's and Marshall's backs for a 1942 invasion of North Africa. At London in July 1942, Marshall failed to persuade Churchill and Brooke of the merits of an early cross-Channel assault, because the Britons knew from Hopkins that Roosevelt privately wanted an attack on the northern coast of Africa. At Casablanca (Symbol) in January 1943, Marshall rightly suspected that Roosevelt was on Churchill's and Brooke's side, and so accepted the Sicily operation *faute de mieux*. At the Third Washington (Trident) Conference, Churchill wrongly thought Brooke and Marshall were conspiring against him, when in fact they were just fighting each other openly, while Roosevelt supported Marshall and won a definite date – 1 May 1944 – for the cross-Channel assault.

At Algiers in May 1943, Marshall simply blocked Churchill and Brooke over the invasion of mainland Italy, because he knew he now had the support of Roosevelt. The result was that tens of thousands of Germans escaped Sicily unnecessarily. At the First Quebec (Quadrant) Conference, Roosevelt and Churchill dashed Brooke's hopes of commanding Overlord, seemingly at the time in favour of Marshall. Over the northern Norway, Sumatran, Andaman, Dodecanese and several other operations, Brooke blocked Churchill's schemes, usually with – but sometimes without – the support of Roosevelt and Marshall. The First Cairo (Sextant) Conference saw another mutual blocking operation, as a result of which Stalin was the only true victor at Teheran (Eureka), while at the Second Cairo Conference Brooke and Roosevelt dashed Marshall's plans for invading the Andaman Islands and Roosevelt told Churchill that Eisenhower rather than Marshall would command Overlord. On the invasion of the south of France, Marshall and Roosevelt argued hard with Brooke and Churchill, but finally overruled them. The same happened with the remit for Eisenhower's supreme command. As for the hare-brained plan to capture Vienna via the Ljubljana Gap, Brooke swiftly changed his mind and joined Roosevelt and Marshall in opposing Churchill.

At the Malta (Cricket) Conference in February 1945, Marshall simply employed *force majeure* to silence Brooke over Montgomery's criticisms of Eisenhower's broad-front strategy, while Roosevelt as good as ignored Churchill. By Yalta the grand strategy of the European denouement was already decided; Marshall approved it and Brooke made the best of it, while Churchill and Roosevelt could find no arguments against the stark fact of a vast Red Army sprawling over Poland and East Prussia, its vanguard only 40 miles from Berlin.

Although the complicated minuet between the Masters and Commanders stopped with the death of Roosevelt, it had in effect ended at Malta on 1 February 1945, when Marshall had laid down his law with aggression, and Brooke could do nothing about it. Britain was in dire financial straits, and the United States was contributing far more on the ground in Europe in terms of men and matériel than the British Commonwealth. Brooke was therefore reduced to expostulating – and occasionally ranting – to his diary, but he could no longer significantly affect the course of the campaign. Once ashore in northern France after June 1944, the Americans could enjoy the dominance over strategy to which their provision of roughly two-thirds of the troops entitled them. Perhaps if the atomic bombs had not worked, and the United States had needed to call on Britain more in the war against Japan, the dynamic between Marshall and Brooke might have shifted again, but by then Roosevelt was dead, Churchill had lost office, and the two soldiers were both exhausted.

On 28 June 1944, in acknowledging defeat over the south of France operation, Brooke and Churchill told Marshall and Roosevelt that they would leave it 'for history to judge' whether they or the Americans had been right, sooner than split the alliance. The answer today is one that probably none would have liked: Brooke's and Churchill's campaign up to the Po Valley was largely a waste of effort after Rome, while Marshall's and Roosevelt's Operation Anvil/Dragoon was also largely a waste of time. The geometry lessons about which Roosevelt wrote to Churchill should have told him that the distance between Marseilles and anywhere else of value – especially Paris nearly 500 miles distant – was too far to make Dragoon a particularly worthwhile use of four hundred thousand men. If anywhere needed capturing in order to shorten the war in the west, it was not Toulon or Marseilles or Milan or Trieste or Istria or even Vienna, but instead the estuary of the Scheldt, thereby freeing up the great port of Antwerp and drastically shortening the long Allied lines of communication that still went via Normandy. Yet that did not happen for another two months after the Dragoon landings. The trouble with leaving things 'for history to judge' is that the verdict might not go your way.

If Marshall and Brooke had trusted the strategic judgements of their political Masters more, they might have taken up the lustrous commands offered them, of supreme commander in Europe and commander-in-chief of the Eighth Army respectively. If they had, there is little doubt that they would be as revered as Eisenhower and Montgomery are today. It is a tribute to their greatness that they didn't accept those posts out of ambition or vanity or because, as Marshall put it, every general longs for a field command. They contributed far more to ultimate victory by staying behind their desks and forfeiting any hopes of global fame and glory. (So little recognized was Brooke that, when he was invited to inspect a parade of Jewish ex-servicemen at Horse Guards Parade in 1950, a corporal of the Life Guards allowed his car through the main archway only on the basis that if challenged he would pretend that he had thought his name was Lord Beaverbrook.)[5]

As they had warned Marshall in April 1942 – if not quite explicitly or early enough – Churchill and Brooke vetoed any cross-Channel operation until they were satisfied that both the Indian Ocean and the Middle East were safe, with the United States expected to get fully committed in bringing about that seemingly Utopian state of affairs. By the time of Trident in May 1943, Roosevelt saw that those preconditions had now been met, and that it was time for the British to commit to the cross-Channel operation. The debate over whether an earlier assault would have been successful will last for ever, but at Trident Roosevelt sensed that the summer of 1944 would be ideal for the assault, and so it happened then.

It was fortuitous that the tipping point in influence between America and Britain came at much the same time that a cross-Channel attack was at last genuinely becoming viable, and just as the British were over-concentrating on the Mediterranean. 'Brooke and his colleagues got their way to a remarkable extent in 1942 and 1943,' notes the historian Brian Bond, and it was just as well at that stage in the war.[6] But George Marshall – who was wrong to press for the invasion of France in 1942 or 1943 – was absolutely right to demand one for 1944. As for Brooke's comment to Forrest Pogue in 1947 that the Americans had insisted on the cross-Channel attack mainly because they 'had created a large army and that was the only place where it could be impressively used', Marshall rightly said: 'I didn't know he felt that way, and that wasn't true.'[7]

The central aims for the Mediterranean theatre agreed at the Casablanca, First Quebec and Teheran conferences had all been achieved by the end of 1943, including clearing the Mediterranean for Allied shipping. Italy had surrendered, forty-five German divisions had been drawn into operating in that country and in the Balkans, Allied aircraft from the Foggia Plains were bombing central and south-eastern European targets (especially Ploesti), and Yugoslav partisans were receiving their supplies. That was about as much as the Mediterranean campaign could ever realistically have achieved south of the Alps. Although it was true that division for division the Germans tied down as many Allies as themselves in Italy, and it was far more difficult for the Allies to resupply their forces there than it was for the Germans, proportionately the Third Reich could ill afford the men and matériel involved, while the United States was deploying ever larger numbers. With far greater Allied populations and productive capacities, ratios of 1:1 made no sense to Germany on a lesser front like Italy.

Great Britain lost 397,700 killed in the war, which, as Alex Danchev has noted, was 'historically a high figure, but by the standards of this gargantuan war proportionately and comparatively very light'.[8] We can now see that whereas Churchill and Brooke were right to call for the invasion of Italy up to Naples in 1943, they were wrong to continue to demand fighting on up to Rome, let alone further north to the Gothic Line. 'Group-think' and 'mission-creep' are ugly modern words for long-established, dangerous phenomena. Brooke's unnecessary post-war claims to total omniscience were also unsustainable.

If a moment can be pinpointed when the British started to get strategy wrong, and the Americans started to get it right, it came in the fortnight after Tuesday 19 October 1943, when Churchill successfully persuaded Brooke to join him in attempting to postpone Overlord. Small wonder that Roosevelt and Marshall lost their patience, with the 150,000 men of the first wave already in full-scale training for the operation. It is indicative that Churchill, who rarely forsook the opportunity to blow his own trumpet when the events justified it, never explicitly claimed credit for having postponed the invasion of France until June 1944. 'Was his silence a kind of escape clause or insurance in case posterity held him responsible for prolonging the war?' asked Lord Moran.[9]

Yet Churchill did not prolong the war, despite General Albert Wedemeyer's denunciation of the Mediterranean campaign as 'a trap which prolonged the war in Europe by a year' and 'a side show, which cost many unnecessary lives'.[10] In fact the African, Sicilian and Naples campaigns weakened the Third Reich before the decisive blow was struck in France, not least in terms of morale. At the point that concentration on the Mediterranean might indeed have threatened to lengthen the war, the Americans insisted upon Overlord taking place no longer than five weeks later than scheduled.

When Henry Stimson visited Britain soon after Overlord to see for himself 'the colossal scope of the undertaking', visiting Bradley's headquarters in the process, Churchill said of the operation, 'It is wonderful, a great triumph. But we could never have done it last year.' Though unconvinced, the US Secretary of War 'did not see any need to quarrel' over the Prime Minister's intentionally provocative remark. In his auto-biography, he acknowledged that Churchill had tried to avoid 'the bloody futility of the Western Front in 1915', and asked: 'If the Americans had suffered similar losses in the First World War or faced similar succeeding dangers, would they have felt differently?'[11]

As the Normandy landings and their immediate aftermath showed, numerical superiority and complete control of the air were necessary for victory, which even then in view of the Germans' formidable capacity for counter-attack was not a foregone conclusion. However unpalatable it might be to admit it, the statistics allow no doubt: soldier for soldier the German fighting man and his generals outperformed Britons, Americans and Russians both offensively and defensively by a significant factor virtually throughout the Second World War. The consequences of Overlord failing were too serious to permit any undue risks being taken over its timing. Colonel Ian Jacob wrote after the war: 'Our thinking was probably too limited; we were too obsessed with the memories of the First World War. On the other hand, the Americans were certainly too optimistic and too expansive in their conception.'[12] That was true of the Americans until mid-1943.

Churchill had a healthy and well-founded (if necessarily private) respect for the quality of the Wehrmacht. In mid-July 1942, he spoke to Major-General John Kennedy about 'the Germans being better than our troops', which prompted Brooke to say that he 'must speak to Winston about this', as he feared that these remarks did harm when uttered in front of politicians, 'as they so often were. It was a case of giving a dog a bad name.' Comments from the Prime Minister such as 'If Rommel's army were *all* Germans, they would beat us,' were probably intended as a joke against the Italians, but Kennedy also saw it as 'a dig at the British Army (which unfortunately he can't resist)'.[13] Yet both Churchill and Brooke always knew that eventually the Western Allies needed to confront the Wehrmacht head on in France and Germany, and by capturing the Ruhr and Saar deny Hitler the ability to fight on indefinitely.

Victory over the Germans in North Africa and Italy was the necessary prelude, in order to give the Western Allies the belief that the Wehrmacht could be defeated on the European Continent. A landing in 1942 would have suffered from far lower morale, as the Germans had not been defeated on land anywhere in the west before El Alamein. As Admiral Cunningham told Forrest Pogue in the Athenæum Club in 1947, there was 'never any doubt' in British minds 'that they would do the operation ... They were always convinced they would go across the Channel.'[14] With the V-weapon offensive about to open against London, the Allies would be able to overrun the launch sites.

'Under their hearty and friendly manner one feels there is suspicion and contempt,' noted John Kennedy of the Americans attending the Casablanca Conference, 'although a few individuals – and those in the highest plane – are true friends. I should put Roosevelt and Marshall in this class.'[15] Britain was fortunate that, despite the undoubted presence of some Anglophobes in the higher reaches of the American Army and Navy Departments, the President and US Army Chief of Staff were not of their number. One who was, Albert Wedemeyer, told interviewers that 'Many American planners and even a few members of the Joint Chiefs of Staff' – he cited Admiral King – 'maintained that the British were insincere' about ever wanting to cross the Channel, and he concluded: 'The diversions into the Mediterranean area would never have been possible had it not been for the so-called charisma of the prime minister.'[16] The influence of personality, in this case Churchill's charisma, on grand strategy has rarely been so ably put. Yet it was Roosevelt who appreciated that in fact the British were indeed sincere about crossing the Channel, but only after the Germans had been softened up by defeat in North Africa and the Mediterranean.

Sir Michael Howard is right when he points out that the US Army Planning Staff always 'drew a sharp distinction between the UK itself, which they were prepared to defend, and "British interests" – particularly British imperial interests – which they were not'.[17] This was perfectly reasonable; the defence of the British mainland was a vital American interest while British possession of a vast global empire manifestly was not. No amount of Churchillian rhetoric or charisma could alter that. Yet as Stimson concluded in his autobiography: 'When all the arguments have been forgotten, this central fact will remain. The two nations fought a single war, and their quarrels were the quarrels of brothers.'[18]

For the first half of the period between Pearl Harbor and the President's death in April 1945, Roosevelt, Churchill and Brooke got their way in matters of grand strategy, and it was fortunate that they did, considering Marshall's plans for an over-hasty assault on France. In the second half of the period, from Trident onwards, Roosevelt supported Marshall and together they established hegemony over decision-making, by which time it was just as fortunate that they had, considering British over-emphasis on the Italian campaign and its projected aftermaths.

Roosevelt was the ultimate arbiter between the competing strategies of Marshall, Churchill and Brooke. The Second World War was an intensely political conflict, and he was a superlatively talented politician, with unrivalled insight into what Americans wanted and would allow. He understood the political as well as military strategies that were needed to keep the Western democracies' morale high. When Marshall spoke of Operation Sledgehammer as 'a sacrificial play' in order to keep the Russians in the war, for example, Roosevelt realized that that was un-acceptable.[19] Similarly his intuition told him that American troops needed to be fighting Germans on the ground during the year 1942, partly in order to save the Germany First policy. Of the four Masters and Commanders of the Western Allies, therefore, the man who most influenced the course of the war was the one who openly acknowledged that he knew the least about grand strategy: Franklin Delano Roosevelt.

On Saturday 27 June 1942 Adolf Hitler dined with Dr Otto Dietrich, the Reich press chief, at the Berghof in Berchtesgaden. The conversation – after the subjects of Julius Caesar's viaducts and the future of Belgium had been exhausted – got round to the eight days of discussions that had taken place between Churchill and Roosevelt in Washington. 'When two people are in general agreement,' the Führer opined, 'decisions are swiftly taken. My own conversations with the Duce have never lasted longer than an hour and a half, the rest of the time being devoted to ceremonies of various kinds ... To harness to a common purpose a coalition composed of Great Britain, the United States, Russia and China demands little short of a miracle.'[20] As the struggles described in this book have shown, harnessing solely Great Britain and the United States to a common purpose, let alone the other powers, demanded even more of a miracle than the Führer could have guessed. Yet, through their rows, standoffs, fist-shaking, charm offensives, hard-fought compromises and occasional tantrums, the Masters and Commanders performed that miracle, and won victory in the west.

*

The disagreements between the Masters and Commanders which this book has chronicled were often deep and sometimes bitter. But were the enmities permanent? Did the opinions that Roosevelt, Churchill, Brooke and Marshall expressed so forcefully about each other, and which reflected such fundamental differences about how the war should be conducted, endure? Eight years after Yalta, the three survivors each performed a prominent role at the Coronation of Queen Elizabeth II at Westminster Abbey on 2 June 1953. General George C. Marshall, representing the United States, was seated in the front row of the Choir, very close to the centre of events and with a superb view of the proceedings only a few feet away.[21] He wore white tie and a black tailcoat, adorned with the insignia of a Knight Grand Cross of the Most Honourable Order of the Bath. Two of the Great Officers of State, appointed for the day, were the Lord High Constable, Viscount Alanbrooke – the title Brooke took when raised to the peerage – and the Lord High Steward, Viscount Cunningham of Hyndehope, who bore St Edward's crown. Near by stood Viscount Portal of Hungerford, bearing the Queen's sceptre. At the rear of the procession of eight Commonwealth premiers entering the Abbey was Sir Winston Churchill.[22] As he drew level with the Choir, Churchill stopped, shook Marshall's hand, and resumed his stately progress.

Appendix A
The Major Wartime Conferences

1941

August	Placentia Bay Conference (Riviera)
December–	
January 1942	First Washington Conference (Arcadia)

1942

April	London visit by Marshall and Hopkins (Modicum)
June	Second Washington Conference (Argonaut)
July	London visit by Marshall, Hopkins and King
August	First Moscow Conference (Bracelet)

1943

January	Casablanca Conference (Symbol)
May	Third Washington Conference (Trident)
August	First Quebec Conference (Quadrant)
November	First Cairo Conference (Sextant)
November	Teheran Conference (Eureka)
December	Second Cairo Conference

1944

June	London visit by US Joint Chiefs of Staff
September	Second Quebec Conference (Octagon)
October	Second Moscow Conference (Tolstoy)

1945

January–	
February	Malta Conference (Cricket)
February	Yalta Conference (Argonaut)
July	Potsdam Conference (Terminal)

Appendix B
Glossary of Codenames

Ajax Autumn 1941 plan to attack Trondheim
Anakim Anglo-American amphibious assault on Burma
Anvil Allied landing in the south of France, August 1944 (later Dragoon)
Arcadia First Washington Conference, December 1941–January 1942
Argonaut Second Washington Conference, June 1942, and Yalta Conference, February 1945
Autumn Mist See Herbstnebel
Avalanche Amphibious Salerno landings, September 1943
Barbarossa German invasion of USSR, June 1941
Baytown Montgomery's landings near Reggio, September 1943
Bolero Build-up of American forces in Britain prior to invasion of western Europe
Bracelet First Moscow Conference, August 1942
Breastplate Allied plan for amphibious attack on Tunisia from Malta, November 1942
Brimstone Proposed invasion of Sardinia, 1943
Buccaneer Planned attack on the Andaman Islands
Capital Advance into central Burma, 1944 (previously Champion)
Champion Previous name for Capital
Citadel German offensive against the Kursk salient, July 1943
Cricket Malta Conference, January–February 1945
Culverin Proposed attack on northern Sumatra
Diadem Offensive against the Gustav Line, 1944
Dracula Combined-operations attack on Rangoon, May 1945
Dragoon Allied landings in the south of France, August 1944 (previously Anvil)
Eureka Teheran Conference, November 1943
Freehold Churchill's visit to Athens, February 1945
Gymnast American invasion of Morocco (later Super-Gymnast, later Torch)
Habbakuk British scheme to build floating aircraft stations out of ice and wood pulp
Herbstnebel German offensive in the Ardennes, December 1944–January 1945
Husky Allied invasion of Sicily, July 1943
Iceberg American capture of the Ryukyu Islands, April 1945
Ironclad British Commonwealth attack on Madagascar, March 1942
Jubilee Raid on Dieppe, August 1942
Jupiter Proposed operations in northern Norway
Kutuzov Marshal Zhukov's successful counter-attack during the battle of Kursk
Mandibles Proposed operations against the Dodecanese Islands
Manhattan Engineer District Development of the atomic bomb
Manna British movement into Greece after German withdrawal, October 1944
Market Garden Attempt to seize the Rhine bridges, September 1944
Modicum Marshall's visit to London, April 1942

Mohican   Allied attacks along the North African coast, similar to Super-Gymnast
Neptune   Naval and air support for Overlord, June 1944
Octagon   Second Quebec Conference, September 1944
Orange   US pre-war contingency plans against the eventuality of war versus Japan
Overlord   Invasion of Normandy, 1944 (previously Roundup)
Plunder   Allied crossing of the Rhine, March 1945
PLUTO   Pipeline Under The Ocean, Normandy 1944
Pointblank   Allied strategic bombing offensive against Germany
Quadrant   First Quebec Conference, August 1943
Rankin   Plan for Allied occupation of post-war western Europe
Riviera   Meeting between Roosevelt and Churchill at Placentia Bay, August 1941
Roundup   Anglo-American invasion of Europe (changed to Overlord)
Sealion   German plan for invasion of Britain, from 1940
Sextant   First Cairo Conference, November 1943
Shingle   Anzio landings, January 1944
Slapstick   Landings at Taranto, September 1943
Sledgehammer   Small-scale attack on the Cotentin Peninsula, 1942 or 1943
Supercharge   Final phase of the battle of El Alamein, November 1942
Super-Gymnast   Attack on Morocco and Algeria (later Torch)
Symbol   Casablanca Conference, January 1943
Tarzan   Attack in north Burma to open the Burma Road
Tolstoy   Second Moscow Conference, October 1944
Torch   Allied invasion of French North Africa in November 1942 (previously Gymnast and Super-Gymnast)
Trident   Third Washington Conference, May 1943
Ultra   The signals-interception project based at Bletchley Park
Veritable   21st Army Group Rhine operations, February 1945
Victor   Anti-invasion exercises in 1940
Winter Storm (Wintergewitter)   General von Manstein's attempt to relieve Stalingrad, December 1942

Appendix C
The Selection of Codenames

After the US bombing attack on the Ploesti oil fields in Roumania on 1 August 1943, codenamed Operation Soapsuds, Churchill minuted to Ismay that military operations 'ought not to be described by codenames which imply a boastful and over-confident sentiment' or equally ones 'which are calculated to invest the plan with an air of despondency'. Furthermore, 'They ought not to be names of frivolous character' or be 'ordinary words' and 'Names of living people should be avoided.' He had already spoken to Marshall on this subject, and Soapsuds was duly rechristened with the altogether more macho name of Tidalwave. Churchill thought that codenames should be taken from 'heroes of antiquity, figures from Greek and Roman mythology, the constellations and stars, famous racehorses and the names of British and American heroes'.

Racehorse names perhaps betrayed Churchill's English aristocratic upbringing, but it does indeed seem astonishing that operations in which men's lives were at stake were often given light-hearted and sometimes downright flippant codenames, but war often throws up such strange phenomena. They almost seem like fey, light-hearted jokes deliberately designed to contrast with the lethal reality of the operations they masked. Among such frivolous codenames of the Second World War were Operations Bingo, Boozer, Bunghole, Cabaret, Cellophane, Chastity, Chatanooga Choo-Choo, Corkscrew, Duck, Grapefruit, Haddock, Hats, Horlicks, Infatuate, Jockey, Juggler, Lilo, Loincloth, Mallard, Manhole, Market Garden, Modified Dracula, Mutton, Nest Egg, Pancake, Pantaloon, Peanut, Puddle, Pumpkin, Raincoat, Razzle, Rhubarb, Rhumba, Sardine, Saucy, Seaslug, Skinflint, Spinach, Squid, Teacup, Wowser and Zipper.

It cannot have been easy for parents to discover after the war that their sons' lives had been lost on Operation Slapstick, Toenails or Maggot, in comparison to the far more martial-sounding Retribution, Mailfist, Supercharge or Musketeer. Churchill's fears about creating 'an air of despondency' were surely justified by the codenames for Operations Orphan and Batty (remote-controlled B-17 bombers), Moonshine (the naval operation to pick up supplies from Sweden in 1945), Penitent (an attack on Yugoslav ports in 1945), Blot (British air operations in Europe), Grubworm (the air transport of the Chinese Army from Assam to China), Hasty (parachute drops east of Rome in June 1944), Deficient (the advance of the Indian 10th Division in 1941), Frantic (bombing raids by 8th and 15th US Air Forces in 1944), Lost (the SAS raid to the Serent after D-Day), Rockbottom (special operations over the Hump in 1943), Ratweek (a Balkan bombing offensive), Stalemate (invasion of the Palau Islands in 1944) and especially Taxable (the British radar deception operation). Nor did the French take to heart Churchill's words about not naming operations after living people: the airborne operation by the 2ème Regiment des Chasseurs Parachutistes to harass the German retreat from Normandy near Corrèze in August 1944 was codenamed Marshall.

Notes

ABBREVIATIONS

ALAB	Papers of Lord Alanbrooke
ASTL	Papers of Joan Bright Astley
AVON	Papers of Sir Anthony Eden, later 1st Earl of Avon
BEAV	Papers of Lord Beaverbrook
BLAC	Archive of Lord Black of Crossharbour
BRGS	Papers of Lawrence Burgis
CAB	Cabinet Papers at the National Archives, Kew
CHAR	Chartwell Papers at Churchill Archives Centre, Cambridge
CHUR	Churchill Papers at Churchill Archives Centre, Cambridge
CUNN	Papers of Admiral Lord Cunningham
DILL	Papers of Sir John Dill
DUPO	Papers of Sir Dudley Pound
FDR	Papers of Franklin D. Roosevelt
FRUS	Foreign Relations of the United States
JACB	Papers of Sir Ian Jacob
KENN	Papers of Sir John Kennedy
LEAH	Papers of Admiral William Leahy
LH	Papers of Sir Basil Liddell Hart
MAR	Papers of George C. Marshall
MHI	USA Military History Institute, Carlisle, Pennsylvania
NA	British National Archives
PEAK	Papers of Sir Charles Peake
PORT	Papers of Lord Portal

PREFACE

1. Lees-Milne, *Enigmatic Edwardian*, p. 229
2. *Ibid.*, p. 330
3. BRGS 1/1/46
4. Leasor, *War at the Top*, pp. 40–41
5. BRGS 1/1
6. BRGS 1/2
7. BRGS 1/1/44
8. BRGS 1/2
9. CAB 69/4/38
10. KENN 4/2/3–5

INTRODUCTION

1. eds Danchev and Todman, *Diaries*, p. 268
2. JACB 1/14/B
3. eds Danchev and Todman, *Diaries*, p. 268
4. ALAB 11/9
5. Churchill, *Hinge of Fate*, p. 344

CHAPTER 1: FIRST ENCOUNTERS: 1880–JUNE 1940

1. Churchill, *The World Crisis: 1915*, p. 166
2. The phrase is from William F. Buckley Jr
3. ed. Kimball, *Correspondence*, II, p. 355; Beschloss, *Kennedy and Roosevelt*, p. 200; ed. Smith, *Hostage to Fortune*, p. 411; Black, *Roosevelt*, p. 91; Ward, *First-Class Temperament*, p. 393
4. Robbins, 'The Atlantic Charter', p. 18; Churchill, *Gathering Storm*, p. 345
5. Halifax Diary, 30/11/1941
6. Marshall, *Together*, p. 110
7. Moran, *Struggle for Survival*, p. 67
8. ed. Bland, *Interviews*, p. 40
9. MAR Secretary of Defense Papers Box 207/12
10. *Ibid.*
11. ed. Stephen, *Sir Victor Brooke, passim*
12. Lord Carver in *Times Literary Supplement*, 7/5/1982, p. 504
13. Layton, *Mr Churchill's Secretary*, p. 51
14. *Oxford Dictionary of National Biography*, VII, p. 863
15. Russell, *Churchill*, pp. 49–50
16. Churchill, *My Early Life*, p. 57
17. Major Buckley recollections in Astley Papers
18. ed. Wheeler-Bennett, *Action This Day*, p. 194
19. Halifax Diary, 27/11/1941
20. JACB 4/8
21. Jenkins, *Roosevelt*, p. 5
22. Black, *Roosevelt*, pp. 65–6
23. *Ibid.*, pp. 646, 715
24. *Ibid.*, p. 21
25. Fraser, *Alanbrooke*, p. 93
26. *Ibid.*, pp. 64–5; ALAB 11/9; ALAB 11/55
27. Greenwood, *Auchinleck*, p. 51
28. Ronald Lewin reviewing Fraser, *Alanbrooke*, in *The Times*, 13/5/1982
29. KENN 4/2/5
30. Interview with General Sir David Fraser, 8/6/2006
31. MHI Pogue Notes for *Supreme Command*, 26/8/1958
32. MAR McCarthy Papers B-17/f-18, MAR GCM Library Xerox 2256
33. Russell, *Churchill*, pp. 288, 297; Churchill, *My Early Life*, p. 327
34. Cray, *General of the Army*, p. 110
35. *Ibid.*, pp. 114–15
36. Pogue, *Education*, p. 353
37. *Ibid.*, p. 340; ed. Bland, *Interviews*, p. 109
38. ed. Bland, *Interviews*, p. 109

39. MHI Handy Interview 1974 section 4, p. 12
40. *Ibid.*; MAR William M. Spencer Interview with Marshall, 9/7/1949
41. Morgan, *FDR*, p. 589
42. MHI Hull Interview 1974 section 5, p. 29
43. MHI Hull Interview Session 6, pp. 24ff., interview with Larry Bland, 28/11/2006; MAR Richard DeMartino interview with Merill Pascoe, 11/11/1997, p. 7
44. MAR Pentagon Papers Boxes 80 and 81; ed. Bland, *Papers, passim*
45. MAR Pentagon Papers Box 71/20
46. Pogue, *Ordeal and Hope*, p. 22
47. MAR Pentagon Papers Box 80/28
48. Pogue, *Education*, pp. 341–2
49. ed. Bland, *Interviews*, p. 329
50. LH 15/15/1
51. Stimson and Bundy, *Active Service*, p. 238
52. ed. Kimball, *Correspondence*, I, p. 421
53. Churchill, *Great Contemporaries*, p. 302
54. LH 15/15/1
55. ed. Bond, *Chief of Staff*, I, p. 367
56. *Oxford Dictionary of National Biography*, VII, p. 865
57. 1959 BBC Interview of Lord Alanbrooke, videotape no. VC338475; *Sunday Times*, 27/1/1946
58. Gilbert, 'Churchill and D-Day', p. 24
59. ed. Macleod, *Ironside Diaries*, p. 351
60. eds Danchev and Todman, *Diaries*, p. 79
61. *Ibid.*, p. 80
62. *Ibid.*, p. 81
63. *Ibid.*
64. ALAB 11/9
65. eds Danchev and Todman, *Diaries*, p. 273
66. ALAB 11/9

CHAPTER 2: COLLECTING ALLIES: JUNE 1940–DECEMBER 1941

1. ALAB 11/59
2. *Finest Hour*, no. 130, Spring 2006, pp. 34–6
3. Moran, *Struggle for Survival*, p. 192
4. KENN 4/2/3, 21/11/1941
5. Best, *Churchill and War*, p. 176
6. David Freeman in *Finest Hour*, no. 115, Summer 2002, p. 34
7. Colville, *Footprints*, p. 189
8. Simpson, *Stark*, p. 144
9. Colville, *Fringes*, p. 283
10. ed. Parker, *Churchill*, p. 89
11. Reynolds, *Creation*, p. 167
12. Gilbert, *Churchill War Papers*, III, p. 55
13. eds Danchev and Todman, *Diaries*, p. 133
14. Gilbert, *Churchill War Papers*, III, p. 397
15. eds Danchev and Todman, *Diaries*, p. 154
16. MAR Pentagon Papers Box 80/29
17. ed. Danchev, *Anglo-American Alliance*, pp. 10–11
18. Morgan, *FDR*, p. 610

19. Papers of Jo, Countess of Onslow
20. BRGS 2/12, 30/3/1942
21. BLAC 4/8/1941
22. Robbins, 'The Atlantic Charter', p. 18
23. Morgan, *FDR*, pp. 596–7
24. BLAC
25. Morgan, *FDR*, pp. 596–7
26. BEAV D/491
27. Reynolds, *In Command*, p. 257; Danchev, 'Dilly-Dally', p. 21
28. ALAB 11/64
29. KENN 4/2/3
30. CHAR 20/22/273
31. eds Danchev and Todman, *Diaries*, p. 206
32. Howard, *Mediterranean Strategy*, p. 18
33. Jackson and Bramall, *The Chiefs*, p. 214
34. Danchev, 'Waltzing', p. 221; Fraser, *Alanbrooke*, p. 233
35. Jackson and Bramall, *The Chiefs*, p. 214
36. ALAB 11/74
37. CUNN Add MSS 52578/28
38. BRGS 2/22
39. CUNN Add MSS 52577/95
40. *Ibid.*
41. CAB 65/44/66
42. MAR Pentagon Papers Box 116
43. Marshall, *Together*, p. 145
44. KENN 4/2/3; *Sunday Times*, 27/1/1946
45. KENN 4/2/3
46. *Ibid.*
47. *Ibid.*
48. BRGS 2/10
49. KENN 4/2/3
50. *Ibid.*
51. eds Danchev and Todman, *Diaries*, p. 209; KENN 4/2/4, p. 301

CHAPTER 3: EGOS IN ARCADIA: DECEMBER 1941–FEBRUARY 1942

1. Churchill, *Hinge of Fate*, p. 290
2. CHUR 4/225A/138; Ismay, *Memoirs*, p. 243
3. Eisenhower, *Crusade in Europe*, p. 30
4. BRGS 2/10
5. Stimson and Bundy, *Active Service*, p. 207
6. Gilbert, *Winston S. Churchill*, VII, pp. 10–13
7. LH 15/15/1
8. Best, *Churchill and War*, p. 93
9. Stimson and Bundy, *Active Service*, p. 214
10. Morgan, *FDR*, p. 637
11. MAR Pentagon Papers Box 80/30
12. Stimson and Bundy, *Active Service*, p. 212
13. MHI Handy Interview 1974 section 3, p. 42
14. JACB 1/13/16
15. *Ibid.*

16. *Ibid.*
17. MAR Robinett Papers Box 21, p. 357
18. MHI Handy Interview 1974 section 4, pp. 10–11
19. MAR Robinett Papers Box 21, p. 363
20. Danchev, *On Specialness*, p. 19
21. ed. Kimball, *Correspondence*, I, p. 293
22. MHI Pogue Interview, 20/12/46, p. 3
23. Pogue, *Ordeal and Hope*, p. 272; Ismay, *Memoirs*, p. 244
24. ed. Bland, *Interviews*, pp. 413–14
25. *Ibid.*
26. Danchev, 'Very Special Relationship', p. 5
27. MAR Robinett Papers Box 21, p. 365
28. Bercuson and Herwig, *One Christmas*, p. 178
29. ed. Bland, *Interviews*, p. 601
30. Fenby, *Alliance*, p. 53
31. Gilbert, *Churchill War Papers*, III, p. 1696; Reynolds, *In Command*, p. 40
32. JACB 1/13/20
33. Moran, *Struggle for Survival*, p. 20
34. ed. White, *Stillwell Papers*, p. 44
35. *Ibid.*, pp. 44–5
36. *Ibid.*, p. 47
37. ed. Bland, *Interviews*, pp. 592–3
38. Halifax Diary, 18/2/1942
39. MAR Pentagon Papers Box 80/31
40. Dupuy and Dupuy, *Collins Encyclopedia*, p. 1309
41. MAR Pentagon Papers Box 71/20
42. MAR Robinett Papers Box 21, p. 356
43. JACB 1/13/40
44. BRGS 2/11, 18/1/1942
45. *Ibid.*
46. *Ibid.*
47. CAB 69/4/23
48. BRGS 2/11, 2/2/1942
49. AVON 20/1/22
50. KENN 4/2/4
51. Danchev, *On Specialness*, p. 23
52. MHI Charles Donnelly Autobiography, p. 634
53. ed. Danchev, *Anglo-American Alliance*, p. 13
54. Danchev, 'Very Special Relationship', p. 6
55. ed. Danchev, *Anglo-American Alliance*, p. 8
56. MHI Handy Interview 1974 section 4, p. 18
57. ed. Bland, *Interviews*, pp. 623–4; Parrish, *Roosevelt and Marshall*, p. 251
58. Stimson and Bundy, *Active Service*, p. 213
59. MHI Caraway Interview 1971 section 6, pp. 50–51
60. Hastings, *Nemesis*, p. 21; MHI Handy Interview 1974 section 4, p. 10
61. Jackson and Bramall, *The Chiefs*, p. 226
62. Harmon, 'Alanbrooke and Churchill', p. 36
63. Colville, *Footprints*, p. 190
64. Tully, *FDR*, p. 262
65. ed. Bland, *Interviews*, p. 436
66. eds Danchev and Todman, *Diaries*, p. 228
67. Danchev, *On Specialness*, p. 18
68. ALAB 6/2/12/8A
69. Kennedy, *Business of War*, pp. 104–5

CHAPTER 4: BROOKE AND MARSHALL ESTABLISH DOMINANCE: FEBRUARY–MARCH 1942

1. NA WO 193/334, 4/10/1942
2. Brodhurst, *Churchill's Anchor*, p. 208
3. *Ibid., passim*; David Freeman in *Finest Hour*, no. 115, Summer 2002, p. 35
4. Brodhurst, *Churchill's Anchor*, pp. 211–12
5. *The Times*, 24/4/1971
6. ALAB 11/64
7. KENN 4/2/4
8. ed. Wheeler-Bennett, *Action This Day*, p. 61
9. *Ibid.*, p. 165
10. Churchill, *Finest Hour*, p. 20
11. LH 15/15/1
12. ALAB 6/4/21
13. Howarth, *Intelligence Chief Extraordinary*, pp. 143, 164–5
14. BRGS 1/1/47 and 1/2
15. *Ibid.*, p. 166
16. CAB 69/4–1–4
17. AVON 20/1/22
18. Interview with General Sir David Fraser, 8/6/2006
19. MHI Pogue Interview, 17/12/1946, p. 5
20. Brendon, *Churchill*, p. 163
21. ed. Taylor, *Churchill*, p. 199
22. KENN 4/2/4, 18/3/1942
23. Moran, *Struggle for Survival*, p. 35
24. Nel, *Mr Churchill's Secretary*, p. 74
25. MHI Lincoln Papers letter of 5/1/1954
26. ed. Bland, *Interviews*, pp. 610–11

CHAPTER 5: GYMNAST FALLS, BOLERO RETUNED: FEBRUARY–APRIL 1942

1. For some reason the money was paid in 1969 to St Paul's School, Athens
2. BRGS 2/11, 10/2/1942
3. KENN 4/2/4
4. *Ibid.*
5. Fraser, *Alanbrooke*, p. 235
6. Mosley, *Backs to the Wall*, p. 245; ed. Dilks, *Cadogan*, p. 447
7. ed. Dilks, *Cadogan*, p. 446
8. MAR Pentagon Papers Box 80/32
9. MHI Hull Interview 1973 section 4, p. 38
10. ed. Kimball, *Correspondence*, I, p. 392
11. Gilbert, *Churchill: A Life*, p. 719; Gilbert, 'Churchill and D-Day', pp. 24–5
12. ed. Kimball, *Correspondence*, I, pp. 392–3
13. Gilbert, *Churchill: A Life*, p. 719
14. Gilbert, *Winston S. Churchill*, VII, p. 74; Gilbert, 'Churchill and D-Day', pp. 24–5
15. ed. Kimball, *Correspondence*, I, pp. 398–9
16. KENN 4/2/4
17. *Ibid.*

18. Sherwood, *White House Papers*, II, p. 523
19. *Ibid.*, pp. 523–4
20. ed. Kimball, *Correspondence*, I, p. 421
21. Butler, *Grand Strategy*, III Part II, p. 573
22. BRGS 2/12, 23/3/1942
23. Stimson and Bundy, *Active Service*, p. 214
24. Parrish, *Roosevelt and Marshall*, p. 254
25. Stimson and Bundy, *Active Service*, p. 215
26. ed. Kimball, *Correspondence*, I, p. 437
27. MAR Pentagon Papers Box 71/19
28. Parrish, *Roosevelt and Marshall*, pp. 261–3
29. AVON 20/1/22
30. KENN 4/2/4
31. MHI Caraway Interview 1971 section 6, p. 29
32. KENN 4/2/4

CHAPTER 6: MARSHALL'S MISSION TO LONDON: APRIL 1942

1. ASTL Major Buckley's Recollections
2. eds Danchev and Todman, *Diaries*, p. 249
3. Pogue, *Ordeal and Hope*, p. 314
4. Morison, *American Contributions*, p. 15
5. ALAB 9/3/8
6. Pogue, *Ordeal and Hope*, p. 310
7. Sherwood, *Roosevelt and Hopkins*, p. 528
8. *Ibid.*, p. 523; Pogue, *Ordeal and Hope*, p. 308; Bryant, *Turn of the Tide*, p. 284
9. MHI Handy Interview 1974 section 4, p. 11
10. Moran, *Struggle for Survival*, p. 35
11. Pogue, *Ordeal and Hope*, p. 309
12. ed. Bland, *Interviews*, p. 434
13. Pogue, *Ordeal and Hope*, p. 315; Butler, *Grand Strategy*, III Part II, Appendix III
14. CAB 79/56 COS Cmtee No. 112 (42), 9/4/1942
15. Pogue, *Ordeal and Hope*, p. 317; Steele, *First Offensive*, pp. 117–18
16. Fenby, *Alliance*, p. 115
17. eds Danchev and Todman, *Diaries*, p. 246
18. *Ibid.*
19. Ismay, *Memoirs*, p. 251
20. LH 15/15/2
21. *Ibid.*
22. Wedemeyer, *Wedemeyer Reports!*, pp. 105–6
23. *Ibid.*
24. LH 15/15/1
25. *Ibid.*
26. Mosley, *Marshall*, p. 202
27. eds Danchev and Todman, *Diaries*, p. 247
28. Pogue, *Ordeal and Hope*, p. 313
29. Sherwood, *Roosevelt and Hopkins*, pp. 530–31
30. Marshall Papers Verifax 631–660
31. MAR Pentagon Papers Box 80/32
32. ed. Kimball, *Correspondence*, I, p. 449
33. eds Danchev and Todman, *Diaries*, p. 247

34. CAB 79/56 COS Cmtee No. 118, 14/4/1942
35. Gilbert, *Winston S. Churchill*, VII, p. 89
36. Sherwood, *Roosevelt and Hopkins*, p. 533
37. CAB 69/4/59
38. *Ibid.*, pp. 61–2
39. CAB 69/4, Defence Committee no. 10 (1942), 14/4/1942
40. Gilbert, *Winston S. Churchill*, VII, p. 89
41. Churchill, *Hinge of Fate*, pp. 289–90
42. Pogue, *Ordeal and Hope*, pp. 319–20
43. LH 15/15/1
44. eds Danchev and Todman, *Diaries*, p. 248
45. Howard, *Mediterranean Strategy*, p. 29
46. eds Danchev and Todman, *Diaries*, p. 248
47. Morgan, *FDR*, p. 638
48. eds Danchev and Todman, *Diaries*, p. 249
49. Matloff and Snell, *Strategic Planning*, p. 191
50. eds Danchev and Todman, *Diaries*, p. 249
51. Morison, *American Contributions*, p. 19
52. eds Danchev and Todman, *Diaries*, *passim*; KENN 4/2/4
53. Mosley, *Marshall*, p. 202
54. ALAB 12/8/8; ed. Bland, *Interviews*, pp. 429, 436, 608
55. eds Danchev and Todman, *Diaries*, p. 250
56. ed. Kimball, *Correspondence*, I, p. 523
57. *Ibid.*, pp. 458–9
58. Kimball and Rose, 'Churchill and D-Day: Another View', p. 31
59. Mosley, *Marshall*, pp. 204–5
60. AVON 20/1/22
61. Dimbleby and Reynolds, *An Ocean Apart*, pp. 141–2
62. MHI Wedemeyer 1972 Interview section 4, p. 31
63. MHI Pogue Interview, 17/12/46, p. 4

CHAPTER 7: THE COMMANDERS AT ARGONAUT: APRIL–JUNE 1942

1. Historian Ronald Lewin reviewing Fraser, *Alanbrooke* in *The Times*, 13/5/1982
2. KENN 4/2/4
3. *Ibid.*
4. Gilbert, 'Churchill and D-Day', p. 24
5. Gilbert, *Winston S. Churchill*, VII, p. 86
6. ed. Eade, *Secret Session Speeches*, p. 74
7. Halifax Diary, 13/6/1942
8. FDR President's Secretary's File Box 83
9. *Ibid.*
10. KENN 4/2/4
11. ed. Kimball, *Correspondence*, I, p. 497
12. ed. Ferrell, *Eisenhower Diaries*, pp. 59–60
13. KENN 4/2/4
14. Sherwood, *White House Papers*, II, p. 567
15. *Ibid.*
16. FRUS, 1942, III, p. 594
17. LH 15/15/1
18. ed. Kimball, *Correspondence*, I, p. 503

19. Brinkley, *Washington Goes to War*, p. 75
20. BRGS 1/2. According to Martin Middlebrook and Chris Everitt, *The Bomber Command War Diaries: An Operational Reference Book, 1939–1945*, 1985, this raid on Cologne involved 1,103 bombers, of which 43 were lost
21. BRGS 2/12, 1/6/1942; KENN 4/2/4
22. AVON 20/1/22
23. BRGS 2/12, 8/6/1942
24. Black, *Roosevelt*, p. 745; Ziegler, *Mountbatten*, pp. 183–5
25. Wedemeyer, *Wedemeyer Reports!*, pp. 132–6
26. Ziegler, *Mountbatten*, pp. 183–5
27. KENN 4/2/4, 10/6/1942
28. CAB 195/1 WM (42) 74th
29. *Ibid.*
30. KENN 4/2/4
31. AVON 19/1/11
32. Stimson and Bundy, *Active Service*, p. 216
33. MAR Pentagon Papers Box 80/33
34. Sandys, *Chasing Churchill*, p. 150
35. Halifax Diary, 18/6/1942
36. Stimson and Bundy, *Active Service*, pp. 217–19
37. *Ibid.*
38. MHI Wedemeyer 1972 Interview section 4, p. 7
39. CAB CCS 27th Meeting, 19/6/1942
40. ed. Danchev, *Anglo-American Alliance*, p. 158
41. JACB 1/14/B
42. eds Danchev and Todman, *Diaries*, p. 267
43. JACB 1/14/C
44. CAB CCS 28th Meeting, 23/6/1942 section 2, American–British Conversations, p. 334
45. MHI Wedemeyer 1972 Interview section 4, p. 15
46. JACB 1/14/C
47. eds Freedman and others, *War, Strategy*, p. 180

CHAPTER 8: THE MASTERS AT ARGONAUT: JUNE 1942

1. Moran, *Struggle for Survival*, p. 33
2. ed. Kimball, *Correspondence*, I, p. 515
3. *Ibid.*; JACB 1/14/E
4. MHI Wedemeyer 1973 Interview section 5, p. 6
5. Tully, *FDR*, p. 304
6. *Ibid.*
7. Colville, *Fringes*, p. 382
8. CAB 65/30, pp. 83–90
9. Ismay, *Memoirs*, p. 255
10. Churchill, *Hinge of Fate*, pp. 342–3
11. ALAB 9/3/8
12. KENN 4/2/4
13. JACB 1/14/F
14. Churchill, *Hinge of Fate*, pp. 344–5
15. CHUR 4/277 War Cabinet Paper No. 278 (42)
16. Gilbert, *Winston S. Churchill*, VII, p. 129; Churchill, *Hinge of Fate*, pp. 344–5
17. Fraser, *Alanbrooke*, p. 232

18. Holmes, *World at War*, pp. 362–3
19. Attlee, *As It Happened*, p. 123
20. ed. Bland, *Interviews*, p. 622
21. Morgan, *FDR*, p. 635
22. Stimson and Bundy, *Active Service*, p. 220
23. Butler, *Grand Strategy*, III Part II, p. 627
24. eds Danchev and Todman, *Diaries*, p. 270
25. Pogue, *Ordeal and Hope*, p. 334; Stimson and Bundy, *Active Service*, 22/6/1942
26. MAR Pentagon Papers Box 80/33; Parrish, *Roosevelt and Marshall*, p. 287; Pogue, *Ordeal and Hope*, p. 334; ed. Bland, *Papers*, III, p. 249
27. MAR Pentagon Papers Box 80/33
28. Moran, *Struggle for Survival*, p. 39
29. JACB 1/14/H and I; Pogue, *Ordeal and Hope*, p. 333
30. Pogue, *Ordeal and Hope*, p. 333
31. Eichelberger, *Jungle Road to Tokyo*, pp. 22–4
32. eds Danchev and Todman, *Diaries*, p. 271
33. Ismay, *Memoirs*, p. 257
34. Bryant, *Turn of the Tide*, pp. 331–2
35. eds Danchev and Todman, *Diaries*, p. 271
36. Stimson and Bundy, *Active Service*, p. 216
37. LH 15/15/1, 31/4/1948 and 14/4/1948
38. MHI Handy Interview 1974 section 4, p. 3, and Hull Interview 1973 section, 4, p. 42
39. Churchill, *Hinge of Fate*, p. 309
40. ALAB 9/3/8
41. Wedemeyer, *Wedemeyer Reports!*, p. 134
42. ALAB 11/73
43. MHI Handy Interview 1974 section 4, pp. 5–6
44. Mosley, *Marshall*, p. 204
45. Ismay, *Memoirs*, p. 250
46. Churchill, *Closing the Ring*, p. 514
47. MHI Hull Interview 1973 section 4, p. 53
48. ed. Bland, *Interviews*, p. 588
49. *Evening Standard*, 28/8/1982, p. 15; *Scotsman*, 21/8/1982
50. eds Danchev and Todman, *Diaries*, pp. 272–3; KENN Papers 4/2/4
51. eds Danchev and Todman, *Diaries*, p. 273
52. ALAB 11/9
53. Wedemeyer, *Wedemeyer Reports!*, p. 141
54. KENN 4/2/4

CHAPTER 9: TORCH REIGNITED: JULY 1942

1. KENN 4/2/4, p. 180B
2. ed. Bland, *Interviews*, p. 580
3. Black, *Roosevelt*, p. 798
4. Keegan, *Second World War*, p. 312
5. ALAB 11/9
6. Eisenhower, *Crusade in Europe*, p. 79
7. Fraser, *Alanbrooke*, p. 240
8. ed. Bland, *Interviews*, 5/10/1956, p. 590
9. Overy, *Why the Allies Won*, pp. 44–5
10. Stoler, *Allies and Adversaries*, p. 116; MHI Hull Interview 1974 section 5, p. 15
11. Wedemeyer, *Wedemeyer Reports!*, p. 164; Fenby, *Alliance*, p. 129
12. Gilbert, *Winston S. Churchill*, VII, p. 137

13. Hansard, 2/7/1942, col. 589
14. AVON 20/1/22
15. CAB 195/1 WM (42) 86th
16. BRGS 2/12, 6/7/1942
17. ALAB 6/2/6/7
18. CAB 65/31, p. 9
19. ed. Dilks, *Cadogan*, p. 461
20. KENN 4/2/4
21. ed. Kimball, *Correspondence*, I, p. 518
22. CAB 65/31, p. 14
23. Danchev, 'Very Special Relationship', p. 6
24. Stimson and Bundy, *Active Service*, p. 220
25. ed. Harvey, *War Diaries*, p. 139
26. Pogue, *Ordeal and Hope*, p. 340
27. ed. Bland, *Interviews*, p. 593; MAR Microfilm 322
28. Stoler, *Politics of the Second Front*, p. 55
29. Mosley, *Marshall*, p. 205
30. ed. Kimball, *Correspondence*, I, p. 529
31. Pogue, *Ordeal and Hope*, pp. 340, 477 n. 1
32. CHAR 20/78/5-6
33. *Ibid.*
34. Robertson, *Soldiers and Statesmen*, I, 96
35. *Ibid.*
36. Ismay, *Memoirs*, p. 252
37. MHI Handy Interview 1974 section 4, p. 5
38. KENN 4/2/4, p. 181A
39. Pogue, *Ordeal and Hope*, p. 342; Stimson Diary, 12/7/1942
40. Stimson and Bundy, *Active Service*, p. 220
41. *Ibid.*, p. 211
42. Parrish, *Roosevelt and Marshall*, p. 292
43. Reynolds, *In Command*, pp. 318, 588 n. 13; CHUR 4/277A/7; Churchill, *Hinge of Fate*, pp. 396-7
44. Reynolds, *In Command*, passim
45. Wedemeyer, *Wedemeyer Reports!*, p. 167
46. Gilbert, *Winston S. Churchill*, VII, p. 149
47. Sherwood, *White House Papers*, II, pp. 604-7
48. Butcher, *Three Years*, p. 20
49. eds Danchev and Todman, *Diaries*, pp. 281-2
50. CAB 79/56 COS Cmtee No. 75 (Operations), 18/7/1942
51. LH 15/15/1
52. CAB 65/31, p. 33
53. Moran, *Struggle for Survival*, p. 44
54. eds Danchev and Todman, *Diaries*, p. 284
55. Gilbert, *Winston S. Churchill*, VII, p. 151
56. KENN 4/2/4, p. 186
57. CAB 65/31, p. 27
58. *Ibid.*, p. 28
59. *Ibid.*, p. 30
60. eds Danchev and Todman, *Diaries*, p. 284
61. Butcher, *Three Years*, p. 24
62. Fenby, *Alliance*, p. 134
63. KENN 4/2/4, p. 186
64. *Ibid.*
65. *Ibid.*

66. *Ibid.*, p. 187
67. eds Danchev and Todman, *Diaries*, p. 284
68. *Ibid.*
69. ed. Bland, *Interviews*, p. 581
70. LH 15/15/1
71. KENN 4/2/4, p. 194
72. Butler, *Grand Strategy*, III Part II, Appendix V, pp. 635, 684–5
73. Howard, *Grand Strategy*, IV, pp. 191–2
74. eds Danchev and Todman, *Diaries*, p. 285
75. Howard, *Grand Strategy*, IV, p. xxi
76. AVON 20/1/22
77. Haffner, *Churchill*, p. 128
78. MHI Handy Interview 1974 section 4, p. 9
79. Fraser, *Alanbrooke*, p. 235
80. eds Danchev and Todman, *Diaries*, p. 406

CHAPTER 10: THE MOST PERILOUS MOMENT OF THE WAR: JULY–NOVEMBER 1942

1. ALAB 11/9
2. Howard, *Mediterranean Strategy*, p. 33
3. MHI Hull Interview 1973 section 4, p. 42
4. CHAR 20/78/124
5. Howard, *Mediterranean Strategy*, pp. 32–3
6. CHAR 20/78/79–81, 102
7. *Ibid.*, pp. 100–101; ed. Kimball, *Correspondence*, I, p. 551
8. FDR Hopkins Papers Container 190
9. KENN 4/2/4, p. 193
10. AVON 20/1/22
11. Tedder, *With Prejudice*, p. 321
12. KENN 4/2/4, p. 215
13. eds Danchev and Todman, *Diaries*, p. 290
14. *Ibid.*, pp. 292–3
15. *Ibid.*, p. 293
16. KENN 4/2/5, 20/11/1943
17. Churchill, *Hinge of Fate*, p. 413
18. JACB 1/17/10
19. BRGS 1/3 and 2/13; CAB 65/31, p. 80
20. Stimson and Bundy, *Active Service*, p. 221
21. ed. Kimball, *Correspondence*, I, pp. 510, 562
22. JACB 1/17/Minutes; JACB 1/17/45
23. KENN 4/2/4, p. 302
24. JACB 1/17/10
25. ed. Kemper, *Resolution*, p. 92
26. Van Der Vat, *D-Day*, p. 21
27. Mosley, *Backs to the Wall*, p. 288
28. eds Danchev and Todman, *Diaries*, p. 317
29. KENN 4/2/4, pp. 206–7
30. *Ibid.*
31. Dupuy, *Genius for War*, p. 254
32. KENN 4/2/4, p. 215, 24/8/1942
33. CAB 195/1 WM (42) 118th

34. ed. Kimball, *Correspondence*, I, p. 578
35. KENN 4/2/4, p. 224
36. ed. Ferrell, *Eisenhower Diaries*, p. 78
37. Payne, *Franco and Hitler*, p. 91
38. Clark, *Calculated Risk*, p. 57
39. Eisenhower, *At Ease*, p. 365
40. Clark, *Calculated Risk*, p. 58
41. MAR Pentagon Papers Box 71/20
42. ed. Kimball, *Correspondence*, I, pp. 581–2
43. CAB 195/1 WM (42) 119th
44. ed. Kimball, *Correspondence*, I, p. 592
45. ed. Ferrell, *Eisenhower Diaries*, p. 78
46. ed. Kimball, *Correspondence*, I, p. 592
47. CAB 65/31, p. 126
48. *Ibid.*, p. 128
49. *Ibid.*, p. 135
50. Stoler, *Allies and Adversaries, passim*; Stoler, *Politics of the Second Front*, pp. 75, 114, 167
51. ed. Kimball, *Correspondence*, I, pp. 603–5
52. KENN 4/2/4, p. 239
53. Overy, *Why the Allies Won*, p. 44
54. ed. Kimball, *Correspondence*, I, p. 611
55. *Ibid.*, p. 612
56. Moran, *Struggle for Survival*, p. 71
57. AVON 20/1/22
58. ed. Harvey, *War Diaries*, p. 165
59. KENN 4/2/4, 9/10/1942
60. MAR Pentagon Papers Box 80/36
61. *Ibid.*
62. BRGS 2/13, 26/10/1942
63. *Ibid.*
64. KENN 4/2/4, p. 255
65. ed. Bland, *Interviews*, p. 589
66. Stimson and Bundy, *Active Service*, p. 222
67. MAR Pentagon Papers Box 80/37; FDR Hopkins Papers Container 180
68. FDR Hopkins Papers Container 180
69. CAB 195/1 WM (42) 151st
70. BRGS 2/13, 9/11/1942
71. eds Danchev and Todman, *Diaries*, p. 340

CHAPTER 11: THE MEDITERRANEAN GARDEN PATH: NOVEMBER 1942–JANUARY 1943

1. KENN 4/2/4, p. 280
2. *Ibid.*
3. ed. Kemper, *Resolution*, p. 87
4. ed. Hart-Davis, *King's Counsellor*, p. 75
5. ed. Blum, *The Price of Vision*, p. 132
6. Interview with Lt-Gen. Sir Ian Jacob, 28/10/1988
7. MAR Pentagon Papers Box 80/37
8. Black, *Roosevelt*, p. 798
9. Gilbert, 'Churchill and D-Day', p. 26

10. AVON 20/1/22
11. ed. Taylor, *Churchill*, p. 193
12. KENN 4/2/4, p. 280
13. BRGS 2/13, 16/11/1942
14. KENN 4/2/4, p. 280
15. ed. Kemper, *Resolution*, p. 87
16. CUNN Add MSS 52570/123
17. BRGS 2/13, 30/11/1942
18. KENN 4/2/4, p. 302
19. *Ibid.*, p. 307
20. *Ibid.*, pp. 307–8
21. Danchev, 'Waltzing', p. 221
22. eds Danchev and Todman, *Diaries*, pp. 347–8
23. MAR Pentagon Papers Box 58/30
24. KENN 4/2/5, 24/5/1943
25. CAB 195/2 WM (42) 168th
26. CHAR 20/85/13
27. ed. Harvey, *War Diaries*, p. 200
28. Gilbert, *Winston S. Churchill*, VII, pp. 281–2
29. KENN 4/2/4, pp. 311–12
30. Brodhurst, *Churchill's Anchor*, p. 261
31. CAB 195/2 WM (42) 170th
32. FDR Personal File 8277
33. CAB 195/2 WM (42) 171st; CAB 195/2 WM (43) 1st
34. CAB 195/2 WM (42) 171st
35. AVON 20/1/22
36. *Ibid.*
37. *Ibid.*
38. Morison, *American Contributions*, pp. 14–15
39. CAB 195/2 WM (43) 4th
40. Tedder, *With Prejudice*, Preface
41. eds Danchev and Todman, *Diaries*, p. 357
42. Nicolson, *Alex*, p. 171
43. LEAH Leahy Diary reel 4, p. 71
44. eds Danchev and Todman, *Diaries*, p. 358

CHAPTER 12: THE CASABLANCA CONFERENCE: JANUARY 1943

1. Berlin, *Mr Churchill*, pp. 37–8
2. Astley, *Inner Circle*, p. 87
3. JACB 1/21/32
4. JACB 1/21/33–4
5. KENN 4/2/4, p. 291, 27/11/1942
6. JACB 1/21/35
7. JACB 1/21/52
8. JACB 1/21/37
9. ed. Huston, *American Airpower*, I, p. 463
10. eds Danchev and Todman, *Diaries*, p. 359
11. KENN 4/2/5, p. 12
12. *Ibid.*
13. Brian Bond in eds Freedman and others, *War, Strategy*, p. 182

14. Wedemeyer, *Wedemeyer Reports!*, p. 178
15. *Ibid.*, p. 181
16. ed. Sherwood, *White House Papers*, II, p. 672
17. MHI Wedemeyer 1973 Interview section 5, p. 28
18. Black, *Roosevelt*, p. 93
19. Roosevelt, *As He Saw It*, p. 83
20. Macmillan, *War Diaries*, p. 8
21. LH 15/15/1
22. eds Danchev and Todman, *Diaries*, p. 360
23. *Ibid.*, p. 361
24. Moran, *Struggle for Survival*, pp. 720–21
25. FRUS, *Washington, 1941–1942 and Casablanca, 1943*, p. 594
26. ed. Huston, *American Airpower*, I, p. 463
27. Howard, *Mediterranean Strategy*, p. 35
28. *Ibid.*, p. 37
29. eds Danchev and Todman, *Diaries*, p. 361
30. ALAB 6/1/1, p. 236
31. *Ibid.*, p. 239; Hayes, *History of the Joint Chiefs*, pp. 286–8
32. Wedemeyer, *Wedemeyer Reports!*, p. 167
33. KENN 4/2/5, p. 25
34. eds Danchev and Todman, *Diaries*, p. 362
35. Ibid.; Danchev, 'Very Special Relationship', p. 6
36. Mosley, *Marshall*, pp. xi, 227
37. Black, *Roosevelt*, p. 799
38. JACB 1/2/44
39. Gilbert, 'Churchill and D-Day', p. 26
40. JACB 1/21/50
41. MHI Handy Interview 1974 section 3, p. 42
42. MHI Hull Interview 1973 section 4, p. 12
43. Macmillan, *War Diaries*, p. 8
44. JACB 1/21/25
45. MHI Wedemeyer 1973 Interview section 5, pp. 27–8
46. MHI Wedemeyer 1972 Interview section 4, p. 3
47. ed. Bland, *Interviews*, p. 608; MAR GCM Library Xerox 2256
48. KENN 4/2/5, p. 38
49. eds Danchev and Todman, *Diaries*, p. 364
50. JACB 1/21/44
51. eds Danchev and Todman, *Diaries*, p. 366
52. JACB 1/21/44
53. KENN 4/2/5, p. 44
54. *Ibid.*
55. eds Danchev and Todman, *Diaries*, p. 367
56. ALAB 6/1/1, p. 236
57. Arnold, *Global Mission*, p. 204
58. Black, *Roosevelt*, pp. 805–6
59. PEAK Correspondence, 26/12/1944
60. Wedemeyer, *Wedemeyer Reports!*, p. 179
61. *Ibid.*, p. 181; MHI Wedemeyer 1972 Interview section 4, p. 16
62. BLAC
63. Jackson and Bramall, *The Chiefs*, p. 239
64. LH 15/15/1

CHAPTER 13: THE HARD UNDERBELLY OF EUROPE: JANUARY–JUNE 1943

1. *Sunday Telegraph* review of Fraser, *Alanbrooke*, 2/5/1982
2. JACB 1/21/42
3. ALAB 9/3/8
4. Coombs, *Life through his Paintings*, p. 163
5. BBC interview in 1959, Videotape no. VC338475
6. Sandys, *Chasing Churchill*, p. 160; BRGS 1/3
7. Hassett, *Off the Record*, pp. 151–2
8. Beevor, *Stalingrad*, pp. 387ff.
9. ed. Kimball, *Correspondence*, II, 133
10. KENN 4/2/5, p. 38
11. Moran, *Struggle for Survival*, p. 88
12. CAB 195/2 WM (43) 34th
13. eds Danchev and Todman, *Diaries*, p. 386
14. CAB 195/2 WM (43) 37th
15. *Ibid.*
16. CAB 195/2 WM (43) 40th
17. CAB 195/2 WM (43) 48th, 6/4/1943
18. BRGS 2/16, 8/3/1943
19. CAB 195/2 WM (43) 44th
20. BRGS 2/16, 29/3/1943
21. Dupuy, *Genius for War*, pp. 253–5
22. CAB 65/38, p. 6
23. KENN 4/2/3
24. CAB 195/2 WM (43) 53rd
25. Alex Danchev in eds Reynolds and Kimball, *Allies at War*, p. 18
26. eds Danchev and Todman, *Diaries*, p. 394
27. ed. Hart-Davis, *King's Counsellor*, p. 129
28. Howard, *Mediterranean Strategy*, p. 39
29. LEAH Reel 4, pp. 109–10
30. Moran, *Struggle for Survival*, p. 96
31. eds Danchev and Todman, *Diaries*, p. 402
32. Moran, *Struggle for Survival*, p. 96
33. LEAH Reel 4, p. 110; Leahy, *I Was There*, p. 191
34. eds Danchev and Todman, *Diaries*, pp. 403–4
35. Tuchman, *Stilwell*, p. 371
36. Astley, *Inner Circle*, p. 95
37. MAR Pentagon Papers Box 116
38. Black, *Roosevelt*, p. 832
39. MAR D736/1B3
40. MAR Pentagon Papers Box 116
41. ALAB 11/73, 23/6/1955
42. eds Danchev and Todman, *Diaries*, p. 405
43. LEAH Reel 4, p. 111
44. eds Danchev and Todman, *Diaries*, p. 406
45. Jackson and Bramall, *The Chiefs*, p. 243
46. ALAB 9/3/8
47. Howard, *Grand Strategy*, IV, p. 432
48. Brian Bond in eds Freedman and others, *War, Strategy*, p. 183
49. Leahy, *I Was There*, p. 193

50. FRUS, *Washington and Quebec, 1943*, pp. 34–48
51. Leahy, *I Was There*, p. 194
52. LEAH Reel 4, p. 113
53. eds Danchev and Todman, *Diaries*, p. 410
54. Stimson and Bundy, *Active Service*, p. 223
55. Blum, *Years of War*, p. 63
56. ed. Bland, *Papers*, III, p. 708
57. ed. Bland, *Interviews*, 20/11/1956, pp. 552–3
58. Churchill, *Hinge of Fate*, pp. 812–13
59. Ismay, *Memoirs*, p. 253; ed. Soames, *Speaking for Themselves*, p. 483
60. Moran, *Struggle for Survival*, p. 272
61. Roosevelt, *As He Saw It*, pp. 144, 184
62. Eisenhower, *At Ease*, p. 264
63. ed. Bland, *Papers*, IV, pp. 3–4
64. Morgan, *FDR*, p. 632
65. Moran, *Struggle for Survival*, p. 79
66. *Ibid.*, p. 103
67. MHI Typescript for John R. Deane's *Strange Alliance*, 1946
68. CAB 195/2, WM (43) 81st
69. *Ibid.*
70. BRGS 2/17, 7/6/1943

CHAPTER 14: THE OVERLORDSHIP OF OVERLORD: JUNE–AUGUST 1943

1. MHI Charles Donnelly Autobiography, p. 640
2. eds Danchev and Todman, *Diaries*, p. 420
3. Leahy, *I Was There*, p. 195
4. Roosevelt, *Rendezvous with Destiny*, p. 338
5. ed. Kimball, *Correspondence*, II, pp. 303–4
6. KENN 4/2/5
7. ed. Dilks, *Cadogan*, p. 545
8. CAB 195/2 WM (43) 94th, 7/7/1943
9. eds Danchev and Todman, *Diaries*, p. 427
10. BRGS 2/17, 12/7/43
11. Howard, *Mediterranean Strategy*, p. 40
12. *Ibid.*
13. Stimson and Bundy, *Active Service*, p. 222
14. *Ibid.*, p. 226
15. *Ibid.*
16. Howard, *Mediterranean Strategy*, p. 45
17. KENN 4/2/5
18. Nicolson, *Alex*, p. 206
19. Butcher, *Three Years*, p. 330
20. Nicolson, *Alex*, p. 207
21. ALAB 9/3/8; KENN 4/2/5
22. AVON 20/1/23
23. Brodhurst, *Churchill's Anchor*, p. 281
24. Astley, *Inner Circle*, pp. 103–4
25. MHI Pogue Interview, 26/2/1947, p. 1
26. Astley, *Inner Circle*, p. 109
27. Jackson and Bramall, *The Chiefs*, p. 247

28. Cozzens, *Guard of Honor*, p. 395
29. Stimson and Bundy, *Active Service*, pp. 228–9
30. Morgan, *Overture to Overlord*, pp. 159, 181, 188
31. Howard, *Grand Strategy*, IV, pp. 563, 569
32. eds Danchev and Todman, *Diaries*, pp. 441–2
33. Moran, *Struggle for Survival*, p. 720
34. Churchill, *Closing the Ring*, p. 76
35. CHUR 4/319A/7
36. *Ibid.*
37. Stimson and Bundy, *Active Service*, pp. 228–9
38. ALAB 11/64
39. KENN 4/2/5, 13/9/1943
40. MHI Pogue Interview, 17/12/46, pp. 1–2

CHAPTER 15: FROM THE ST LAWRENCE TO THE PYRAMIDS: AUGUST–NOVEMBER 1943

1. eds Danchev and Todman, *Diaries*, p. 447
2. *Ibid.*, p. 442
3. ed. Bland, *Interviews*, 5/10/1956, p. 590
4. eds Danchev and Todman, *Diaries*, p. 443
5. Danchev, *On Specialness*, p. 27
6. MHI Pogue Interview, 26/2/1947, p. 4
7. Liddell Hart in ed. Taylor, *Churchill*, p. 194
8. eds Danchev and Todman, *Diaries*, p. 445
9. *Ibid.*, pp. 444–5
10. *Ibid.*, pp. 445–6; Leahy, *I Was There*, p. 213; MHI Donnelly Autobiography, pp. 633–4
11. ed. Bland, *Papers*, IV, pp. 91–4; FRUS, *Washington and Quebec, 1943*, pp. 1024–5
12. eds Danchev and Todman, *Diaries*, p. 709; Rose, *Churchill*, p. 323
13. MHI Wedemeyer 1973 Interview section 5, p. 35
14. eds Danchev and Todman, *Diaries*, pp. 447–8; Brian Bond in eds Freedman and others, *War, Strategy*, p. 184
15. ASTL Major Buckley Recollections
16. eds Danchev and Todman, *Diaries*, p. 449
17. KENN 4/2/5
18. Nicolson in *Financial Times*, 24/4/1982
19. KENN 4/2/5
20. *Ibid.*
21. See Howard, *Captain Professor* for a first-hand account of the Italian campaign
22. Howard, *Mediterranean Strategy*, p. 42
23. Warlimont, *Inside Hitler's Headquarters*, p. 387
24. AVON 20/1/23
25. ed. Kimball, *Correspondence*, II, p. 506
26. Brian Bond in eds Freedman and others, *War, Strategy*, p. 185
27. ed. Kimball, *Correspondence*, II, pp. 498–9; Ehrman, *Grand Strategy*, V, p. 95
28. Morison, *American Contributions*, p. 32; ed. Bland, *Interviews*, p. 622
29. ed. Bland, *Interviews*, p. 321
30. Reynolds, *In Command*, pp. 357–8
31. KENN 4/2/5
32. ed. Bland, *Papers*, IV, pp. 128–9
33. Butcher, *Three Years*, p. 360
34. BRGS 2/18

35. MAR Pentagon Papers Box 81/2
36. *Ibid.*
37. AVON 20/1/23
38. AVON 20/1/24, 18/11/1944
39. KENN 4/2/5
40. eds Danchev and Todman, *Diaries*, p. 386
41. ed. Hart-Davis, *King's Counsellor*, p. 176
42. Reynolds, *In Command*, pp. 379–80
43. DUPO 2/3; Pawle, *The War and Colonel Warden*, p. 253
44. Colville, *Footprints*, p. 188
45. Jackson and Bramall, *The Chiefs*, p. 250
46. Winton, *Cunningham*, p. 211
47. Frank McLynn in *Literary Review*, September 1998, p. 17
48. Winton, *Cunningham*, p. 337
49. ed. Kimball, *Correspondence*, II, pp. 555–7
50. CHAR 20/122/22
51. eds Danchev and Todman, *Diaries*, p. 462
52. *Ibid.*, pp. 462–3
53. Brian Bond in eds Freedman and others, *War, Strategy*, p. 186
54. Stoler, *Politics of the Second Front*, p. 131; ed. Dilks, *Cadogan*, p. 571
55. Howard, *Mediterranean Strategy*, pp. 35, 56
56. MAR Pentagon Papers Box 81/4
57. Moran, *Struggle for Survival*, p. 562
58. KENN 4/2/5
59. *Ibid.*
60. eds Danchev and Todman, *Diary*, p. 465
61. Howard, *Mediterranean Strategy*, p. 47
62. ed. Bland, *Papers*, IV, pp. 127, 176–8
63. ed. Hart-Davis, *King's Counsellor*, p. 176
64. CHAR 20/124/5

CHAPTER 16: *EUREKA!* AT TEHERAN: NOVEMBER–DECEMBER 1943

1. Arnold, *Global Mission*, p. 219
2. Ehrman, *Grand Strategy*, V, p. 111; Brian Bond in eds Freedman and others, *War, Strategy*, p. 187
3. Leighton, 'Overlord Revisited'
4. MHI Sidney Matthews Box 3, 15/11/1943
5. KENN 4/2/5
6. eds Danchev and Todman, *Diaries*, pp. 472–3
7. MHI Sidney Matthews Box 3, 19/11/1943
8. MAR Pentagon Papers Box 81/5
9. MHI Sidney Matthews Box 3, 19/11/1943
10. *Ibid.*
11. MHI Charles Donnelly Autobiography, p. 672
12. Douglas, *Years of Command*, p. 229
13. Cunningham, *Odyssey*, p. 586
14. ASTL Major Buckley's Recollections
15. MHI Charles Donnelly World War II Conferences Box
16. eds Danchev and Todman, *Diaries*, pp. 477, 479; ed. Bland, *Papers*, IV, p. 192
17. eds Danchev and Todman, *Diaries*, p. 478

18. Arnold, *Global Mission*, p. 220
19. Ronald Lewin in *The Times*, 13/5/1982
20. ed. White, *Stilwell Papers*, pp. 230–32
21. Douglas, *Years of Command*, p. 188
22. Pogue, *Ordeal and Hope*, pp. 330–31
23. Black, *Roosevelt*, p. 799
24. ed. White, *Stilwell Papers*, pp. 230–32
25. eds Danchev and Todman, *Diaries*, pp. 478–80
26. Tuchman, *Stilwell*, pp. 403–4
27. MHI Wedemeyer 1973 Interview section 6, p. 7
28. ed. Bland, *Papers*, IV, p. 192
29. MHI Charles Donnelly Autobiography, p. 669
30. *Ibid.*, p. 670
31. Arnold, *Global Mission*, p. 221
32. ed. Dilks, *Cadogan*, p. 578
33. Stoler, *Allies and Adversaries*, chapters 6 and 7
34. ed. Dilks, *Cadogan*, p. 539
35. Moran, *Struggle for Survival*, p. 131
36. *Ibid.*
37. *Ibid.*, p. 132
38. Ingersoll, *Top Secret*, p. 48
39. eds Danchev and Todman, *War Diaries*, p. 481
40. Cunningham, *Odyssey*, p. 587
41. FRUS, *Cairo and Tehran, 1943*, pp. 150–51, 363–5
42. eds Danchev and Todman, *Diaries*, p. 476,
43. ASTL Major Buckley Recollections
44. eds Danchev and Todman, *Diaries*, p. 483
45. FRUS, *Cairo and Tehran, 1943*, pp. 499–501, 505–7
46. Arnold, *Global Mission*, p. 224
47. ASTL Major Buckley Recollections
48. Perkins, *The Roosevelt I Knew*, pp. 70–71
49. Lunghi, 'Troubled Triumvirate', pp. 17–18
50. Churchill, *Closing the Ring*, pp. 377, 378
51. Leahy, *I Was There*, p. 191
52. Sherwood, *White House Papers*, II, p. 778
53. *Ibid.*
54. Letter to author from Hugh Lunghi, 2/1/2008; Brian Bond in eds Freedman and others, *War, Strategy*, p. 187
55. MHI Wedemeyer 1973 Interview section 5, p. 35
56. Deane, *Strange Alliance*, p. 42
57. ed. Bland, *Papers*, IV, p. 194
58. Harriman and Abel, *Special Envoy*, p. 271
59. *Ibid.*
60. ALAB 3/8
61. Bohlen, *Witness to History*, pp. 149–50
62. Ehrman, *Grand Strategy*, V, p. 189
63. Howard, *Mediterranean Strategy*, p. 57 n. 3
64. Holmes, *Footsteps*, p. 263
65. ed. Bland, *Interviews*, p. 344
66. Stimson and Bundy, *Active Service*, pp. 222–3
67. *Ibid.*
68. MHI Handy Interview 1974 section 4, p. 30
69. Roosevelt, *As He Saw It*, p. 210

CHAPTER 17: ANZIO, ANVIL AND CULVERIN: DECEMBER 1943–MAY 1944

1. Churchill, *Finest Hour*, p. 20
2. KENN 4/2/5, 9/12/1943
3. ed. Kimball, *Correspondence*, II, pp. 622–3
4. Hastings, *Overlord*, p. 32
5. eds Heiber and Glantz, *Hitler and his Generals*, p. 316
6. CHAR 20/130/39
7. CHAR 20/179/37
8. MAR GCM Library Xerox 2256, note 14; Howard, *Mediterranean Strategy*, p. 58
9. Clark, *Calculated Risk*, p. 348
10. Howard, *Mediterranean Strategy*, p. 51
11. ed. Bland, *Interviews*, p. 339
12. eds Danchev and Todman, *Diaries*, p. 459
13. Warlimont, *Inside Hitler's Headquarters*, p. 411
14. eds Reynolds and Kimball, *Allies at War*, p. 13
15. LH 15/15/2
16. MAR Pentagon Papers Box 80/31
17. BRGS 2/19
18. Eisenhower, *At Ease*, p. 273
19. ed. Chandler, *Papers*, III, p. 1707
20. ALAB 6/2/3/B
21. ed. Nicolson, *Diaries and Letters*, p. 348
22. BRGS 2/19
23. MAR Pentagon Papers Box 81/9
24. ed. Chandler, *Papers*, III, pp. 1732–6
25. ed. Bland, *Papers*, IV, pp. 313–14
26. BRGS 2/19
27. Ehrman, *Grand Strategy*, V, pp. 230–31
28. Marshall, *Together*, p. 188; Meacham, *Franklin and Winston*, p. 274
29. Howard, *Mediterranean Strategy*, p. 59
30. eds Danchev and Todman, *Diaries*, pp. 529, 530
31. ed. Bland, *Papers*, IV, p. 342
32. Dupuy, *Genius for War*, p. 254
33. Farrell, *Basis and Making*, II, p. 777
34. Matloff, *Strategic Planning*, chapter 14
35. CHAR 20/188A/64–5
36. *Ibid.*, p. 66
37. *Ibid.*, p. 68
38. LH 15/15/2
39. Reynolds, *In Command*, pp. 403–4
40. ALAB 6/3/10
41. Halifax Diary, 20/2/1942
42. MHI Pogue Interview, 17/12/46, p. 4
43. BRGS 2/20
44. ALAB 6/3/9
45. CHAR 20/188B/128; ALAB 6/3/9
46. CHAR 20/188B/132
47. ALAB 6/3/10
48. ALAB 11/9; Smith, *Government*, p. 208
49. ALAB 6/3/10

50. Halifax Diary, 31/3/1944
51. Matloff, *Strategic Planning*, pp. 424–5
52. MHI, Caraway Interview 1971 section 6, p. 59
53. DILL 3/1
54. ed. Bland, *Papers*, IV, p. 404
55. CUNN Add MSS 52577/7
56. ed. Bland, *Papers*, IV, p. 405
57. *Ibid.*
58. Reynolds, *In Command*, p. 393
59. CHAR 20/163/15
60. CHAR 20/163/59
61. eds Danchev and Todman, *Diaries*, p. 551
62. ed. Kemper, *Resolution*, p. 86
63. eds Danchev and Todman, *Diaries*, p. 544
64. CUNN Add MSS 52577/26
65. Hastings, *Overlord*, pp. 33–4
66. BRGS 2/20
67. CUNN Add MSS 52577/8
68. Cray, *General of the Army*, pp. 449–50
69. ed. Bland, *Papers*, IV, p. 468

CHAPTER 18: D-DAY AND DRAGOON: MAY–AUGUST 1944

1. ed. Kemper, *Resolution*, p. 83
2. ed. Kimball, *Correspondence*, III, p. 153
3. Djilas, *Conversations with Stalin*, p. 61
4. CUNN Add MSS 52577/8
5. BRGS 2/21
6. *Ibid.*
7. Butcher, *Three Years*, p. 486
8. Brian Bond in eds Freedman and others, *War, Strategy*, p. 190
9. Ehrman, *Grand Strategy*, V, pp. 279–80
10. eds Reynolds and Kimball, *Allies at War*, p. 18
11. MHI Charles Donnelly Autobiography, p. 689
12. MHI Handy Interview 1974 section 3, pp. 46–7
13. Morison, *American Contributions*, p. 29
14. ed. Bland, *Interviews*, p. 589
15. Bryant, *Triumph in the West*, p. 223
16. CUNN Add MSS 52577/31
17. ed. Bland, *Papers*, IV, p. 477
18. MHI Handy Interview 1974 section 3, p. 47
19. ALAB 9/3/8
20. LH 15/15/1
21. ed. Taylor, *Churchill*, p. 196
22. ed. Hart-Davis, *King's Counsellor*, p. 233
23. eds Danchev and Todman, *Diaries*, p. 560
24. ed. Chandler, *Papers*, III, pp. 1938–9
25. ed. Bond, *Chief of Staff*, II, p. 177
26. CUNN Add MSS 52577/36; ed. Kimball, *Correspondence*, III, p. 199
27. Danchev, 'Very Special Relationship', p. 4
28. Howard, *Mediterranean Strategy*, p. 61

29. CUNN Add MSS 52577/38
30. *Ibid.*
31. ed. Bland, *Papers*, IV, pp. 497–8
32. ed. Kemper, *Resolution*, p. 83
33. Macmillan, *War Diaries*, p. 476
34. ed. Kimball, *Correspondence*, III, pp. 227–9
35. ed. Hart-Davis, *King's Counsellor*, p. 208
36. Holmes, *Footsteps*, p. 263
37. CUNN Add MSS 52577/42
38. BRGS 2/21
39. Gilbert, *Winston S. Churchill*, VII, p. 843
40. eds Danchev and Todman, *Diaries*, p. 566
41. CUNN Add MSS 52577/42
42. AVON 20/1/24
43. *Ibid.*
44. CUNN Add MSS 52577/40
45. AVON 20/1/24
46. ed. Bland, *Papers*, IV, p. 524
47. CUNN Add MSS 52577/51
48. ALAB 6/3/6
49. CUNN Add MSS 52577/55
50. Gilbert, *Winston S. Churchill*, VII, p. 877
51. AVON 20/1/24
52. CUNN Add MSS 52577/57
53. *Ibid.*, p. 58
54. *Ibid.*, p. 94

CHAPTER 19: OCTAGON AND TOLSTOY: AUGUST–DECEMBER 1944

1. ed. Davenport-Hines, *Letters from Oxford*, pp. 33–4
2. CUNN Add MSS 52577/65
3. MAR Pentagon Papers Box 81/21
4. CUNN Add MSS 52577/67
5. *Ibid.*, p. 68
6. Blum, *Years of War*, pp. 362–3
7. eds Danchev and Todman, *Diaries*, p. 590
8. Pickersgill and Forster, *Mackenzie King Record*, II, p. 67
9. *Ibid.*
10. Ehrman, *Grand Strategy*, V, pp. 510–12
11. FRUS, *Quebec, 1944*, p. 303
12. Bryant, *Triumph in the West*, p. 283n
13. MHI Charles Donnelly Autobiography, p. 697
14. MHI Pogue Interview, 12/2/1947, p. 1
15. LH 15/15/1
16. Howard, *Mediterranean Strategy*, p. 66
17. Morison, *American Contributions*, p. 34
18. ALAB 6/1/5, p. 236
19. FRUS, *Quebec, 1944*, pp. 40–41
20. *Ibid.*, pp. 238, 313
21. ALAB 6/1/5, p. 240
22. Cunningham, *Odyssey*, p. 611; ALAB 6/1/5, p. 241; FRUS, *Quebec, 1944*, pp. 312ff.

23. Buell, *Master of Sea Power*, pp. 470–71
24. *Ibid.*, p. 536
25. *Ibid.*, p. 471
26. FRUS, *Quebec, 1944*, pp. 330–35
27. Arnold, *Global Mission*, p. 243
28. Buell, *Master of Sea Power*, p. 471
29. ed. Huston, *American Airpower*, II, p. 212 n. 23
30. FRUS, *Quebec, 1944*, p. 335
31. Cunningham, *Odyssey*, p. 597
32. Kimball, *Forged in War*, p. 276
33. CUNN Add MSS 52577/73
34. BRGS 2/22
35. Deane, *Strange Alliance*, p. 155
36. eds Barnes and Nicholson, *Empire at Bay*, p. 1018
37. CUNN Add MSS 52577/110
38. Danchev, 'Very Special Relationship', p. 1
39. CUNN Add MSS 52577/110, p. 96
40. CHAR 20/174/92
41. Montgomery, *Memoirs*, pp. 534–5
42. PORT Box A File V
43. Eisenhower, *General Ike*, p. 124
44. BRGS 2/22, 29/11/1944
45. CUNN Add MSS 52577/109

CHAPTER 20: AUTUMN MIST: DECEMBER 1944–FEBRUARY 1945

1. CUNN Add MSS 52577/109
2. Hobbs, *Dear General*, p. 183
3. MHI Pogue Interview, 14/2/1947, p. 1
4. PORT Box A File V
5. CUNN Add MSS 52578/4
6. MHI Pogue Interview, 8/5/1947, p. 3
7. CHAR 20/197A/92
8. eds Barnes and Nicholson, *Empire at Bay*, p. 1026
9. FRUS, *Malta and Yalta, 1945*, pp. 22–3
10. BRGS 2/23
11. CUNN Add MSS 52578/5
12. DILL 3/1
13. ed. Bland, *Interviews*, p. 550
14. BRGS 2/23
15. CUNN Add MSS 52578/13; eds Danchev and Todman, *Diaries*, p. 651
16. FRUS, *Malta and Yalta, 1945*, p. 473
17. ed. Soames, *Speaking for Themselves*, pp. 511–12
18. CUNN Add MSS 52578/15
19. MAR GCM Library Microfilm Reel 322 Part I, 25/7/1949
20. eds Danchev and Todman, *Diaries*, p. 653
21. ed. Bland, *Interviews*, p. 400; MAR GCM Library Microfilm Reel 322 Part II, 25/7/1949
22. MAR GCM Library Microfilm Reel 322 Part II, 25/7/1949
23. *Ibid.*
24. Black, *Nixon*, pp. 265, 323

25. Schlesinger, *A Thousand Days*, p. 16
26. FRUS, *Malta and Yalta, 1945*, p. 543
27. Moran, *Struggle for Survival*, p. 348
28. MHI Charles Donnelly Autobiography, p. 713
29. ASTL Joan Bright to Ismay, 10/2/1945
30. MHI Hull Autobiography chapter 18, p. 7
31. Sir Stewart Crawford's Diary: Yalta section

CHAPTER 21: YALTA REQUIEM:
FEBRUARY–MAY 1945

1. ed. Dilks, *Cadogan*, p. 432
2. ed. Bland, *Interviews*, p. 415
3. ed. Kimball, *Correspondence*, III, p. 524
4. ed. Bland, *Interviews*, pp. 400, 406
5. ed. Hart-Davis, *King's Counsellor*, p. 297
6. MHI Charles Donnelly Autobiography, p. 721
7. MHI Hull Autobiography chapter 18, p. 7
8. Sir Stewart Crawford's Diary, Yalta section
9. FRUS, *Malta and Yalta, 1945*, pp. 564–7; Farrell, *Basis and Making*, II, p. 778
10. ed. Bland, *Papers*, V, p. 44
11. CUNN Add MSS 52578/15–16
12. FRUS, *Malta and Yalta, 1945*, pp. 650–53
13. CUNN Add MSS 52578/17; Rose, *Churchill*, p. 313
14. BRGS 1/2
15. Ranfurly, *To War*, p. 328
16. ed. Parrish, *Simon and Schuster Encyclopedia*, p. 520
17. MHI Charles Donnelly Autobiography, p. 721
18. *Ibid.*
19. CUNN Add MSS 52578/18
20. MHI Hull Interview Session 7, p. 4
21. CUNN Add MSS 52578/26
22. Reynolds, *Summits*, p. 102; Cowles, *Churchill*, p. 339
23. BRGS 2/23
24. *Ibid.*
25. CAB 65/51/77–9
26. BRGS 1/2
27. ALAB 6/3/5
28. MHI Pogue Interview, 13/2/1947, p. 2
29. ALAB 6/3/5
30. ALAB 9/3/8
31. ed. Kimball, *Correspondence*, III, p. 574
32. CUNN Add MSS 52578/36
33. Butcher, *Three Years*, p. 663
34. ed. Chandler, *Papers*, IV, p. 2589
35. CUNN Add MSS 52578/39
36. CHAR 20/213A/101–14
37. CUNN Add MSS 52578/40
38. CHAR 20/213A/101–14
39. MAR Pentagon Papers Box 81/32
40. ed. Bland, *Papers*, V, p. 114 n. 1
41. Cowles, *Churchill*, p. 350; ed. Kimball, *Correspondence*, III, p. 608

42. MAR Pentagon Papers Box 81/32
43. CUNN Add MSS 52578/45
44. Below, *At Hitler's Side*, p. 234; Kershaw, *Hitler*, pp. 791–2
45. Colville, *Fringes*, p. 587
46. ed. Hart-Davis, *King's Counsellor*, p. 313
47. Moran, *Struggle for Survival*, p. 347
48. CAB 195/3 WM (45) 44th
49. Best, *Churchill and War*, p. 135; Jenkins, *Churchill*, p. 785
50. Churchill, *Triumph and Tragedy*, pp. 413–17
51. CUNN Add MSS 52578/47
52. ed. Bland, *Papers*, v, p. 159
53. CUNN Add MSS 52578/52
54. eds Danchev and Todman, *Diaries*, pp. 686–7
55. Astley, *Inner Circle*, p. 209
56. CUNN Add MSS 52578/56
57. ed. Bland, *Papers*, v, p. 171 n. 1
58. eds Danchev and Todman, *Diaries*, p. 688; CUNN Add MSS 52578/56

CONCLUSION: THE RIDDLES OF THE WAR

1. eds. Danchev and Todman, *Diaries*, p. 680
2. Moran, *Struggle for Survival*, p. 110
3. Berlin, *Mr Churchill*, pp. 37–8
4. Moran, *Struggle for Survival*, p. 326
5. ALAB 11/64
6. *London Review of Books*, 2–15/9/1982, p. 16
7. ed. Bland, *Interviews*, p. 589
8. Alex Danchev in eds Reynolds and Kimball, *Allies at War*, p. 18
9. Moran, *Struggle for Survival*, p. 45
10. Wedemeyer, *Wedemeyer Reports!*, p. 168
11. Stimson and Bundy, *Active Service*, p. 235
12. LH 15/15/1
13. KENN 4/2/4, pp. 180–82B
14. MHI Pogue Interview, 12/2/1947
15. KENN 4/2/5, p. 6
16. MHI Wedemeyer 1973 Interview section 5, pp. 1–2
17. Howard, *Mediterranean Strategy*, p. 24
18. Stimson and Bundy, *Active Service*, p. 237
19. MAR GCM Library Microfilm Reel 322 Part 1, 25/7/1949; MAR GCM Library Xerox, p. 2256
20. ed. Trevor-Roper, *Hitler's Table Talk*, pp. 538–9
21. My thanks to Paul Courtenay for this paragraph
22. Major-General David Dawnay, *Wish Stream*, vol. 7 no. 2, October 1953

Bibliography

ARCHIVES AND PRIVATE PAPERS

General Sir Ronald Adam (Liddell Hart Centre for Military Archives, King's College London)
Field Marshal Lord Alanbrooke (Liddell Hart Centre for Military Archives, King's College London)
General H. H. Arnold (Library of Congress, Washington DC)
Mrs Joan Bright Astley (by kind permission of Mrs Astley)
Lord Avon (Birmingham University Archives)
Lord Beaverbrook (Parliamentary Archives, Palace of Westminster)
General Charles L. Bolte (Military History Institute, Carlisle, Pennsylvania)
General Omar N. Bradley (Military History Institute, Carlisle, Pennsylvania)
Lawrence Burgis (Churchill Archives Centre, Cambridge University)
Sir Alexander Cadogan (Churchill Archives Centre, Cambridge University)
General Paul Caraway (Military History Institute, Carlisle, Pennsylvania)
Neville Chamberlain (Birmingham University Archives)
Sir Winston Churchill (Churchill Archives Centre, Cambridge University)
General Mark W. Clark (Military History Institute, Carlisle, Pennsylvania)
General Richard Collins (Military History Institute, Carlisle, Pennsylvania)
Francis P. Corrigan (Franklin D. Roosevelt Presidential Library, Hyde Park)
Sir Stewart Crawford (by kind permission of Mr Michael Crawford)
Admiral of the Fleet Lord Cunningham (British Library)
F. H. N. Davidson (Liddell Hart Centre for Military Archives, King's College London)
J. C. C. Davidson (Parliamentary Archives, Palace of Westminster)
General Jacob L. Devers (Military History Institute, Carlisle, Pennsylvania)
Field Marshal Sir John Dill (Liddell Hart Centre for Military Archives, King's College London)
Colonel Charles H. Donnelly (Military History Institute, Carlisle, Pennsylvania)
General Sir David Fraser (by kind permission of Sir David Fraser)
Lord Halifax (Churchill Archives Centre, Cambridge University)
General Thomas T. Handy (Military History Institute, Carlisle, Pennsylvania)
William D. Hassett (Franklin D. Roosevelt Presidential Library, Hyde Park)
Harry L. Hopkins (Franklin D. Roosevelt Presidential Library, Hyde Park)
General John E. Hull (Military History Institute, Carlisle, Pennsylvania)
General Lord Ismay (Liddell Hart Centre for Military Archives, King's College London)
Lieutenant-General Sir Ian Jacob (Churchill Archives Centre, Cambridge University, and private collection by kind permission of the late Sir Ian Jacob)
Major-General Sir John Kennedy (Liddell Hart Centre for Military Archives, King's College London)
Admiral Ernest J. King (Library of Congress, Washington DC)

BIBLIOGRAPHY

Admiral William D. Leahy (Library of Congress, Washington DC)
General Raymond E. Lee (Military History Institute, Carlisle, Pennsylvania)
Sir Basil Liddell Hart (Liddell Hart Centre for Military Archives, King's College, London)
Lawrence J. Lincoln (Military History Institute, Carlisle, Pennsylvania)
David Lloyd George (Parliamentary Archives, Palace of Westminster)
Isador Lubin (Franklin D. Roosevelt Presidential Library, Hyde Park)
Colonel Frank McCarthy (George C. Marshall Foundation, Lexington, Virginia)
Admiral John L. McCrea (Franklin D. Roosevelt Presidential Library, Hyde Park)
General Carter Magruder (Military History Institute, Carlisle, Pennsylvania)
General George C. Marshall (George C. Marshall Foundation, Lexington, Virginia)
Sidney Matthews (Military History Institute, Carlisle, Pennsylvania)
Lowell Mellett (Franklin D. Roosevelt Presidential Library, Hyde Park)
Lady Alexandra Metcalfe (by kind permission of Mr David Metcalfe)
Ronald Middleton (by kind permission of Mr Mark Katzenellenbogen)
Henry Morgenthau (Franklin D. Roosevelt Presidential Library, Hyde Park)
Jo, Countess of Onslow (Churchill Archives Centre, Cambridge University)
Oleg Pantuhoff a.k.a. Colonel John L. Bates (Military History Institute, Carlisle, Pennsylvania)
Sir Charles Peake (private collection)
Marshal of the RAF Lord Portal (Christ Church, Oxford University)
Admiral of the Fleet Sir Dudley Pound (Churchill Archives Centre, Cambridge University)
General Matthew B. Ridgway (Military History Institute, Carlisle, Pennsylvania)
Colonel Paul M. Robinett (George C. Marshall Foundation, Lexington, Virginia)
Eleanor Roosevelt (Franklin D. Roosevelt Presidential Library, Hyde Park)
Franklin D. Roosevelt (Franklin D. Roosevelt Presidential Library, Hyde Park)
Wellington Alexander Samouce (Military History Institute, Carlisle, Pennsylvania)
General William H. Simpson (Military History Institute, Carlisle, Pennsylvania)
General Lucian K. Truscott (George C. Marshall Foundation, Lexington, Virginia)
General Orlando W. Ward (Military History Institute, Carlisle, Pennsylvania)
General Albert C. Wedemeyer (Military History Institute, Carlisle, Pennsylvania)
R. W. W. Wilmot (Liddell Hart Centre for Military Archives, King's College London)
John Winant (Franklin D. Roosevelt Presidential Library, Hyde Park)
Sir Evelyn Wrench (British Library)

BOOKS

(All published in London unless otherwise stated)

Alldritt, Keith, *The Greatest of Friends: Franklin D. Roosevelt and Winston Churchill 1941–1945*, 1995
eds Ancell, R. Manning, and Miller, Christine M., *The Biographical Dictionary of World War II Generals and Flag Officers*, Westport, Connecticut 1996
Arnold, Henry H., *Global Mission*, New York 1949
Astley, Joan Bright, *The Inner Circle: A View of War at the Top*, 1971
Attlee, C. R., *As It Happened*, 1954
eds Barnes, John, and Nicholson, David, *The Empire at Bay: The Leo Amery Diaries 1929–1945*, 1988
Barnett, Correlli, *The Desert Generals*, 1960
Beitzell, Robert, *The Uneasy Alliance*, New York 1972
Beevor, Antony, *Crete: The Battle and the Resistance*, 1991
——, *Stalingrad*, 1998
——, *Berlin: The Downfall 1945*, 2002
Bell, Coral, *The Debatable Alliance: An Essay in Anglo-American Relations*, 1964

Below, Nicolaus von, *At Hitler's Side: The Memoirs of Hitler's Luftwaffe Adjutant 1937–1945*, 2001

Ben-Moshe, Tuvia, *Churchill as Historian and Strategist*, 1992

Bennett, Ralph, *ULTRA and the Mediterranean Strategy 1941–45*, 1989

Bercuson, David J., and Herwig, Holger H., *One Christmas in Washington: Churchill and Roosevelt Forge the Grand Alliance*, 2005

Berlin, Isaiah, *Mr Churchill in 1940*, 1949

Berthon, Simon, *Allies at War: The Bitter Rivalry among Churchill, Roosevelt, and De Gaulle*, 2001

Beschloss, Michael R., *Kennedy and Roosevelt: The Uneasy Alliance*, 1987

Best, Geoffrey, *Churchill: A Study in Greatness*, 2001

——, *Churchill and War*, 2006

Birkenhead, The Earl of, *Halifax: The Life of Lord Halifax*, 1965

Black, Conrad, *Franklin Delano Roosevelt: Champion of Freedom*, 2003

——, *Richard Milhous Nixon: The Invincible Quest*, 2007

ed. Bland, Larry I., *The Papers of George Catlett Marshall*, vols III, IV and V, 1996–2003

——, *George C. Marshall: Interviews and Reminiscences for Forrest C. Pogue*, Lexington 1996

Blum, John Morton, *Years of War 1941–1945: From the Morgenthau Diaries*, Boston 1967

ed. Blum, John Morton, *The Price of Vision: The Diary of Henry A. Wallace 1942–1946*, Boston 1973

Blumenson, Martin, *Mark Clark*, 1985

Bohlen, Charles E., *Witness to History 1929–1969*, 1973

ed. Bond, Brian, *Chief of Staff: The Diaries of Lieutenant-General Sir Henry Pownall*, 2 vols, 1972, 1974

Brendon, Piers, *Winston Churchill: A Brief Life*, 2001

Brinkley, David, *Washington Goes to War*, 1988

Brodhurst, Robin, *Churchill's Anchor: The Biography of Admiral of the Fleet Sir Dudley Pound*, 2000

——, *Triumph in the West*, 1959

Browne, Anthony Montague, *Long Sunset: Memoirs of Winston Churchill's Last Private Secretary*, 1995

Bryant, Arthur, *The Turn of the Tide*, 1957

Buell, Thomas B., *Master of Sea Power: A Biography of Admiral Ernest J. King*, 1995

Burns, James MacGregor, *Roosevelt: The Soldier of Freedom*, 1970

Butcher, Harry C., *Three Years with Eisenhower*, 1946

Butler, J. R. M., *Grand Strategy*, vol. II, 1957

——, *Grand Strategy*, vol. III Parts I and II, 1964

ed. Butler, Susan, *My Dear Mr Stalin: The Complete Correspondence between Franklin D. Roosevelt and Joseph V. Stalin*, 2005

Cannadine, David, *In Churchill's Shadow*, 2002

Casey, Lord, *Personal Experience*, 1962

Chadwick, Owen, *Britain and the Vatican during the Second World War*, 1986

ed. Chandler, Alfred D., *The Papers of Dwight David Eisenhower*, vols I to V, Baltimore 1970

Chant, Christopher, *The Encyclopedia of Codenames of World War II*, 1986

Charmley, John, *Churchill: The End of Glory*, 1993

——, *Churchill's Grand Alliance: The Anglo-American Special Relationship 1940–1957*, 1995

Chisholm, Anne, and Davie, Michael, *Beaverbrook: A Life* 1992

Churchill, Winston S., *The World Crisis*, 2 vols, 1938

——, *My Early Life*, 1944

——, *Their Finest Hour*, 1949

——, *The Hinge of Fate*, 1951

——, *Closing the Ring*, 1952
——, *Triumph and Tragedy*, 1954
——, *Great Contemporaries*, 1959
Clark, Mark W., *Calculated Risk*, 1951
ed. Cockett, Richard, *My Dear Max: The Letters of Brendan Bracken to Lord Beaverbrook*, 1990
Collins, John M., *Grand Strategy: Principles, Practices and Historical Perspectives*, 2002
Collins, R. J., *Lord Wavell*, 1947
Colville, John, *Man of Valour*, 1972
——, *Footprints in Time*, 1976
——, *Fringes of Power: Downing Street Diaries 1939–1955*, 1985
Coombs, David, *Sir Winston Churchill's Life through his Paintings*, 2003
Cooper, Artemis, *Cairo in the War 1939–1945*, 1989
Cowles, Virginia, *Churchill: The Era and the Man*, 1953
Cozzens, James G., *Guard of Honor*, 1949
Cray, Ed, *General of the Army: George C. Marshall, Soldier and Statesman*, 2000
Cunningham, Andrew Browne, *A Sailor's Odyssey*, 1951
Danchev, Alex, *Very Special Relationship: Field Marshal Sir John Dill and the Anglo-American Alliance*, 1986
——, *On Specialness: Essays in Anglo-American Relations*, 1998
ed. Danchev, Alex, *Establishing the Anglo-American Alliance: The Second World War Diaries of Brigadier Vivian Dykes*, 1990
eds Danchev, Alex, and Todman, Daniel, *War Diaries 1939–1945: Field Marshal Lord Alanbrooke*, 2001
ed. Daniel, Jonathan, *Complete Presidential Press Conferences of Franklin D. Roosevelt*, vols XXI and XXII, New York 1972
ed. Davenport-Hines, Richard, *Letters from Oxford: Hugh Trevor-Roper to Bernard Berenson*, 2006
Davidson, Edward, and Manning, Dale, *Chronology of World War Two*, 1999
Davies, Vernon E., *The History of the Joint Chiefs of Staff in World War II*, 2 vols, Washington DC 1972
Davis, Kenneth S., *FDR: The War President*, 2000
eds Deakin, William, and others, *British Political and Military Strategy in Central, Eastern and Southern Europe in 1944*, 1988
Deane, John R., *The Strange Alliance: The Story of our Efforts at Wartime Co-operation with Russia*, 1947
D'Este, Carlo, *Patton*, 1996
——, *Eisenhower*, 2002
ed. Dilks, David, *The Diaries of Sir Alexander Cadogan 1938–1945*, 1971
Dimbleby, David, and Reynolds, David, *An Ocean Apart*, 1988
Djilas, Milovan, *Conversations with Stalin*, 1962
Douglas, Sholto, *Years of Command*, 1966
Dupuy, R. Ernest, and Dupuy, Trevor N., *The Collins Encyclopedia of Military History*, 4th edn, 1993
Dupuy, Trevor, *A Genius for War*, 1977
ed. Eade, Charles, *Secret Session Speeches*, 1946
Ehrman, John, *Grand Strategy*, vols. V and VI, 1956
Eichelberger, Robert L., *Jungle Road to Tokyo*, New York 1957
Eisenhower, Dwight, *Crusade in Europe*, 1948
——, *At Ease: Stories I Tell to Friends*, 1967
Eisenhower, John S. D., *General Ike: A Personal Reminiscence*, 2003
Ellis, John, *The Sharp End of War: The Fighting Man in World War II*, 1980
——, *Brute Force*, 1999
ed. Evans, Trefor, *The Killearn Diaries 1934–1946*, 1970

Farrell, Brian, *The Basis and Making of British Grand Strategy 1940–1943: Was There a Plan?*, 2 vols, 1998

Fenby, Jonathan, *Alliance: The Inside Story of How Roosevelt, Stalin and Churchill Won One War and Began Another*, 2006

ed. Ferrell, Robert H., *The Eisenhower Diaries*, 1981

Foreign Relations of the United States, *The Conferences at Washington, 1941–1942 and Casablanca, 1943*, Washington DC 1968

——, *1942*, vol. III, Washington DC 1961

——, *The Conferences at Washington and Quebec, 1943*, Washington DC 1970

——, *The Conferences at Cairo and Tehran, 1943*, Washington DC 1961

——, *The Conference at Quebec, 1944*, Washington DC 1972

——, *The Conferences at Malta and Yalta, 1945*, Washington DC 1955

Fraser, David, *Alanbrooke*, 1997

eds Freedman, Lawrence, and others, *War, Strategy and International Politics*, 1992

Garrett, Stephen A., *Ethics and Airpower in World War II: The British Bombing of German Cities*, 1997

Gilbert, Martin, *Winston S. Churchill*, vols VI and VII, 1983, 1986

——, *Churchill: A Life*, 1991

——, *The Churchill War Papers*, vols I, II and III, 1993–2000

——, *In Search of Churchill*, 1994

——, *Churchill and America*, 2006

Gimlette, John, *The Theatre of Fish: Travels through Newfoundland and Labrador*, 2005

Goodwin, Doris Stearns, *No Ordinary Time*, 1994

Greenfield, Kent Roberts, *American Strategy in World War II: A Reconsideration*, Baltimore 1963

ed. Greenfield, Kent Roberts, *Command Decisions*, 1960

Greenwood, Alexander, *Field Marshal Auchinleck*, 1991

Grigg, John, *1943: The Victory that Never Was*, 1999

Guderian, General Heinz, *Panzer Leader*, 1950

Guedalla, Philip, *Mr Churchill*, 1941

Gwyer, J. M. A., *Grand Strategy*, vol. III Part I, 1964

Haffner, Sebastian, *Churchill*, 2003

Halifax, The Earl of, *Fulness of Days*, 1959

ed. Handel, Michael, *Intelligence and Strategy in the Second World War*, 1995

ed. Hardy, Henry, *Isaiah Berlin: Flourishing: Letters 1928–1946*, 2005

Harriman, W. Averell, and Abel, Elie, *Special Envoy to Churchill and Stalin 1941–1946*, 1976

Harris, Sir Arthur, *Bomber Offensive*, 1947

Harrison, Gordon A., *Cross-Channel Attack*, 1951

ed. Hart-Davis, Duff, *King's Counsellor: Abdication and War: The Diaries of Sir Alan Lascelles*, 2006

ed. Harvey, John, *The War Diaries of Oliver Harvey 1941–1945*, 1978

Harvey, Robert, *American Shogun: MacArthur, Hirohito and the American Duel with Japan*, 2006

Hassett, William D., *Off the Record with FDR 1942–1945*, 1960

Hastings, Max, *Bomber Command*, 1979

——, *Overlord: D-Day and the Battle for Normandy*, 1984

——, *Armageddon: The Battle for Germany 1944–45*, 2004

——, *Nemesis: The Battle for Japan 1944–45*, 2007

Hayes, Grace Person, *The History of the Joint Chiefs of Staff in World War II: The War against Japan*, Annapolis 1982

eds Heiber, Helmut, and Glantz, David M., *Hitler and his Generals: Military Conferences 1942–1945*, 2002

Higgins, Trumbull, *Winston Churchill and the Second Front 1940–1943*, 1957

Hinsley, Harry, *British Intelligence in the Second World War*, 1993
Hobbs, Joseph, *Dear General: Eisenhower's Wartime Letters to Marshall*, Baltimore 1971
Holland, James, *Together We Stand: North Africa 1942–1943*, 2006
——, *Italy's Sorrow: A Year of War 1944–1945*, 2008
Hollis, General Sir Leslie, *One Marine's Tale*, 1956
Holmes, Richard, *In the Footsteps of Churchill*, 2006
——, *The World at War: The Landmark Oral History*, 2007
Horner, D. M., *High Command: Australia and Allied Strategy: 1939–45*, 1983
Howard, Michael, *The Mediterranean Strategy in the Second World War*, 1968
——, *Grand Strategy*, vol. IV, 1972
——, *The Causes of War and Other Essays*, 1983
——, *Captain Professor: A Life in War and Peace*, 2007
Howarth, David, *Dawn of D-Day*, 2004
Howarth, Patrick, *Intelligence Chief Extraordinary: The Life of the Ninth Duke of Portland*, 1966
Hull, Cordell, *The Memoirs of Cordell Hull*, 2 vols, New York 1948
ed. Hunter, Ian, *Winston and Archie: The Letters of Sir Archibald Sinclair and Winston S. Churchill 1915–1960*, 2005
Hussert, Stewart W., *George C. Marshall*, 2004
ed. Huston, John W., *American Airpower Comes of Age: General Henry H. 'Hap' Arnold's World War II Diaries*, 2 vols, Maxwell Air Force Base, Alabama 2002
Ingersoll, Ralph, *Top Secret*, 1946
Ismay, Lord, *The Memoirs of General the Lord Ismay*, 1960
Jackson, Sir William, and Bramall, Lord, *The Chiefs: The Story of the United Kingdom Chiefs of Staff*, 1992
ed. Jacobsen, Hans-Adolf, and Smith, Arthur L., *World War Two Policy and Strategy*, 1979
James, Robert Rhodes, *Churchill Speaks 1897–1963: Collected Speeches in Peace and War*, 1981
Jenkins, Roy, *Churchill*, 2001
——, *Franklin Delano Roosevelt*, 2003
Jones, R. V., *Most Secret War: British Scientific Intelligence 1939–1945*, 1978
Keegan, John, *The Second World War*, 1990
——, *Churchill*, 2002
ed. Keegan, John, *Churchill's Generals*, 1991
ed. Kemper, R. Crosby, *Winston Churchill: Resolution, Defiance, Magnanimity, Good Will*, Columbia, Missouri 1996
Kennedy, Sir John, *The Business of War*, 1957
Kersaudy, François, *Norway 1940*, 1990
Kershaw, Ian, *Hitler 1936–1945: Nemesis*, 2000
Kimball, Warren F., *Forged in War: Churchill, Roosevelt and the Second World War*, 1997
ed. Kimball, Warren F., *Churchill and Roosevelt: The Complete Correspondence*, 3 vols, 1984
King, Ernest J., and Whitehill, Walter Muir, *Fleet Admiral King: A Naval Record*, 1953
Kolinsky, Martin, *Britain's War in the Middle East: Strategy and Diplomacy 1936–42*, 1999
Lamb, Richard, *Churchill as War Leader*, 2003
eds Lane, Ann, and Temperley, Harold, *The Rise and Fall of the Grand Alliance 1941–45*, 1995
Langer, William L., and Gleason, S. Everett, *The Challenge to Isolation: 1937–1940*, New York 1952
Lash, Joseph P., *Roosevelt and Churchill 1939–1941*, 1977
Lavery, Brian, *Churchill Goes to War*, 2007
Layton, Elizabeth, *Mr Churchill's Secretary*, 1958
Leahy, William D., *I Was There*, 1950
Leasor, James, *War at the Top: Based on the Experiences of General Sir Leslie Hollis*, 1959

Lees-Milne, James, *The Enigmatic Edwardian: The Life of Reginald 2nd Viscount Esher*, 1988
Leutze, James R., *The London Journal of General Raymond E. Lee 1940–1941*, Boston 1971
——, *Bargaining for Supremacy: Anglo-American Naval Collaboration 1937–1941*, Chapel Hill, North Carolina 1977
Lewin, Ronald, *Churchill as Warlord*, 1973
——, *Slim: The Standard-Bearer*, 1976
ed. Lewis, William Roger, *More Adventures with Britannia*, 1998
——, *Yet More Adventures with Britannia*, 2005
Liddell Hart, Sir Basil, *The History of the Second World War*, 1970
Lilienthal, David E., *The Diary of David E. Lilienthal*, vol. II, 1964
Lovat, Lord, *March Past*, 1978
Lukacs, John, *Churchill: Visionary. Statesman. Historian*, 2002
McIntyre, Vice-Admiral Ross, *Twelve Years with Roosevelt*, 1948
McJimsey, George, *Harry Hopkins*, 1987
Maclachlan, Donald, *In the Chair: Barrington-Ward of The Times 1927–1948*, 1971
ed. Macleod, Colonel Roderick, *The Ironside Diaries 1937–1940*, 1962
Macmillan, Harold, *War Diaries: Politics and War in the Mediterranean January 1943–May 1945*, 1984
Maney, Patrick J., *The Roosevelt Presence*, 1992
Marshall, George C., *Biennial Report of the Chief of Staff of the United States Army*, 1941
——, *Biennial Report of the Chief of Staff of the United States Army*, 1943
——, *General Marshall's Report: The Winning of the War in Europe and the Pacific*, 1945
Marshall, Katherine Tupper, *Together: Annals of an Army Wife*, New York 1946
Martin, Sir John, *Downing Street: The War Years*, 1991
Matloff, Maurice, *Strategic Planning for Coalition Warfare 1943–44*, Washington DC 1959
Matloff, Maurice, and Snell, E. M., *Strategic Planning for Coalition Warfare 1941–42*, Washington DC 1953
Meacham, Jon, *Franklin and Winston: Portrait of a Friendship*, 2003
Montgomery, Bernard Law, *The Memoirs of Field-Marshal Montgomery*, 1958
ed. Moon, Penderel, *Wavell: The Viceroy's Journal*, 1973
Moran, Lord, *Winston Churchill: The Struggle for Survival 1940–1965*, 1966
Morgan, Sir Frederick E., *Overture to Overlord*, New York 1950
——, *Peace and War: A Soldier's Life*, 1961
Morgan, Ted, *FDR: A Biography*, New York 1985
Morison, S. E., *American Contributions to the Strategy of World War II*, 1958
Morton, H. V., *Atlantic Meeting*, 1943
Mosier, John, *The Blitzkrieg Myth: How Hitler and the Allies Misread the Strategic Realities of World War Two*, 2003
Mosley, Leonard, *Backs to the Wall: London under Fire 1939–45*, 1982
——, *Marshall: Organizer of Victory*, 1971
ed. Muggeridge, Malcolm, *Ciano's Diary 1939–1943*, 1947
Nel, Elizabeth, *Mr Churchill's Secretary*, 1958
Nevins, Brigadier-General Arthur S., *Gettysburg's Five-Star Farmer*, New York 1977
Nicolson, Nigel, *Alex: The Life of Field Marshal Earl Alexander of Tunis*, 1974
ed. Nicolson, Nigel, *Harold Nicolson: Diaries and Letters 1939–1945*, 1967
Nisbet, Robert, *Roosevelt and Stalin: The Failed Courtship*, Washington DC 1988
Overy, Richard, *Why the Allies Won*, 1995
ed. Parker, R. A. C., *Winston Churchill: Studies in Statesmanship*, 1995
Parrish, Thomas, *Roosevelt and Marshall: Partners in Peace and War*, 1989
ed. Parrish, Thomas, *The Simon and Schuster Encyclopedia of World War II*, New York 1978
Pawle, Gerald, *The War and Colonel Warden*, 1963
Payne, Robert, *General Marshall: A Study in Loyalties*, 1952

Payne, Stanley G., *Franco and Hitler: Spain, Germany and World War II*, 2008
Perkins, Frances, *The Roosevelt I Knew*, 1947
Pickersgill, J. W., and Forster, D. F., *The McKenzie King Record*, 2 vols, 1960, 1968
Pogue, Forrest C., *The Supreme Command*, Washington DC 1954
——, *George C. Marshall: Education of a General 1880–1939*, 1964
——, *George C. Marshall: Ordeal and Hope 1939–42*, 1965
——, *George C. Marshall: Organizer of Victory 1943–45*, 1973
Porch, Douglas, *Hitler's Mediterranean Gamble*, 2004
Powell, Anthony, *The Military Philosophers*, 1968
Probert, Henry, *Bomber Harris: His Life and Times*, 2006
Ramsden, John, *Man of the Century: Winston Churchill and his Legends since 1945*, 2002
Ranfurly, Hermione, Countess of, *To War with Whitaker: The Wartime Diaries of the Countess of Ranfurly 1939–1945*, 1994
Reynolds, David, *The Creation of the Anglo-American Alliance 1937–41*, 1981
——, *In Command of History: Churchill Fighting and Writing the Second World War*, 2004
——, *Summits: Six Meetings that Shaped the Twentieth Century*, 2007
eds Reynolds, David, and Kimball, Warren, *Allies at War*, 1994
Richards, Denis, *Portal of Hungerford*, 1978
Richardson, Charles, *From Churchill's Secret Circle to the BBC: The Biography of Lieutenant General Sir Ian Jacob*, 1991
Roberts, Andrew, *The Holy Fox: A Life of Lord Halifax*, 1991
——, *Eminent Churchillians*, 1994
Robertson, Sir William, *Soldiers and Statesmen*, 2 vols, 1926
Rooney, David, *Stilwell the Patriot*, 2005
Roosevelt, Eleanor, *This I Remember*, 1950
Roosevelt, Elliott, *As He Saw It*, New York 1946
——, *The Roosevelt Letters 1928–1945*, vol. III, 1952
——, *A Rendezvous with Destiny*, 1975
Roosevelt, Franklin D., *The Public Papers and Addresses of Franklin D. Roosevelt*, New York 1941
——, *The War Messages of Franklin D. Roosevelt*, 1945
Roosevelt, James, *Affectionately, FDR*, New York 1959
——, *A Differing View*, 1977
Rose, Norman, *Churchill: An Unruly Life*, 1994
Roskill, Stephen, *Churchill and the Admirals*, 1977
Ross, Stephen, *American War Plans 1941–45*, 1997
Russell, Douglas S., *Winston Churchill: Soldier*, 2005
Sainsbury, Keith, *The Turning Point: Roosevelt, Stalin, Churchill and Chiang Kai-shek, 1943*, 1986
——, *Churchill and Roosevelt at War*, 1994
Sandys, Celia, *Chasing Churchill: The Travels of Winston Churchill*, 2003
Schlesinger, Arthur, *A Thousand Days*, 1965
Sebag Montefiore, Simon, *Stalin: The Court of the Red Tsar*, 2003
Sherwood, Robert E., *The White House Papers of Harry L. Hopkins*, vol. II, 1949
——, *Roosevelt and Hopkins: An Intimate History*, 1950
Simpson, B. Mitchell, *Admiral Harold R. Stark: Architect of Victory 1939–1945*, Columbia, South Carolina 1989
Slessor, Sir John, *The Central Blue*, 1956
Slim, Field Marshal Viscount, *Defeat into Victory*, 1956
ed. Smith, Amanda, *Hostage to Fortune: The Letters of Joseph P. Kennedy*, New York 2001
Smith, Kevin, *Conflict over Convoys: Anglo-American Logistics Diplomacy in the Second World War*, 1996
Smith, Paul, *Government and the Armed Forces 1856–1990*, 1990
Smith, Walter Bedell, *Eisenhower's Six Great Decisions*, 1956

Soames, Mary, *Clementine Churchill*, 2002
ed. Soames, Mary, *Speaking for Themselves: The Personal Letters of Winston and Clementine Churchill*, 1988
Stafford, David, *Churchill and Secret Service*, 1997
——, *Roosevelt and Churchill*, 2000
Steele, Richard W., *The First Offensive 1942: Roosevelt, Marshall and the Making of American Strategy*, Bloomington, Indiana 1973
ed. Stephen, Oscar, *Sir Victor Brooke, Sportsman and Naturalist*, 1894
Stimson, Henry L., and Bundy, McGeorge, *On Active Service in Peace and War*, 1949
Stoler, Mark, *The Politics of the Second Front: American Military Planning and Diplomacy in Coalition Warfare, 1941–1943*, Westport, Connecticut 1977
——, *George C. Marshall: Soldier–Statesman of the American Century*, 1989
——, *Allies and Adversaries: The Joint Chiefs of Staff, the Grand Alliance and US Strategy in World War II*, Chapel Hill, North Carolina 2000
Strong, Kenneth, *Intelligence at the Top*, 1968
Summersby, Kay, *Eisenhower was my Boss*, 1949
ed. Sweetman, John, *Sword and Mace: Twentieth-Century Civil–Military Relations in Britain*, 1986
ed. Taylor, A. J. P., *Churchill: Four Faces and the Man*, 1969
——, *Off the Record: Political Interviews 1933–1943*, 1973
Tedder, Lord, *Air Power in War*, 1947
——, *With Prejudice*, 1966
Thomas, Evan, *Sea of Thunder: Four Commanders and the Last Great Naval Campaign 1941–1945*, 2006
Thompson, R. W., *Churchill and Morton*, 1976
Thorne, Christopher, *Allies of a Kind*, 1978
ed. Trevor-Roper, Hugh, *Hitler's Table Talk 1941–1944*, 2000
Tuchman, Barbara, *Stilwell and the American Experience in China 1911–1945*, 2001
Tully, Grace, *FDR: My Boss*, New York 1949
Tunis, Field-Marshal Earl Alexander of, *The Alexander Memoirs 1940–1945*, 1962
Van Der Vat, David, *D-Day*, 2003
Ward, Geoffrey, *A First-Class Temperament: The Emergence of Franklin D. Roosevelt*, New York 1989
ed. Ward, Geoffrey C., *Closest Companion: The Unknown Story of the Intimate Friendship between Franklin Roosevelt and Margaret Suckley*, New York 1995
Warlimont, Walter, *Inside Hitler's Headquarters*, 1964
Wedemeyer, General Albert C., *Wedemeyer Reports!*, New York 1958
Weigley, Russell F., *Eisenhower's Lieutenants: The Campaign of France and Germany 1944–1945*, Bloomington, Indiana 1981
Weinberg, Gerhard L., *Visions of Victory: The Hopes of Eight World War Leaders*, New York 2006
Weintraub, Stanley, *Fifteen Stars: Eisenhower, MacArthur, Marshall*, New York 2007
ed. Wheeler-Bennett, John, *Action This Day*, 1968
——, *Special Relationships*, 1975
ed. White, Theodore, *The Stilwell Papers*, New York 1948
Wills, Matthew B., *FDR: A Diminished President*, New York 2003
Wilmot, Chester, *The Struggle for Europe*, 1952
Wilson, Theodore, *The First Summit: Roosevelt and Churchill at Placentia Bay*, 1991
Winton, John, *Cunningham: The Greatest Admiral since Nelson*, 1998
Ziegler, Philip, *Mountbatten*, 1985
ed. Ziegler, Philip, *Personal Diary of Admiral the Lord Mountbatten 1943–1946*, 1988

ARTICLES, PAMPHLETS AND DISSERTATIONS

Ben-Moshe, Tuvia, 'Winston Churchill and the "Second Front": A Reappraisal', *Journal of Modern History*, December 1990

Danchev, Alex, 'Very Special Relationship: Field Marshal Sir John Dill and the Anglo-American Alliance', SHAFR Lecture, August 1984

——, ' "Dilly-Dally", or Having the Last Word: Field Marshal Sir John Dill and Prime Minister Churchill', *Journal of Contemporary History*, vol. 22, 1987

——, 'The Grand Alliance Revisited', *RUSI Journal*, vol. 137 no. 2, April 1992

——, 'Waltzing with Winston: Civil–Military Relations in Britain in the Second World War', *War in History*, no. 2, 1995

——, 'The Birdwatcher Who Saved Britain', in ed. Lewis, *Yet More Adventures with Britannia*, 2005

Gilbert, Sir Martin, 'Churchill and D-Day', *Finest Hour*, no. 122, Spring 2004

Harmon, Christopher C., 'Alanbrooke and Churchill', *Finest Hour*, no. 112, Autumn 2001 *Journal of the Royal Artillery*, vol. 25 no. 2, November 1963

Kimball, Warren, and Rose, Norman, 'Churchill and D-Day: Another View', *Finest Hour*, no. 124, Autumn 2004

Kurtz, Megan Kathleen, *Leap-Frogging to the Bomb: The Allied Strategy against Japan 1941–1945*, George C. Marshall Research Foundation Scholarship Paper

Leighton, Richard, 'Overlord Revisited', *American Historical Review*, July 1963

Liddell Hart, Basil, 'Western War Strategy', *RUSI Journal*, vol. 105, February 1960

Lunghi, Hugh, 'The Troubled Triumvirate', *Finest Hour*, no. 135, Spring 2006

Parker, Sally Lister, 'Attendant Lords: A Study of the Joint Staff Mission in Washington 1941–1945', University of Maryland dissertation 1984

Pogue, Forrest, 'Why Eisenhower's Forces Stopped at the Elbe', *World Politics*, April 1952

Rigby, David Joseph, 'The Combined Chiefs of Staff and Anglo-American Strategic Coordination in World War II', Brandeis University dissertation 1996

Robbins, Ron Cynewulf, 'The Atlantic Charter 1941–2001', *Finest Hour*, no. 112, Autumn 2001

Stagner, Elizabeth, *Politics and Diplomacy, Not Strategy: The Second Front in World War II*, George C. Marshall Research Foundation Scholarship Paper

Stoler, Mark A., 'The "Pacific First" Alternative in American World War II Strategy', *International History Review*, no 2, 3 July 1980

Strange, Joseph L., 'The British Rejection of Operation Sledgehammer: An Alternative Motive', *Military Affairs*, vol. 46 no. 1, February 1982

Index